AF575424

Product Cost Controlling with SAP S/4HANA®

John Jordan, Janet Salmon

Product Cost Controlling with SAP S/4HANA®

Editor Rachel Gibson
Acquisitions Editor Emily Nicholls
Copyeditor Kim Hannel
Cover Design Graham Geary
Photo Credit Shutterstock: 119117617/© Design Stock
Layout Design Vera Brauner
Production Kelly O'Callaghan
Typesetting SatzPro, Germany
Printed and bound in Canada

ISBN 978-1-4932-2534-7

1st edition 2024

Library of Congress Cataloging-in-Publication Data
Names: Jordan, John, 1959- author. | Salmon, Janet, author.
Title: Product cost controlling with SAP S/4HANA / by John Jordan and Janet Salmon.
Description: 1st edition. | Bonn ; Boston : Rheinwerk Publishing, [2024] | Includes index.
Identifiers: LCCN 2024019989 | ISBN 9781493225347 (hardcover) | ISBN 9781493225354 (ebook)
Subjects: LCSH: SAP HANA (Electronic resource) | Cost control--Data processing.
Classification: LCC HD47.3 .J673 2024 | DDC 658.15/52--dc23/eng/20240510
LC record available at https://lccn.loc.gov/2024019989

Contents at a Glance

Contents

12 Overhead 425

13 Work in Process 431

Preface

Welcome to the latest edition of *Product Cost Controlling with SAP*, now titled *Product Cost Controlling with SAP S/4HANA*. The best-selling previous editions, published in 2008, 2012, and 2016, are still universally accepted as the only comprehensive work on the topic. By reading this book, you will gain a deep understanding of product cost controlling with SAP S/4HANA, empowering you to make informed decisions.

This new edition is not just a revision but a comprehensive update tailored to your needs. It still contains the complete, easy-to-follow SAP product costing guide you want to read and keep as a reference. In addition, the long-awaited SAP S/4 HANA content in all chapters sets it apart. See the "Structure of This Book" section for an overview of the new material.

For this latest edition, I've invited Janet Salmon from SAP as coauthor to provide not only the latest SAP S/4HANA Cloud content but also the direction of future SAP developments on the horizon.

Concept of This Book

Most manufacturing and service companies are interested in accurately apportioning their operational and other costs to products or services provided. Comparing these costs with revenues allows you to analyze your company's profitability.

Accurately determining the profitability of products, product lines, and services is a complex task, which is magnified when you consider the details that SAP requires to set up product cost controlling correctly. For instance, the seemingly straightforward task of determining the cost of purchased components involves multiple factors, including supplier quotations, planned, and purchased quantities, and costs. This is just a glimpse into the intricacies of using highly integrated SAP S/4HANA.

When departments are allowed to determine the costs of processes manually in spreadsheets, there is less scrutiny, questioning, and transparent access to the details behind many of these important figures and calculations. The highly integrated nature of SAP S/4HANA increases discipline in data accuracy and validity and, most importantly, improves communication within and between departments.

The complicated task of apportioning labor and general overhead costs to individual products or services can compromise the accuracy required and the time and effort necessary to gain more accuracy. For example, it's possible to either manually enter the time to perform each activity when manufacturing an assembly or automatically enter the average time required for each activity. You gain more accurate information by recording actual activity time, but you incur the extra cost of manually entering the time.

This book aims to increase your understanding of the detailed reporting and analysis capability possible with SAP S/4HANA. This allows you to make informed decisions when designing processes and the level of detail you input into the system. The level of costing and profitability detail you input is determined by the output reporting detail you require.

Much detailed documentation on setting up data and processes in product cost controlling is available online. This book presents many process overviews, examples, and case scenarios to assist you in understanding how the details fit together, the overall processes, and leveraging your investment in the integrated SAP S/4HANA system.

Target Audience

This book is designed for users, managers, consultants, and anyone interested in gaining a greater understanding of the product costing process. It contains easy-to-understand process overviews and detailed master data and configuration setup. You can use this book as a reference, referring to specific sections when needed. For example, you can refer to a specific chapter on master data during master data setup, or you may refer to the chapter on costing sheets when configuring overhead.

The screenshots, menu paths, and SAP Fiori apps in this book are taken from an SAP S/4HANA 2023 system. Manufacturing order is an umbrella term for production and process orders throughout.

Structure of This Book

This book is structured to help you quickly navigate to the sections of particular interest depending on your role and interests.

Initial Planning

Initial planning is the first step in planning and controlling costs for future fiscal periods. You create sales plan quantities in margin analysis or sales and operations planning and convert the quantities to a production plan that you transfer to demand management. Long-term planning accesses these requirements to determine work center loads and purchasing requirements based on production master data. Scheduled activity requirements are transferred to cost center accounting, where plan activity requirements are determined.

Note

Sales and operations planning will be gradually replaced by SAP Integrated Business Planning for Supply Chain (SAP IBP).

In **Chapter 1**, we introduce product cost controlling by converting the sales plan into a production plan and, from there, determining component procurement and cost center activity quantities. We follow a typical best practice initial planning scenario, as described above. Included are detailed instructions on planning scenarios and long-term planning runs, including exception messages.

Product Cost Planning

Product cost planning involves planning procurement and production costs and setting prices for materials and services. You typically determine the purchase price for externally procured items first and then the manufactured cost for assemblies. The following master data and configuration chapters describe the setup required for cost estimates.

Master Data

Master data stays relatively constant over time. In these three chapters, we examine master data relevant to product cost planning, including many examples and case scenarios:

- **Chapter 2**
  We discuss controlling master data, including accounts, cost elements, cost centers, activity types, and statistical key figures. You'll also learn about cost of sales cost elements, swapping the cost center standard hierarchy, and activity type groups.
- **Chapter 3**
  We look at material master views, including material requirements planning (MRP), Controlling, and financial accounting. We also examine the ABC indicator, fixed lot size, discontinued parts, Material Ledger documents, and the actual cost component split.
- **Chapter 4**
  We examine logistics master data, including bills of materials (BOMs), routings, product cost collectors, and purchasing information records (purchasing info records). More specific topics include recursive BOMs, product cost collector lists, plant-specific purchasing info records, and source lists.

Configuration

Most configuration settings are made during the initial implementation. All configuration settings are closely controlled and monitored because they have a major impact on the system process design. Configuration settings are discussed in the following four chapters:

- **Chapter 5**
  We examine costing sheets and how you use them to calculate overhead. This chapter includes sections on dependency, overhead groups, and keys, default costing sheets, and quantity-based overhead.
- **Chapter 6**
  We analyze cost components and structures and how they group similar costs, including material, labor, and overhead, for reporting. We also look at how cost component groups can improve your reporting by making cost components available in standard cost estimate lists.
- **Chapter 7 and Chapter 8**
  We examine how you set up costing variants with the configuration to create cost estimates. Detailed information on purchased material delivery costs is included.

After setting up master data and configuration, we're ready to create cost estimates, as discussed in the next two chapters.

Cost Estimates

Cost estimates plan procurement and manufacturing costs. Different types of cost estimates are used for different purposes, as discussed in the following chapters:

- **Chapter 9**
  We discuss how you create, mark, and release standard cost estimates to update the standard price and revalue inventory. We examine costing runs, which mass process standard cost estimates. You learn about marking allowances and optimizing costing runs.

  We look at how unit cost estimates are designed for use during the product development phase. They are easily changed because they are in a spreadsheet format, ideal for a development environment. We also analyze cost estimate user exits.
- **Chapter 10**
  We analyze how preliminary cost estimates calculate the planned costs for manufacturing orders and product cost collectors.

Cost Object Controlling

In the first ten chapters of this book, we focused on plan costs, which eventually allowed us to create, mark, and release standard cost estimates. In the next ten chapters, we'll examine actual costs and compare them with standard costs. This comparison provides a variance that we'll analyze in margin analysis in **Chapter 11**. During simultaneous costing, actual costs, including components and activities, are posted to production orders. Default activities and operation sequence, margin analysis, and detailed reports are also covered in this chapter.

Planned and Actual Costs

In **Chapter 12**, we determine how to plan and post actual costs to cost objects. We analyze how preliminary cost estimates plan costs for production orders and product cost collectors based on order type configuration.

Period-End Processing

During period-end processing, you carry out overhead, work in process, variance calculation, and settlement, as discussed in the following seven chapters:

- **Chapter 13**
  We look at how work in process (WIP) is configured and how the period-end step is executed.
- **Chapter 14**
  We look at production variance configuration, period-end processing, analysis, and examples of variance calculation types. We also discuss how SAP S/4HANA improves variance calculation performance.
- **Chapter 15**
  We analyze settlement configuration and period-end processing and discuss settlement and processing types. We also examine new transactions available with SAP S/4HANA.
- **Chapter 16**
  Actual costing values all goods movements within a period at the standard price. At the end of the period, the price and exchange rate differences are used to calculate an actual weighted average price for the period, called the periodic unit price.
- **Chapter 17**
  Sales order controlling scenarios involve customer orders that require you to procure components or assemblies manufactured specifically for individual customers or orders.
- **Chapter 18**
  Subcontracting, delivery, and cross-company-code costs are examined.
- **Chapter 19**
  Event-based processing involves goods movements and confirmations representing events that trigger overhead calculation according to the costing sheet. Depending on the order status, this triggers either the posting of a journal entry for the WIP or the cancellation of any existing WIP and the calculation of production variances.

Information System

In the last chapter, **Chapter 20**, we discuss the standard reporting available with SAP S/4HANA. We examine margin analysis, the variance split to assign each variance category to an account, and the cost component split to assign each cost component to an account. We examine standard SAP S/4HANA reports, drill-down and summarization reports, and line-item reports. We also discuss the production order information system and standard cost center reports.

Looking Ahead

After reading this book, you will have a clear understanding of the product cost controlling process, including initial planning, product cost planning, cost object controlling, and reporting. Although much detailed information is available in online documentation, this book provides an overview of how the many product costing details fit together in the integrated system.

This book will help you take advantage of the many aspects of controlling costs by following the product costing process flow from start to finish and explaining in detail the many analysis and reporting options available. This information can increase the profitability of your company by helping you analyze and understand the flow of costs and the reasons for variances between planned and actual costs. Companies that regularly analyze production variances and take immediate corrective action gain a competitive advantage over companies that react only after increased costs or inefficient processes have already decreased profitability.

Readers do not need a detailed knowledge of accounting or production processes to understand the concepts and details discussed in this book.

You can contact the authors at *jjordan@erpcorp.com* and *janet.dorothy.salmon@sap.com*.

Acknowledgments

I'd like to acknowledge the support and encouragement of the SAP Controlling community. Thanks for your many requests for a new edition based on the latest version of SAP S/4HANA. We welcome your feedback. I'd also like to thank Csilla, Hapci, and Tejfol.

—**John Jordan**

Thank you to John for getting me involved in what was already the classic work on product cost controlling, and thank you to my husband, Nick, and children, Martin and Lucy, for their endless patience when I am writing.

—**Janet Salmon**

We would also like to thank Rachel Gibson and Emily Nicholls for their encouragement and hard work in coordinating and editing this book's manuscript.

—**John Jordan and Janet Salmon**

Introduction

What is product cost controlling, and why do we need it? It addresses the requirements of different stakeholders as follows:

- Engineers responsible for designing the product
- Product managers responsible for the market success of the product
- Plant controllers who ensure the product is manufactured efficiently and create standard costs
- Inventory accountants who manage the inventory accounts of the goods in stock

Inventory accountants manage the subledger, decide whether to work with standard or actual costs, and handle the inventory balance sheet at year-end.

While product costing might appear to focus on manufacturing, a production plant is integrated within an organization and with external business partners. Plants can sell products to distribution centers and to selling companies in the market. This introduces intercompany relationships and the need to represent values from a corporate point of view and legal entities in the supply chain.

We'll begin our journey by setting scenarios, defining some terms, and sharing SAP S/4HANA screens. In Chapter 1, we begin planning for standard cost estimates with initial planning 6–12 months before we release standard cost estimates and revalue inventory.

Sales creates a 12-month sales plan, which you convert into a production plan. The production plan is key to the following:

- Planning component quantities required for the next 12 months so purchasing can enter supplier quotations in purchasing information records.
- Plan activity quantities are required at each work center to assemble components into products. The cost center plan costs divided by the plan activity quantity provide the plan activity rate.

In Chapter 2 through Chapter 5, we explain how master data is a prerequisite to initial planning. We discuss general ledger accounts, cost centers, activity types, material masters, bills of materials (BOMs), work centers, operations, and routings.

The next prerequisite for creating cost estimates, discussed in Chapter 6 through Chapter 9, is configuring cost components, costing sheets, and costing variants. These provide the system with instructions on how to obtain the purchase price for components and calculate work in process (WIP), overhead, and variances.

In the first half of the book, Chapter 1 through Chapter 10, we discuss plan costs and how to set up standard cost estimates. Then, in the second half, Chapter 11 through

Chapter 20, we discuss actual costs and how you compare plan and actual to gauge purchasing and manufacturing performance. Of course, there are many details to discuss, and we break down the process flow into easy-to-follow chapters.

This book is designed for end users, managers, consultants, and in-house experts responsible for configuration or design decisions, such as whether to value inventory at standard or actual costs. We provide examples and recommendations to guide you on your journey.

This book explains how to achieve your requirements with SAP S/4HANA and SAP S/4HANA Cloud. We provide different configurations depending on whether you're running on-premise or in the cloud based on best practices and standard configuration settings provided by SAP.

SAP S/4HANA: Deployment Options

We previously published a product cost controlling book based on the on-premise edition of SAP ERP and SAP S/4HANA. There are now also two cloud approaches: SAP S/4HANA Cloud Private Edition and SAP S/4HANA Cloud Public Edition.

In this book, when we explain transactions and describe the configuration settings, the screens are identical for SAP S/4HANA Cloud Private Edition and on-premise SAP S/4HANA. We don't specifically explain SAP S/4HANA Cloud Private Edition settings.

When we explain how to run SAP S/4HANA Cloud Public Edition, the business transactions are via SAP Fiori (a web interface) rather than the SAP GUI. The system configuration is delivered via best practices, with only minor changes possible with self-service configuration user interfaces (SSCUI).

When we discuss SAP S/4HANA Cloud, we mean SAP S/4HANA Cloud Public Edition. The SAP Fiori user interface is role-based, and we'll begin instructions with the application's default role and then explain how to find the tile to run the application.

If you use SAP S/4HANA Private Edition or on-premise SAP S/4HANA, you can choose whether you use SAP GUI transactions or SAP Fiori. We give you the transaction code and the menu path for every SAP GUI application. For example, to create a material cost estimate, you can either enter Transaction CK11N, as shown in Figure 1, or follow the menu path **Accounting • Controlling • Product Cost Controlling • Product Cost Planning • Material Costing Cost Estimate with Quantity Structure • Create**. Figure 1 shows both the transaction code and the menu path. Both options will take you to the **Create Material Cost Estimate with Quantity Structure** screen shown in Figure 2, and from then on, the instructions are identical.

You'll quickly follow how to work with the SAP Easy Menu, and we typically show the initial screen after you enter the transaction code or follow the menu path to reach **Create Material Cost Estimate with Quantity Structure** in this example. We'll then provide instructions on how to complete the fields and access the cost estimate.

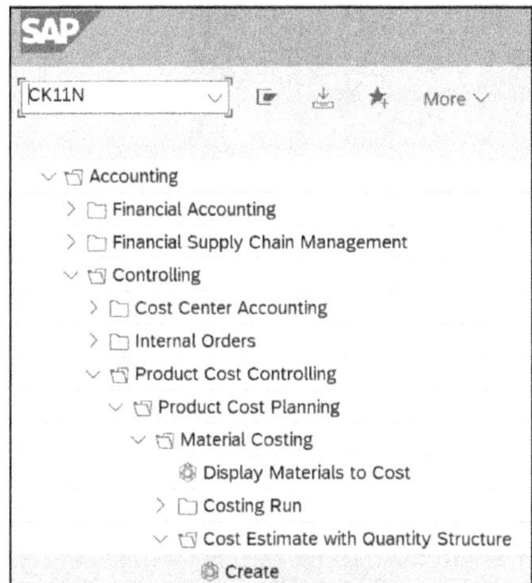

Figure 1 SAP Easy Menu: Create Cost Estimate with Quantity Structure

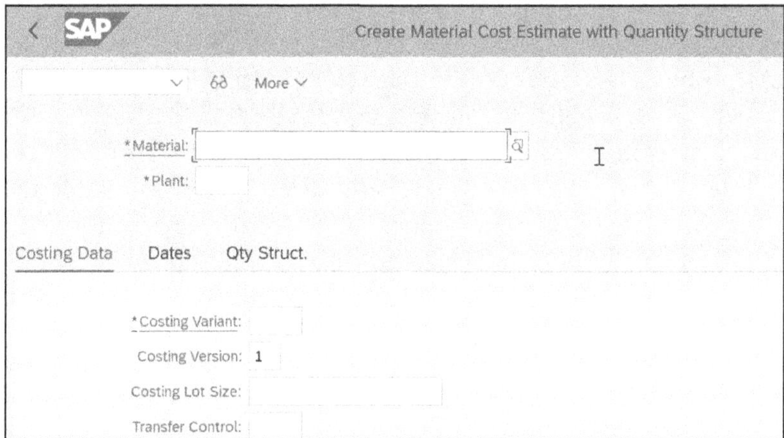

Figure 2 Initial Screen - Create Material Cost Estimate with Quantity Structure

With SAP Fiori, transactions are delivered as *tiles* within a *role*. The product costing delivered roles are:

- Cost Accountant—Inventory
- Cost Accountant—Production
- Cost Accountant—Overhead
- Cost Accountant—Sales

Figure 3 shows the tiles for the sample applications in the SAP Fiori launchpad for the Cost Accountant—Inventory role. An inventory accountant aims to create cost estimates, update material prices, and value inventory at period close. Each application is represented as a tile, and you access each app by clicking on the tile. In this example, let's select the **Manage Material Valuations** app.

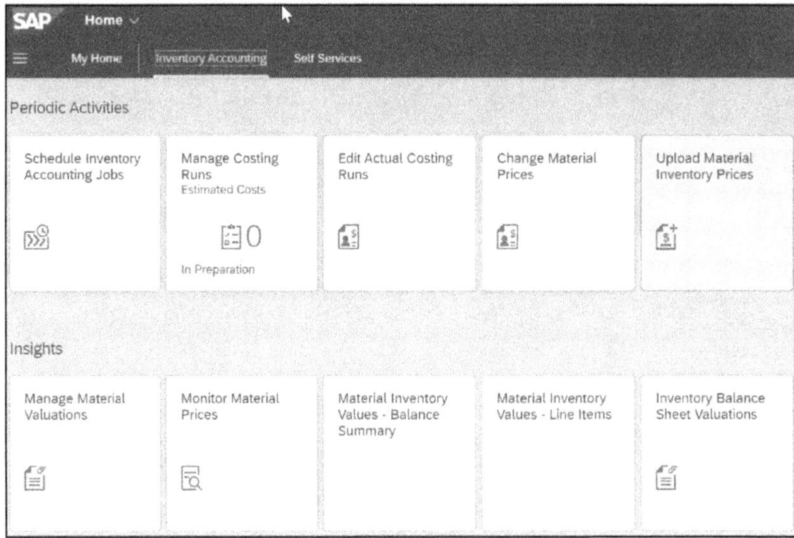

Figure 3 Sample SAP Fiori Apps for Inventory Accounting

Figure 4 shows the results after you select the **Manage Material Valuations** app (SAP Fiori ID F2680), complete the **Material** and **Plant** fields, and select **Go**. The material valuations are displayed. You can navigate to **Material Price Analysis**, **Change Future Prices with Reference**, **Release Planned Price Changes**, and **Delete Future Prices** for the selected material by choosing the navigation options at the top of the result list.

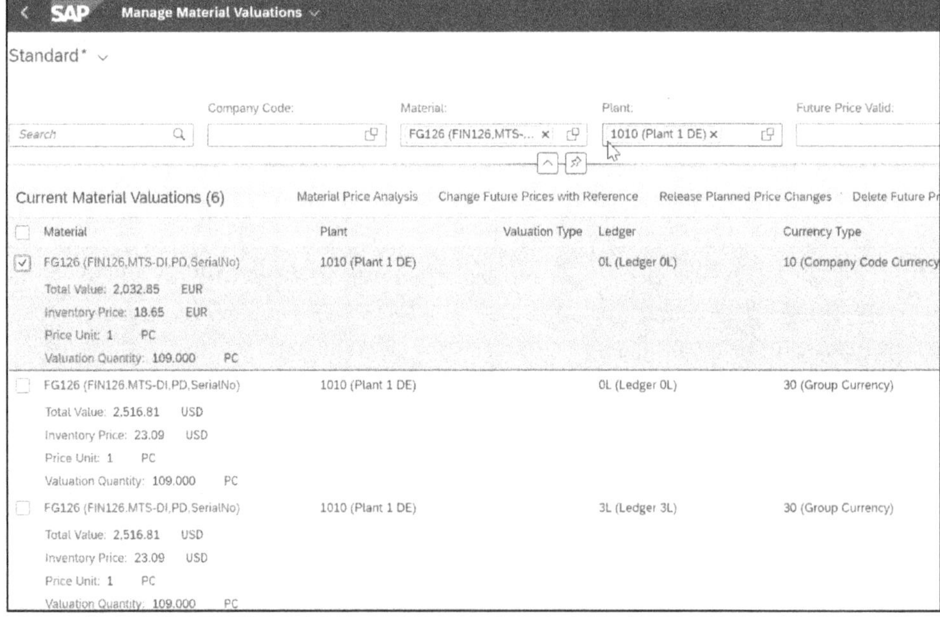

Figure 4 Manage Material Valuations App

You can also access the applications by entering the transaction code in the search box shown in Figure 5.

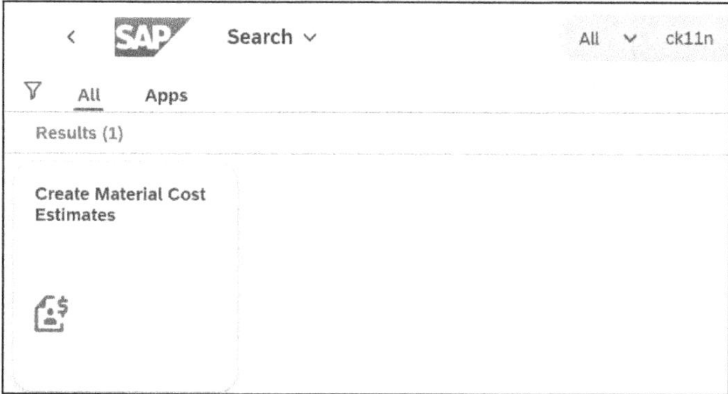

Figure 5 Enter the Transaction Code to Search for an SAP Fiori App

We list the SAP Fiori apps when available with the GUI transaction. There isn't always a one-to-one relationship between SAP GUI transactions and SAP Fiori apps. The Manage Material Valuations app covers several SAP GUI transactions. The tile in Figure 5 calls up a classic transaction with only minor changes in the user interface.

Material Prices and the Material Ledger

The material master is fundamental in SAP S/4HANA. There are many stakeholders. It must exist for all materials purchased, manufactured, or sold and is referenced by purchase orders, production orders, and sales orders. There can be only one material master for each material in a plant.

Important fields for costing include price and valuation class (link to account determination). Figure 6 shows a material master with tabs of interest to us: **Accounting 1**, **Accounting 2**, **Costing 1**, and **Costing 2**. You display a material master with Transaction MM03 or via the menu path **Logistics • Material Management • Material Master • Material • Display • Display Current**, enter the material, choose the **Accounting 1** view, and choose the plant. You can display all material price information and key parameters for assigning the material to a general ledger account and a profit center during goods movements. We'll return to this screen in Chapter 3 and explain how to set material prices for raw materials and how to include them in cost estimates. In Chapter 9, we'll explain how to create a costing run and set the standard costs for assemblies and finished goods.

Material valuation in SAP S/4HANA is based on the Material Ledger. The **ML Act.** (Material Ledger active) checkbox in Figure 6 is always selected. Every journal entry for a goods movement includes the following:

- The latest stock value for the material
- The source of the material (purchase from a supplier, in-house production, or subcontracting)
- How it is used (next level of production, sale to a customer, sale to an affiliated company)

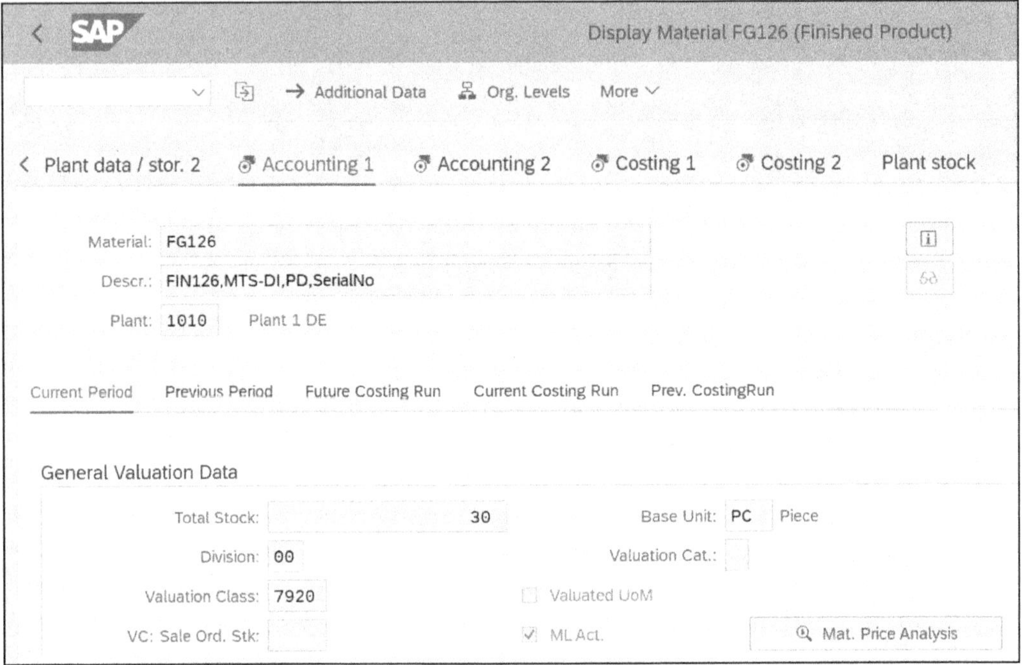

Figure 6 Material Master with Active Material Ledger

Every journal entry that can result in a price change, such as a supplier invoice or production order settlement, shows the price change in price and exchange rate variances.

You see the details by clicking the **Mat. Price Analysis** (material price analysis) button shown in Figure 6 to display the screen shown in Figure 7.

In Figure 7, **30** pieces of material **FG126** are currently in **Inventory** with a **PrelimVal** (preliminary valuation) of **150 EUR**. **15** were in stock at the start of the period (**Beginning Inventory**), and **15** were received during the period (**Receipts**). To see the source of the **15** pieces, select **Receipts • Production • BOM/Routing** to show that the materials were produced internally using one production order for 10 pieces and one for 5 pieces with a preliminary valuation of EUR 10.00 per piece. (**100.00/10** and **50.00/5**).

Production variances of **90,00-** were incurred on production order **1000118663** and **30,00-** on production order **1000117342**. These variances provide the basis for calculating the *actual costs* for each material at period close. Actual costing is optional in SAP S/4HANA. In Figure 7, since the material is flagged as **Not Relevant for Settle** (not relevant for settlement), actual costing is inactive for this material in this plant.

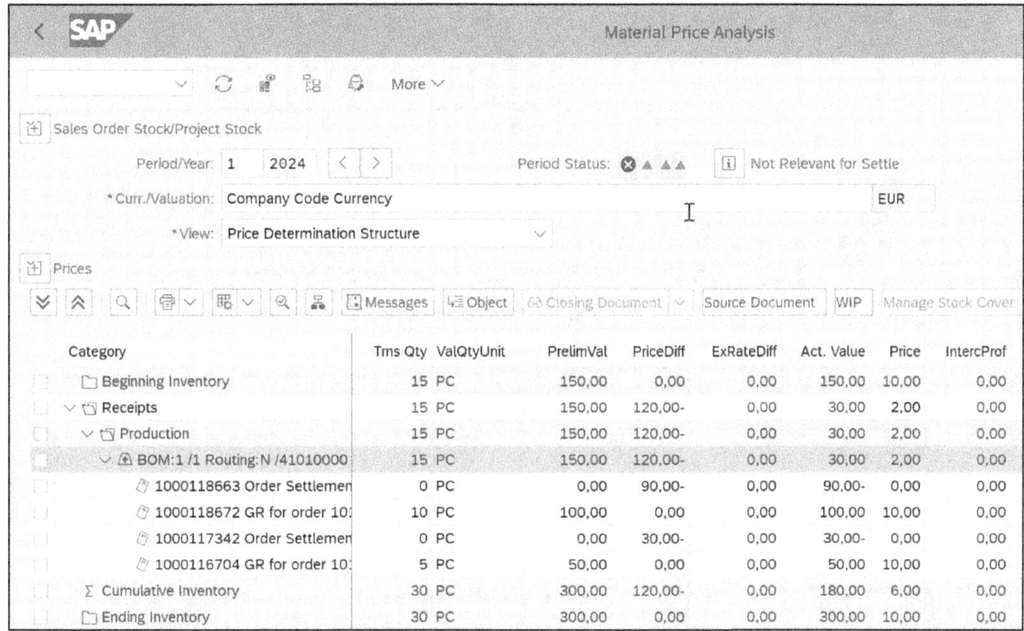

Category	Trns Qty	ValQtyUnit	PrelimVal	PriceDiff	ExRateDiff	Act. Value	Price	IntercProf
Beginning Inventory	15	PC	150,00	0,00	0,00	150,00	10,00	0,00
Receipts	15	PC	150,00	120,00-	0,00	30,00	2,00	0,00
Production	15	PC	150,00	120,00-	0,00	30,00	2,00	0,00
BOM:1 /1 Routing:N /41010000	15	PC	150,00	120,00-	0,00	30,00	2,00	0,00
1000118663 Order Settlemen	0	PC	0,00	90,00-	0,00	90,00-	0,00	0,00
1000118672 GR for order 10	10	PC	100,00	0,00	0,00	100,00	10,00	0,00
1000117342 Order Settlemen	0	PC	0,00	30,00-	0,00	30,00-	0,00	0,00
1000116704 GR for order 10	5	PC	50,00	0,00	0,00	50,00	10,00	0,00
Σ Cumulative Inventory	30	PC	300,00	120,00-	0,00	180,00	6,00	0,00
Ending Inventory	30	PC	300,00	0,00	0,00	300,00	10,00	0,00

Figure 7 Material Price Analysis, Showing Relevant Goods Movements

Notice also the **Curr./Valuation:** field in Figure 7. The material ledger supports multiple currencies, and you can scroll from the **Company Code Currency** (**EUR**) through the currencies in which the values for the material are stored.

The example shows every goods movement and price change recorded for the material, and you can access the accounting documents by selecting any of the documents shown in Figure 7. Let's choose material ledger document **1000118672 GR for order** and explore the fields in the accounting document. Figure 8 shows the journal entry for the goods receipt from the production order with a local currency value of **100,00 EUR**. To see the other currencies, either click the **Display Currency** button or adjust the report layout to include the other currencies, as we'll explain in the next section. Let's examine the relationship between the accounting document and the material ledger document in the material master, which summarizes the stock levels and prices.

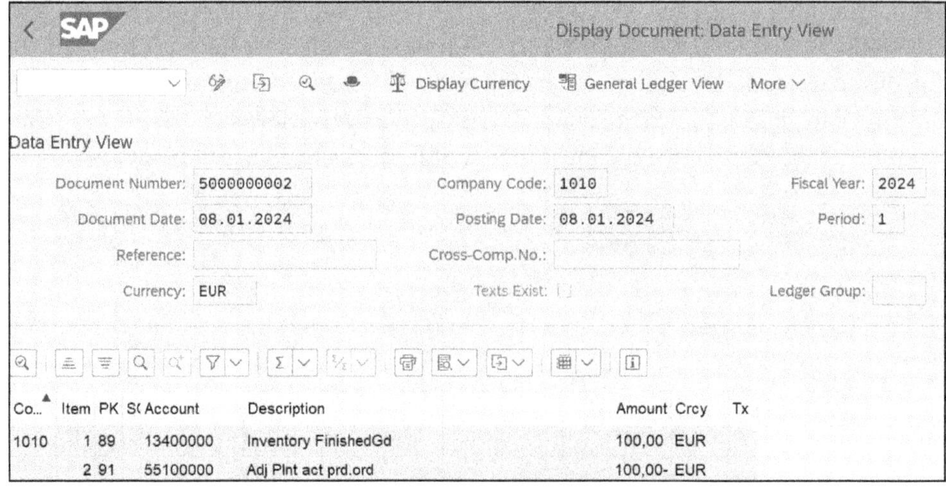

Figure 8 Accounting Document for Goods Movement (Data Entry View)

Accounting Documents

If you've worked with SAP R/3 or SAP ERP, the accounting document shown in Figure 8 will look familiar. Behind the scenes, however, there has been a change in SAP S/4HANA: financial and management accounting are merged into a single accounting document via the Universal Journal. The new journal entry includes the fields from financial accounting and many more previously stored in separate documents in Controlling, profit center accounting, and special ledgers.

In many cases, you can display these fields directly in the accounting document instead of separate applications. To do this, choose the **Manage Layouts** grid icon (second from right) in Figure 8 and add fields to the current layout via the **Change Layout** screen in Figure 9. The fields in Figure 8 are shown under **Displayed Columns** on the left, and many additional options are available under **Column Set** on the right.

To add **Material** to the layout, select **Material** from the list on the right, use the arrows in the middle to move it into the list of **Displayed Columns** on the left, and then use the up/down arrows to move the **Material** into the required sequence as shown in Figure 10. If you use this layout regularly, click the **Save As** button and save it under your chosen name.

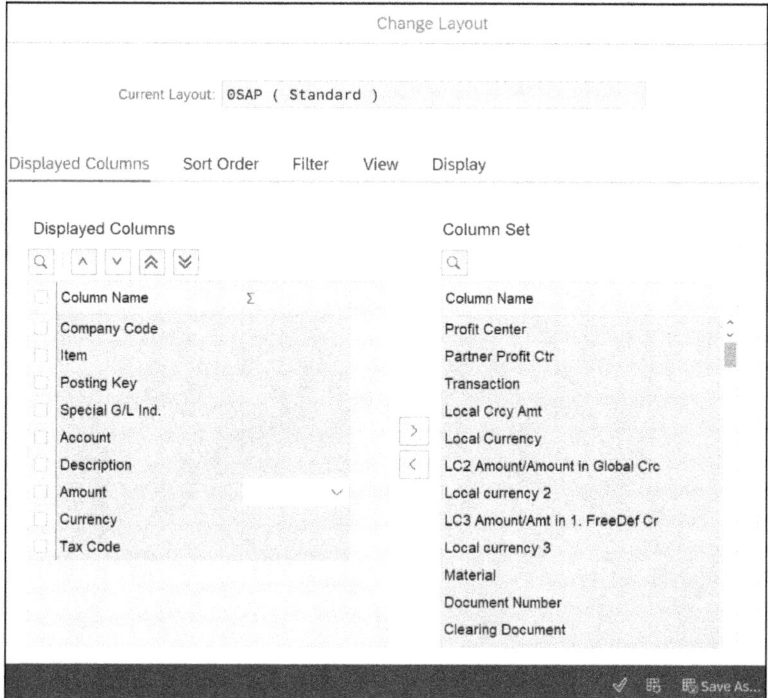

Figure 9 Fields in Document Layout

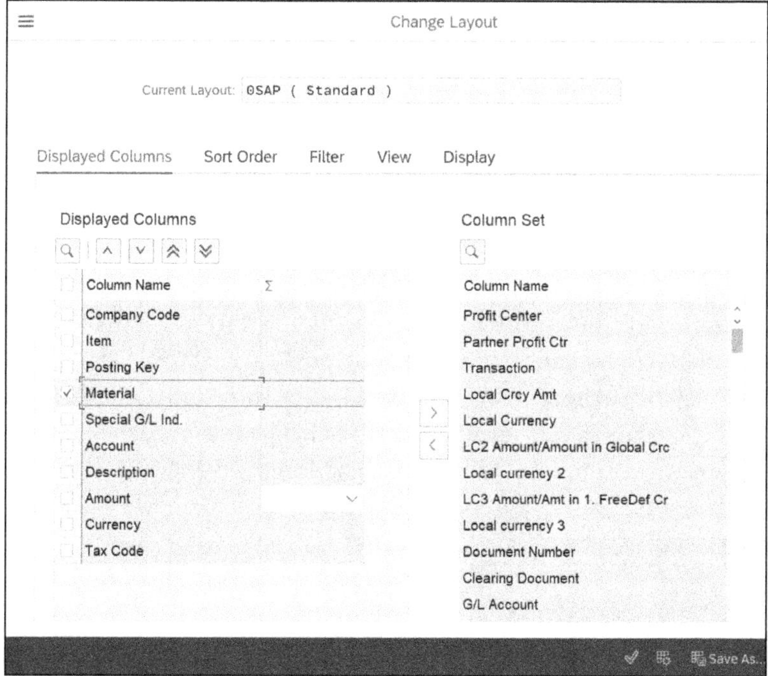

Figure 10 Adding the Material Field to the Layout

When you close the **Change Layout** dialog and return to the accounting document, the result is shown in Figure 11 where the **Material** field is now shown. This process represents the basis for all financial reporting. You can display all the fields you might expect in the trial balance and financial statements and drill down to the associated fields for the material, such as asset, cost center, order, and others.

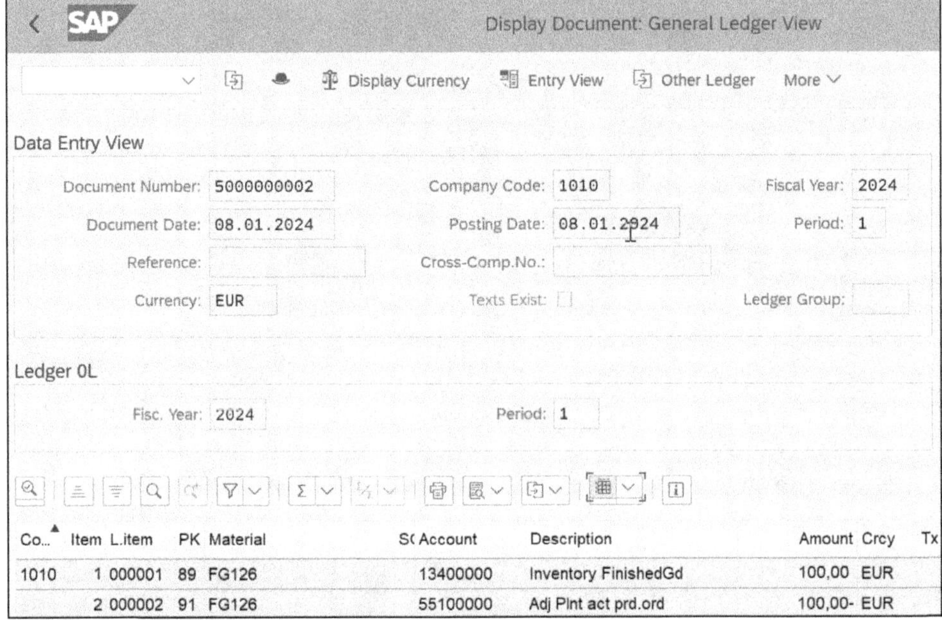

Figure 11 General Ledger View of Accounting Document, Showing Material Received

The goods movement journal entry document is displayed in spreadsheet format. In the same document, you can view information from the goods movement, financial accounting, Controlling, profit center accounting, cost of goods sold (COGS) reporting, and group reporting. The Universal Journal, containing all the fields needed for financial reporting, is one of the key benefits of SAP S/4HANA.

Figure 12 shows the same document with **Order** (production order) added to illustrate how the document now includes information previously only available in the Controlling document. Figure 13 shows the same document, this time with **Profit Center** added. You can create as many layouts as you need. You can use **Change Layout** to choose from over 250 fields, though only a subset is updated for a goods movement. We'll return to this merge in Chapter 2, when we discuss controlling master data and how general ledger accounts and cost elements merge in SAP S/4HANA.

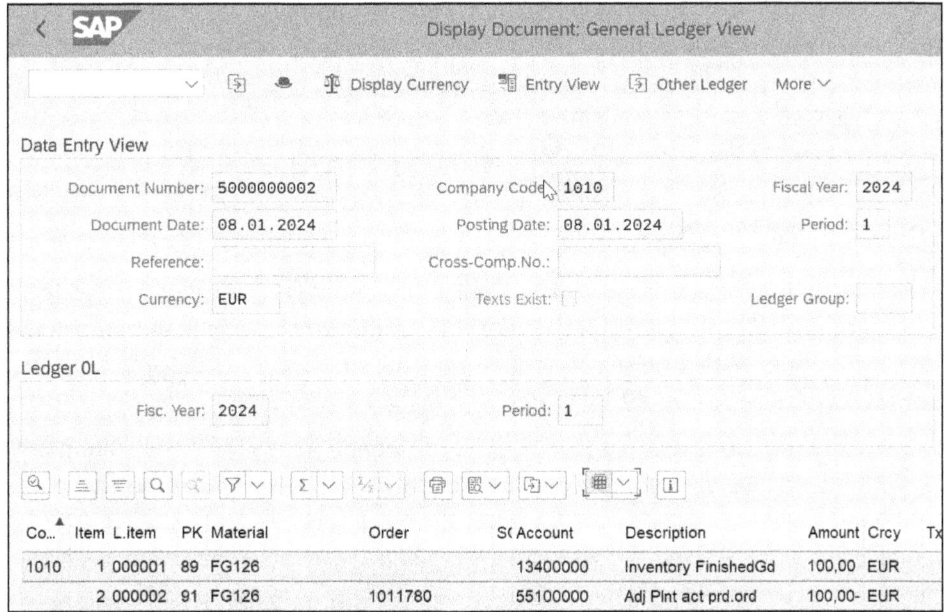

Figure 12 General Ledger View of Accounting Document, with Production Order

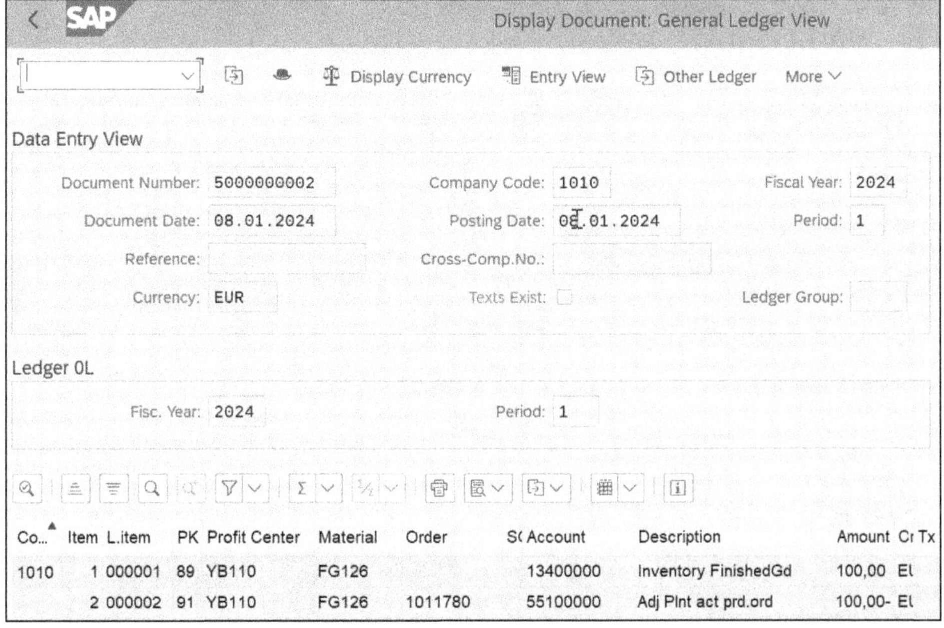

Figure 13 General Ledger View of Accounting Document, with Profit Center

When you work with the Universal Journal, the default view of the document is the **Data Entry View** (refer back to Figure 8), but many fields you might want to analyze are included in the **General Ledger View** (refer back to Figure 11, Figure 12, and Figure 13).

The **Data Entry View** includes those fields provided by the delivering application, while the **General Ledger View** includes derived fields, such as the profit center, partner profit center, and functional area. The **General Ledger View** contains the fields of the Universal Journal and is the basis for all financial reporting.

Contribution Margins

In SAP S/4HANA, you can merge financial and management accounting into one document. The production order is part of the Universal Journal, along with the material and profit center. When we sell a product, we are interested in its costs and the customer and sales fields, like the sales organization and the division.

We want to compare the COGS with the revenue. Figure 14 shows the **Product Profitability with the Production Variances** app (SAP Fiori ID W0182), which displays the profitability information in the Universal Journal, the billed revenue, and the COGS (**COGS - Variable** and **COGS – Fixed**) for sales document **289925**. The revenue is captured when the customer is invoiced, and the COGS is captured when the goods are delivered and is derived from the standard cost estimate.

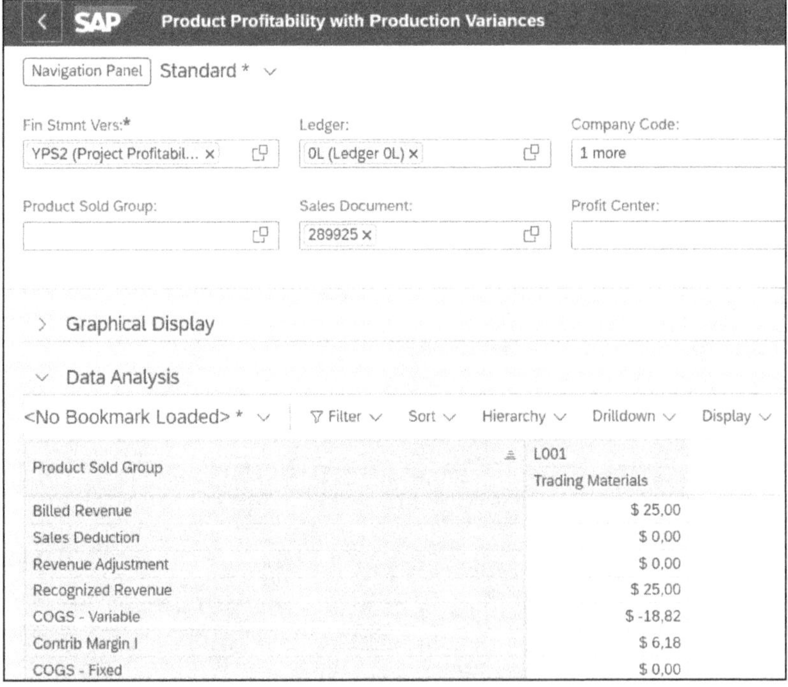

Figure 14 Product Profitability with Production Variances App

In Figure 14, the **Sales Document** and the **Product Sold Group** are in the selection screen at the top. The **Product Sold Group** filters the report, while many fields are derived from the sales document and included in the profitability segment.

Figure 15 shows a profitability segment for a sales document, where the **Customer**, the **Product**, the **Company Code**, and the **Plant** deliver a reporting view from the Universal Journal. SAP S/4HANA Cloud includes a fixed set of fields for margin analysis, which an administrator or key user can extend with the Custom Fields and Logic app (SAP Fiori ID F1481). The fields that comprise the profitability segment in SAP S/4HANA are part of the operating concern, which is defined in the configuration. The Universal Journal contains placeholders for the fields added as part of the operating concern and uses cloud extensibility so the fields can be used in profitability reporting.

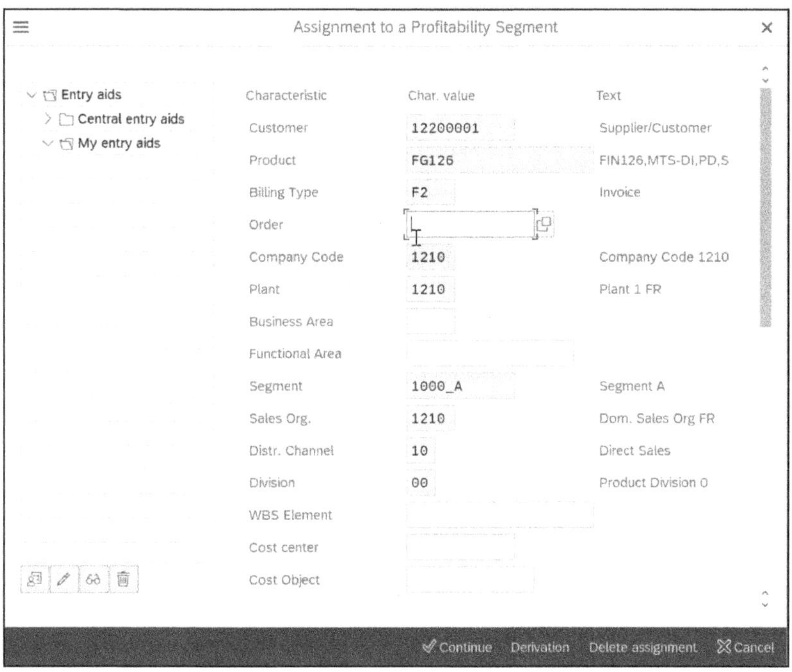

Figure 15 Assignment of Sales Order to Profitability Segment

In SAP Fiori, the **Customer** and the **Product Sold** are shown directly in the Display Line Items–Margin Analysis app (SAP Fiori ID F4818), as shown in Figure 16. You can add the remaining profitability segment fields with the **Settings** button to select further fields.

The Universal Journal also offers account assignment types to select accounting documents. In Figure 16, the **Account Assignment Type EO** represents the profitability segment and the fields in Figure 15. In the document shown in Figure 11 through Figure 13, you choose **Account Assignment Type** OR for order.

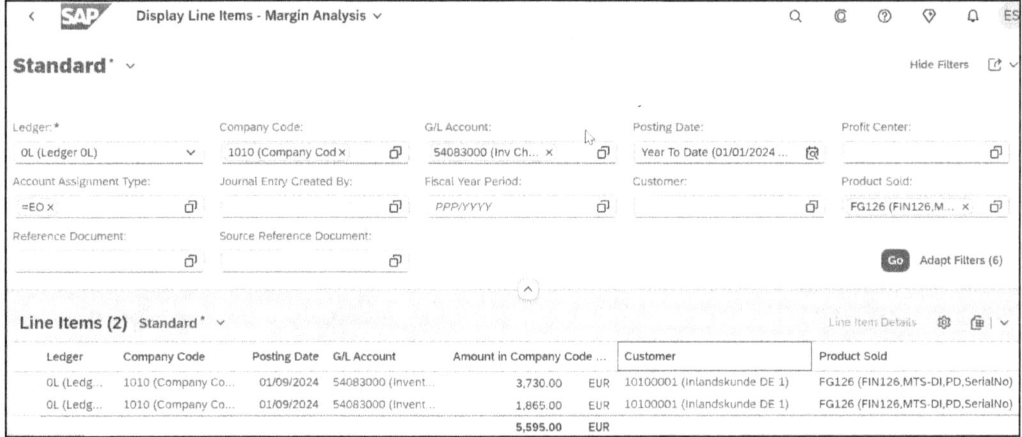

Figure 16 Display Line Items - Margin Analysis App, Showing Customer and Product

Cost Estimates and Product Profitability

To find the COGS in Figure 14, return to the material master previously shown in Figure 6 and navigate to the **Costing 2** tab, as shown in Figure 17. Click the **Current** button to display the current cost estimate, as shown in Figure 18.

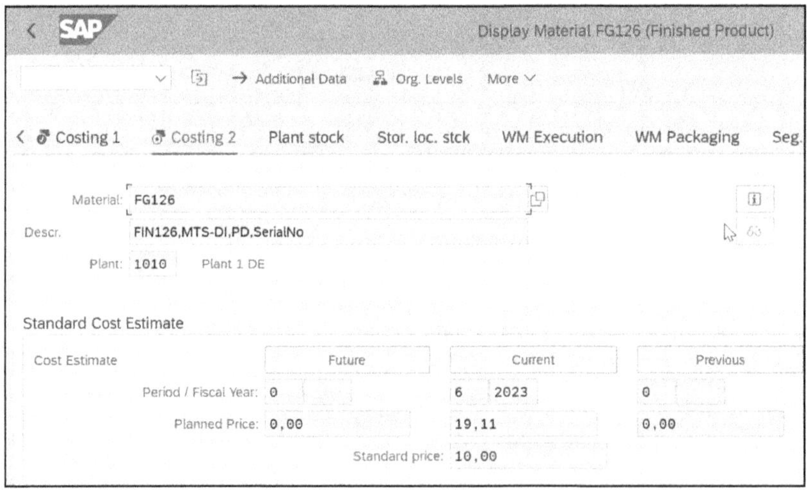

Figure 17 Costing View of Material Master

Figure 17 displays a material master **Costing 2** view. With this screen, you can quickly locate standard cost estimates by clicking the **Future** (marked standard cost estimates), **Current**, or **Previous** (previous standard cost estimate) buttons.

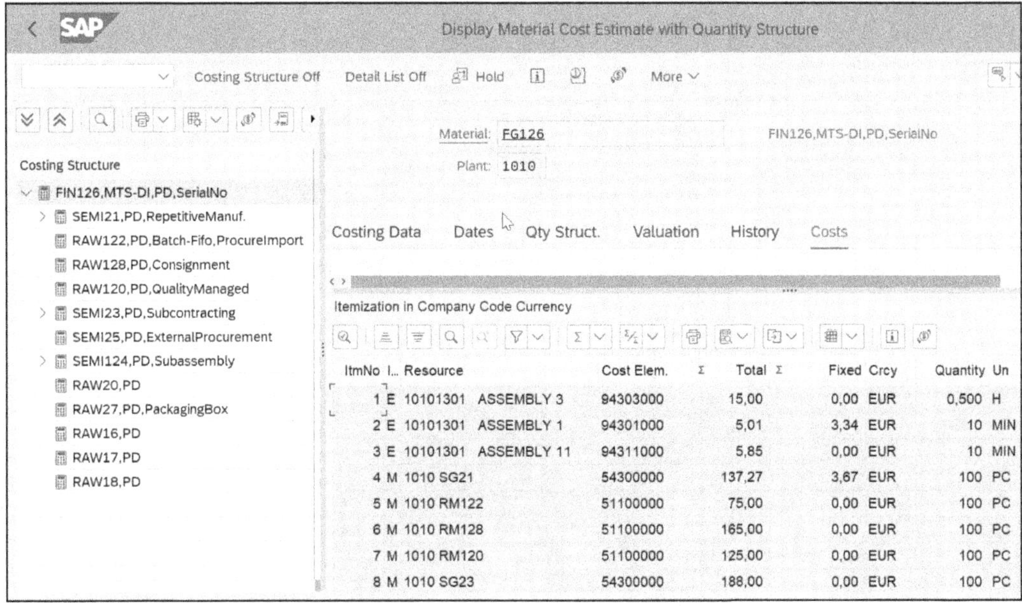

Figure 18 Material Standard Cost Estimate with Quantity Structure

This is the cost estimate for **Material FG126** in **Plant 1010**. You can also access the cost estimate via the menu path shown in Figure 1, the **Manage Material Valuations** SAP Fiori tile shown Figure 3, and the transaction search field in Figure 5. In the **Costing Structure** on the left, you see the BOM structure with the raw materials and semi-finished products used during manufacture. Click on each item in the structure to show the detailed costs for the chosen component on the right of the screen. In the **Itemization in Company Code Currency** section, you see the materials used and the activities performed to convert the raw materials into the finished product **FG126**.

The standard cost delivers the first contribution margin, as shown previously in Figure 14. During manufacturing, variances are assigned to the product for an additional contribution margin.

Having set the scene, let's look at the initial planning process in Chapter 1.

Chapter 1
Initial Planning

Initial planning allows you to plan production and procurement quantities and costs based on planned sales quantities.

With initial planning, you can take advantage of the fully integrated SAP S/4HANA. You enter a sales plan and determine a production plan. Together with cost center planning, this allows you to calculate plan activity rates and standard cost estimates, which plan the cost of manufacture for each product. This process allows you to plan costs based on sales quantities, which is a best practice in manufacturing.

One of the main advantages of integrated planning is that you can compare planned costs with actual costs and determine the reason for the differences, which is the basis for variance analysis. You can then use this as an iterative process to improve your period-by-period and year-by-year sales and production planning.

Many alternatives for entering and processing plan data are available in SAP S/4HANA. In this chapter, we'll examine entering sales data in profitability analysis and sales and operations planning and then follow a typical flow to long-term planning and to cost center accounting. A sales manager typically enters a sales plan into either margin analysis or sales and operations planning and analyzes multiple sales scenarios. You convert a preferred sales plan into a production plan, which you then transfer to long-term planning.

Long-term planning accesses a *bill of materials (BOM)* and routing to determine component procurement and cost center capacity requirements. A BOM is a hierarchical structure of components and subassemblies, while a *routing* lists operations and standard values required to manufacture an assembly.

Activity-scheduled quantities are then transferred from long-term planning to cost center accounting, where, together with cost center planning data, you calculate activity and overhead rates.

To calculate the plan activity price, divide the cost center plan's activity-dependent costs by the plan activity quantity. You can enter the plan activity price manually or automatically calculate it with Transaction KSPI. Cost center under/over absorption will result as the actual activity price and quantity differ from the plan activity price and quantity.

We'll start initial planning by entering a sales plan in costing-based profitability analysis.

1.1 Margin Analysis

Sales managers can enter the quantity of finished products they expect to sell in future periods with Transaction KEPM or by following the menu path **Accounting • Controlling • Profitability Analysis • Planning • Edit Planning Data**. Enter the operating concern (costing-based profitability analysis or margin analysis) and press Enter to display the planning levels. Expand the planning levels and double-click a planning package to display the planning methods. Expand the Enter planning data node and double-click a planning package to enter planning data.

After you have entered sales data in costing-based profitability analysis or margin analysis, you can do one of the following:

- Transfer quantities to sales and operations by following the menu path **Accounting • Controlling • Profitability Analysis • Planning • Integrated Planning**.
- Create a planning scenario with Transaction MS31 and access the costing-based profitability analysis data directly with long-term planning, which we'll discuss further in Section 1.3.

We'll follow an example of entering data in sales and operations planning so you can see this functionality.

1.2 Sales and Operations Planning

You enter a sales plan for future periods and fiscal years directly into sales and operations planning, or you can transfer the data from other components, such as margin analysis. You enter the sales plan for a product group, which you then disaggregate to lower materials or enter directly for individual materials. The production plan is determined from the sales plan and then transferred to long-term planning. If you determine the production plan from the sales plan on a spreadsheet, you can enter it manually into planned independent requirements in demand management.

You enter a sales plan for a material and a plant into sales and operations planning with Transaction MC88 or via the menu path **Logistics • Production • SOP • Planning • For Material • Change**. Enter **Material** and **Plant** and select a version. The data entry screen in Figure 1.1 is displayed.

You enter sales **Plan** quantities in the **Sales** row and production **Plan** quantities in the **Production** row. Figure 1.1 shows an example of the production plan offset one month forward in time from the sales plan to help ensure that sales plan delivery dates are met. The production plan may differ from the sales plan due to known capacity requirements.

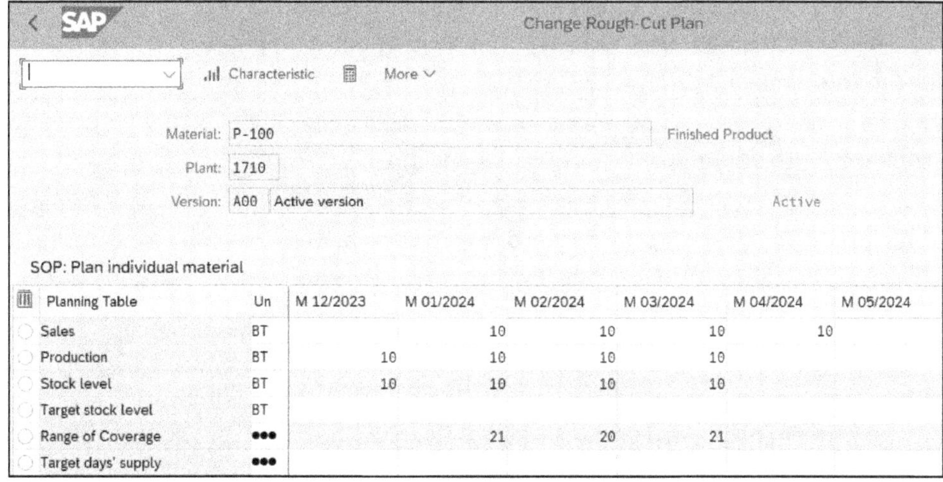

Figure 1.1 Sales and Operations Planning Entry Screen

Example: Adjust Production for Sales Plan

If the sales plan is to sell 40 in month 04/2024, you may adjust the production plan to manufacture a quantity of 10 in the four preceding months.

After you determine the production plan, you transfer it to demand management with Transaction MC74 or via the menu path **Logistics • Production • SOP • Planning • For Material • Transfer Material to Demand Management**. The screen shown in Figure 1.2 is displayed.

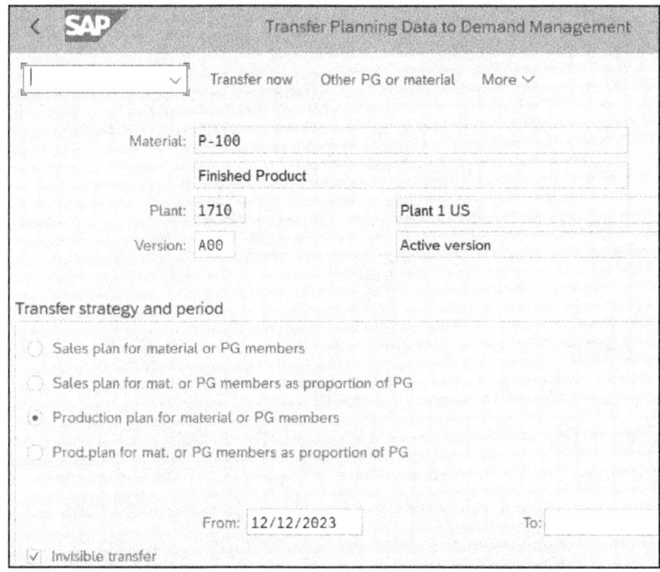

Figure 1.2 Transfer Production Plan to Demand Management

You can transfer the **Sales plan** or **Production plan** for a **material** or product group (**PG**) **members** with your selection in the **Transfer strategy and period** section and then clicking the **Transfer now** button at the top.

SAP Integrated Business Planning for Supply Chain

Sales and operations planning is being replaced by SAP Integrated Business Planning for Supply Chain (SAP IBP), which supports all sales and operations planning features. Sales and operations planning is an interim solution to allow a smooth transition from SAP ERP to SAP S/4HANA and SAP IBP. See SAP Note 2268064 for more details.

1.3 Long-Term Planning

Now that we've converted the sales plan into a production plan and transferred the production plan to demand management, let's start working with the information in long-term planning.

Long-term planning simulates future material demands and allows you to enter medium- to longer-term production plans. Medium-term production plans involve quantities between three months and three years into the future, while longer-term production plans can plan production quantities as far into the future as you need.

The production plan represents planned independent requirements that meet two downstream prerequisites to create cost estimates:

- Planned independent requirements generate requirements for purchased items. With these, you request vendor quotes, negotiate raw material prices, and ensure that purchasing info records are current. Cost estimates use purchasing info records to determine the purchase price of components.
- You transfer scheduled activity requirements to cost centers. Cost-center planned costs, divided by scheduled activity requirements, provide the planned activity price, which cost estimates use to determine activity costs.

Let's walk through the procedure in the following sections to generate planned independent requirements that meet the two prerequisites for creating cost estimates.

1.3.1 Transfer the Production Plan

You transfer the production plan from sales and operations planning to generate planned independent requirements, as previously discussed in Section 1.2, or you can enter the production plan directly with Transaction MD62 or via the menu path **Logistics • Production • Production Planning • Long-Term Planning • Planned Independent Requirements • Change**. The data entry screen shown in Figure 1.3 is displayed.

Change Planned Independent Requirements: Initial Screen

User Parameters More

Planned Independent Requirements for

Material: P-100

Product group:

Requirements Plan:

Ext. Req. Plan:

MRP Area:

Plant: 1710

Selection Parameters

Requirements type:

Selected version 00 Requirements Plan

All active versions

All active/inactive versions

Planning Horizon

From: 12/13/2023 To: 01/16/2025 Planning period: M Month

Figure 1.3 Planned Independent Requirements Initial Screen

Complete the **Material** and **Plant** fields, select **All active versions**, and enter the **Planning Horizon From** and **To** dates. You typically set the planning horizon at one to three years; however, this depends on your planning process. You typically set the **Planning period** to **M Month**. Press Enter to display the planned independent requirements planning table shown in Figure 1.4.

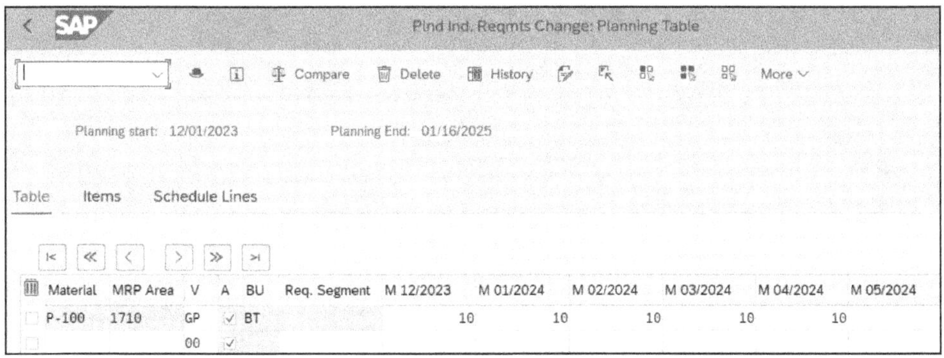

Figure 1.4 Change Planned Independent Requirements

The requirements displayed in Figure 1.4 correspond with the production plan transferred from sales and operations planning that we saw earlier in Figure 1.1. You can change the requirements or enter additional requirements directly. The **A** (version active) checkbox shown in Figure 1.4 determines if the requirements are relevant to operative *material requirements planning (MRP)*. If relevant to operative MRP, requirements will generate planned orders, which you can convert into production orders for in-house production, purchase requisitions, or purchase orders for external procurement. The system also explodes the BOM for in-house assemblies and generates dependent requirements for material components.

To generate dependent requirements, we must create a planning scenario combining all the long-term planning parameters as follows.

1.3.2 Create Planning Scenario

In long-term planning, you first define a planning scenario, which controls how you carry it out. To compare different versions of planned independent requirements, you create scenarios and compare the planning results. You create a planning scenario with Transaction MS31 or via the menu path **Logistics • Production • Production Planning • Long-Term Planning • Scenario • Create**. Give the planning scenario a name and description, and then select the long-term planning radio button. Press Enter to display the screen shown in Figure 1.5.

Figure 1.5 Planning Scenario Details

Complete the planning scenario details as follows:

1. The **Start** date is the beginning of the **Planning period for indep. requirements** for long-term planning.
2. The **End D:** (end date) is the end of the planning period. You use the planned start and finish dates to limit the long-term planning period.
3. You typically set the **Opening stock** at **1 Safety stock as opening stock.** Press F4 for more options.
4. Select **Dep. Reqmts for reorder point materials** to plan bulk materials in MRP.
5. The **Consider sales orders** checkbox determines whether you consider sales orders in long-term planning. Sales orders are automatically copied from operative planning, and you cannot change them in long-term planning. The customer requirements from operative planning consume planned independent requirements in long-term planning. If you don't select this checkbox, you only consider planned independent requirements. You can use this checkbox for scenario planning.
6. Select **Switch off planning time fence** to switch off the planning time fence in long-term planning for materials with a planning time fence maintained and active in MRP.
7. If you plan materials with the special procurement key, the system creates a collective order. If you do not select this indicator in long-term planning, the system does not create a collective order—it only displays individual planned orders, even if the materials have a special procurement key.
8. Select **Use make-to-order** to include make-to-order planning and project planning in long-term planning. The system automatically copies customer requirements from operative planning.
9. Select the **Switch off scrap calculation** checkbox to ensure scrap is not calculated in long-term planning. No assembly scrap is planned in the net requirements calculation, and component or operation scrap is not planned during the BOM explosion. Deactivating scrap calculation is useful for gross requirements planning, as you would like to plan the pure requirements due to cost evaluation and budgeting.
10. Select **Use gross lot size** so the lot size defined in the planning scenario is used for the lot size calculation in long-term planning for all materials, irrespective of the material-specific settings.

You assign planned independent requirements versions by clicking the **Planned Independent Requirements** button, and plants by clicking the **Plants** button. After you have checked the **Control parameters**, release the planning scenario for planning by clicking the **Release + Save** button.

After you release the planning scenario, you can only change the allocation of the planned independent requirements versions. You can only use released planning scenarios in a long-term planning run.

Click any checkbox on this screen and press F1 for detailed information.

Government Contracts

Government defense contracts can extend many years into the future, long before you create sales orders. You can determine production planning requirements by entering planned independent requirements in place of future sales orders. If you select the **Consider sales orders** checkbox in Figure 1.5, sales orders that you create in the future automatically consume the planned independent requirements that you manually entered. This avoids removing planned independent requirements as you create sales orders.

1.3.3 Long-Term Planning

Now that we've created a planned scenario, we are ready to run and evaluate long-term MRP and send the results to the purchasing information system.

You can carry out a long-term planning run either for an individual material or collectively for all materials in a plant. All planned independent requirements and sales orders within the planning scenario time frame, depending on the settings shown in Figure 1.5, are analyzed. The dependent requirements for all manufactured and purchased items are determined from the operative BOM. You can also carry out long-term planning with a separate BOM. You create simulative dependent requirements and simulative planned orders for all components during a long-term planning run that are for planning purposes only.

Production Versions Are Mandatory

Production versions are mandatory with SAP S/4HANA. They allow you to define which BOM alternative goes together with which routing alternative. They simplify MRP sourcing since there is only one option to determine BOM and routing alternatives to manufacture a material. See SAP Note 2267880 for more details.

You carry out a collective long-term planning run with Transaction MS01 for online processing, with Transaction MSBT for background processing, or by following the menu path **Logistics • Production • Production Planning • Long-Term Planning • Long-Term Planning • Planning Run**. The screen shown in Figure 1.6 is displayed.

The regenerative planning processing key **NEUPL** disregards whether planning-relevant changes were made to the material. All materials with an MRP-relevant planning type in the **MRP 1** view are processed.

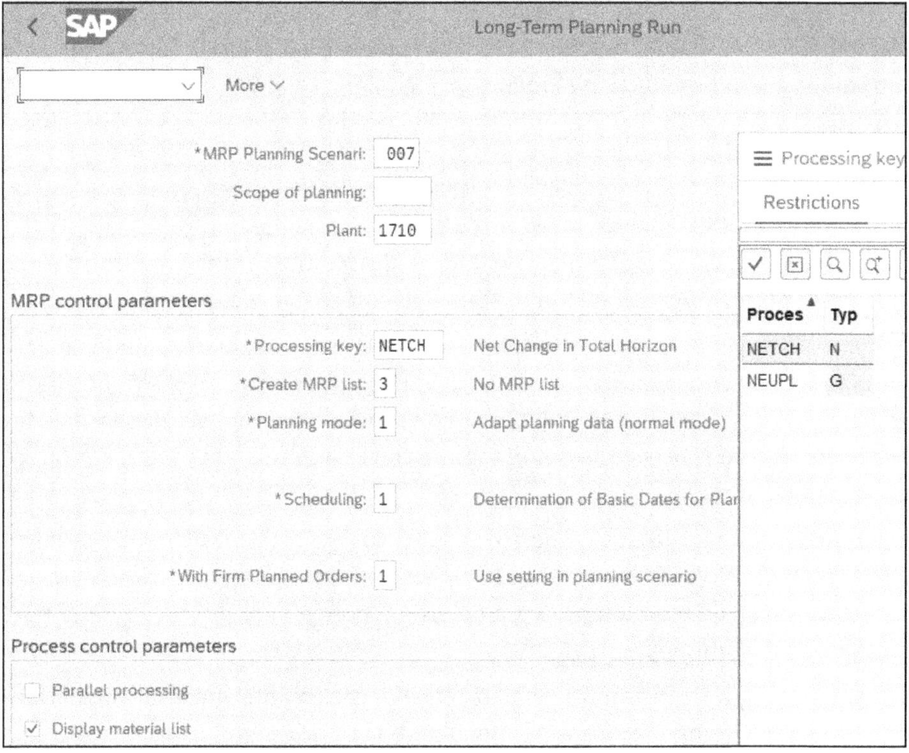

Figure 1.6 Collective Long-Term Planning Run

Planning Inside the Planning Horizon

The processing key **NETPL** is no longer supported with SAP S/4HANA, and planning inside a planning horizon is no longer available. Planning inside the planning horizon improves MRP runtime because fewer planned orders are created, although with a more complicated planning procedure. In SAP S/4HANA, it's no longer necessary to improve the MRP runtime, so MRP always covers all material demand. See SAP Note 2270241 for more details.

An **MRP list** displays the results of the last planning run for a material. It remains unchanged until the next MRP run. The MRP list contains *exception messages* that assist the planner. Exception messages indicate, for example, whether you should postpone an order or that the safety stock has been exceeded.

After you've reviewed all the planning run parameters, press Enter to start the planning run. When the planning run is complete, you'll see a results screen with its statistics and parameters. If you choose to create an **MRP list** in the selection screen shown in Figure 1.6, you'll also see a list of materials included at the top of the planning run results screen.

You can display the long-term MRP list at any time with Transaction MS05 for an individual material, by using Transaction MS06 for a collective display, or by following the menu path **Logistics • Production • Production Planning • Long-Term Planning • Evaluations**.

1.3.4 Transfer Requirements to Purchasing

Now that we've executed a planning run and evaluated the results, let's transfer the results to the purchasing information system and cost center accounting.

Long-term MRP generates simulative planned orders based on planned independent requirements. You don't convert simulative planned orders into purchase requisitions or production orders; they are for planning only. You can transfer simulative data for external procurement to the purchasing information system and evaluate for future purchasing requirements.

This information can be used to generate vendor *requests for quotations (RFQs)*, negotiate component prices, and ensure that purchasing information records are current. Cost estimates can access the updated purchasing information records to determine purchased component prices.

You transfer long-term planning data to the purchasing information system with Transaction MS70 or by following the menu path **Logistics • Production • Production Planning • Long-Term Planning • Evaluations • Purchasing Information System • Set Up Data**. The screen shown in Figure 1.7 is displayed.

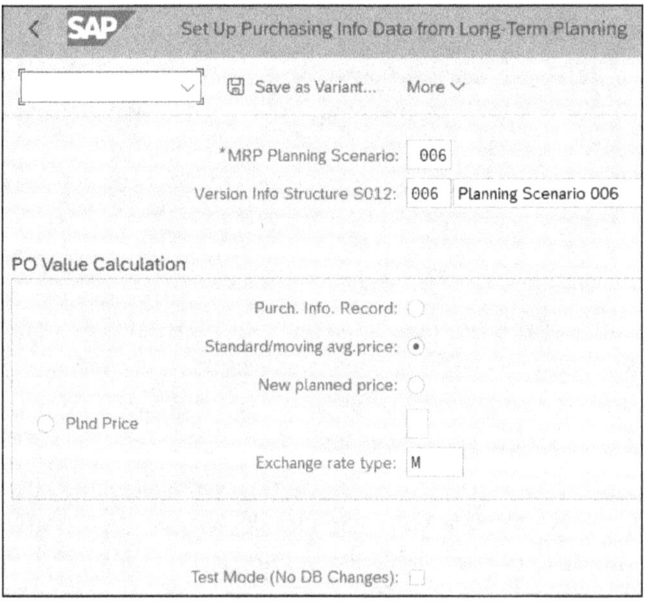

Figure 1.7 Set Up Purchasing Info Data from Long-term Planning

Version Info Structure S012 allows you to determine the receiving plan version of the purchasing plan data. If you don't enter a version, the system uses the planning scenario as the planning version number. You can also choose how you calculate the purchase order value in the **PO Value Calculation** section. Complete the selection screen as follows:

1. Complete the **MRP Planning Scenario** and **Version Info Structure S012** fields.
2. Select **Standard/moving avg.price**. You choose this option because there may not be any **Purch. Info. Record** (purchasing info records) or a **New planned price** at an early planning stage.
3. Deselect **Test Mode (No DB Changes)**.
4. Press F8 to execute the transaction.

You can also run a report on long-term planning purchasing data with Transaction MCEC or by following the menu path **Logistics • Production • Production Planning • Long-Term Planning • Evaluations • Purchasing Information System • Material**. The selection screen shown in Figure 1.8 is displayed.

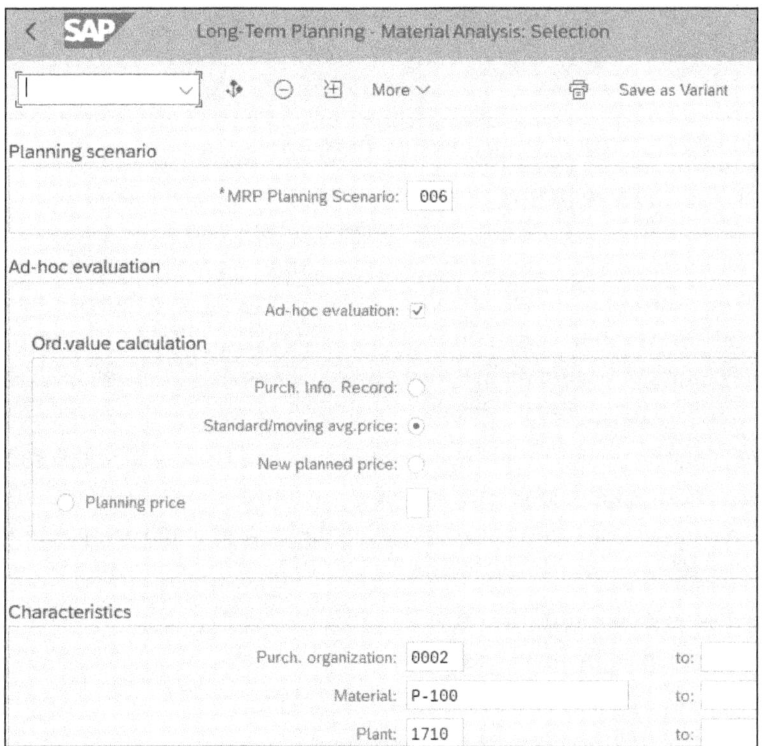

Figure 1.8 Purchasing Information System Selection Screen

The **Planning Scenario** field allows you to choose the long-term planning scenario on which to base the analysis. You can choose how the purchase order value is calculated in the **Ord.value calculation** section. Complete the selection screen as follows:

1. Complete the **MRP Planning Scenario** field.
2. Select **Ad-hoc evaluation** if you require one data selection for analysis of the standard analysis.
3. You may choose **Standard/moving avg.price** because there may not be purchasing info records or a new planned independent requirement at this early planning stage.
4. Complete the **Purch. Organization**, **Material**, and **Plant** fields.
5. Press F8 to start the transaction.

This report provides information on future purchasing requirements.

Activated planned independent requirements are also visible in operative MRP. In addition to data transferred to the purchasing information system, the purchasing department can view activated planned independent requirements through planned orders generated by operative MRP and purchase requisitions converted from planned orders. This can also be the basis for updating purchasing info records, which cost estimates access to determine the price of purchased materials.

1.3.5 Transfer Activity Quantities to Cost Center Accounting

In addition to ensuring that purchasing info records are up to date, you can transfer long-term planning activity quantities to cost centers. Long-term MRP generates requirements for all lower-level components and work centers from the production plan for products. You transfer the activity requirements to corresponding cost centers with Transaction KSPP or via the menu path **Logistics • Production • Production Planning • Long-Term Planning • Environment • Activity Requirement • Transfer to Cost Centers**. The screen shown in Figure 1.9 is displayed.

This selection screen allows you to enter the **Parameters** of the activity quantities to send to cost centers. Because we're interested in activity quantities sent to cost centers per activity, we select the corresponding radio button in the **Level of detail: Output lists** section. Complete the selection screen as follows:

1. Complete the **Version**, **Period**, and **Fiscal Year** fields.
2. Select the **Execute period adjustment** checkbox.
3. Select the **Cost center/activity type** radio button.
4. Click the **Transfer control** button at the top or run Transaction OMIK.

Figure 1.10 shows the next screen that is displayed.

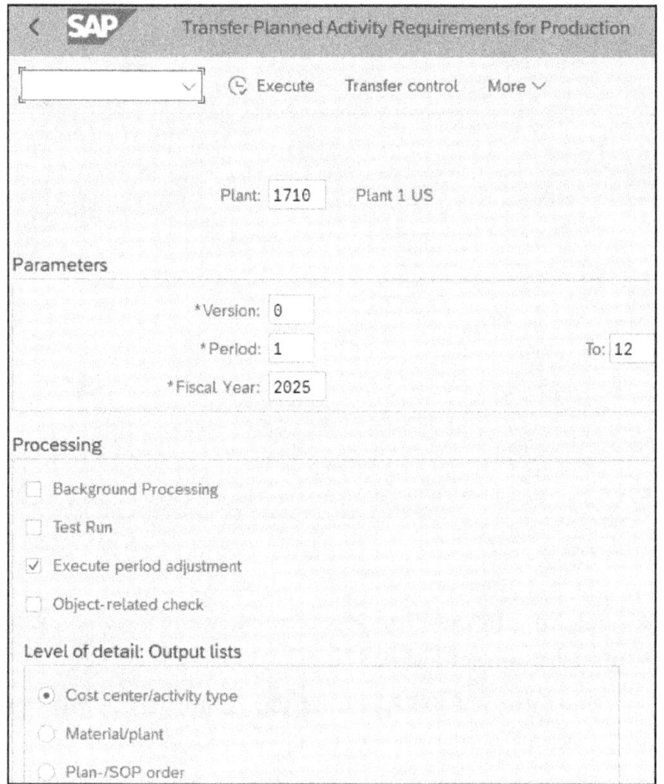

Figure 1.9 Transfer Planned Activity Requirements Selection Screen

Change View "Control: Transferring Activity Reqs to Cost Center Planni

New Entries More Display Exit

CoAr	Version	Version Name	Fiscal Year
0001	0	Plan/Actual Version	2023
AEZ1	0	Plan/Actual Version	2023

Figure 1.10 Transfer Controls for Activity Requirements

Each line in Figure 1.10 corresponds to a controlling **Version**. You use versions to carry out scenario testing with different cost center plans, activity prices, and any other parameters in cost center planning. You can create as many versions as you like, but only version 0 contains the plan and actual data. To change transfer control settings, select a row and click the **Details** (magnifying glass) icon. You can also double-click on a row to display the details. The screen shown in Figure 1.11 is displayed.

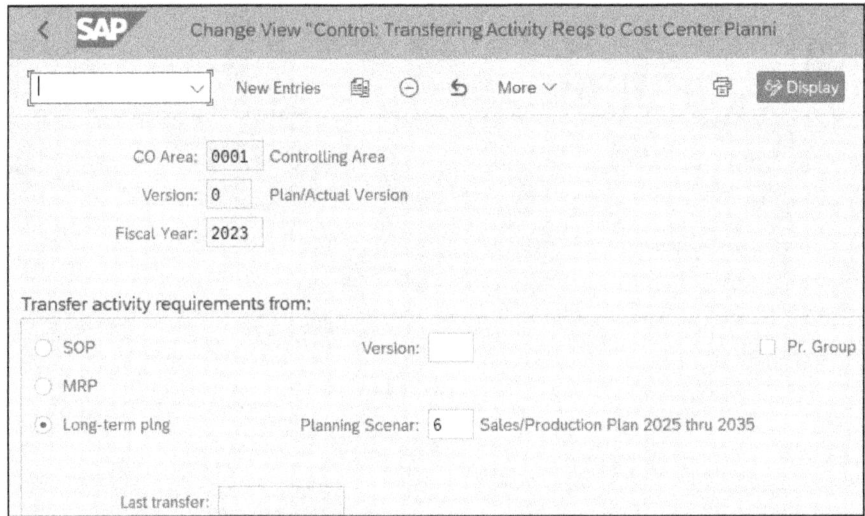

Figure 1.11 Transfer Control Definition

You transfer sales and operations planning, MRP, or long-term planning activity quantities to cost center accounting by doing the following:

1. Choose the appropriate radio button—**SOP**, **MRP**, or **Long-term plng**—in the **Transfer activity requirements from:** section.
2. Select **Pr. Group** (product group) to transfer **SOP** information based on product group.
3. Press F3 twice.
4. Click the **Execute** button shown earlier in Figure 1.9 to start the transaction.

You can create only one transfer control per version. In long-term planning, we determined the following:

- The component purchasing requirements and then transferred them to the purchasing information system.
- The scheduled activity requirements and then transferred them to cost center planning.

The next step in initial planning is to carry out cost center primary planning and then, together with scheduled quantities transferred from long-term planning, calculate the planned activity rate required by cost estimates to determine activity costs.

1.4 Cost Center Planning

Cost center planning meets two requirements for variance analysis:

- First, primary cost planning allows you to compare plan costs against actual costs as they occur. This allows you to measure a cost center manager's performance.
- Second, dividing the cost center plan's activity-dependent primary costs by the planned activity quantity provides an estimate of the planned activity rate, which cost estimates use to determine activity costs.

Plan costs also allow you to determine target costs, which are plan costs adjusted for actual output or operating rate. For example, if the plan is to deliver 100 hours, and 50 hours have been delivered, then target costs will be 50% of the plan costs.

Determining the planned workload (activity quantities) of production cost centers for the following fiscal year is a best-practice prerequisite for cost center planning. Activity quantities are necessary to determine variable costs such as wages and energy. Planned activity quantities are determined from work center loads resulting from the production plan, which are determined from the sales plan. You can transfer scheduled activity quantities from sales and operations planning, MRP, or long-term planning to cost center planning. You can then convert the scheduled activity quantities into planned activity quantities with plan reconciliation. We'll walk through each of these tasks in this section.

SAP Analytics Cloud

SAP is moving financial planning to SAP Analytics Cloud for planning, where there is dedicated business content to support the integrated financial planning process. Integration with scheduled activity planning is not yet supported at the time of writing, so we'll explain the classic method in SAP S/4HANA in this section. See SAP Note 2270407 for more details.

1.4.1 Primary Costs

You enter the plan for primary costs by primary cost element, corresponding to a general ledger expense account. Examples are plan payroll and depreciation costs against corresponding cost elements for each cost center.

You enter a primary cost plan for a cost center with Transaction KP06 or by following the menu path **Accounting • Controlling • Cost Center Accounting • Planning • Cost and Activity Inputs • Change**. The selection screen in Figure 1.12 displays.

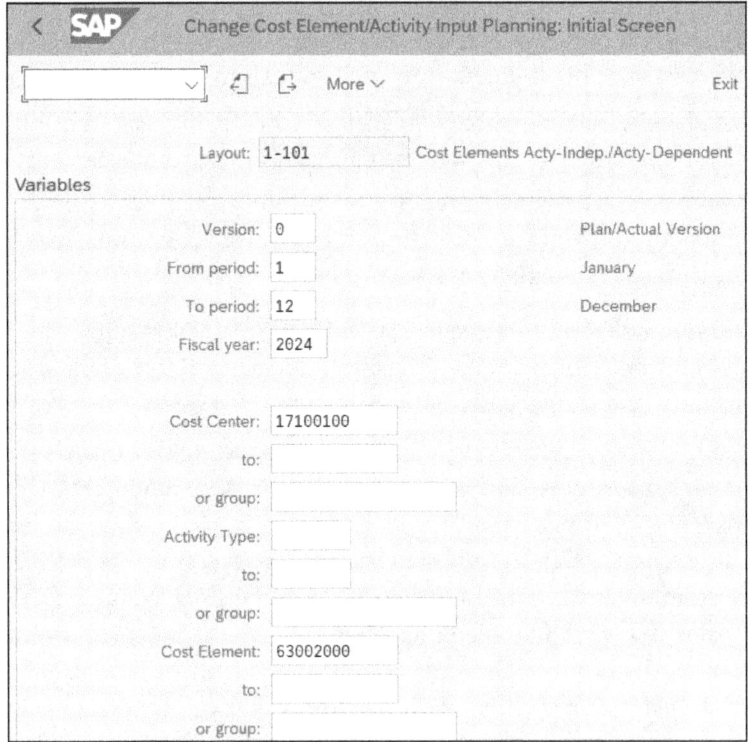

Figure 1.12 Cost Element Planning Selection Screen

You may see different fields depending on the planning **Layout**. You can scroll through available planning layouts with the left- and right-pointing arrow icons at the top of the screen.

You can create many versions, and you can enter planning data for each of them. In this example, we'll use **Version 0**. Actual costs post to **Version 0**, which is the version compared with target costs during variance analysis. Complete the selection screen as follows:

1. Complete the **Version**, **From period**, **To period**, and **Fiscal year** fields.
2. Leave **Activity Type** blank to plan for activity-independent costs.
3. Complete **Cost Center** and **Cost Element and** click the **Overview Screen** button (not shown) or press F5 to display the screen shown in Figure 1.13.

In this screen, you enter **Cost Center Plan Fixed Costs** per **Cost Element**. These are activity-independent costs because we didn't enter an activity in the **Activity Type** field in the screen shown in Figure 1.12. To carry out primary cost planning, enter the plan cost in the **Plan Fixed Costs** column. Click the **Period Screen** button (not shown) or press F6 to plan per period.

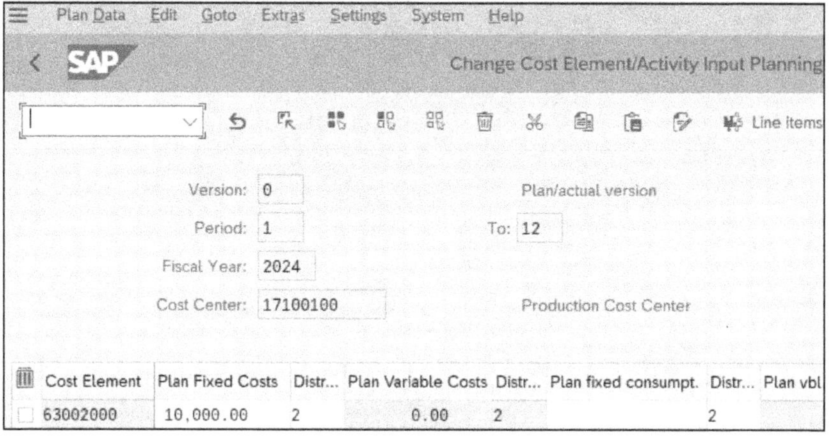

Figure 1.13 Cost Element Planning Screen for Cost Centers—Fixed Costs

If you enter an activity type in the selection screen in Figure 1.12, you can plan both fixed and variable costs, as shown in Figure 1.14.

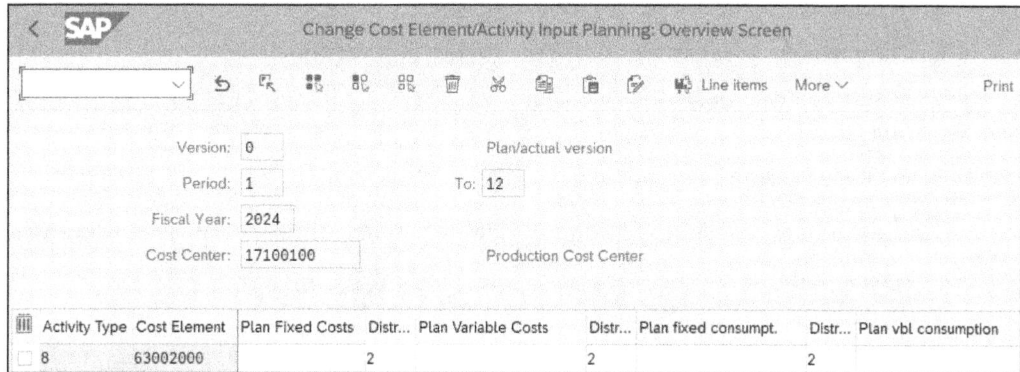

Figure 1.14 Cost Element Planning Screen for Cost Centers—Fixed and Variable

You can enter **Plan Variable Costs** because we entered an **Activity Type** in the selection screen in Figure 1.12.

Several reports are available to view planning data. One such report can be viewed with, Transaction KSBL or via menu path **Accounting • Controlling • Cost Center Accounting • Information System • Reports for Cost Center Accounting • Planning Reports • Cost Centers: Planning Overview**. A selection screen is displayed, as shown in Figure 1.15.

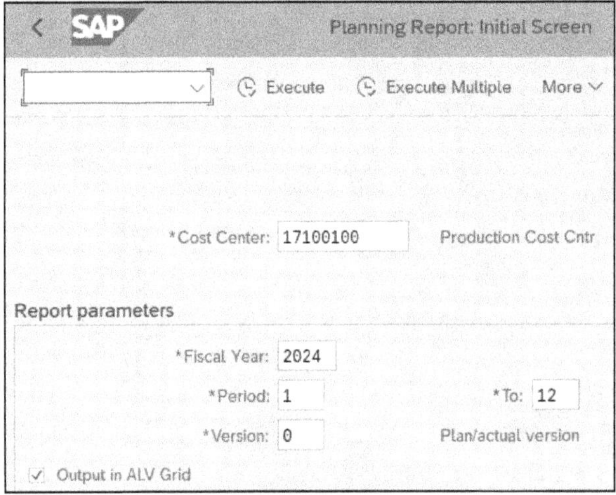

Figure 1.15 Cost Center Planning Report Selection Screen

You make entries in the **Report parameters** section to filter the values in the output screen. Complete the selection screen as follows:

1. Complete the **Cost Center**, **Fiscal Year**, **Period**, and **Version** fields.
2. Ensure that the **Output in ALV** checkbox is selected for a user-friendly display format.
3. Click the **Execute** button or press F8 to start the transaction.

Figure 1.16 shows a resulting summary view of planned primary costs for a cost center.

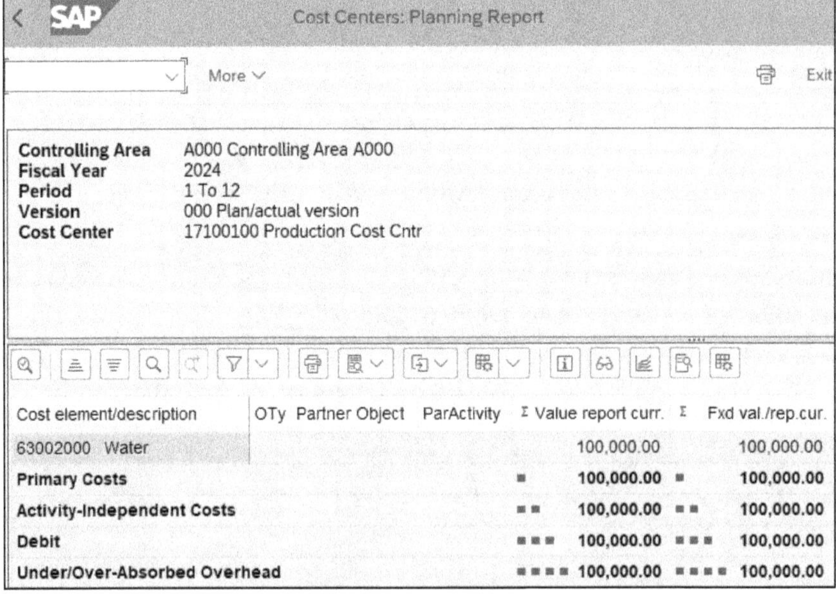

Figure 1.16 Cost Center Planning Overview Report

This screen displays a summary view of planned primary costs for **Cost Center 17100100**. Now let's examine how to calculate and enter activity rates.

1.4.2 Activity Price Planning

Determining the planned workload (activity quantities) of production cost centers for the next fiscal year is a prerequisite for cost center planning to calculate the planned activity price automatically. Activity quantities are necessary to determine variable costs such as wages and energy use. Planned activity quantities are determined from work center loads resulting from the production plan, which is determined from the sales plan. You can transfer scheduled activity quantities from sales and operations planning, MRP, or long-term planning to cost center planning. You can then convert the scheduled activity quantities into planned activity quantities with plan reconciliation.

After planning cost element costs for the next fiscal year (as discussed in the previous section), you can manually calculate and enter activity rates, or the system can automatically calculate them. Many companies calculate and enter planned activity rates manually for the first couple of years after system implementation, which allows for fine-tuning of master data and plan costs.

You can enter plan activity prices for a cost center with Transaction KP26 or by following the menu path **Accounting • Controlling • Cost Center Accounting • Planning • Activity Output/Prices • Change**. A selection screen is displayed, as shown in Figure 1.17.

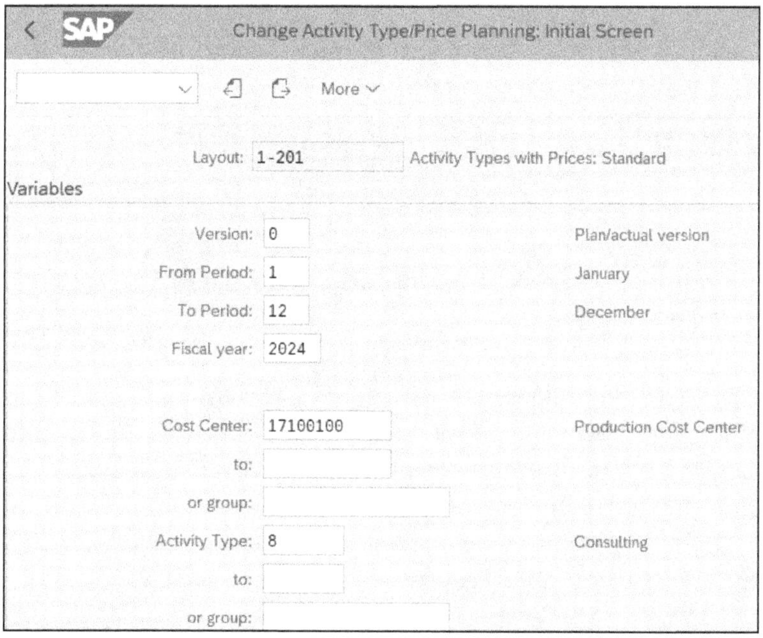

Figure 1.17 Planned Activity Price Selection Screen

This selection screen lets you enter the version, periods, cost center, and activity type you want to plan. Complete the selection screen as follows:

1. Complete the **Version**, **From Period**, **To Period**, and **Fiscal year** fields.
2. Complete the **Cost Center** and **Activity Type** fields.
3. Click the **Overview Screen** button or press F5 to display the screen in Figure 1.18.

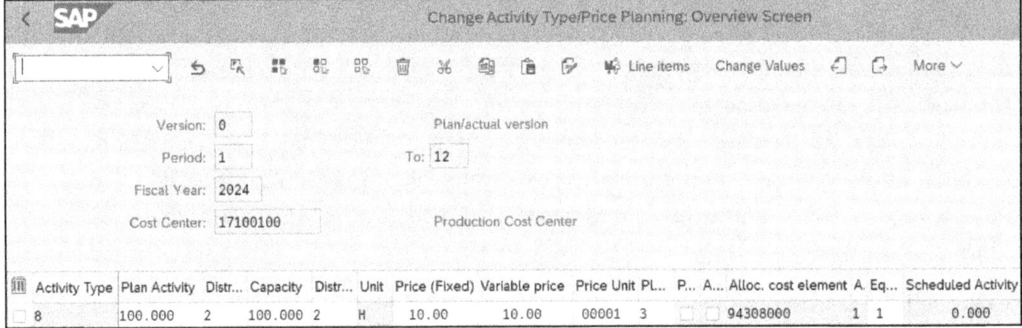

Figure 1.18 Planned Activity Price Entry Screen

In this screen, you enter the plan activity quantity, capacity quantity, and plan fixed and variable activity prices. To do this, follow these three steps:

1. Complete the **Plan Activity** and **Capacity** quantity fields.
2. Complete the **Price (Fixed)** and/or **Variable Price** fields.
3. Save your entries.

The **Plan activity** quantity, entered in the second column in Figure 1.18, is required to calculate the planned activity price automatically.

Advantage of Planned Activity Quantity

Another benefit of entering the planned activity quantity is that it appears at the bottom of the standard cost center report S_ALR_87013611 (Cost Centers: Actual/Plan/Variance). You can then compare the plan and actual activity quantities in the cost center report to analyze cost center variance.

The **Scheduled Activity** column in Figure 1.18 is transferred from sales and operations planning, MRP, or long-term planning, as we discussed in Section 1.3.5. You cannot adjust this field manually. You can use it to overwrite the planned activity quantity (the **Plan Activity** column in Figure 1.18) with Transaction KPSI or by following the menu path **Accounting • Controlling • Cost Center Accounting • Planning • Planning Aids • Plan Reconciliation**. A selection screen is displayed, as shown in Figure 1.19.

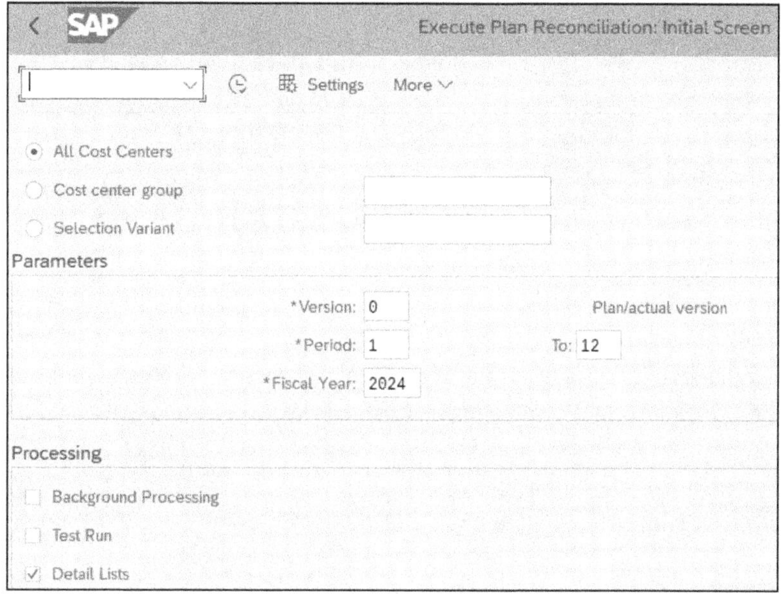

Figure 1.19 Plan Reconciliation Screen

This selection screen allows you to enter the parameters to choose which cost centers and periods you'll update with the scheduled activity quantity transferred from long-term planning. Complete the screen as follows:

1. Select either the **All Cost Centers** or **Cost center group** radio button.
2. Complete the **Version**, **Period**, and **Fiscal Year** fields.
3. Deselect the **Test Run** checkbox and select the **Detail Lists** checkbox.
4. Click the **Execute** icon or press F8 to start the transaction.

The following points describe the columns shown earlier in Figure 1.18:

- **Total plan activity** corresponds to the **Plan activity** column.
- **New plan activity** corresponds to the last column.
- **Activity difference** is the difference between the two columns.

When you execute plan reconciliation, you automatically overwrite the planned activity manually entered in the **Plan activity** column in Figure 1.18 with the scheduled activity from the last column in Figure 1.18.

Now that we've carried out cost element and activity type planning, you can automatically calculate the planned activity price with Transaction KSPI or by following the menu path **Accounting • Controlling • Cost Center Accounting • Planning • Allocations • Price Calculation.** In the selection screen, you enter the version, time frame, and cost centers and then press F8 or click the **Execute** icon.

Next, we'll discuss a method of uploading plan data directly from Microsoft Excel.

1.4.3 Uploading Plan Data from Microsoft Excel

You often enter controlling plan data manually in spreadsheet-style screens. You also have the option of copying and pasting from Excel into the planning screens. With a little-known third option, you can upload planning data directly into SAP from a spreadsheet file.

To upload planning data directly from a file, follow the three steps detailed in the following sections.

Define Planning Layout

You carry out manual activity type price planning with Transaction KP26 or by following this menu path **Accounting • Controlling • Cost Center Accounting • Planning • Activity Output/Prices**. The selection screen shown in Figure 1.20 is displayed.

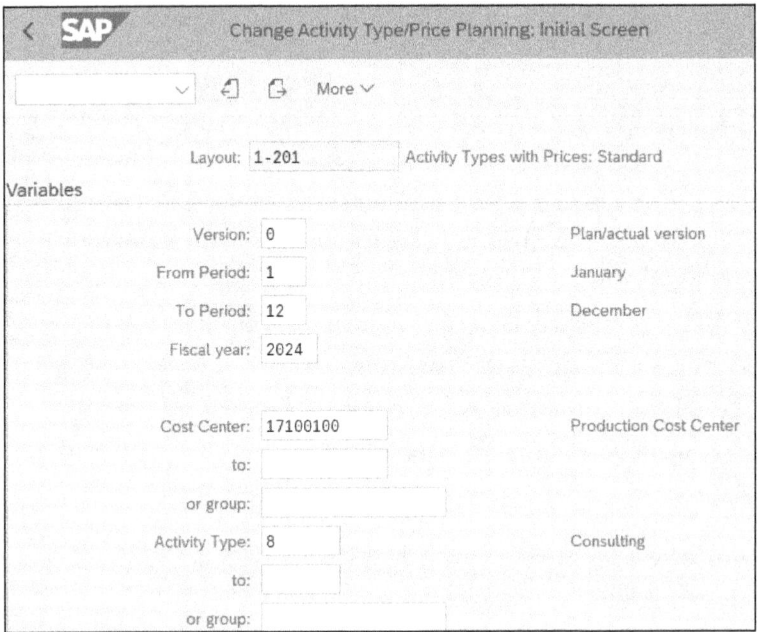

Figure 1.20 Activity Type Price Planning Selection Screen

Complete the selection screen as follows:

1. Complete the **Version**, **From Period**, **To Period**, and **Fiscal year** fields.
2. Complete the **Cost Center** and **Activity Type** fields.
3. Click the **Overview Screen** button or press F5 to display the planning screen in Figure 1.21.

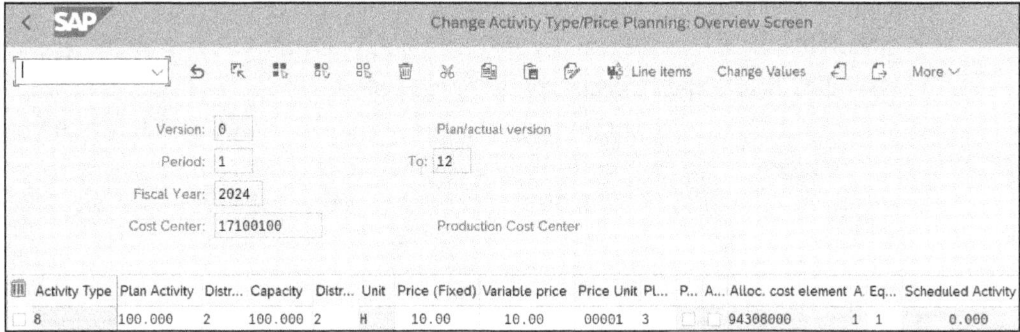

Figure 1.21 Activity Type Price Planning Screen

Layout 1-201, shown earlier in the selection screen in Figure 1.20, determines which fields appear in the following overview screen shown in Figure 1.21. Lead columns (**Activity type,** in this example) appear on the left in the overview screen, and you cannot change them. Layout general selection parameters together with lead columns define the selection screen parameters shown in the initial screen.

To examine the definition of Layout 1-201, run Transaction KP77. Double-click **layout 1-201** and select **Edit • Gen. data selection • Gen. data selection** from the menu bar to display the screen shown in Figure 1.22.

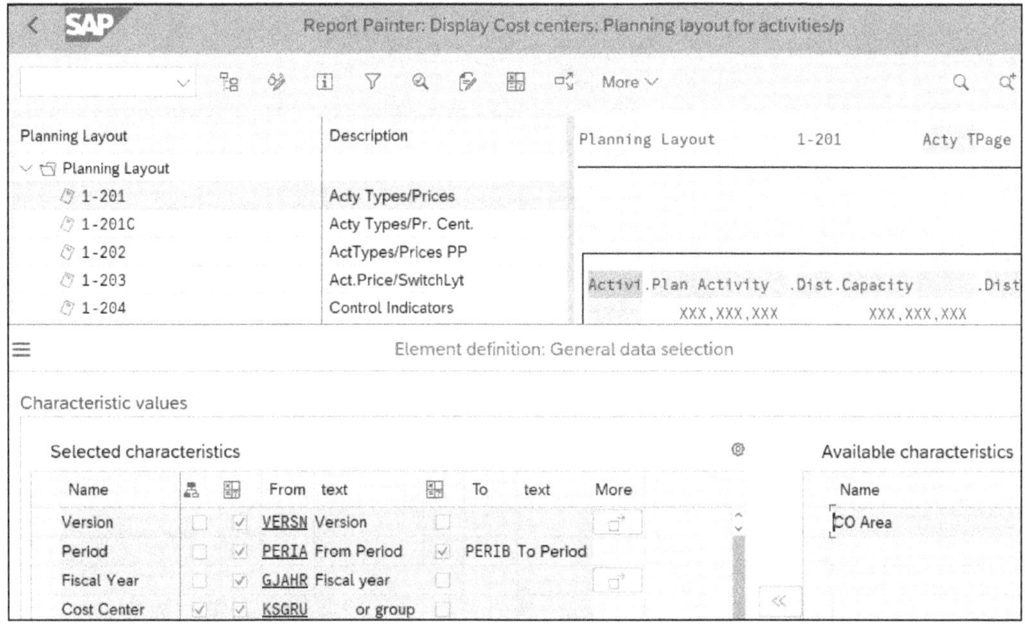

Figure 1.22 General Data Selection for Layout 1-201

The general data selection **Characteristics** shown in Figure 1.22 determine which selection fields appear in the screen in Figure 1.20. Lead columns (such as **Activity type,** in this example) also appear in the screen shown in Figure 1.20.

With the **Cost Center** as part of the general data selection (as shown in Figure 1.22), you can only upload data for one cost center at a time. To upload a plan for multiple cost centers in one file, remove it from the general data selection and include it as a lead column.

Because you cannot change standard SAP layouts, create a new layout with Transaction KP75 while using **Layout 1-201** as a reference. You can remove the cost center from the general data selection shown in Figure 1.22 by selecting **Edit • Gen. data selection • Gen. data selection** from the menu bar. Then, add the cost center as a lead column by selecting **Edit • Columns • New lead column** from the menu bar. Double-click the new **Cost Center** column and complete the subsequent popup screens.

Define Planner Profile

Create a new planner profile with Transaction KP34. Select **Edit • New Entries** from the menu bar to display the screen shown in Figure 1.23.

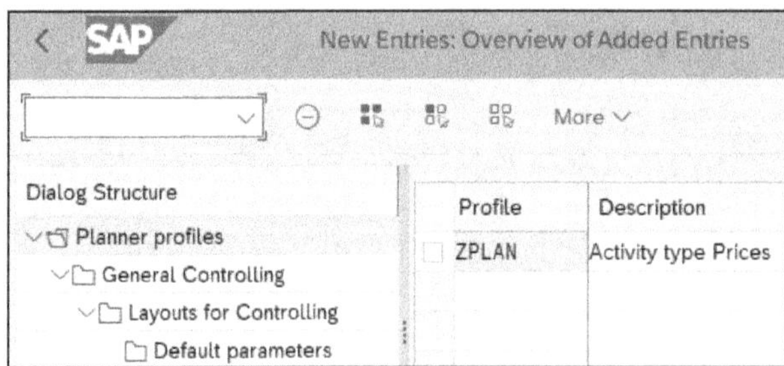

Figure 1.23 Define Planner Profile

Type in your new profile and description, select the row, and double-click on **General Controlling**. On the next screen, choose a planning area to link with your new planner profile, select the line, and double-click **Layouts for Controlling** to display Figure 1.24.

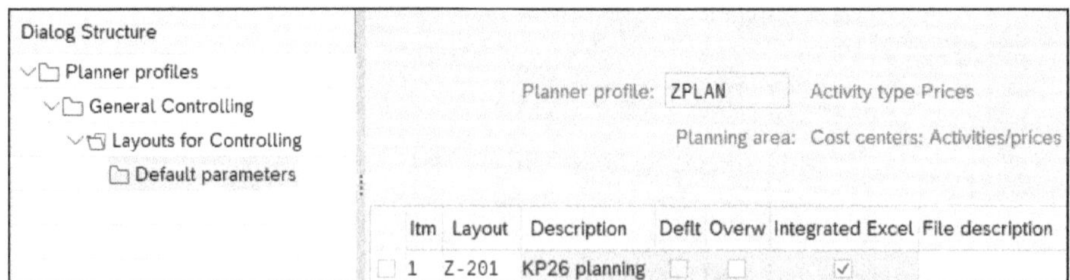

Figure 1.24 Associate Planning Layout with Planner Profile

Type in a layout, select the **Integrated Excel** checkbox, select the line, and double-click **Default parameters** to display the screen shown in Figure 1.25.

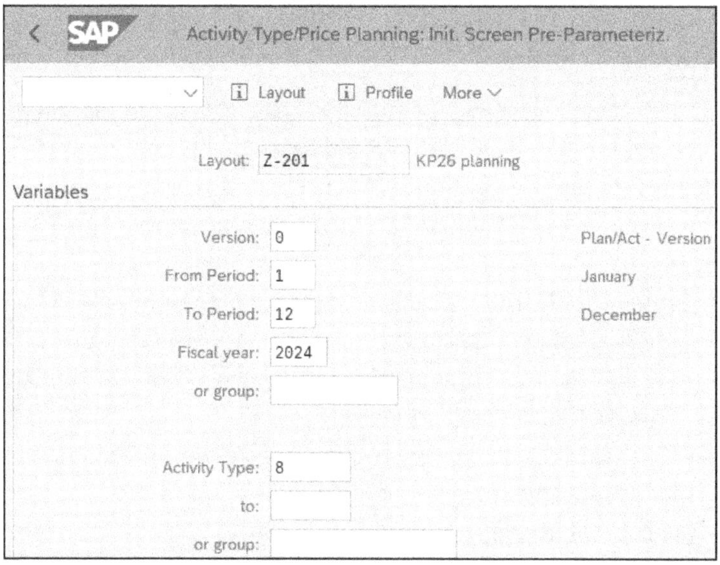

Figure 1.25 Default Parameters Screen

Complete the selection screen as follows:

1. Complete the **Version**, **From Period**, **To Period**, and **Fiscal year** fields.
2. Complete the **Activity Type** field.
3. Click the **Overview Screen** button or press F5.

You may need to enable macros, depending on your security settings. Click the **Generic file name** button (not shown), type in a generic name (such as "KP26*.TXT") and select **File description • Save all** from the menu bar. You can save the Excel file to your local drive by selecting **File • Save copy as** from the menu bar. Save the Excel file as a tab-delimited text file, which you can then use as a template for creating upload files.

Upload the Plan Data

Enter the plan data using a copy of the Excel file you saved. Run Transaction KP26 and select **Extras • Excel Planning • Upload** from the menu bar. Navigate to the Excel file that contains your plan data and execute. A green checkmark on a subsequent screen indicates that the upload was successful.

Now that we've completed cost center planning, the integrated planning cycle has some more steps. We'll discuss these in the next section on final planning.

1.5 Final Planning

You can take some more planning steps to help determine the accuracy of your initial planning in costing-based profitability analysis:

- **Calculate standard costs**
  We've estimated component procurement quantities in long-term planning, and we can access component quantities and prices in the purchasing information system.

  We've also estimated future activity quantities in long-term planning, transferred this information to cost center accounting, and calculated planned activity prices by dividing cost center activity-dependent costs by activity quantities.

  This information, together with the master data and configuration setup, allows you to create standard cost estimates, which allow you to estimate projected procurement and manufacturing costs for components and finished goods and update inventory valuation based on the planned costs.

  After you have created standard cost estimates, you can transfer this information to profitability analysis for detailed margin analysis and reporting.
- **Transfer standard costs to margin analysis**
  You can transfer standard cost estimate and cost component information to margin analysis using the valuation functionality. You can display this data with standard margin analysis reports by running Transaction KE30 or by following the menu path **Accounting • Controlling • Profitability Analysis • Information System • Execute Report**. You can choose an existing report or create your own to compare planned sales and cost information.

Transactions KP06 and KP26

In on-premise SAP S/4HANA, the legacy planning Transactions KP06 and KP26 that we've discussed are still available for compatibility even though they have been removed from the SAP Easy Menu. This allows you to transition smoothly from legacy planning to SAP Analytics Cloud. Planning table ACDOCP can exchange data with legacy planning tables such as COSP and COSS, which can also be accessed by SAP Business Planning and Consolidation (SAP BPC) optimized for SAP S/4HANA. See SAP Note 2977560 for details.

1.6 Summary

In this chapter, we discussed the initial planning scenarios. First, we looked at creating a sales plan in margin analysis. We then considered creating a sales plan directly in sales and operations planning or transferring the plan from margin analysis. We converted the sales plan into a production plan that we transferred to demand planning as planned independent requirements.

We then ran long-term planning and generated planned dependent requirements based on BOM and routing information. We analyzed purchasing requirements in the purchasing information system. We transferred work center loads to cost center accounting.

Cost center activity quantities transferred from long-term planning, together with cost element planning, allowed us to calculate planned activity prices. We then considered the final planning steps of creating standard cost estimates and using valuation in margin analysis for margin reporting on planned sales and cost information.

Now that we've discussed initial planning, let's next examine the following:

- Controlling master data in Chapter 2
- Material master data in Chapter 3
- Logistics master data in Chapter 4

This information is a prerequisite for creating standard cost estimates, which we'll discuss in detail in Chapter 9.

Chapter 2
Controlling Master Data

Controlling master data identifies the cost type and provides information for the responsible department, project, person, and internal management reporting.

Master data, such as accounts and activity types, stays relatively constant over long periods. Some examples of master data include the following:

- **Controlling master data**
  Provides information for the department, project, or person responsible for costs and identifies the type of costs.
- **Material master data**
  Provides all the information required to manage a material.
- **Logistics master data**
  Provides information on how materials are procured and manufactured.

Even though master data is relatively stable, companies that want to remain competitive in rapidly changing environments constantly assess whether it's most cost-effective to manufacture assemblies in-house, procure externally, or outsource. Changing procurement methods can significantly affect variance calculation and require constant master data and purchasing information maintenance. These effects may also influence the frequency of price updates with standard cost estimates and costing runs.

This chapter discusses controlling master data relevant to product costing and how it influences cost estimates and reporting, walking you through accounts, cost centers, activity types, and statistical key figures. In the following two chapters, we'll look at material and logistics master data. The first Controlling master data items we analyze are accounts and cost elements.

2.1 Accounts and Cost Elements

You identify all costs posted in Controlling by selecting accounts linked with cost elements that indicate the cost type. In this section, we'll explain the basic concepts used to differentiate costs arising outside the organization from costs charged within the organization. We'll also explain the concepts of primary and secondary costs and the settings needed to ensure that costs are captured on the correct accounts.

2.1.1 Basic Concepts

Controlling costs fall into two groups, based on posting origin:

- Postings to Controlling from external business transactions (profit and loss [P&L]) accounts with primary cost elements)
- Business transactions within Controlling between cost objects (accounts with secondary cost elements)

We'll examine primary and secondary costs further in the following sections.

2.1.2 Primary Costs

All costs in Controlling originate from journal entries identified by accounts with primary cost elements. You create general ledger expense accounts during implementation and with primary cost elements. Accounts with a primary cost element require all postings to include a mandatory cost object. Typical cost objects are cost centers, internal orders, product cost collectors, and work breakdown structure (WBS) elements. This ensures that all general ledger expenses include cost objects. You can see an example of such a posting in Figure 8 in the Introduction, and we'll now explore the process in more detail. You can see an example of a primary expense account posting in Figure 2.1.

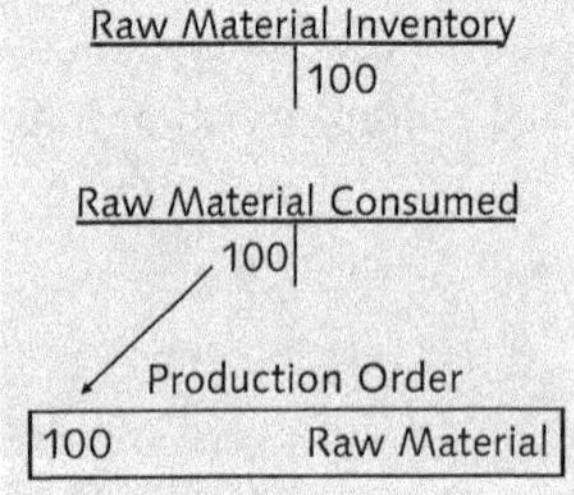

Figure 2.1 Goods Issue Debits to Production Order

Components valued at 100 are issued from the raw material inventory (balance sheet) to a production order (cost object). The raw material consumed general ledger expense account receives a debit of 100, resulting in a production order debit of 100.

You identify and control all costs in SAP S/4HANA by the general ledger account master, which you maintain with the Manage G/L Account Master Data app (SAP Fiori ID F0731A or SAP GUI Transaction FS00).

General Ledger Accounts and Cost Elements Merge in SAP S/4HANA

Before SAP S/4HANA, general ledger accounts and cost elements were separate master data in the financial accounting and Controlling modules. With SAP S/4HANA and the Universal Journal, general ledger accounts and cost elements are combined. This merger eliminates reconciliation issues and simplifies reporting.

Now that we've discussed how primary costs occur, let's examine how to create and maintain accounts with primary cost elements. You create accounts with Transaction FS00 or via the menu path **Accounting • Financial Accounting • General Ledger • Master Records • G/L Accounts • Individual Processing • Centrally**. The selection screen is displayed in Figure 2.2. Notice the **G/L Account Type P** for primary costs. The same account type is used for external revenues linked with market segments rather than cost objects.

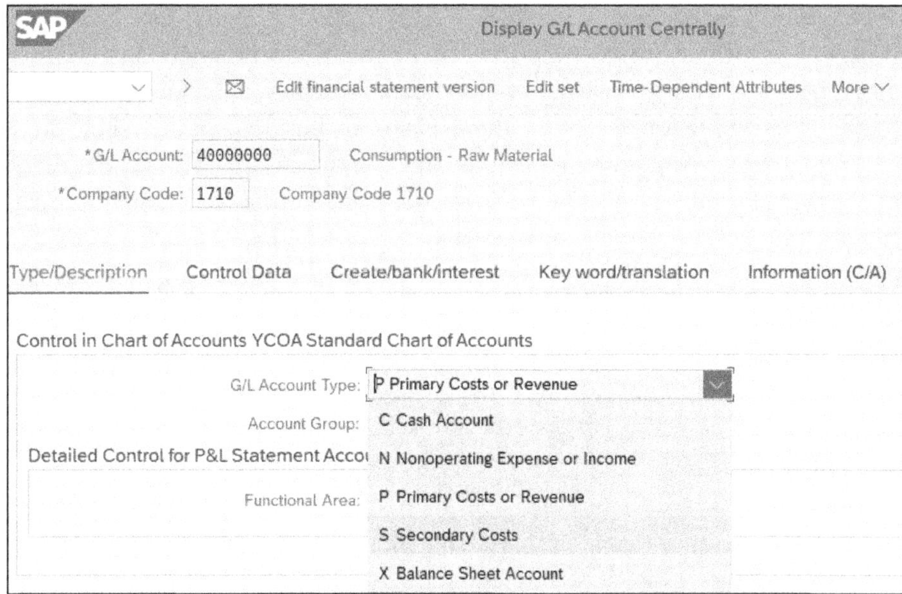

Figure 2.2 General Ledger Account Type in the General Ledger Account Master

Secondary Cost Element Account Types

You create accounts with secondary cost elements by selecting **G/L Account Type: Secondary Costs**, as shown in Figure 2.2. This changes the list of available cost element categories, as shown later in Figure 2.7.

In the following sections, we'll discuss the general ledger account settings: account type and cost element category. We'll also discuss cost object assignments required by cost element categories.

Secondary Cost Elements Are Now General Ledger Accounts

Secondary cost elements were previously independent master data in the Controlling module. SAP S/4HANA integrates secondary cost elements with general ledger accounts, ensuring primary and secondary journal entries are posted to the Universal Journal.

General Ledger Account Type

The **G/L Account Type** is the first setting in the general ledger account master. In Controlling, we're interested in the following three options:

- **P Primary Costs or Revenue**
  Primary costs include, for example, material expenses, salaries, asset depreciation, and energy and are mostly captured with reference to a cost center but also with reference to the production order, as we saw above. You post revenue by selling products and services. If you are using margin analysis, the cost object for the revenue and the cost of goods sold (COGS) is the market segment (the combination of characteristics, such as product, customer, region, and sales office, by which you want to analyze the market success of your products).
- **S Secondary Costs**
  With these general ledger accounts, you post allocations between cost objects. Examples are allocations of costs from overhead cost centers to production cost centers; from production orders to customer projects, production orders, or market segments; and from internal projects to cost centers and market segments.
- **N Nonoperating Expense or Income**
  These are P&L accounts that are not related to the company's products and services but belong in the financial statement, such as gains or losses from a foreign exchange market.

Cost Element Category

The next setting of interest in the general ledger account master is the cost element category. The cost element category determines the processes for which you can use general ledger accounts. You'll find the **CElem category** field in the **Control Data** tab of the general ledger account master, as shown in Figure 2.3.

Click in the **CElem category** (cost element category) field and press F4 to display a list of possible entries, as shown in Figure 2.3. This screen shows typical cost element category entries for **Primary** cost elements. There is a different list of possible cost element types for secondary cost elements, as shown later in Figure 2.7. The more commonly used primary cost element categories are listed below:

- **1: Primary costs/cost-reducing revenues**
  Created for primary expense general ledger accounts and COGS accounts.
- **11: Revenues**
  Created for revenue general ledger accounts.
- **12: Sales deduction**
  Created for sales deduction accounts, sales discounts, and rebates accounts.
- **22: External settlement**
  Created for external settlement. Typically, costs are posted to cost objects through postings to primary general ledger accounts with corresponding primary cost elements. It's also possible to settle costs from Controlling cost objects back to general ledger accounts via category **22** cost elements. This might be the case for assets under construction that are first expensed to a project and then settled.

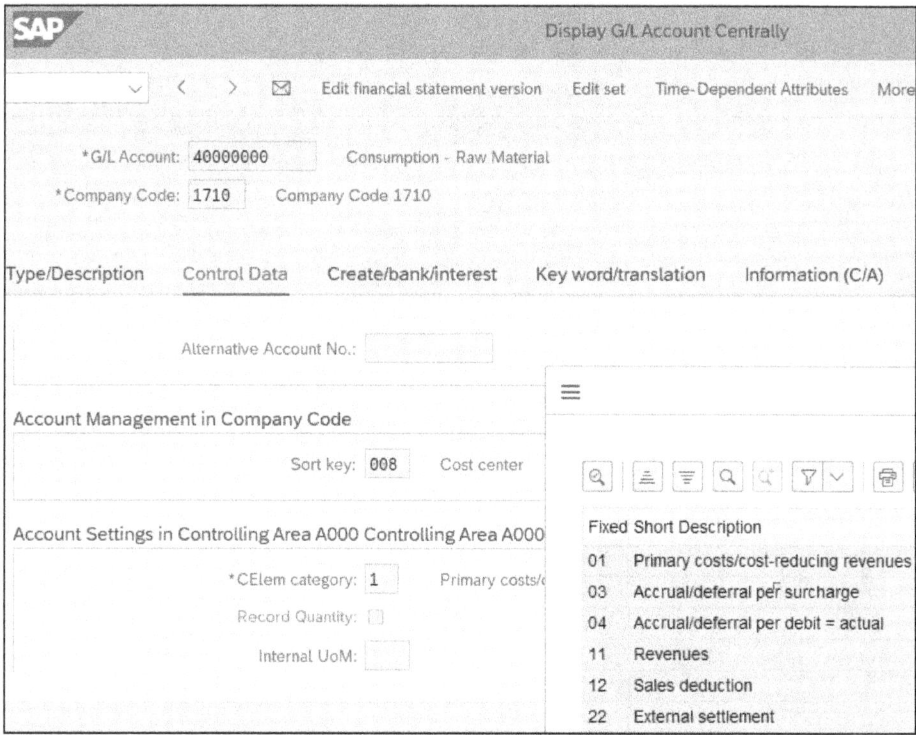

Figure 2.3 Cost Element Categories Available for Primary Costs or Revenue

You cannot change the cost element category after either plan or actual transactional data exists for a cost element during the current fiscal year. The following case study explains when to change the cost element category.

Case Study: Purchase Price Variance Cost Element Category

You can create category **12** (**Sales deduction**) cost elements for difference accounts, such as purchase price variance (PPV) and physical inventory adjustment, as defined here:

- PPV postings occur when the purchase order/invoice prices differ from the standard price.
- Physical inventory adjustment postings occur when you adjust inventory quantities during a physical inventory count and reconciliation with inventory quantities.

Some companies create these cost elements as category **12** because you don't need to set up automatic account determination, configuring which cost centers are posted to automatically. Revenue and sales deduction postings don't require automatic account assignment to cost centers but rather to market segments and also post to profit centers. However, you lose the cost analysis with cost center reports.

It's not possible to change account cost element categories after transactional data has been posted, so you will need to create new accounts with an expense cost element, which requires cost center assignment. You must set up automatic account assignment to assign cost centers automatically. We'll discuss automatic account assignment in detail in Section 2.1.6.

The field status group in the general ledger account (Transaction FS00) determines which cost objects are available during transactions. Because the **Cost Center** wasn't required when the cost element was category **12**, the **Cost Center** field wasn't available. You can change the field status group to allow the **Cost Center** to be populated by automatic account assignment, as described in the next section.

2.1.3 Field Status Group

The field status group is in the general ledger account master (Transaction FS00), as shown in Figure 2.4.

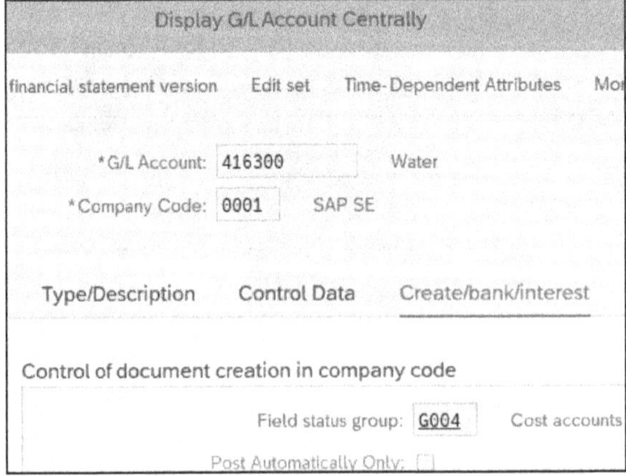

Figure 2.4 General Ledger Account Field Status Group

You'll find the **Field status group** in the **Create/bank/interest** tab. Selecting **Post Automatically Only** ensures no manual entries are allowed, which can assist during reconciliation. For example, you may set this indicator for inventory or PPV accounts, which are posted automatically. Double-click the field status group in Figure 2.4 to display an overview as shown in Figure 2.5.

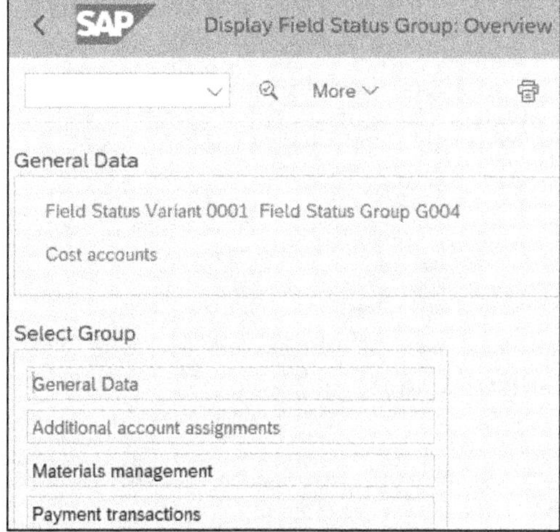

Figure 2.5 Field Status Group Overview

The **Field Status Variant** in the **General Data section** lists field status groups in the **Select Group** section. Double-click the group **Additional account assignments** to display individual fields, as shown in Figure 2.6.

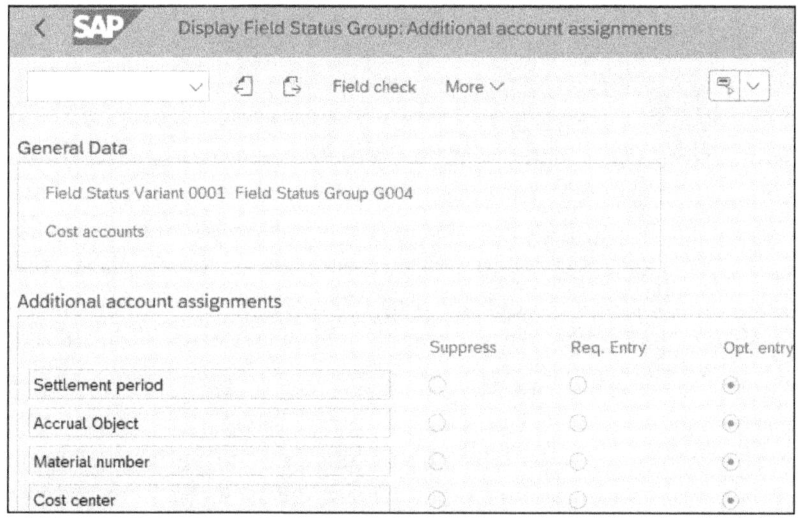

Figure 2.6 Field Status Group Additional Account Assignments

You may need to scroll down to display more entries across several pages.

This field status group shown in Figure 2.6 contains the fields we're interested in Controlling. In this example, the **Cost center** is an **Opt. entry (optional entry).** You can also **Suppress (hide)** fields or make them a **Req. entry** (mandatory).

You maintain field status groups via the IMG menu path **Financial Accounting • Financial Accounting Global Settings • Ledgers • Fields • Define Field Status Variants.** If you work with SAP S/4HANA Cloud or use the best practice settings with SAP S/4HANA, these settings are delivered as part of scope item J45 (financial accounting).

Now that we've discussed primary costs and external revenues, let's look at secondary costs.

2.1.4 Secondary Costs

The possible cost element categories when you choose secondary costs as the general ledger account type in Figure 2.3 (shown earlier) are shown in Figure 2.7.

Fixed	Short Description
21	Internal settlement
31	Order/project results analysis
41	Overhead Rates
42	Assessment
43	Internal activity allocation
50	Project-related incoming orders: Sales revenue
51	Project-related incoming orders: Other revenues
52	Project-related incoming orders: Costs
61	Earned value
66	Reporting Cost Element CO-PA

Figure 2.7 Cost Element Categories for Secondary Costs

After external primary costs post to overhead cost centers, you move the costs within Controlling to production cost centers. These secondary flows allow you to progressively allocate costs from overhead cost centers to production cost centers via **Assessment** with cost element category **42** and then on to manufacturing orders via **Internal activity allocation** (confirmation) with cost element category **43.** This ensures each assembly takes its share of overhead costs in the COGS via the released standard cost estimate that sets the standard price of the assemblies.

In the example shown in Figure 2.8, the production order is debited, and a production cost center is credited during activity confirmation. To increase the rate of distribution of overhead to production cost centers and orders, you can either increase the planned activity rate or increase the standard activity quantity in routings. We'll discuss routings in Chapter 4.

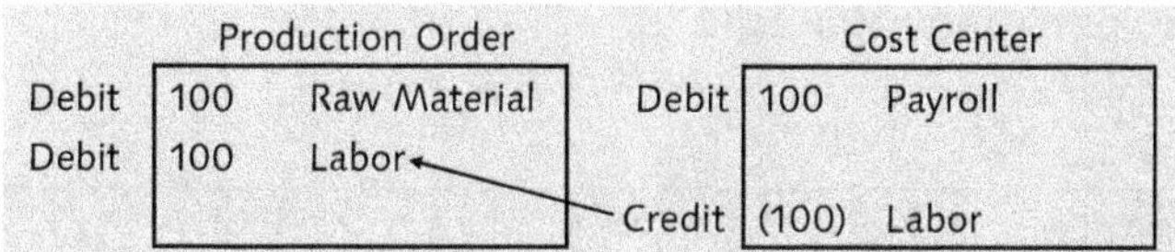

Figure 2.8 Production Cost Center Allocation during Activity Confirmation

Cost of Sales Account Cost Element Category

You typically create category **1** primary cost elements for expense general ledger accounts. Category **1** ensures that all primary expenses post to cost objects and are available for analysis with standard reports. In margin analysis, you create cost of sales (COS) accounts with category **1** cost elements.

Let's now discuss how the system determines which general ledger accounts to post to during goods movements. The general ledger accounts must be determined automatically because you cannot always manually enter general ledger accounts during goods movements and settlements.

2.1.5 Automatic Account Determination

The valuation class in the **Costing 2** view of the material master determines which general ledger accounts are posted to via automatic accounting assignment configuration Transaction OBYC or the IMG menu path **Materials Management • Valuation and Account Assignment • Account Determination • Account Determination Without Wizard • Configure Automatic Postings**. We discuss the **Costing 2** view in detail in Chapter 3 and automatic account determination in Chapter 13.

2.1.6 Automatic Account Assignment

Certain primary general ledger accounts that receive automatic postings, such as exchange rate differences and PPV, require automatic assignment of a cost center because they aren't inherently assigned to a cost object. You maintain automatic account assignment with configuration Transaction OKB9 or via the IMG menu path **Controlling • Cost Center Accounting • Actual Postings • Manual Actual Postings • Manage Default Account Assignments**. The screen in Figure 2.9 is displayed.

Complete the following steps to proceed to a screen where you can assign a cost center for each plant:

1. Type "1" in the **Acct assignmt detail** column.
2. Select the row of the required **Cost Elem.** (cost element).

3. Double-click **Detail per business area/valuation area**.
4. Click the **New Entries** button.

Change View "Default account assignment": Overview

New Entries More Display

Dialog Structure
- Default account assignment
  - Detail per business area/valuation area
  - Detail per profit center

CoCd	Cost Elem.	Cost Ctr	Order	PrfSeg	Profit Ctr	Acct assignmt detail	Acct assignmt detail
0005	2000001	51-5000000				1	Valuation area is mandatory
0005	2100001	51-6000000					
0005	2111111	51-6000000					

Figure 2.9 Automatic Account Assignment Configuration

The screen shown in Figure 2.10 is displayed.

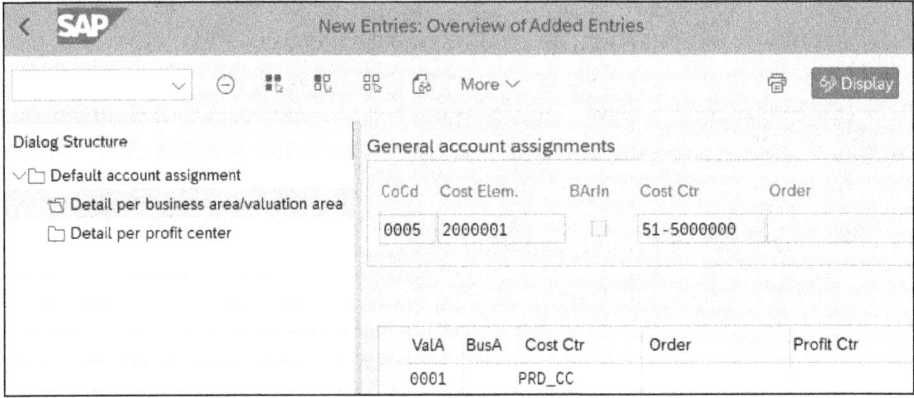

Figure 2.10 Default Cost Center per Plant

Complete the **ValA** (valuation area) and **Cost Ctr** (cost center) fields and save your work to make the cost center assignment. After you assign a cost center to a plant (**ValA**), only the **Cost Ctr** field is changeable. You can add more plants and assign cost centers, or you can delete rows as required.

Default Account Assignment

Before SAP S/4HANA, you could enter a default cost center directly in cost element master data. While this was convenient, it was applied per controlling area, not per plant. Cost element master data does not exist in SAP S/4HANA, and you must configure automatic account assignment with Transaction OKB9, as described in Section 2.1.5.

Now that we've examined the purpose of accounts and cost elements and how they are set up and maintained, let's examine cost element groups and cost centers.

2.1.7 Cost Element Groups

Cost element groups are hierarchical groups of cost elements with similar characteristics that allow you to create flexible subtotals along the lines of a report. You maintain cost element groups with Transaction KAH2.

2.2 Cost Center

Cost centers identify where expenses occur by geographic location, area of responsibility, or both. Most organizations also assign budgets at this level to ensure their cost center managers control their spending accordingly. In this section, we'll introduce the basic concepts, including the need for budget control and hierarchical reporting. We'll then explain the various fields in the cost center master.

2.2.1 Basic Concepts

In manufacturing, products are assembled at work centers with activities supplied by cost centers. The work center master data includes a mandatory cost center field, and as activities at a work center are confirmed, the cost center is credited, and the production order is debited. The activity cost is determined by multiplying the activity quantity by the plan activity rate.

A responsible person assigned to a cost center analyzes and explains reasons for the cost center variance and under/over absorption. The reasons can include:

- The difference between plan and actual primary expenses
- The difference between plan and actual secondary debits and credits from internal activity allocation

Cost center managers can use budget *availability control* to monitor and control their cost center budget. To set up availability control, create a budget planning profile by following the IMG menu path **Controlling • Cost Center Accounting • Budget Management • Define Budget Planning Profiles**. Double-click a budget profile to display the details, as shown in Figure 2.11.

In the **Time Frame** section, the **Past** (years allowed in the past) specifies how many years into the past you can plan/budget for. The start year is the reference point. The **Future** (years allowed into the future) specifies how many years into the future you can plan/budget for. The start year is the reference point.

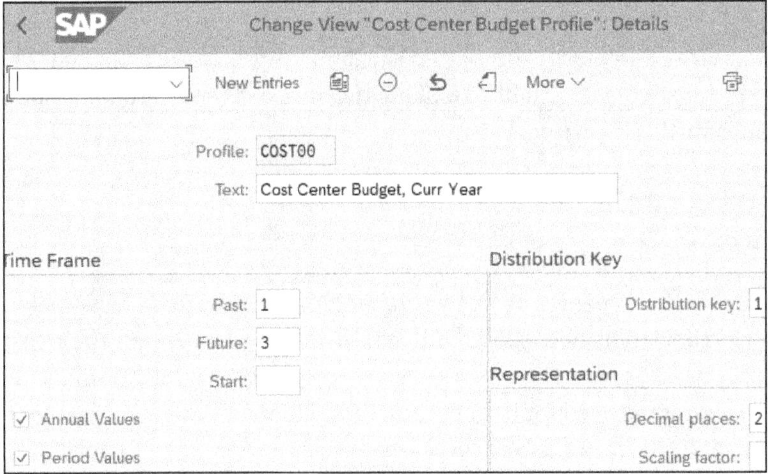

Figure 2.11 Cost Center Budget Profile

You then create a budget availability control profile via the IMG menu path **Controlling • Cost Center Accounting • Budget Management • Budget Availability Control for Cost Centers • Maintain Budget Availability Control Profile for Cost Centers.**

You assign a budget profile for account groups to the cost center. Then you activate budget availability control for the cost center in the Manage Cost Centers app (SAP Fiori ID F1443A) by setting the indicator **Budget Availability Control is Active** and entering the **Budget Availability Control Profile** as shown in Figure 2.12. When costs are posted to the cost center, the budget availability control calculates the available budget and issues a warning or error message as actual spend approaches the budget limit in accordance with the rules set in the profile.

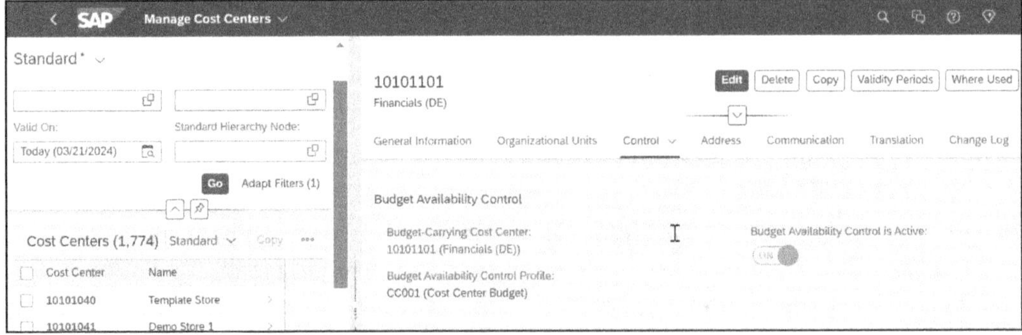

Figure 2.12 Manage Cost Centers, Showing Settings for Budget Availability Control

Now that we've discussed the purpose of cost centers, let's look at cost center groups. Cost centers are grouped together in decision, control, and responsibility units in a hierarchical structure known as the *standard hierarchy*, which we'll now examine.

2.2.2 Standard Hierarchy

A standard hierarchy node is a mandatory field in all cost centers. You must create the standard hierarchy and assign it to a controlling area before creating any cost centers. This guarantees that the standard hierarchy contains all cost centers in a controlling area.

Standard Hierarchy Groups

The standard hierarchy is a special cost center group that contains all the cost centers in a Controlling area.

Each node of the standard hierarchy is a cost center group. You can use this technique of creating a hierarchy by creating groups within groups in other areas of SAP; for example, in bills of materials (BOMs). SAP breaks down what could otherwise be complicated structures into simple concepts such as master data groups.

You maintain alternative cost center hierarchies with the Manage Global Hierarchies app (SAP Fiori ID F2918). Global hierarchies have several advantages over cost center groups, including:

- Validity dates are available, which means they are valid for a limited time only.
- Status management is available with statuses of draft, active, in revision, or no longer in use. Alternative hierarchies can be prepared in draft status before a reorganization. You can change the status to active, available for reporting after the reorganization.

You maintain a standard hierarchy with Transaction OKEON or via the menu path **Accounting • Controlling • Cost Center Accounting • Master Data • Standard Hierarchy • Change**. You can find an example of a standard hierarchy in Figure 2.13. There is not yet an equivalent SAP Fiori app for the standard hierarchy, only for the alternative hierarchies.

You drag and drop cost centers to reassign them to different nodes in the standard hierarchy. You can reassign cost centers by company code, business area, or profit center during a fiscal year if you meet the following three conditions:

- The currency of the old company code is the same as the new one
- You post only plan data in the current fiscal year
- You have not assigned the cost center to a fixed asset, work center, or human resources master record

Cost center **Activation status**, shown in the standard hierarchy in Figure 2.14, changes as you maintain cost centers within the standard hierarchy.

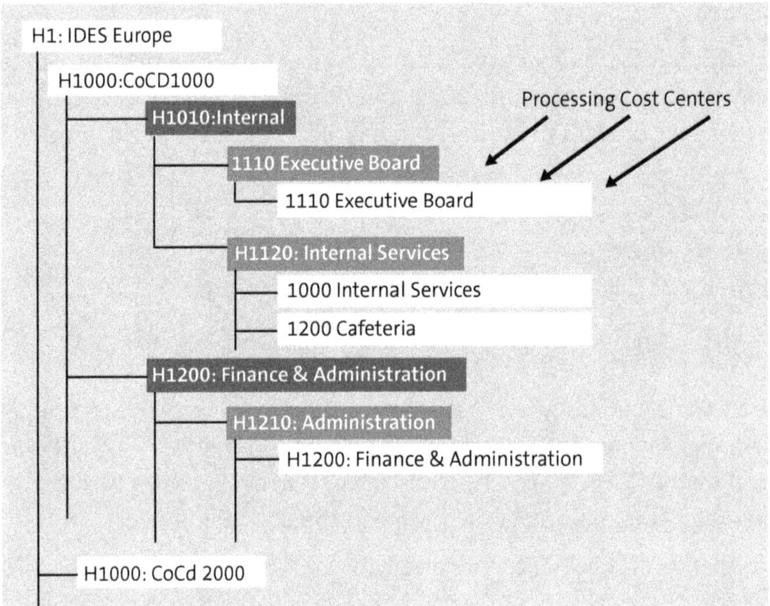

Figure 2.13 Example of a Standard Hierarchy

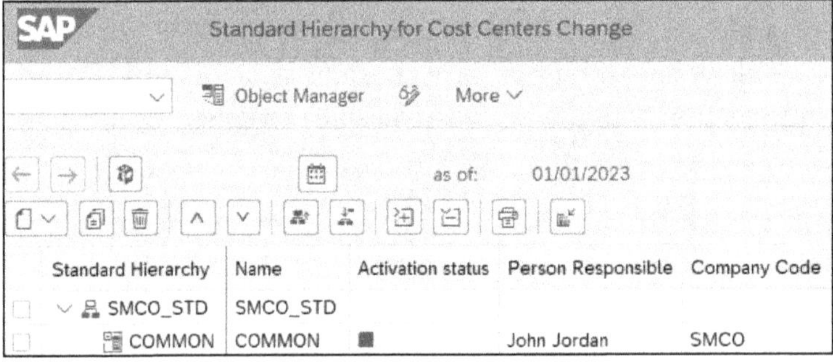

Figure 2.14 Cost Center Status in Standard Hierarchy

A cost center created or maintained in the standard hierarchy is initially assigned inactive (red) status and cannot send or receive costs. You can delete cost centers with an inactive status more quickly while you initially set up and maintain the standard hierarchy because the system doesn't need to check for dependent entries.

Double-click a cost center in a **Standard Hierarchy** shown in Figure 2.14 to display details of the cost center in the lower part of the screen, as shown in Figure 2.15.

To activate a cost center, click the green-and-red traffic light icon in the **Basic data** tab of the cost center to the right of the **Status** field. Assignments are checked during activation, and once activated, the cost center can receive and send costs. Activated cost centers are indicated by a green square traffic light icon in the **Activation status** column in the standard hierarchy shown in Figure 2.14. You can activate cost centers collectively with Transaction KEOA1.

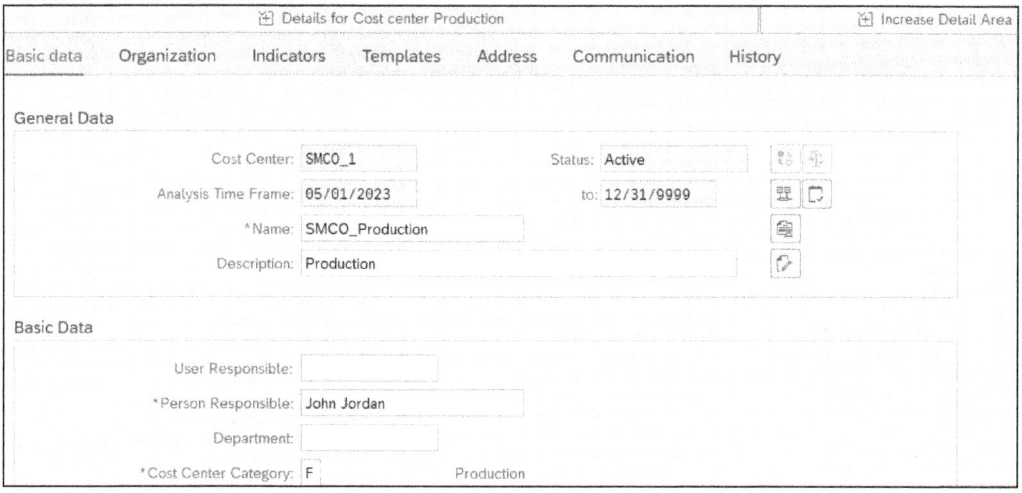

Details for Cost center Production
Increase Detail Area
Basic data | Organization | Indicators | Templates | Address | Communication | History

General Data
Cost Center: SMCO_1
Status: Active
Analysis Time Frame: 05/01/2023 to: 12/31/9999
*Name: SMCO_Production
Description: Production

Basic Data
User Responsible:
*Person Responsible: John Jordan
Department:
*Cost Center Category: F Production

Figure 2.15 Cost Center Details in Standard Hierarchy

You can delete inactive cost centers collectively with Transaction KEOD1 or via the configuration menu path **Controlling • Cost Center Accounting • Master Data • Cost Centers • Delete Inactive Cost Centers**.

2.2.3 Company Structure and Standard Hierarchy

Because the standard hierarchy represents your company structure, you'll likely maintain the standard hierarchy as your company structure changes. If you define your most recent company structure in an alternate hierarchy, it is not immediately obvious how to swap the standard hierarchy with the alternate hierarchy.

You can maintain an alternate hierarchy with the Manage Global Hierarchies app or Transaction KSH2, which has less functionality than the standard hierarchy screen. If you use an alternate hierarchy more often for reporting than the standard hierarchy, you may consider swapping hierarchies.

You can determine the standard hierarchy with Transaction OKKP or by following IMG menu path **Controlling • General Controlling • Organization • Maintain Controlling Area**. Double-click the text **Maintain Controlling Area**, then double-click on a controlling area and scroll down to **Other Settings**. The screen in Figure 2.16 is displayed.

If you attempt to change the standard hierarchy here, you'll receive an error message stating that this can only be done if neither the assigned standard hierarchy nor the proposed hierarchy contains any cost centers. You must still work with the existing hierarchy or risk too much disruption to current reporting.

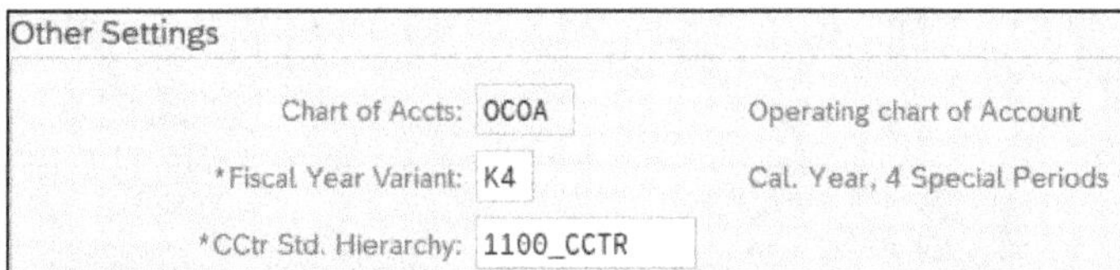

Figure 2.16 Maintain Controlling Area

Copying the alternate hierarchy as a node directly under the standard hierarchy and then deleting other existing nodes would be the simplest solution if it were possible. This would also have the advantage that all existing reports using the alternate hierarchy would still run correctly. However, if you attempt to create the alternate hierarchy as a node directly below the existing standard hierarchy, you'll receive an error message stating that the cost center group already exists.

Follow this procedure to streamline the changeover:

1. **Copy the existing alternate hierarchy to another group.**
   Create a cost center group with Transaction KSH1 or by following the menu path **Accounting • Controlling • Cost Center Accounting • Master Data • Cost Center Group • Create.** Create a new cost center group with reference to your existing alternate hierarchy. Create a new cost center with a similar name by adding a digit at the end.
2. **Delete the existing alternate hierarchy.**
   You can delete a cost center group with Transaction KSH2 or by following the menu path **Accounting • Controlling • Cost Center Accounting • Master Data • Cost Center Group • Change.** Click the highest node, click the Select icon, and then click the trash-can icon to delete the alternate hierarchy.
3. **Create the deleted hierarchy as a node in the standard hierarchy.**
   Now that the original alternate hierarchy doesn't exist, you can create a new alternate hierarchy as a node directly under the top node of the existing standard hierarchy with Transaction OKEON or by following the menu path **Accounting • Controlling • Cost Center Accounting • Master Data • Standard Hierarchy • Change.** Click the new page icon and select **Lower-Level Group** to create the alternate hierarchy name directly under the top node of the standard hierarchy.
4. **Create alternate groups in the standard hierarchy.**
   Now create the alternate hierarchy structure within the standard hierarchy, duplicating any previous steps as necessary, depending on which subnodes of the alternate hierarchy you need to continue reporting on with the same name.
5. **Reassign all cost centers into new groups.**
   Reassign all cost centers to the new groups and delete all the old groups.

2.2.4 Global Hierarchy

In addition to SAP GUI transactions, you can manage cost center alternative hierarchies and other cost center groups in the Manage Global Hierarchies app (SAP Fiori ID F2918). The Manage Global Hierarchies app is used for many types of hierarchies, so to create or change a cost center hierarchy, choose the **Hierarchy Type** cost center hierarchy. You can build up the tree structure node by node and then assign the cost centers to the correct nodes as shown in Figure 2.17 Alternatively, you can use the **Export/Import** button above the list of nodes to download the structure to a spreadsheet and then upload a list of your cost centers. Notice in the header that the cost center hierarchy is valid for a specific timeframe and has a status. The cost center hierarchies that we showed previously did not have a validity period, so it makes sense to investigate the newer options if your organization is in a constant state of flux. It's also possible to create draft versions of a hierarchy. We're focusing on an active version of the hierarchy, but there is a draft version on the left where you can anticipate future changes.

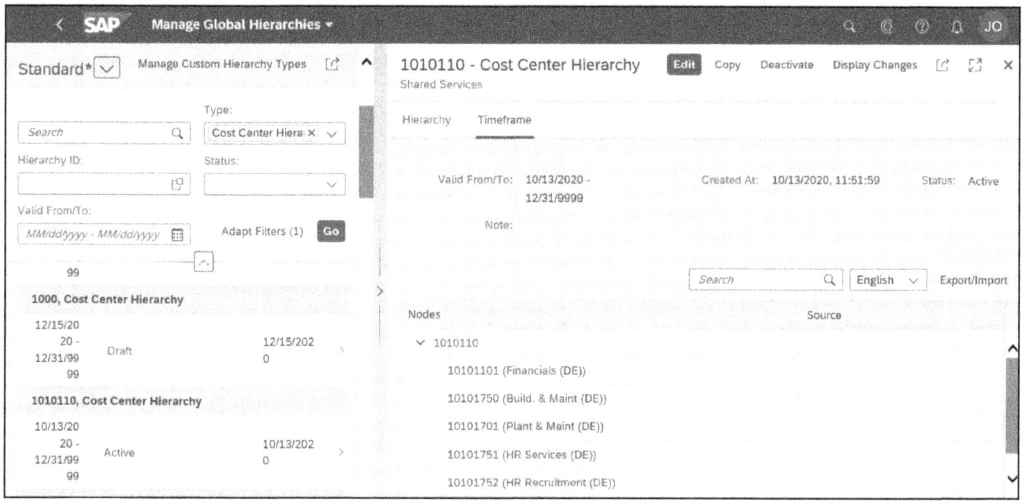

Figure 2.17 Cost Center Hierarchy in the Manage Global Hierarchies App

Now that we've discussed the standard hierarchy, let's examine more cost center master data. We'll first examine fields in the **Basic Data** tab.

2.2.5 Basic Data

You maintain cost centers with Transaction KS02 or by following the menu path **Accounting • Controlling • Cost Center Accounting • Master Data • Cost Center • Individual Processing • Change**. Type in a cost center and press Enter to display the initial cost center screen shown in Figure 2.18. You can also use the Manage Cost Centers app (SAP Fiori ID F1443A).

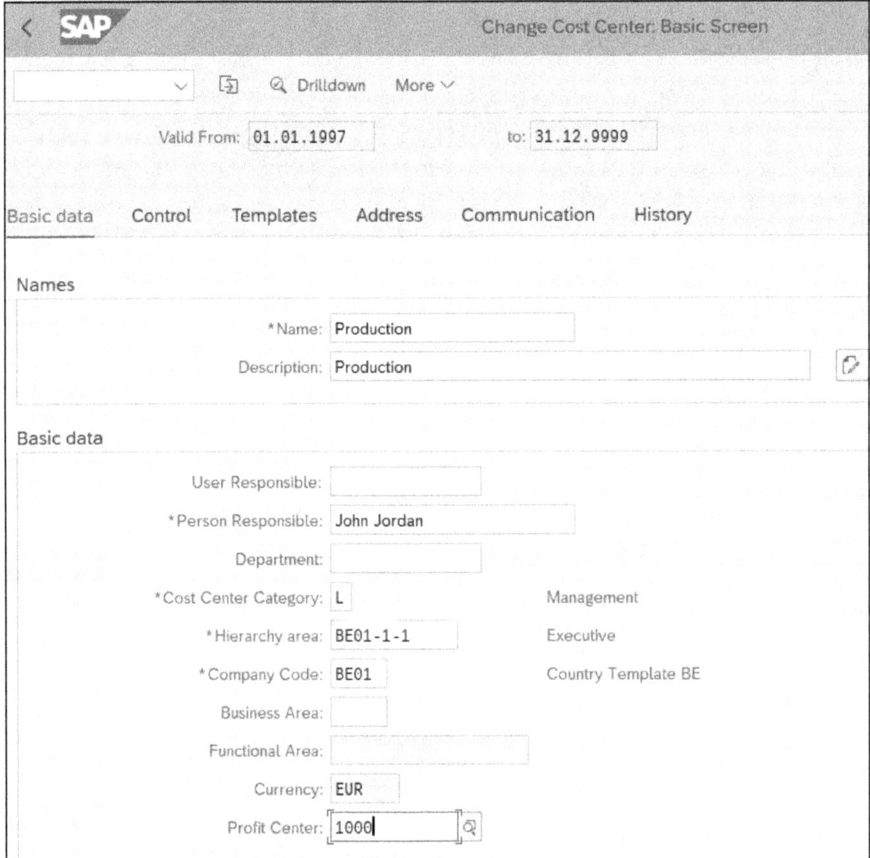

Figure 2.18 Cost Center Basic Data Tab

Let's discuss the **Basic data** tab fields in more detail:

- **Name and Description**
  These fields appear in cost center and order reports, which you can change anytime. You can add long text to the **Description** field by clicking the pencil-and-paper icon to the right of the field.

Cost Center Reports Use Cost Center Text

The standard cost center report S_ALR_87013611 (Cost Centers: Actual Plan/Variance) uses the text from the cost center **Name** field in Figure 2.18.

- **User Responsible**
  You can populate the optional **User Responsible** with the user ID of the person responsible for the cost center.
- **Person Responsible**
  The **Person Responsible** field is a mandatory text field that you populate with the name of the person responsible for controlling the cost center costs and analyzing the cost center planned and actual costs. This should be the name of a real person responsible for the costs and not a high-level manager who receives cost center reports from lower-level managers responsible for explaining the costs.
- **Department**
  Department is an optional text field for reporting purposes only. For example, you can enter department names unique to your company.
- **Cost Center Category**
  Cost Center Category identifies the type of activity a cost center provides, such as administration, production, or sales. You can define your own additional cost center categories with configuration Transaction OKA2 or via configuration menu path **Controlling • Cost Center Accounting • Master Data • Cost Centers • Define Cost Center Categories**. Cost center categories allow you to restrict certain activity types for use with certain cost centers. For example, you can prevent production activities from posting incorrectly to administrative cost centers. We'll examine this further in Section 2.3 when we discuss activity types.
- **Hierarchy area**
  Hierarchy area is a mandatory field that defines the cost center location on the standard hierarchy. We discussed the standard hierarchy in detail in Section 2.2.2.
- **Company Code**
  You can assign a cost center to only one company code in the mandatory **Company Code** field. All postings are assigned to a company code in SAP S/4HANA, and general ledger accounts and cost elements are integrated.

  A company code is the smallest organizational unit for which you can assign a self-contained chart of accounts for external reporting.
- **Business Area**
  You can create financial statements for business areas, and you can also use these statements for internal reporting purposes. The business area is superseded by functional areas and profit centers, as discussed in the following sections.

  With the introduction of the Universal Journal, the separation between internal and external reporting is less relevant, and profit centers have been enhanced to replace the business area in most scenarios. Refer to SAP Note 3256460 for more information.

- **Functional Area**

  A functional area is part of COS accounting that compares sales revenue with the manufacturing costs of an activity for a period. You assign expenses to the cost center to the functional area. Typical examples of functional areas include the following:

  - Research and development
  - General and administration
  - Sales and distribution

- **Cost of sales (COS)**

  Expenses and revenues you cannot assign to functional areas are reported in other P&L items according to expense and revenue type. If no postings exist for this cost center, you can enter or change the **Functional Area**.

  You must activate COS accounting to enter values in the **Functional Area** field for general ledger accounts and Controlling master data. You do this by following the configuration menu path **Financial Accounting • Financial Accounting Global Settings • Functional Area for Cost of Sales Accounting • Activate Cost of Sales Accounting**.

Cost of Sales Reporting

Margin analysis reports based on market segment and functional areas in general ledger accounts and Controlling master data are also available in standard reporting. In the U.S., you must use generally accepted accounting principles (GAAP) to report on functional areas.

You define a financial statement version by assigning functional areas to P&L items. You can create P&L statements according to COS accounting. The functional area logic only works for P&L accounts based on the functional area assignment in the postings, not the assignment in their master data.

- **Currency**

  The cost center **Currency** is determined automatically from the company code currency, which you cannot change in cost centers. You assign company code currency during company code definition with Transaction OX02 or via the IMG menu path **Enterprise Structure • Definition • Financial Accounting • Edit, Copy, Delete, Check Company Code**. You cannot change the company code currency after transactional data exists in a productive company code.

- **Profit center**

  A profit center receives postings made in parallel to a cost center and other cost objects, such as orders. Note that a profit center is not a true cost object but is derived from the underlying cost objects (the cost center here). Profit center accounting (PCA) is part of the Universal Journal and enables reporting based on profit center responsibility.

Profit Center Accounting (PCA) SAP S/4HANA Compatibility Scope

Classical PCA, where the profit center information is stored in a separate set of tables, is part of the SAP S/4HANA compatibility scope, with limited usage rights. For more details on the compatibility scope and its expiry date and links to further information, refer to SAP Note 2993220.

You typically create profit centers in areas that generate revenue and that have a responsible manager assigned. You can analyze operating results for a profit center using the COS approach or period accounting. The COS approach may require maintenance of functional areas by following the IMG menu path **Enterprise Structure • Definition • Financial Accounting • Define Functional Area**. You'll see the screen shown in Figure 2.19.

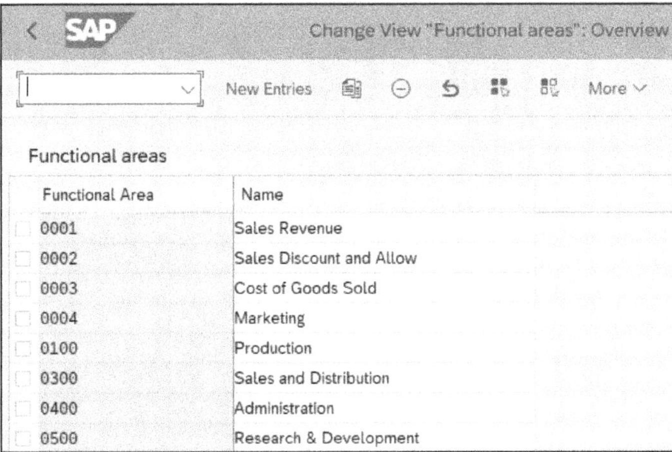

Figure 2.19 Functional Areas Definition Overview

General ledger accounts describe the nature of costs, while functional areas explain the function or area of the expenses.

Profit Center Ledger

Profit centers are contained within the Universal Journal, not a separate ledger.

When PCA is active, you'll receive a warning message if you don't specify a profit center, and all unassigned costs will go to a dummy profit center. You activate PCA during controlling area maintenance with configuration Transaction OKKP or by following the IMG menu path **Controlling • General Controlling • Organization • Maintain Controlling Area.**

Now that we've looked at the fields in the cost center **Basic Data** tab, let's examine the fields in the remaining tabs.

2.2.6 Control

Select the cost center **Control** tab shown earlier in Figure 2.18 to display the screen shown in Figure 2.20.

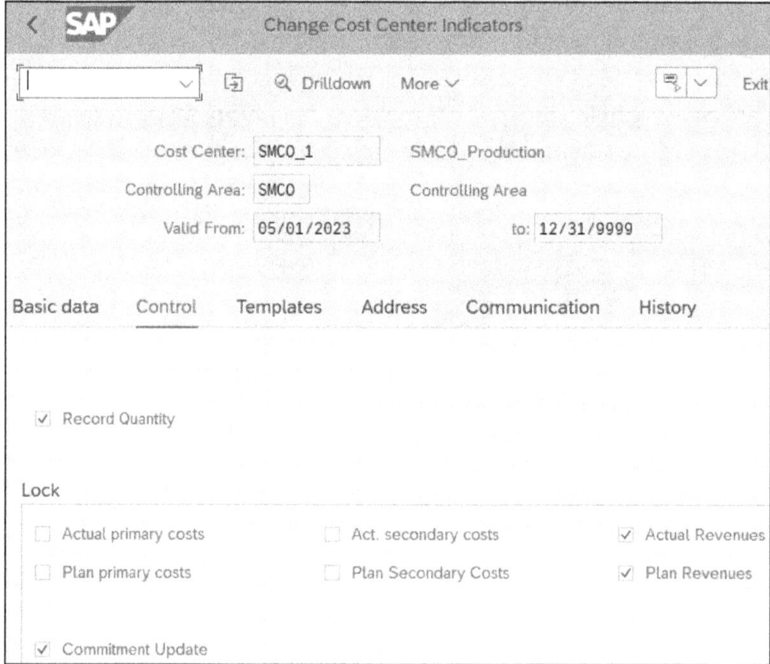

Figure 2.20 Cost Center Control Tab

The following list describes the settings:

- **Record Quantity**

  Select **Record Quantity** to define a message issued when you don't specify a quantity or quantity unit during journal entries. While quantities are typically recorded automatically during activity confirmation and inventory movements, you can also post costs manually via cost elements and cost centers. This may be important if you use advanced cost center reporting with target costs, which depends on quantities.

  This checkbox has no effect on planning or on the identification of quantities in reports. Also, select this checkbox in cost elements to ensure you record quantities for the combination of cost center and cost element.

- **Lock**
  Checkboxes in the **Lock** section in Figure 2.20 allow you to block specific costs and revenue to a cost center. In the example shown in Figure 2.20, the cost center cannot receive **Actual** or **Plan Revenues**. These are typical default settings. You can post revenue statistically to the cost center if you deselect these checkboxes. You can report on statistical postings, but they aren't included in the cost center total costs. Typically, you post revenues directly to a profit center while costs flow from assigned cost centers.
- **Commitment Update**
  If this checkbox is selected, commitments posted in the Universal Journal are assigned to this cost center but are not included in cost center reports.

Example: Block Cost Center Costs

A company structure changes, and you create and assign new cost centers in the standard hierarchy. To block all costs to previous cost centers, select all checkboxes in the **Lock** section of the **Control** tab.

Let's now look at the fields in the **Templates** tab of a cost center.

2.2.7 Templates

Select the **Templates** tab shown in Figure 2.20 to display the screen shown in Figure 2.21.

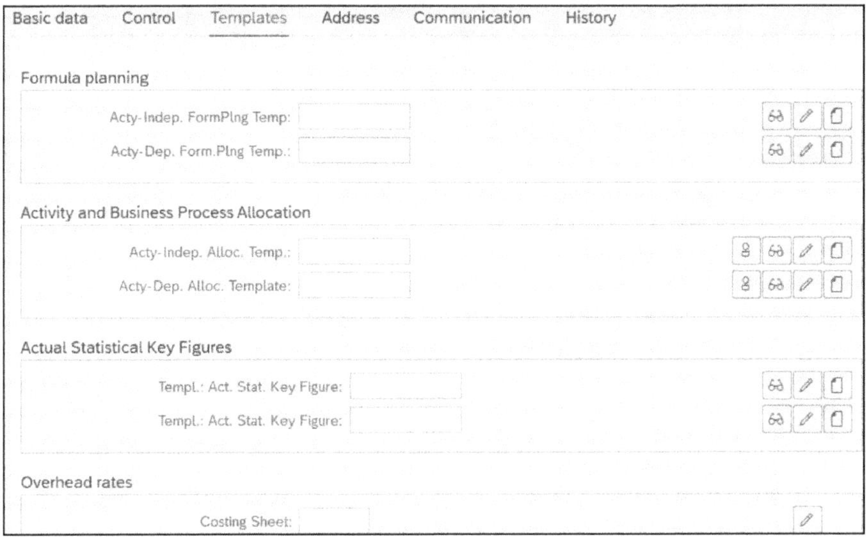

Basic data Control Templates Address Communication History

Formula planning

Acty-Indep. FormPlng Temp:

Acty-Dep. Form.Plng Temp.:

Activity and Business Process Allocation

Acty-Indep. Alloc. Temp.:

Acty-Dep. Alloc. Template:

Actual Statistical Key Figures

Templ.: Act. Stat. Key Figure:

Templ.: Act. Stat. Key Figure:

Overhead rates

Costing Sheet:

Figure 2.21 Cost Center Templates Tab

Templates allow flexibility when allocating overhead costs. Let's discuss the available fields:

- **Activity and Business Process Allocation**
  Templates in the **Activity and Business Process Allocation** section allow you to allocate actual overhead costs between sender and receiver objects dynamically. Enter a template in the fields to assign this cost center as a receiver object. Templates use functions to access data so you can determine the allocation proportions dynamically.

  You can also use formulas to calculate the data accessed by functions. The templates you enter in these fields determine the sending and receiving objects.

Example: Functions and Formulas in Templates

A dairy packaging facility handles large volumes of milk. The cost accountant determines that the most accurate way to allocate overhead costs to process orders is based on the volume of milk processed by an order.

A *function* within a template determines the volume from the material master **Basic Data 1** view, and a *formula* uses this data to calculate the volume of milk processed by the order. The volume of milk is then used to calculate overhead costs allocated to process orders.

- **Overhead rates**
  Overhead rates allow you to debit a cost center with overhead costs based on a **Costing Sheet**. You may be more familiar with debiting a manufacturing order with overhead costs via a costing sheet. You can debit a cost center with overhead costs similarly by entering the costing sheet in this cost center field.

Next, let's discuss the **Address** and **Communication** tabs.

2.2.8 Address and Communication

The **Address** tab contains street address information for the cost center. These fields are for information only and aren't mandatory. This tab also contains a **Tax Jurisdiction** field that you use to determine tax rates in the United States. The **Communication** tab contains language, telephone, fax, and other communication fields.

Now, let's examine the **History** tab fields.

2.2.9 History

Select the **History** tab shown in Figure 2.21 to display the screen shown in Figure 2.22. The **History** tab contains the following information for a cost center:

- **History Data**
  History Data lets you determine which user created the cost center and when.

- **Change document**
  Click the **Change document** button to display a list of changed cost center fields. Double-click a line to display details of changes made to each field.

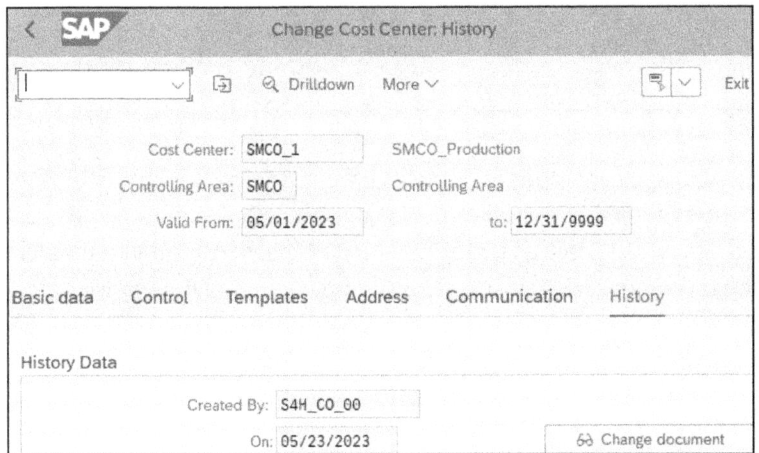

Figure 2.22 Cost Center History Tab

The entries in the **History** tab are determined automatically. You cannot manually change them. If you prefer, you can maintain your cost centers using the Manage Cost Centers app shown in Figure 2.23.

Figure 2.23 Manage Cost Center App

While many organizations continue to use the SAP GUI transactions to maintain their cost center master data or import them from SAP Master Data Governance, some are starting to use the SAP Fiori apps for master data maintenance, and this is the only option in SAP S/4HANA Cloud.

The Manage Cost Center app includes a worklist of available cost centers on the left and tabs for data entry for the selected cost center on the right.

Now that we've examined the purpose of cost centers and how to set up and maintain the master data, let's examine activity types.

2.3 Activity Type

Cost centers are rarely used in isolation for product cost controlling, but always in combination with an activity type that describes the types of work performed by the cost center (manufacturing tasks or service tasks). The activity is used by production orders for manufacturing tasks when time confirmations are recorded, maintenance orders and WBS elements for service tasks, and so on. In this section, we'll look at the two different activity types and explain the basic data that you'll need to maintain to use your cost centers and activity types.

2.3.1 Basic Concepts

Activity types describe activities provided by cost centers and allow you to allocate costs and services to receiving objects such as manufacturing orders, product cost collectors, and other cost centers. As you begin to maintain your activity types, determine which type of activity you want to capture for your cost center: manufacturing or service activities.

Manufacturing Activities

An activity type describes the output of a work center when you confirm manufacturing order activities. Confirmations credit the production cost center and debit each production order with actual activity costs.

During variance analysis, you compare plan and actual activity costs and determine the:

- price difference between the plan and the actual activity rate
- quantity difference between the plan and actual quantity confirmed

You have the option of confirming at standard or actual quantity. If it is difficult to change the standard quantity to actual during confirmation (for example, because you use backflushing), you may decide to confirm at the standard quantity. In this case, there will be no activity quantity variance.

The activity rates for the manufacturing activities can be defined using transaction KP26 or using the Manage Cost Rates—Plan app shown in Figure 2.24. In this example, we've selected all activity types for cost center **10101301**.

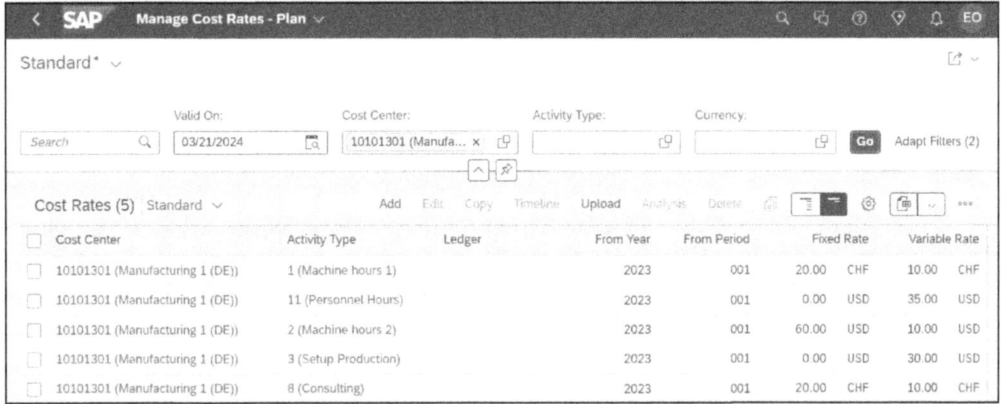

Figure 2.24 Manage Cost Rates—Plan App

Service Activities

Service activities have much in common with manufacturing activities. If performed for a maintenance or service order, they rely on confirmations, just like manufacturing activities. However, where the activity does not take place on the shop floor but rather in a professional service environment, it is recorded using time sheets. In both cases, the service cost center order is credited, and the service order or WBS element is debited with the actual activity costs.

When planning the activity rates for service activities, you can either treat them exactly like manufacturing activities and plan an hourly rate, or you can use the Manage Cost Rates—Professional Services app (SAP Fiori ID F3161) to plan charge rates that take account of the company code or even the WBS element for which the activity is being performed. This is particularly useful where you expect the charge rate to be different depending on where the service will be performed. Figure 2.25 shows the sample cost rates for a senior consultant working in several company codes.

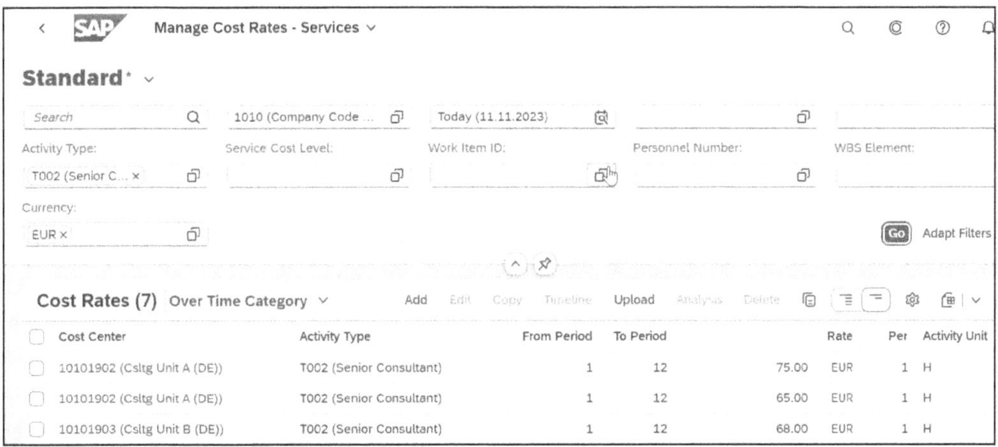

Figure 2.25 Manage Cost Rates—Services App

Now that we've reviewed the overview of activities, let's examine the fields in the **Basic data** tab of the activity type master.

2.3.2 Basic Data

You maintain activity types with Transaction KL02 or by following the menu path **Accounting • Controlling • Cost Center Accounting • Master Data • Activity Type • Individual Processing • Change** or by using the app Manage Activity Types (SAP Fiori ID F1605A). Type in the activity type and press Enter to display the initial activity type screen shown in Figure 2.26.

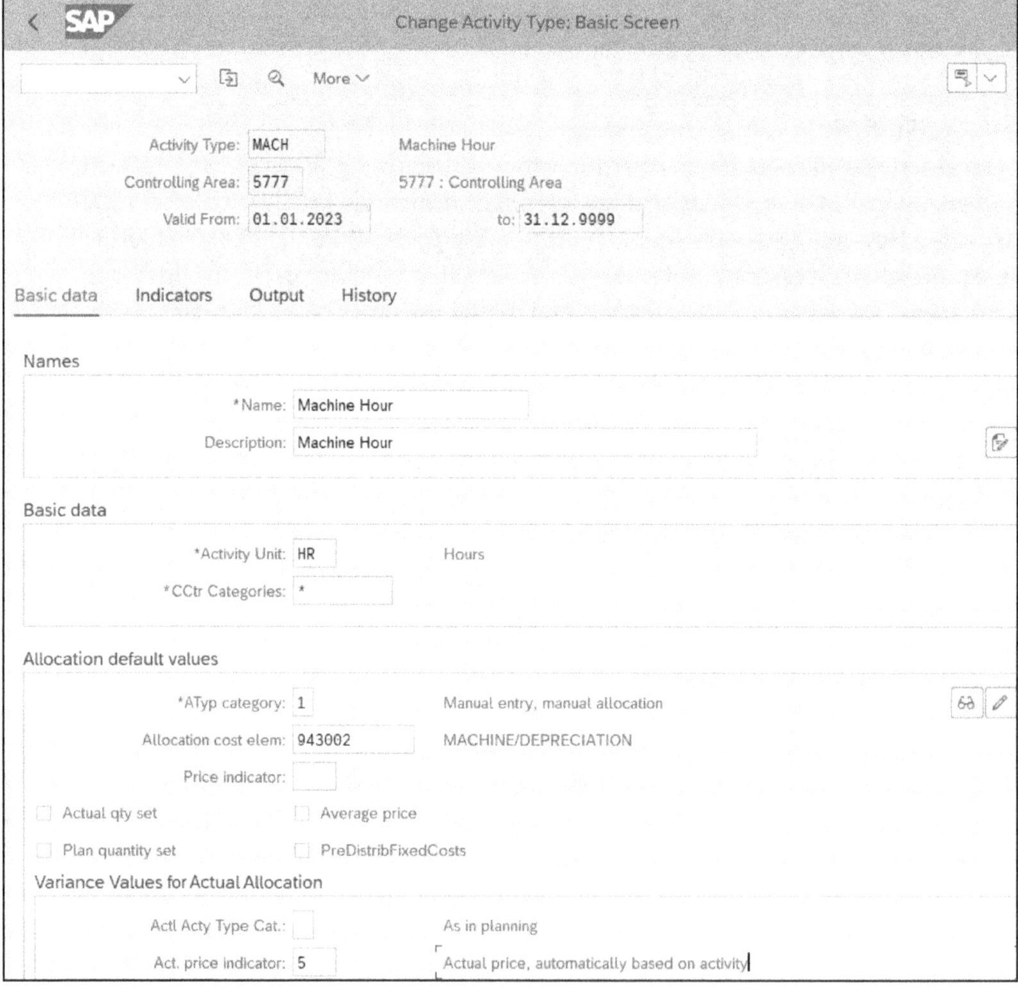

Figure 2.26 Activity Type Basic Data

The fields you maintain in the **Basic data** tab provide basic and control information on how you use the activity type during confirmations and allocations. Let's discuss each field in detail in the following sections.

Name and Description

You can change these fields at any time. They appear in the description in cost center and order reports. You add long text to the **Description** by clicking the pencil-and-paper icon to the right.

> **SAP Cost Center Reports Use Activity Type Text**
>
> Standard cost center report S_ALR_87013611 (Cost Centers: Actual Plan/Variance) uses the **Description** text, which you can change anytime.

Activity Unit

The **Activity Unit** field is the unit that appears in the itemization in cost estimates and order costing. This unit can differ from the unit of measure in work centers and operations in routings and manufacturing orders.

You cannot change the activity unit in the activity type if either plan or actual transactional data exist during the current fiscal year in any version. Even after you have deleted all activity type planning for all versions for the current fiscal year, you may still receive an error message due to dependent data. SAP Note 43230 recommends you use report RKPLNC13 to delete control information stored in table CSSL after you have deleted all other activity type planning for all versions. You may then be able to change the activity unit in the activity type.

Cost Center Categories

You can restrict the use of the activity type to specific types of cost centers by entering the **CCtr categories** (cost center categories) field. You can select any of the standard categories such as administration, production, or sales, or you can define your own with configuration Transaction OKA2 or via IMG menu path **Controlling • Cost Center Accounting • Master Data • Cost Centers • Define Cost Center Categories.** Cost center categories allow you to, for example, prevent production activities from posting incorrectly to administrative cost centers. You can enter multiple categories, up to a maximum of eight, or you can leave the assignment unrestricted by entering an asterisk (*). You can change the cost center category at any time, even if transactional data exists during the current fiscal year.

The **ATyp category** (activity type category) determines whether and how you record and allocate an activity type. For example, for some activity types, you can allow certain activities to be allocated directly based on transactions, while others can be allocated automatically. Right-click in the **ATyp category** field and select possible entries to display the screen shown in Figure 2.27.

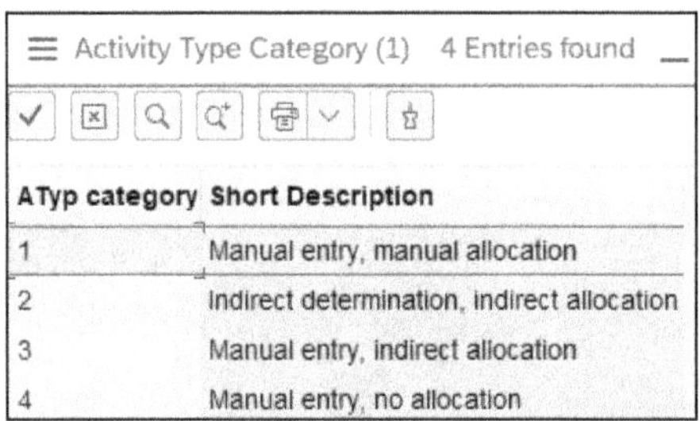

ATyp category	Short Description
1	Manual entry, manual allocation
2	Indirect determination, indirect allocation
3	Manual entry, indirect allocation
4	Manual entry, no allocation

Figure 2.27 Activity Type Category Possible Entries

Each activity type category is explained below:

- **1: Manual entry, manual allocation**
  With this category, you manually plan activity quantities with activity input planning using Transaction KP26.

  You manually allocate actual activity quantities based on business transactions such as activity confirmations. This category should be used for manufacturing activities triggered by order confirmations.
- **2: Indirect determination, indirect allocation; 3: Manual entry, indirect allocation**
  These categories allow you to automatically plan and allocate *quantities*, which is similar to automatically allocating overhead *costs* to receiver cost centers with assessment and distribution cycles and segments. With these methods, you carry out indirect activity allocation with Transaction KSCB, which allocates activity quantities and costs from sender to receiver cost centers based on tracing factors.
- **4: Manual entry, no allocation**
  This category allows you to manually plan activity quantities. You cannot specify receiver objects for this category, but you can calculate target costs.

The category you enter in the activity type is the default value for allocable activity type categories 1, 2, or 3 when planning activity prices and quantities with Transaction KP26. You can change this default value to a different allocable activity type category. You can only change non-allocable activity type category 4 to an allocable activity type category (or the other way around) if no dependent data exists.

Allocation Cost Element

The **Allocation cost elem** (allocation cost element) in Figure 2.26 determines the default value when you enter the planned activity price and quantity with Transaction KP26.

You can overwrite the default cost element when planning for the first time. Once transactions occur, you cannot change this cost element.

Price Indicator and Actual Price Indicator

The **Price indicator** (planned price) and **Act. price indicator** (actual price) fields indicate how the system automatically calculates the price of an activity for a cost center. If you don't set an actual price indicator, the system uses the planned price indicator. Right-click in the **Price indicator** field and select **Possible Entries.** The screen in Figure 2.28 is displayed.

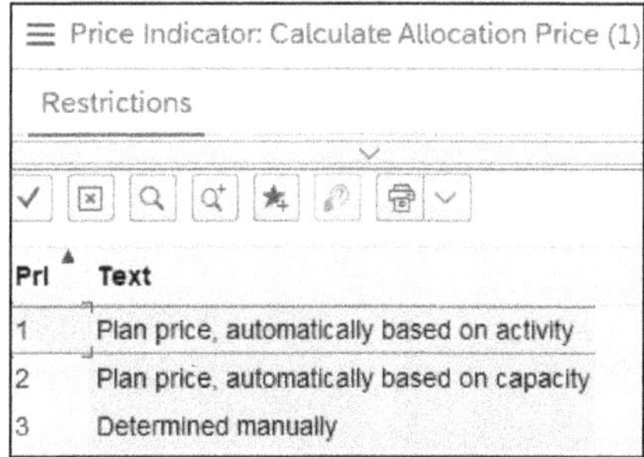

Figure 2.28 Price Indicator Possible Entries

When manually entering the planned activity price with Transaction KP26, you can also enter cost center planned activity and capacity quantities. SAP uses these to calculate planned activity rates automatically. Following are the price indicator possible entries explained:

- **1: Plan price, automatically based on activity**
  Activity price is calculated based on cost center planned activity quantity.
- **2: Plan price, automatically based on capacity**
  Activity price is calculated based on cost center capacity. This setting can lead to underabsorption on the cost center because capacity is usually greater than the planned activity quantity.
- **3: Determined manually**
  This setting indicates that you plan the activity price manually and don't require automatic activity price calculation for this activity type.

When you plan activity prices and quantities with Transaction KP26, the price indicator you enter in the activity type defaults. When planning for the first time, you can overwrite this default value.

SAP Analytics Cloud for Planning

Many settings in the activity type relate to planning with SAP GUI transactions, including Transaction KP26, KP06, KB21N, and others. While these transactions are still available in SAP S/4HANA 2023 without the universal parallel accounting business function activated, SAP future developments in this sphere will focus on SAP Analytics Cloud for planning.

You can also plan activity rates with the Manage Cost Rates—Plan app (SAP Fiori ID F3162).

Actual Quantity Set

When you select the **Actual qty set** checkbox, you must post a manual quantity in addition to the quantity with which the object is credited. You use this when the quantity leading to the object's credit is determined indirectly, but the actual quantity from the sender's view is already known.

If the **Actual qty set** checkbox is set for the activity type with a direct activity allocation with Transaction KB21N, you must manually post the actual activity quantity using Transaction KB51N.

Plan Quantity Set

If you select the **Plan Quantity Set** checkbox, the corresponding checkbox will default as selected when planning activity prices and quantities with Transaction KP26, as shown in Figure 2.29.

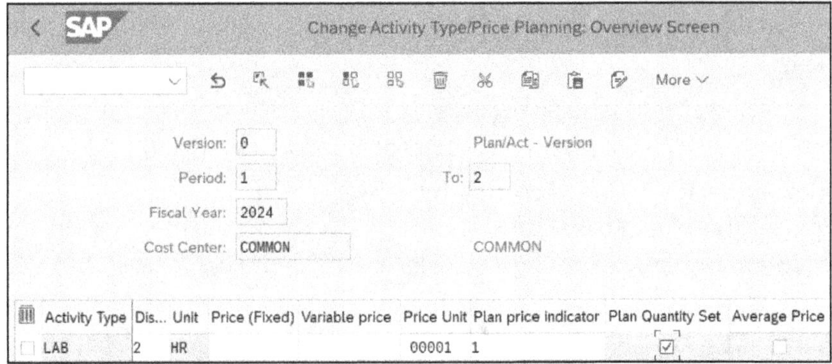

Figure 2.29 Plan Quantity Set Checkbox Activity Price Planning

Select this checkbox if you don't want to manually overwrite, during plan reconciliation, the planned activity quantity entered in the second column of Transaction KP26 (not shown in Figure 2.29) with the scheduled activity quantity automatically calculated. You can transfer scheduled activity quantities from sales and operations planning, material requirements planning (MRP), or long-term planning with Transaction KSPP.

You can change the default setting for this checkbox in Transaction KP26 if you are planning for the first time.

Average Price

If you select the **Average Price** checkbox, the corresponding checkbox will default as selected when planning activity prices and quantities with Transaction KP26, as shown in Figure 2.30.

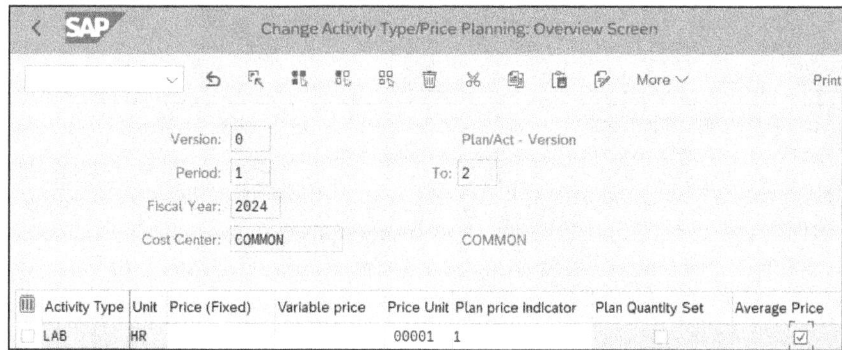

Figure 2.30 Average Price Checkbox Activity Price Planning

Select the **Average price** checkbox if you want the automatically calculated activity price to remain constant for the entire fiscal year, making variance analysis easier. You can still manually change the price per period if necessary. You can change the default setting for this checkbox in Transaction KP26 if you are planning for the first time.

You can indicate that the activity price is to remain constant for the fiscal year, either at the version level or for activity types or cost centers:

- **Version**
  In the **Plan method** field of the fiscal year–dependent version, you can select an average price.
- **Cost center or activity price**
  If you have not defined an average activity price at the version level, you can define an average price for individual cost centers or activity types.

You can change the default setting for this checkbox in Transaction KP26 if you are planning for the first time.

Now that we've reviewed fields and settings in the activity type basic data screen, let's review the **Indicators** tab.

2.3.3 Indicators

Click the **Indicators** tab shown earlier in Figure 2.26 to display the screen shown in Figure 2.31.

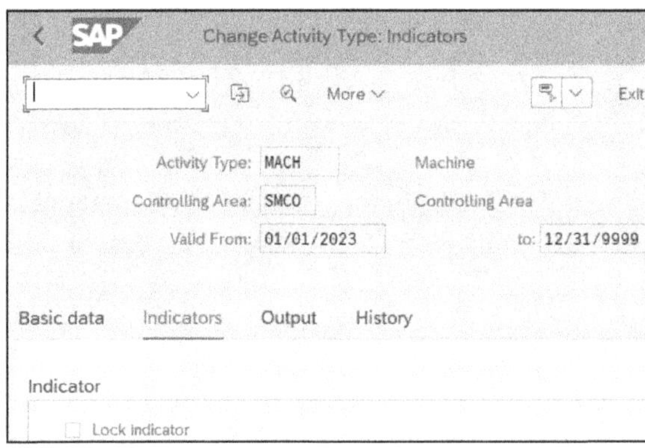

Figure 2.31 Activity Type Indicators Tab

The **Lock indicator** checkbox allows you to block against indirect activity allocation in plan and activity input planning. We discussed indirect activity allocation when looking at the activity type category in Section 2.3.2. Let's examine *activity input planning* next.

Two types of input planning are available for cost centers:

- Basic input planning that most companies use is primary cost planning for cost elements.
- Advanced planning is available with **Cost Element/Activity Input Planning** Transaction KP06 or via the menu path **Accounting • Controlling • Cost Center Accounting • Planning • Cost and Activity Inputs • Change**. You'll see a selection screen like the screen shown in Figure 2.32.

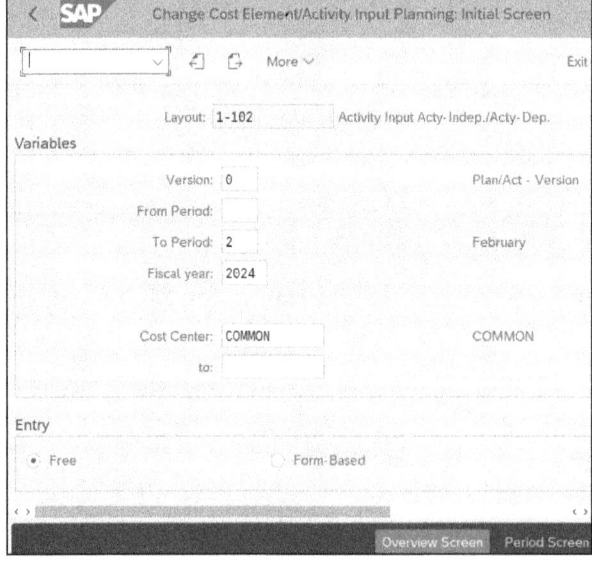

Figure 2.32 Cost Element/Activity Input Planning Selection Screen

To see this screen, first ensure you use planner profile SAPALL with Transaction KP04. You may need to scroll across to **layout 1-102** using the right-pointing arrow icon above the **Layout** field in Figure 2.32. This layout allows you to plan quantities of a sender activity type from a sender cost center to a receiver cost center.

The activity type **Lock indicator** checkbox shown in Figure 2.31 blocks against proceeding with this activity type entered in the **Sender activity type** field. You'll encounter the message shown in Figure 2.33 by clicking on the **Overview Screen** button in Figure 2.32.

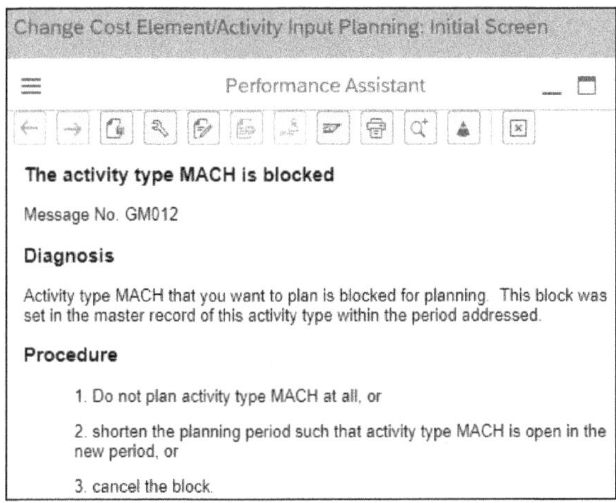

Figure 2.33 Blocked Activity Type Message

Using activity input planning, you'll normally set this indicator following plan reconciliation with Transaction KPSI. Plan reconciliation ensures that the planned activity equals the scheduled activity, and you activate the indicator to ensure that the planned quantity remains unchanged.

With the **Lock indicator** checkbox selected, you can still carry out actual postings, activity price changes, and planning other than activity input planning and indirect activity allocation in the plan.

Now that we've reviewed the **Indicators** tab, let's review the **Output** tab.

2.3.4 Output

Select the **Output** tab shown in Figure 2.31 to display the screen shown in Figure 2.34.

The **Output Unit** is an alternative to the activity unit entered in the **Basic data** tab. Select **Edit • Activity • Output** from the menu bar when entering planned activity rates with Transaction KP26 to change between activity and output units.

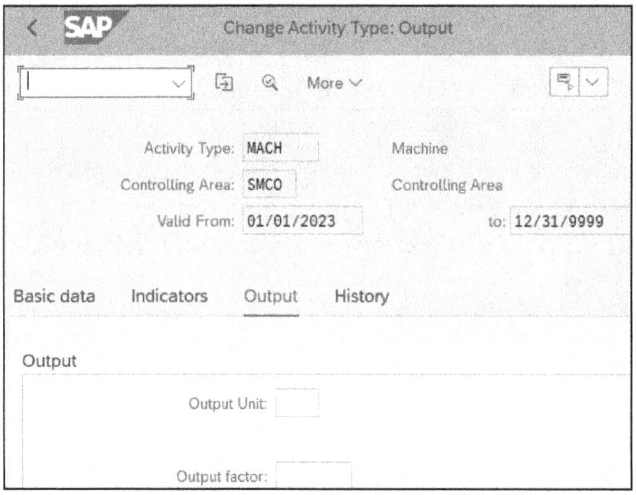

Figure 2.34 Activity Type Output Tab

Output Unit Examples

Output units are useful when you want to use a combination of outputs, such as hours of activity in combination with the number of items produced.

- One hour of activity type "Machining" produces 10 drills. The activity unit is *hours*, and the alternative output unit is *pieces*. You specify an output factor of 10 in this case.
- It takes an average of 30 minutes to process a customer inquiry. The activity unit is *pieces*, and the alternative output unit is *hours*. The alternative output factor is 0.5.

The system can automatically derive the output from the scheduled activity using the alternative output unit and the alternative output factor.

You can also use the alternative output unit with direct internal activity allocation and when you create cycles for indirect internal activity allocation.

Now that we've reviewed the **Output** tab, let's look at the activity type **History** tab.

2.3.5 History

Click the **History** tab shown in Figure 2.34 to display the screen shown in Figure 2.35.

The **History** tab contains the following history and change information:

- **History Data**
  The **History Data** section lets you determine which user created the activity type and when.

- **Change document**
  Click the **Change document** button to display a list of changed activity type fields. Double-click a line to display details of changes made to each field. The entries in the fields in the **History** tab are determined automatically, and you cannot manually change them.

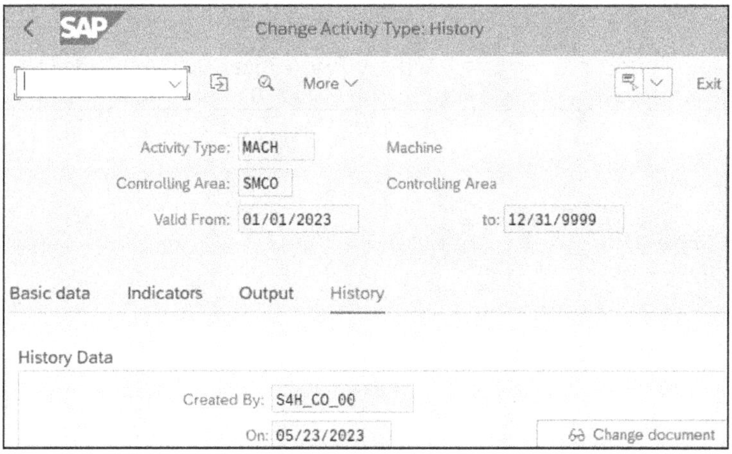

Figure 2.35 Activity Type History Tab

Now that we've examined the data in the **History** tab, let's look at the advantages of creating activity type groups.

2.3.6 Group

Activity type *groups* allow you to limit the list of activity types when planning activity prices with Transaction KP26. You can enter an individual activity type, a range of activity types, or an activity group in the initial selection screen when planning activity prices:

- **Activity type**
  You need to manually enter all your activity types for the group.
- **Range**
  If you don't number your activity types in logical number ranges, you might accidentally include many activity types in the group that you don't require.
- **Group**
  Groups filter in exactly your activity types on the **Plan Activity Price** screen.

Let's examine how to create and use activity type groups. You create an activity type group with Transaction KLH1 or by following the menu path **Accounting • Controlling • Cost Center Accounting • Master Data • Activity Type Group • Create**. Type in the name of your group and save to display the screen shown in Figure 2.36. You can also use the Manage Global Hierarchies app to create activity type groups.

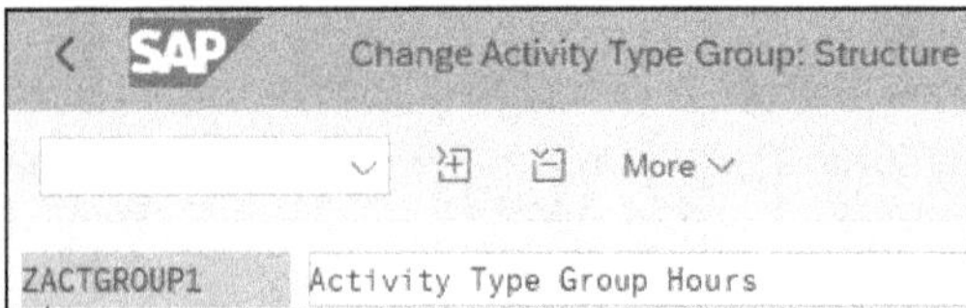

Figure 2.36 Activity Type Group

Type in your group description, choose **Edit • Activity Type • Insert Activity Type** from the menu bar, and enter the activity types in your group to display the screen shown in Figure 2.37.

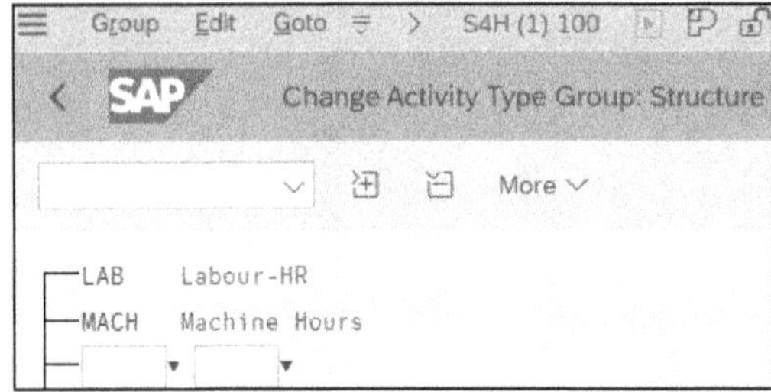

Figure 2.37 Add Activity Type to Activity Type Group

After saving the details in Figure 2.38, you can use your new activity type group during activity price planning. Enter your plan activity prices with Transaction KP26 or by following the menu path **Accounting • Controlling • Cost Center Accounting • Planning • Activity Output/Prices • Change**. Enter the selection screen data (including your activity type group), select the **Form-Based** radio button at the bottom of the screen, and click the **Overview Screen** button to display the plan price screen shown in Figure 2.38.

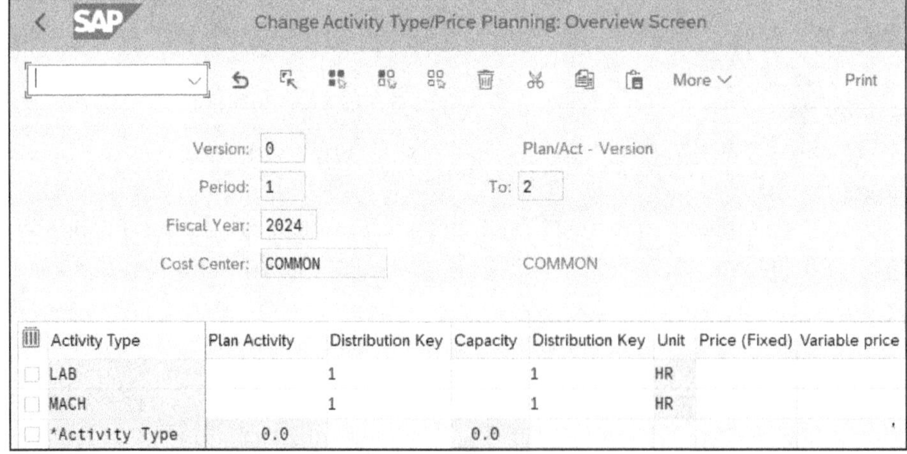

Figure 2.38 Activity Type Price Planning Overview

The screen is restricted to the activity types in your group, and you cannot enter any additional activity types. If you need to plan for more activities, you can either add activity types to your group or select the **Free Entry** radio button at the bottom of the selection screen when you run Transaction KP26. The screen defaults to your last selection of either the **Free Entry** or **Form-Based** radio buttons.

Activity type groups save you time during activity price planning by restricting the planning screen to a list of your activities. You can also group master data in other modules, such as characteristic groups used when editing characteristics or assigning characteristics to classes.

Now that we've examined activity types and their setup and maintenance, let's look at statistical key figures.

2.4 Statistical Key Figures

Statistical key figures define values describing cost centers, profit centers, and overhead orders, such as the number of employees or minutes of long-distance phone calls. You can use statistical key figures as the tracing factor for periodic transactions such as cost center distribution or assessment. You can post both planned and actual statistical key figures.

You maintain statistical key figures with the Manage Statistical Key Figures app (SAP Fiori ID F1603) or Transaction KK02 or via menu path **Accounting • Controlling • Cost Center Accounting • Master Data • Statistical Key Figures • Individual Processing • Change**. Type in the statistical key figure and then press Enter to display the initial statistical key figure screen shown in Figure 2.39.

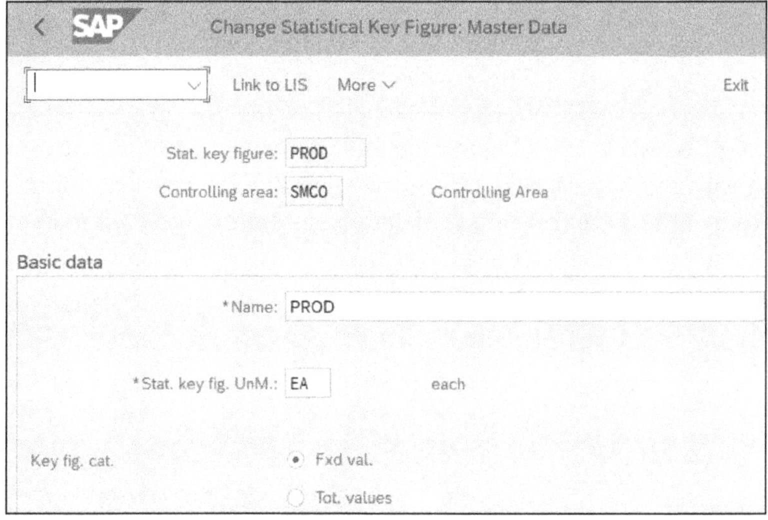

Figure 2.39 Statistical Key Figure

The first fields you maintain are **Name** and **Stat. key fig. UnM.** (statistical key figure unit of measure).

You can define the statistical key figure category as a fixed or total value by selecting **Fxd val.** or **Tot. values**:

- **Fxd val.** (fixed value)
  An example of fixed value, such as employee head count, would carry over from the period in which it's entered to all subsequent periods of the same fiscal year. You need to enter a new posting only when the value changes. The fiscal year total is an average of the period totals.
- **Tot. values** (total value)
  An example of total value, such as "Long-Distance Calls," isn't transferred to the following period but is entered each period. The fiscal year total is the sum of all the period values.

Statistical Key Figures in SAP S/4HANA

Be aware that there are two places to store statistical key figures in SAP S/4HANA: the tables COSR and FINSSKF. Normally you don't need to know this, as Transaction KB31N (update statistical key figures) updates both tables; however, if the contents of the two tables get out of synch, follow the explanation in SAP Note 3289021 to fix the situation. The classical transactions for assessment and distribution read the values for the statistical key figures from table COSR, but universal allocation reads the statistical key figures from table FINSSKF.

Now that we've examined the statistical key figure fields, let's review what we've covered in this Controlling master data chapter.

2.5 Summary

In this chapter, we discussed Controlling master data relevant to product cost controlling, including accounts and cost elements, cost centers, activity types, and statistical key figures. We examined the master data fields and indicators and their roles in identifying, collecting, and allocating Controlling costs.

Now that we've examined Controlling master data in this chapter, we'll look at material master data in Chapter 3 and logistics master data in Chapter 4.

Chapter 3
Material Master Data

A material master contains all the information required to manage a material, including settings that cost estimates access to determine the cost to purchase or manufacture the material.

3

Now that we've examined Controlling concepts and master data, let's look at material master data. Information is stored in *views*, each corresponding to a department or area of business responsibility. Views conveniently group information together for users in different departments—for example, sales and purchasing. The three views of particular interest for product cost controlling are **MRP**, **Costing**, and **Accounting**. These views are plant-specific, so plants can have different values for fields in these views. Material requirements planning (MRP) guarantees material availability by monitoring stocks and generating planned orders for procurement and production.

3.1 Basic Concepts

Cost estimates typically follow MRP settings when calculating the cost to procure or manufacture a material. For example, the components that MRP determines to be necessary to manufacture an assembly are also used by a cost estimate to calculate the material cost of the assembly. This is why this chapter examines the **MRP** as well as the **Costing** and **Accounting** views of the material master.

Material Master Number

A *material master number* uniquely identifies a material within a plant. Material master numbers can be extended from 18 to 40 characters with Transaction OMSL. Refer to SAP Note 2232396 for more information.

Extended material master numbers are helpful for specific industries, such as automotive, which often requires the material number to include the vehicle identification number (VIN), which is 17 characters long.

While most material master views contain some fields of interest to product costing, the **MRP**, **Controlling**, and **Accounting** views contain multiple fields that cost estimates access to determine the cost to purchase or manufacture a material. We'll examine each view in detail, starting with the **MRP 1** view.

3.2 MRP 1 View

You maintain material master views with Transaction MM02 or by following the menu path **Logistics • Production • Master Data • Material Master • Material • Change • Immediately**. Select the **MRP 1** tab to display the screen shown in Figure 3.1.

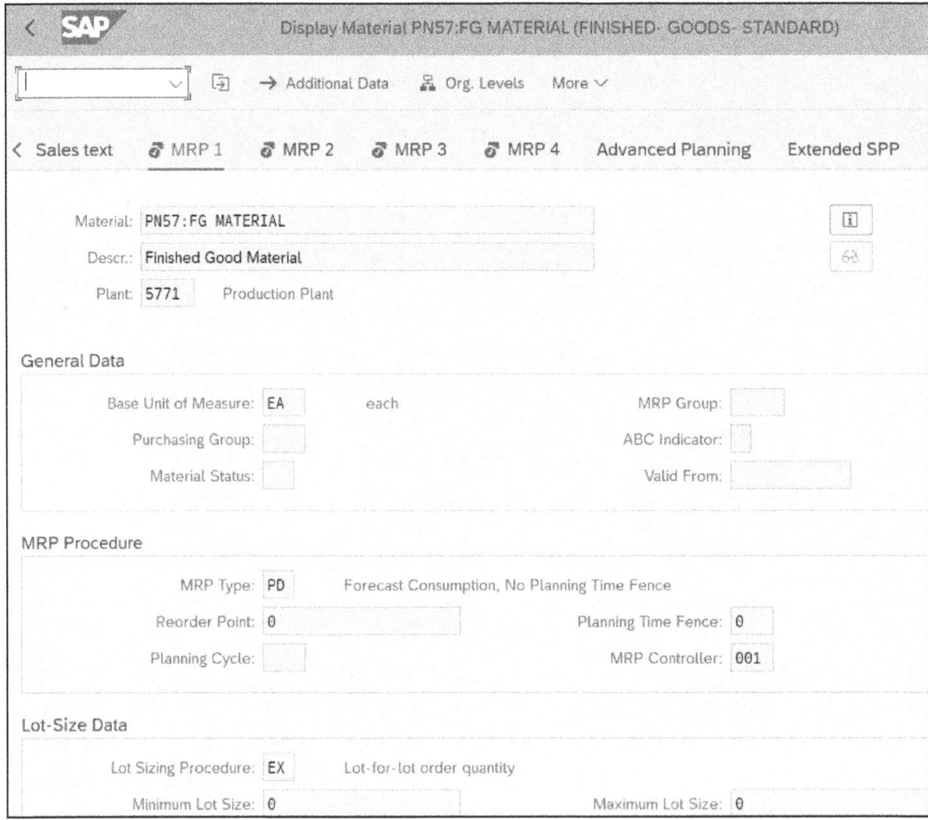

Figure 3.1 MRP 1 View

The fields that are most relevant to costing are in the **General Data** and **Lot Size Data** sections. We'll consider each section in turn.

3.2.1 General Data

In the **General Data** section, let's examine fields relevant to costing.

Base Unit of Measure

You manage stocks of a material in the **Base Unit of Measure**. The system converts all quantities you enter in other units of measure (alternative units of measure) to the base unit of measure. You can use alternative units of measure, for example, for purchasing and sales.

Alternative Units of Measure

You enter alternative units of measure in the material master by clicking the **Additional Data** button in Figure 3.1 and then selecting the **Units of measure** tab.

Material Status

An entry in the **Material Status** (plant-specific material status) field determines if you cost the material. Plant-specific material status can also be used for purposes other than costing, such as to issue a warning or error message when a purchase or production order is created for discontinued materials. You can view the list of possible settings behind plant-specific material status with configuration Transaction OMS4 or via IMG menu path **Controlling • Product Cost Controlling • Product Cost Planning • Material Cost Estimate with Quantity Structure • Settings for Quantity Structure Control • Material Data • Check Material Status.**

Discontinued and Obsolete Materials

You typically set a plant-specific material status for discontinued and obsolete materials:

- Another material will replace discontinued material after the remaining inventory is consumed. You can allow discontinued material to be costed, but you might want to issue a warning or error message when creating a purchase order. The **Discontinued parts** section in the **MRP 4** view allows you to enter the material number that you can use in materials planning to replace the discontinued material as its warehouse stock is depleted.
- Obsolete materials aren't intended for further use. This can arise, for example, if the material is outdated or has a design problem. Typically, you may decide to scrap or sell the remaining inventory. In the case of obsolete material, you can set the plant-specific material status to display either a warning or an error message during costing if the material is contained in the bill of materials (BOM) of an active assembly.

You can also use the plant-specific material status to restrict the use of new parts during different stages of developing or phasing out a product.

Valid From

The **Valid from** field indicates the date the plant-specific material status is valid. For example, if you block a material from purchasing with the material status, you cannot create purchase orders after the valid from date.

ABC Indicator

The **ABC Indicator** determines the importance of the material during cycle counting based on consumption:

- A
  Important part, high consumption value
- B
  Less important part, medium consumption value
- C
  Relatively unimportant part, low consumption value

You can configure the **ABC Indicator** by following the IMG menu path **Logistics – General • Material Master • Settings for Key Fields • Extend ABC Indicator.**

Now, let's discuss the **Lot size data** section of the **MRP 1** view.

3.2.2 Lot Size Data

MRP determines proposed quantities and dates for manufacturing and purchase orders so you can meet the needs of independent requirements such as customer sales orders. MRP uses information from the **Lot size data** fields of the **MRP 1** view to calculate the quantity and date of proposed orders.

Example 1: Lot Size Data

A customer places an order for a quantity of 120 items, due for delivery one year from the date you place the sales order. MRP determines the following proposed orders:

- Requisition or purchase orders for components
- Manufacturing orders for assemblies

MRP also determines the quantity (lot size) and the date to create each order. The quantity and date depend on the quantity of components in stock, work center capacity, and requirements generated by other sales orders. The following are the two proposals in this example:

- One manufacturing order for a quantity of 120 during the next year
- Manufacturing orders with a quantity of 10 each month over the next year

You can manually change system-generated requirements as required.

Let's examine the fields most relevant to costing in the **Lot size data** section of the **MRP 1** view.

Fixed Lot Size

You use the **Fixed lot size** field to enter the quantity that you'll order in the event of a shortage. Two possible scenarios are:

- If the shortage is less than the fixed lot size, you'll order the fixed lot size quantity.
- If the shortage exceeds the fixed lot size, the quantity ordered will be in multiples of the fixed lot size.

Pharmaceutical and food industry companies often use fixed lot size to ensure that the batch size always matches certified quantities. You can also use fixed lot size to determine batches involving pallets or tank contents.

Note

You cannot use the **Fixed lot size** field with **Minimum Lot Size** or **Maximum Lot Size** entries.

Assembly scrap increases production order plan quantity to compensate for the scrap quantity. Assembly scrap in combination with fixed lot size instead reduces the production order yield quantity, while the plan quantity remains fixed.

Example 2: Fixed Lot Size

You define a fixed lot size of 50 pieces for an in-house product. The system can only create production orders with a procurement quantity of 50 pieces each to cover requirements. To cover a requirement of 100 pieces, you must produce 125 pieces altogether, considering an assembly scrap of 20%. The system creates 3 receipts, each with 50 pieces, while the yield of each receipt is 40 pieces.

Assembly Scrap

You can enter a value in the **Assembly scrap (%)** field to indicate the percentage output from manufacturing orders that doesn't meet required production quality standards. You determine this number from your production statistics of scrap rates and update it before each costing run.

A *costing run* is a collective processing of cost estimates, which we'll discuss in detail in Chapter 9. The **Assembly scrap (%)** field only appears in the material master. You can set the **Net ID** indicator on the BOM item **Basic data** tab to ignore assembly scrap for a component.

Note: Assembly Scrap

You can access more detailed information on assembly scrap in *Production Variance Analysis in SAP S/4HANA* (SAP PRESS, 2023) at *https://www.sap-press.com/5629*.

Now that we've discussed the fields and indicators relevant to product cost controlling in the **MRP 1** view, let's examine the **MRP 2** view.

> **Material Master Tabs**
>
> Four **MRP** tabs handle the MRP setup: **MRP 1**, **MRP 2**, **MRP 3**, and **MRP 4**. If all the MRP fields were on one tab, you would need to scroll to see them, hence the reason for the four tabs.

3.3 MRP 2 View

Select the **MRP 2** tab shown in Figure 3.1 to display the screen shown in Figure 3.2.

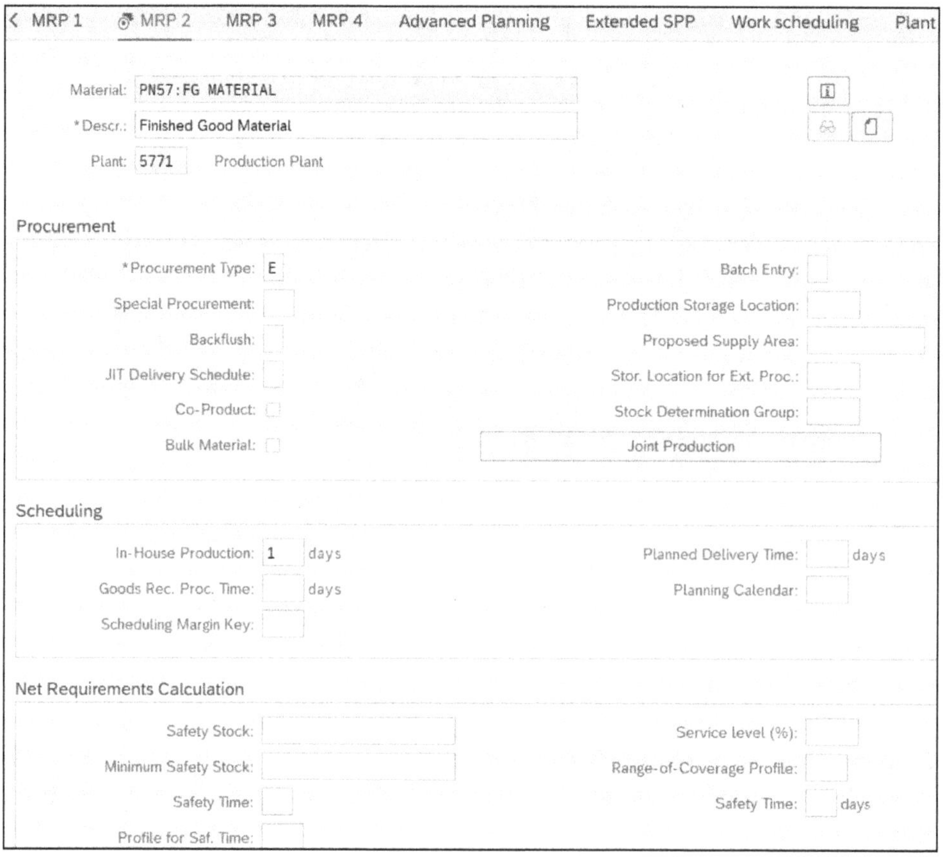

Figure 3.2 MRP 2 View

MRP ensures material availability by monitoring stocks and requirements and by generating planned orders for procurement and production. A cost estimate needs data on how the material is procured as the first step in determining the cost of the material, so let's look at the procurement type, along with other important settings:

- **Procurement Type**
  To display a list of possible entries for the **Procurement Type** field, left-click in the field and press F4. The entries are explained below:
  - **E: In-House Production**
    An entry here means the system will search for production information such as a BOM and routing. A *routing* is a list of tasks containing standard activity times required to perform operations to build an assembly. We'll discuss BOMs and routings in more detail in Chapter 4.
  - **F: External Procurement**
    An entry here means the system will search for purchasing information, usually from a purchasing info record that contains purchasing information for a material from a vendor.
  - **X: Both Procurement Types**
    An entry here means a planned order (proposal) generated by MRP can be converted into either a production or purchase order. A cost estimate is based on a BOM and routing if they are available; that is, the procurement type will behave as in-house production. You can make the cost estimate treat the material as though it's externally procured with an entry in the **Special procurement** type field, as discussed in the next section.
- **Special Procurement Type**
  The **Special Procurement** field found immediately below the **Procurement Type** field in Figure 3.2 more closely defines the procurement type. For example, it might indicate if you produce the item in another plant and transfer it to the plant you are looking at. A cost estimate normally follows the MRP special procurement type when determining costs. However, costing will use an entry in the special procurement type for costing field in the **Costing 1** view instead of the MRP setting.

Special Procurement Type Can Override Procurement Type

You can use the special procurement type to override the procurement type if required. For example, if you enter a special procurement type that contains an external procurement type (F) in configuration Transaction OMD9, the material will behave as if it's externally procured, regardless of the procurement type setting.

- **Backflush**
  You select the **Backflush** checkbox to default the backflush indicator as selected in production orders. *Backflushing* is the automatic goods issue for components after their actual physical issue for use in an order. The goods issue posting for backflushed components is carried out automatically during confirmation.

  You can use backflushing to reduce the amount of work in warehouse management, especially for low-value parts. To use backflushing effectively, assign the material components from the BOM required in the operation to the operations in the routing.

- **Co-Product checkbox**
  Select the **Co-Product** checkbox if this material is a valuated product you produce simultaneously with one or more other products. Selecting this checkbox allows you to assign the proportion of costs this material will receive about other co-products within an apportionment structure. An apportionment structure defines how you distribute costs to co-products. The **Co-Product** checkbox is also located in the **Costing 1** view.
- **Joint Production**
  Joint production involves the simultaneous production of many materials in a single production process. Click the **Joint Production** button to assign apportionment structures, which define how you distribute costs between co-products. You must first select the **Co-product** checkbox to assign the apportionment structures. The **Joint Production** button is also located in the **Costing 1** view.
- **Bulk Material checkbox**
  You select the **Bulk Material** checkbox if you make materials, such as washers or grease, available directly at a work center. Bulk materials aren't relevant for costing in a cost estimate, and you expense them directly instead to a cost center. The cost can be included in the cost of sales (COS) as part of an overhead rate.

You can also maintain the **Bulk Material** checkbox in the BOM item. The indicator in the BOM item has higher priority. (We'll examine BOM items in more detail in Chapter 4.) If you always use the material as a bulk material, select the **Bulk Material** checkbox in the material master. If you only use the material as a bulk material in individual cases, select the checkbox in the BOM item.

Now that we've discussed the fields and indicators relevant to product cost controlling in the **MRP 2** view, let's examine the **MRP 3** view.

3.4 MRP 3 View

Select the **MRP 3** tab to display the screen shown in Figure 3.3.

In the **Planning** section, the **Strategy Group** field includes the following strategies:

- **Make-to-stock production (10)**
  You manufacture to maintain a certain quantity in collective stock.
- **Make-to-order production (20)**
  You manufacture based on individual sales orders for customer stock.

Now, let's look at the **MRP 4** view.

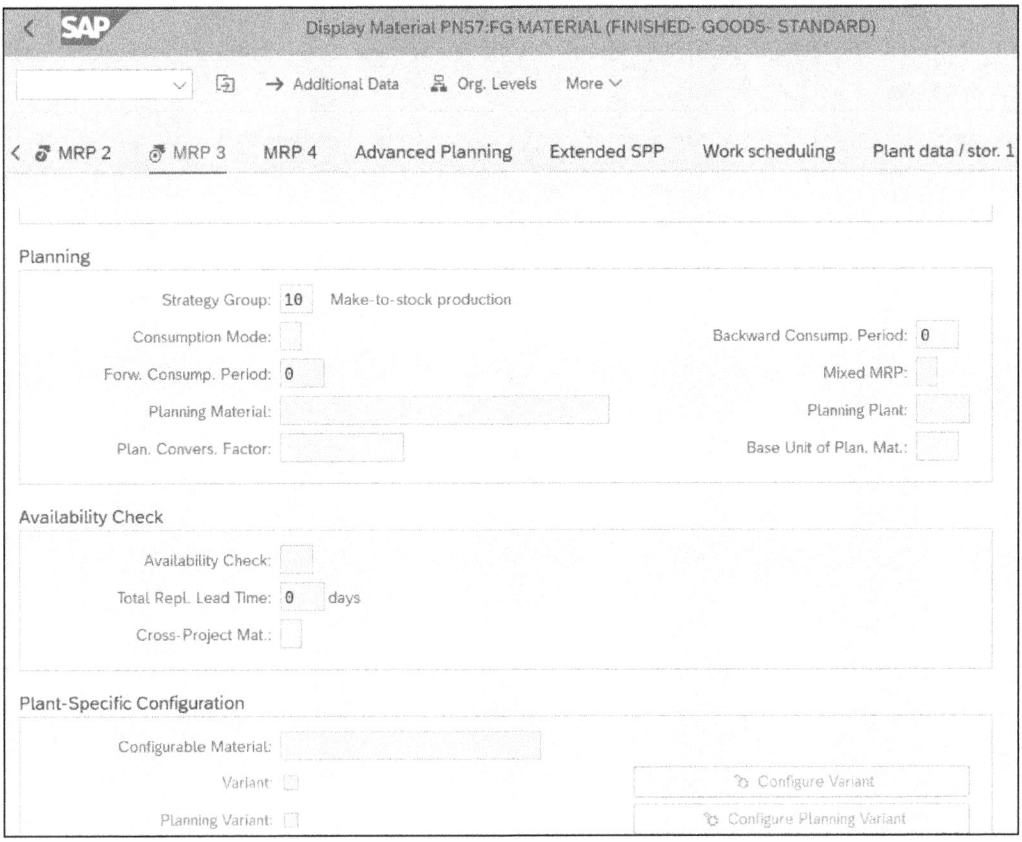

Figure 3.3 MRP 3 View

3.5 MRP 4 View

Select the **MRP 4** tab to display the screen shown in Figure 3.4.

The **MRP 4** view contains further BOM and production version information. A **Production Version** is a unique combination of BOM, routing, and production line. You use a production line in repetitive manufacturing, and a production line typically consists of one or more work centers.

Production Versions Are Mandatory

Production versions are mandatory with SAP S/4HANA. They allow you to define which BOM alternative goes together with which routing alternative. They simplify MRP sourcing because there is only one option to determine BOM and routing alternatives to manufacture a material. See SAP Note 2267880 for more details.

Let's examine the fields of interest in the **MRP 4** view. We'll consider the relevant fields in the **BOM Explosion/Dependent Requirements** and **Discontinued Parts** sections in turn.

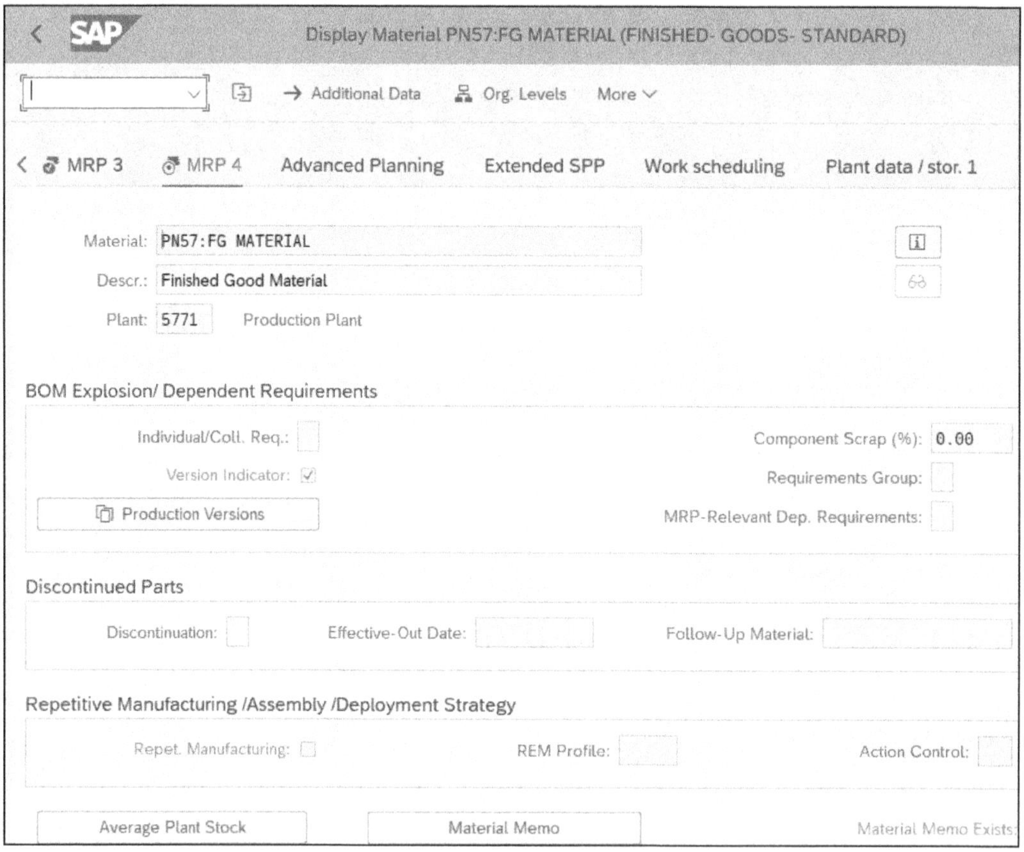

Figure 3.4 MRP 4 View

3.5.1 Bill of Materials Explosion and Dependent Requirements

The fields in this section determine how you select and treat BOM components. The following sections examine the **Individual/Coll. Req.** field, which determines if you consider dependent requirements individually or separately.

Individual and Collective Dependent Requirements

To display a list of possible entries for the dependent requirements for the **Individual/Coll. Req.** (individual and collective requirements) field, as shown in Figure 3.4, left-click in the field and press F4 to display the screen shown in Figure 3.5.

Ind./Coll.	Short Description
	Individual and collective requirements
1	Individual requirements only
2	Collective requirements only

Figure 3.5 Individual and Collective Dependent Requirements

This field determines whether the dependent (lower-level) requirements of an assembly can be grouped together (collective) or must be treated separately (individual). The three possible settings are discussed here:

- **(blank): Individual and collective requirements**
  This setting offers the most flexibility and indicates that the component is planned similarly to the higher-level assembly.
- **1: Individual requirements only**
  This setting means the material is specially manufactured or procured for a sales order. You create an individual requirement only if the higher-level material doesn't create a collective requirement.
- **2: Collective requirements only**
  This setting means the material is produced or procured for various requirements and offers the advantage of economy of scale because you can group together requirements of higher-level assemblies for lower-level assemblies and components.

You can also maintain this indicator when configuring the BOM item basic data with Transaction OS17, which takes priority over the **MRP 4** view setting.

You can ensure that you use the costing lot size of the highest-level material for lower-level materials with individual requirements with the **Pass On Lot Size** field on the **Qty Struct.** tab in the costing variant with configuration Transaction OKKN. We'll look at this indicator in more detail in Chapter 8.

Component Scrap

A value entered in the **Component scrap (%)** field shown earlier in Figure 3.4 indicates the percentage of this component quantity that doesn't meet required production quality standards before being inserted into the production process. You should determine this number from your production statistics of component scrap rates and update this field before each costing run. The **Component scrap (%)** field can also be maintained in the BOM item, which takes priority over an entry in the **MRP 4** view.

Component Scrap

You can find more detailed information on component scrap in the SAP PRESS book *Production Variance Analysis in SAP S/4HANA* at *https://www.sap-press.com/5629/*.

Requirements Group

The **Requirements group** field determines whether you group *dependent requirements* for a material together daily.

Material Requirement Planning-Dependent Requirements

The **MRP-Relevant dep.requirements** indicator controls whether dependent requirements are relevant for MRP for make-to-stock materials and assemblies.

Version Indicator and Production Versions

The **Version Indicator** checkbox and **Production Versions** button are also in the **Costing 1** view. We'll discuss these in more detail in Section 3.6.

Now let's discuss the fields in the **Discontinued parts** section of the **MRP 4** view.

3.5.2 Discontinued Parts

In many industries, discontinuing the use of a material and replacing it with another material is common practice. Here are two typical scenarios:

- You replace one part with a more technically advanced part.
- You replace an expensive part with a less expensive part.

When replacing a component with another, however, the stock of the old material should be used up before the new one is introduced to avoid dead stock. MRP reassigns the dependent requirements for the discontinued component to the follow-up material after you use up the stock of the discontinued part.

The following settings control how the use of the discontinued materials is controlled:

- **Discontinuation indicator**
  The **Discontinuation** field identifies the material as a part to be discontinued. The two possible entries are listed here:
  - **1: Simple Discontinuation**
    One material is to be replaced by one follow-up material. You set this indicator for the main discontinued material in parallel discontinuation.
  - **3: Dependent Parallel Discontinuation**
    A group of materials is to be replaced by another group of materials. You set this indicator for materials dependent on the main discontinued material. You should enter all discontinued parts and all follow-up materials in the BOM with discontinuation and follow-up groups.
- **Effective-out date**
  The effective-out date refers to the date from which the stocks of the material are to be used up. As soon as no more stock exists for this material, the follow-up material will replace it.

- **Follow-up material**
  The follow-up material replaces the discontinued material at the effective-out date. Follow-up materials may cause a production variance if their cost differs from the discontinued material.

Now that we've discussed the fields and indicators relevant to product cost controlling in the **MRP 4** view, let's examine the material master **Costing 1** view fields.

3.6 Costing 1 View

Members of the Controlling or management reporting team generally use the **Costing** views. Select the **Costing 1** tab to display the screen shown in Figure 3.6.

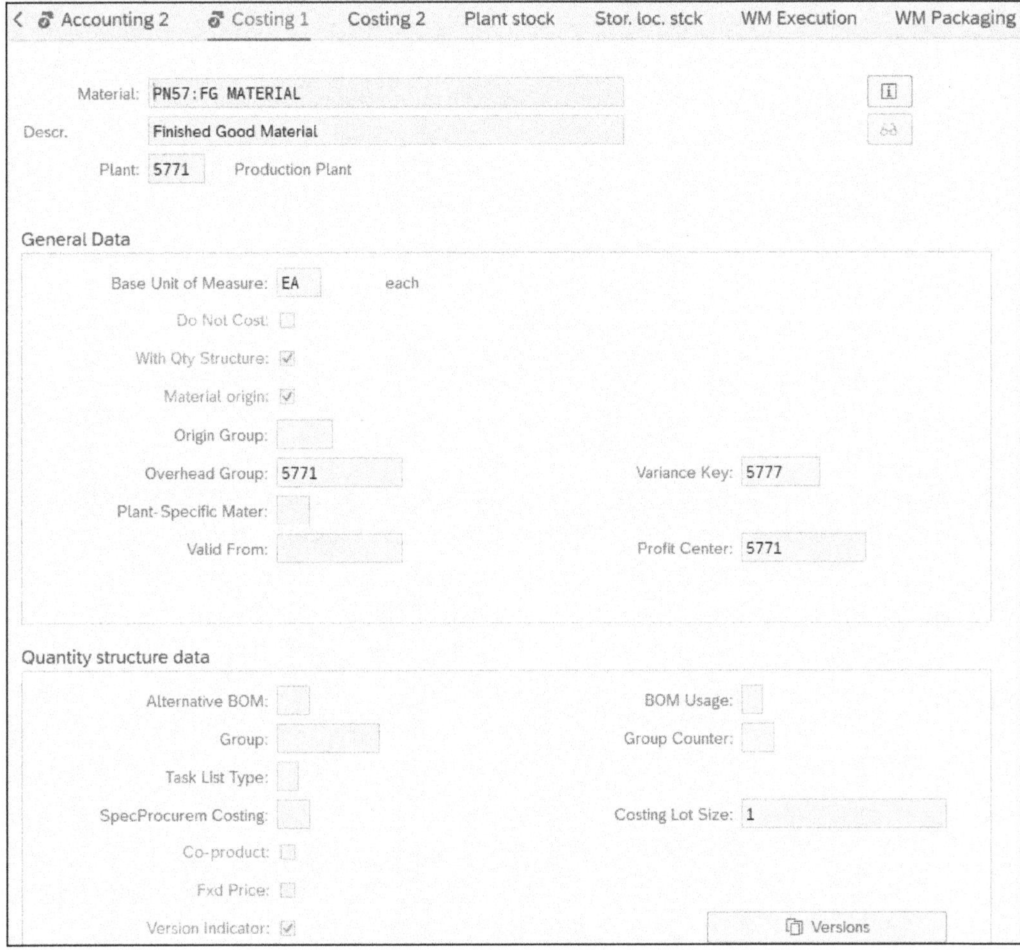

Figure 3.6 Costing 1 View

This screen is divided into two sections: **General data** and **Quantity structure data**. To make the discussion easier to follow, we'll consider the fields in each section individually. First, let's look at the fields in the **General Data** section.

3.6.1 General Data

The **General Data** section contains fields relating to costing, and cost estimates in general, that cannot be included in any of the more specific sections. The following sections describe these fields.

Do Not Cost

The **Do Not Cost** checkbox controls whether you create a cost estimate for the material. With the checkbox selected, the material is excluded from the selection and BOM explosion steps in the costing run selection screen, which you access with Transaction CK40N (we'll examine costing runs in more detail in Chapter 9). If the material is included as a component in another cost estimate, the valuation price is determined through the valuation variant, as we will discuss in Chapter 7.

With Quantity Structure

You should select the **With Qty Structure** checkbox if you create cost estimates with quantity structure (i.e., with BOMs and routings) or procure materials from other plants with a special procurement type.

If you create an individual cost estimate for a material with this checkbox deselected, the outcome depends on the transaction you are using to create the cost estimate, as follows:

- If you create an individual cost estimate with Transaction CK11N, the system will issue a warning message stating that the **With Qty Structure** checkbox is deselected. The system will still create a cost estimate with a quantity structure.
- If you create a cost estimate with costing run Transaction CK40N, the system will not consider the quantity structure and will treat the material as a raw material without quantity structure. The system will search for a price as indicated by the valuation variant search strategy, typically a purchasing info record price or planned price entry in the **Costing 2** view.

Always Select the With Qty Structure Checkbox

If you aren't sure how this checkbox works, it's usually safest to leave it selected, or you may encounter unexpected costing run results, as explained in the previous list.

Selecting the **With Qty Structure** checkbox saves the system time searching for cost estimates without quantity structures (unit costing) when you are initially developing new or improved products and don't yet have a BOM or routing created. *Unit costing* is a method of costing that doesn't use BOMs or routings. If you aren't using unit costing, deselecting the checkbox will slow down the system while it searches for cost estimates that don't have quantity structures.

Select the With Qty Structure Checkbox for Raw Materials

There is an advantage to leaving the **With Qty Structure** checkbox selected for raw materials: Many plants transfer raw materials from plant to plant using the special procurement type in the **MRP 2** view, as discussed in Section 3.3. A costing run considers this plant-to-plant transfer as part of a quantity structure and ignores the special procurement type for materials with the **With Qty Structure** checkbox deselected.

Next, we'll look at the **Material origin** checkbox.

Material Origin

You should select the **Material origin** checkbox for all cost-critical components. This checkbox determines if the material number displays in detailed reports. The material number is one of the most important indicators for providing greater visibility about the causes of production variances. If you have already created material masters without the **Material origin** checkbox selected, you can use report RKHKMATO with Transaction SA38 to select the checkbox automatically.

Origin Group

An **Origin Group** enables you to separately identify materials assigned to the same cost element to assign them to separate cost components. A *cost component* identifies costs of similar types, such as material, labor, and overhead, by grouping together cost elements.

You can also use the origin group to determine the calculation base for overhead. A *calculation base* is a group of cost elements to which overhead is applied and is a component of a costing sheet that summarizes the rules for allocating overhead. We'll discuss calculation bases and costing sheets in more detail in Chapter 5.

You can view origin group configuration settings with Transaction OKZ1 or by following the IMG menu path **Controlling • Product Cost Controlling • Product Cost Planning • Basic Settings for Material Costing • Define Origin Groups**.

Origin Group Example: Active Pharmaceutical Ingredient

The active pharmaceutical ingredient (API) is the highest-cost ingredient in the pharmaceutical industry. If you use the same cost element for all ingredients, it's possible to identify the API as a separate cost component with the following steps:

1. Create an API origin group.
2. Populate all API material masters with the API origin group.
3. Create an additional API cost component that includes the origin group with cost component structure Transaction OKTZ.
4. Subsequent costing runs will report API as a separate cost component.

A cost component structure groups cost elements into cost components.

Overhead Group

An **overhead group** allows you to apply different overhead percentages to individual or groups of materials. You can also use the same overhead percentage rate across all materials in a plant, eliminating the need for overhead groups. This minimizes the maintenance required for regularly updating rates associated with each overhead group.

Example: Overhead Group

A company realized it was possible to assign different overhead percentage rates to individual materials within a plant and created a separate overhead group for each family of materials.

This required setting up and maintaining a large overhead rate table with hundreds of entries in a costing sheet. Within a year, the client realized that any potential gains from increased overhead accuracy were offset by the maintenance time required to calculate the percentages and update the costing sheet. Subsequently, the client changed to a simplified method of allocating overhead using the same percentage across all materials within the plant.

Plant-Specific Material Status and Valid From

The **Plant-Specific.Mater** (plant-specific material status) and **Valid From** fields are also maintained in the **MRP 1** view and were discussed in detail in Section 3.2.

Variance Key

You calculate variances only on manufacturing orders or product cost collectors containing a **Variance Key**. This key defaults from the **Costing 1** view when you create manufacturing orders or product cost collectors. The variance key also determines if you subtract the scrap value from actual costs before production variances are determined.

Profit Center

Profit Center Accounting (PCA) is part of the Universal Journal in SAP S/4HANA and enables reporting from a profit center point of view. You typically create profit centers based on areas in a company that generate revenue with a responsible manager assigned. The **Profit Center** field also appears in the **Sales**, **MRP**, and **Plant Data** views.

You can continue to use classic PCA as part of the compatibility scope. Usage rights are scheduled to expire at the end of 2025. It should only be used as part of the transition to PCA within the Universal Journal. The assignment of the materials to profit centers is the same for both the Universal Journal and classic PCA.

3.6.2 Quantity Structure Data

Now that we've examined the fields in the **General data** section of the **Costing 1** view, we'll consider the fields in the **Quantity structure data** section, as shown in Figure 3.7.

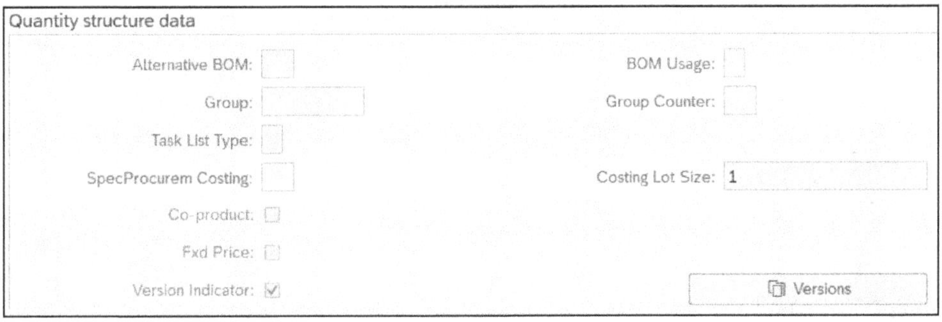

Figure 3.7 Quantity Structure Data Section

A *quantity structure* typically consists of a BOM and a routing. A cost estimate determines the quantity structure from the costing variant configuration, as discussed in Chapter 7, or from the production version, which you access by clicking the **Versions** button.

You typically don't need to make any entries in the screen, except for the mandatory **Costing Lot Size** field, which we'll examine later in this section. You make entries in this screen to direct the cost estimate to use BOMs and routings independently of those determined automatically by MRP or costing configuration for individual materials. Let's examine each field in detail and look at examples to show how these fields are used.

Alternative Bill of Materials

There can be multiple methods of manufacturing an assembly and many possible BOMs. The **Alternative BOM** field allows you to identify one BOM in a group. A BOM group is a collection of BOMs for a product or a number of similar products.

Example: Alternative BOM

A production department sources a particular component from overseas, but soon, you'll source the component locally at a lower cost. For most of the following year, the assembly will be manufactured with the cheaper local component, but the cost estimate would normally set the plan cost based on the present procurement method (with the assembly using the more expensive overseas component) for the entire fiscal year. In this case, you can enter the alternative BOM containing the locally sourced component to ensure that the cost estimate incorporates this cheaper component for the following fiscal year.

Bill of Materials Usage

The **BOM Usage** field determines a specific section of your company, such as production, engineering/design, or costing, for which a BOM is valid.

Case Study: Costing BOM

You use the same material in the production and costing departments in your company. But for each department, you configure a separate BOM controlled by the **BOM Usage** field as follows:

- The engineering BOM contains items used in the early stages of product development.
- The production BOM contains items that are relevant to production. You copy these items to the planned (proposed) order when MRP runs. You generate dependent items, and the planned order is subsequently copied to the production order.
- The costing BOM contains items that are relevant when determining the material costs of a product. You cannot use the costing BOM for production.

You normally copy the costing BOM from the production BOM every year before the main costing run. If a BOM with usage of costing exists, cost estimates will typically refer to this instead of the production BOM. This means that changes in the production BOM during the year will only be reflected in costing after the costing BOM is copied from the production BOM.

It's not common to use costing BOMs because typically you want to see changes in production BOMs reflected in costing if you create new cost estimates during the year.

Group

A task list *group* identifies routings with different operations (production steps) for one material. Just as you can create alternative BOMs to record different hierarchies of components to manufacture an assembly, you can maintain alternative routings containing different production steps. A cost estimate uses an entry in the **Group** field to determine activity costs for this material independently of the task list group used by production.

Group Counter

A *group counter* identifies a unique routing within a task list group.

> **Example: Group Counter**
>
> You can use group counters to identify different lot size ranges. For example, a manufacturer can machine a workpiece either on a conventional machine or on a numerically controlled (NC) machine. An NC machine has a longer setup time than a conventional machine, although machining costs are considerably less. Therefore, whether you use an NC machine will depend on lot size. The different machines can be set up on different routings for the same material, depending on the lot size. Another term for a different sequence of operations that depends on lot size is *alternative sequence*.

Task List Type

The **Task List Type** field classifies a task list according to its functionality. Left-click in the **Task List Type** field shown earlier in Figure 3.7 and then press F4 to display the list shown in Figure 3.8.

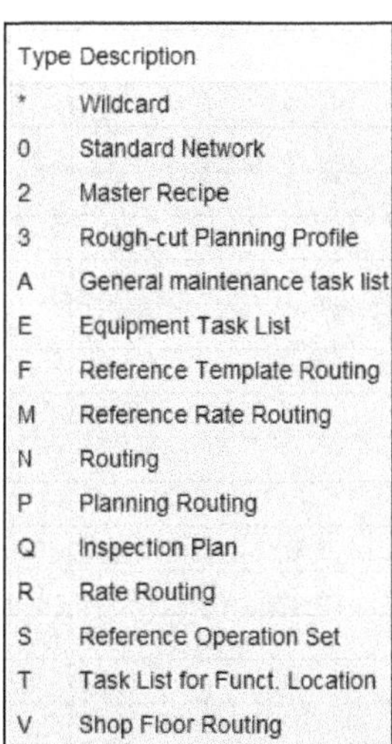

Type	Description
*	Wildcard
0	Standard Network
2	Master Recipe
3	Rough-cut Planning Profile
A	General maintenance task list
E	Equipment Task List
F	Reference Template Routing
M	Reference Rate Routing
N	Routing
P	Planning Routing
Q	Inspection Plan
R	Rate Routing
S	Reference Operation Set
T	Task List for Funct. Location
V	Shop Floor Routing

Figure 3.8 Task List Type Possible Entries

If you enter this field, SAP will search for an existing task list of this type, or a suitable production version containing a task list of this type, when creating a cost estimate. If SAP finds a suitable task list or production version, the cost estimate will use it independently of the production process. If SAP doesn't find a suitable task list or production version, you'll receive a warning message indicating that the material will be costed without reference to a routing or production version. SAP will attempt to cost the material based on the valuation strategy in the costing variant, which we will discuss in detail in Chapter 7.

Task List Types 3, A, E, and T

Task list types **3**, **A**, **E**, and **T** aren't relevant to costing. An error message will result if you attempt to enter them in this field.

Special Procurement Type for Costing

An entry in the **SpecProcurem Costing** (special procurement type for costing) field shown earlier in Figure 3.7 will be used by costing, and an entry in the equivalent field in the **MRP 2** view maintained by production will be ignored. You can use the special procurement type for costing to define the procurement type more closely, which determines whether a material is produced in-house or procured externally.

Special Procurement Type

Assume that the procurement type is set to **Both Procurement Types (X)** in the **MRP 2** view, and you maintain a BOM and routing so the material can be produced in-house as required. The cost estimate in this situation should be based on the price maintained in a purchasing info record because the material is procured externally (purchased) in most cases. To handle this requirement, you can maintain a *special procurement type* with an external procurement type **F** setting in configuration Transaction OMD9 (maintain special procurement type). If you enter this special procurement type in the **Costing 1** view, a cost estimate will first search for a purchasing info record price. Special procurement type **20** supplied as standard contains procurement type **F**.

Costing Lot Size

The **Costing Lot Size** field determines the material quantity on which you base cost estimate calculations. This field is mandatory because a cost estimate must base cost calculations on a quantity. When you create a standard cost estimate, it uses the costing lot size value in the **Costing 1** view by default. You can manually change the costing lot size defaulted from the **Costing 1** view when creating a single cost estimate with Transaction CK11N. You don't get this opportunity when creating cost estimates collectively with costing run Transaction CK40N.

You should set the costing lot size close to the actual purchase or production quantities to reduce lot size variance. Unfavorable variances may result if you create a production order for a quantity less than the costing lot size. You need setup time to prepare equipment and machinery to produce assemblies, and that preparation is generally the same regardless of the quantity produced. Setup time spread over a smaller production quantity increases the unit cost. This also applies to externally procured items because vendors typically quote higher unit prices for smaller quantities.

Equipment Cleaning Time

Setup and teardown time can also represent equipment cleaning time—for example, in the food and pharmaceutical industries. This can be significant to meet cleanliness and regulatory requirements.

A *cost estimate* typically calculates the cost of components at the lowest level in an assembly based on the costing lot size in the **Costing 1** view of each component material master. SAP progressively converts costing results to the costing lot size of the materials of the next-highest level to finally calculate the material costs for the finished product. It's possible to calculate the costs for all materials in a multilevel BOM using the costing lot size of the highest material based on a setting on the **Quantity Structure** tab in the costing variant with configuration Transaction OKKN. You typically use this function in sales order costing, which we discuss in more detail in Chapter 18.

Costing Lot Size and Price Unit

The costing lot size cannot be smaller than the price unit. Because you base costing on the costing lot size, this combination could result in rounding problems when determining a new price. You'll receive an error message if you attempt to enter this combination in the material master.

Co-Product

We already examined the details of the **Co-product** checkbox in the **MRP 1** view in Section 3.2, so let's move on to the **Fxd Price** (fixed price co-product) checkbox.

Fixed Price Co-Product

Select the **Fxd Price** (fixed price co-product) checkbox shown earlier in Figure 3.7 if you don't want a co-product material to be costed using a joint production process. Selecting this checkbox indicates that you'll use a preset price for a co-product instead of determining the costs using an apportionment method. The price for a fixed-price co-product is either taken from a cost estimate without a quantity structure or from the material master. You can only select this checkbox if you have also selected the **Co-product** checkbox for the material. Fixed-price co-products are shown in a cost estimate itemization with category M with negative quantities and values instead of the category A you typically see for co-products.

In the next section, we examine production versions and how they are associated with a material.

Production Version

Click the **Versions** button shown earlier in Figure 3.7 to display a list of production versions, as shown in Figure 3.9.

Production Version Overview

Version	Production Version T	Valid from	Valid To	Repetitive Mfg Allowed
57V1	5777:PRODUCTION VERSION	06/01/2023	12/31/9999	☐

Figure 3.9 Production Version Overview

You'll see a list of production versions assigned to a material in the **Production Version Overview** screen. You can assign many production versions to a material, and you must assign at least one production version if you are using repetitive manufacturing or product cost collectors.

Repetitive manufacturing eliminates the need for production or process orders in manufacturing environments with production lines and long production runs. It reduces the work involved in production control and simplifies confirmations and goods receipt postings. Select the **Repetitive Mfg Allowed** checkbox to use a production version in repetitive manufacturing.

Double-click **Production Version 57V1** as shown in Figure 3.9 to display the **Production Version Details** screen shown in Figure 3.10.

The **Basic data** section contains fields that allow you to lock the production version (if required) so you cannot use it, to set the minimum and maximum lot size, and to enter valid-from and valid-to dates.

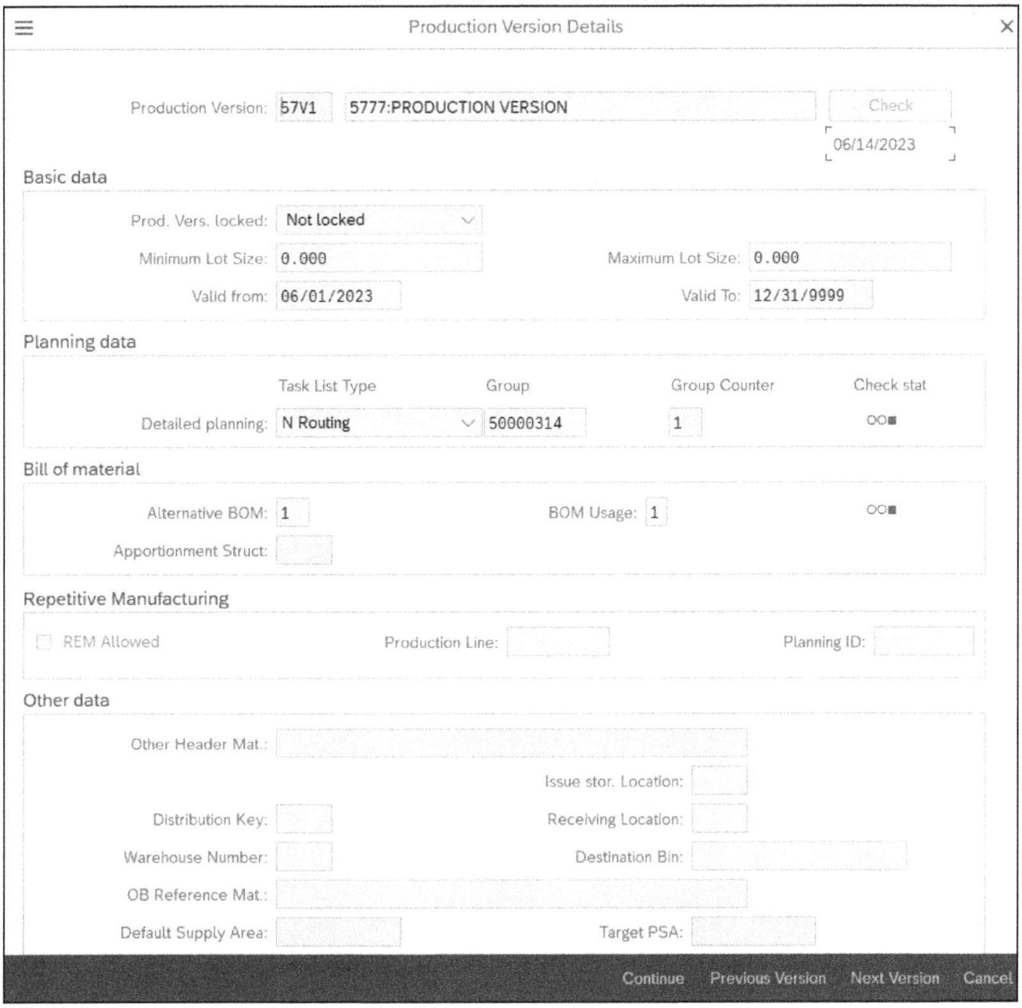

Figure 3.10 Production Version Details Screen

In the **Planning data** section, you can assign each production version to up to three work centers for different planning levels. You carry out operations at a work center, which can represent, for example, machines, production lines, or employees. Your choices in this section are as follows:

- **Detailed planning**
  Lets you determine which routing to use for MRP, product costing, and production orders.
- **Rate-based planning**
  Lets you determine which rate routing is used in repetitive manufacturing.

- **Rough-cut planning**
  Lets you determine in preliminary planning which rough-cut planning profile you use for sales and operations planning. To create capacity requirements during preliminary planning, you should use a routing instead of a rough-cut planning profile.

In the **Bill of material** section, the following fields are available:

- **Alternative BOM**
  Multiple methods of manufacturing an assembly and many possible BOMs can exist. The **Alternative BOM** field allows you to identify one BOM in a group.
- **BOM Usage**
  The **BOM Usage** field determines a specific section of your company, such as production, engineering/design, or costing, for which the BOM is valid.
- **Apportionment Struct**
  An apportionment structure defines how you distribute costs to co-products. The system uses the apportionment structure to create a settlement rule that distributes costs from an order header to the co-products. For each co-product, the system generates a further settlement rule that assigns the costs distributed to the order item to stock.

Now that we've discussed production versions, let's examine the last field in the **Costing 1** view shown earlier in Figure 3.7, the **Production Version** field, which indicates the production version to be costed. A production version entered in this field is used for cost estimates rather than the production version used by the manufacturing process. Costing will determine the production version automatically if there is no entry in this field, if you have configured the appropriate setting in the alternate BOM selection method screen in the **MRP 4** view, as discussed earlier in Section 3.5.

Now that we've examined the fields in the **Costing 1** view, let's look at the fields in the **Costing 2** view.

3.7 Costing 2 View

Select the **Costing 2** tab to display the screen shown in Figure 3.11.

This screen is divided into three sections: **Standard Cost Estimate**, **Planned prices**, and **Valuation Data**. To make the discussion easier to follow, we'll consider the fields in each section individually. Let's first look at the **Standard Cost Estimate** section.

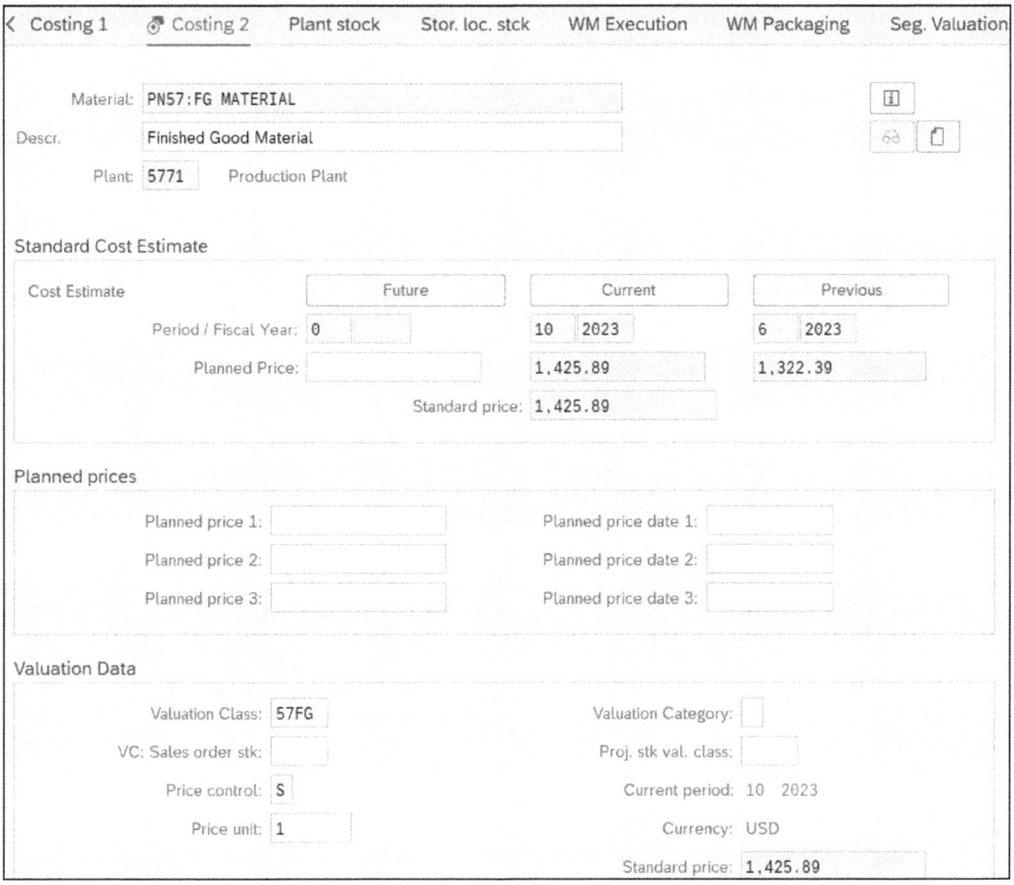

Figure 3.11 Costing 2 View

3.7.1 Standard Cost Estimate

You can click the **Future**, **Current**, and **Previous** buttons to display the corresponding cost estimates, if they exist. If a cost estimate exists, you'll notice an entry in the **Period/Fiscal Year** row under the corresponding **Cost Estimate** button. For example, in Figure 3.11, a current cost estimate was released in **Period 10/Fiscal year 2023**.

> **Display Current Released Cost Estimate**
>
> Many cost estimates can exist for a material. The quickest way to display the current released standard cost estimate is via the **Current Cost Estimate** button in the **Costing 2** view (Transaction MM02).

You can update the **Cost Estimate** fields in the **Standard Cost Estimate** section when a standard cost estimate is marked and released. When you first create and save a standard cost estimate, inventory valuation isn't changed. You populate the **Future Planned price** field when a cost estimate is marked.

A *marked* cost estimate is a proposed standard price that does not affect inventory valuation. You can create new standard cost estimates and mark them as many times as you like if you have not released a standard cost estimate during the period. Marking costs allows you to make adjustments and corrections to cost estimates before you release them, at which time you update the inventory valuation.

Note: Future Planned Price Field

If no marked standard cost estimate exists, manually enter a price in the **Future Planned price** field in standard SAP. If you include this field in the material valuation strategy sequence in the costing variant described in Chapter 7, a component cost estimate can take the value in this field when determining material valuation.

If you subsequently create and mark a standard cost estimate for this material, however, you overwrite any manual entry in this field with the results of the cost estimate, which you cannot change. To avoid overwriting manual entries, you can configure the component cost estimate to search first for a valid purchase info record price, and if unsuccessful, to then search for manual entries in the **Planned price 1**, **Planned price 2**, and **Planned price 3** fields. We'll discuss this procedure in detail in Chapter 9.

You handle the remaining fields in the **Standard Cost Estimate** section as follows:

- The system overwrites the **Current Planned price** and **Current Standard price** fields when you agree with the marked cost estimate proposed standard price and subsequently release the cost estimate. When you release a standard cost estimate, any change in inventory value posts as a financial document.
- The system overwrites the **Previous Planned price** field with the value of the previously released standard cost estimate when you release a marked cost estimate.

Both of these steps occur simultaneously as you release a standard cost estimate.

3.7.2 Planned Prices

We'll now examine the fields in the **Planned prices** section of the **Costing 2** view, as shown in Figure 3.12.

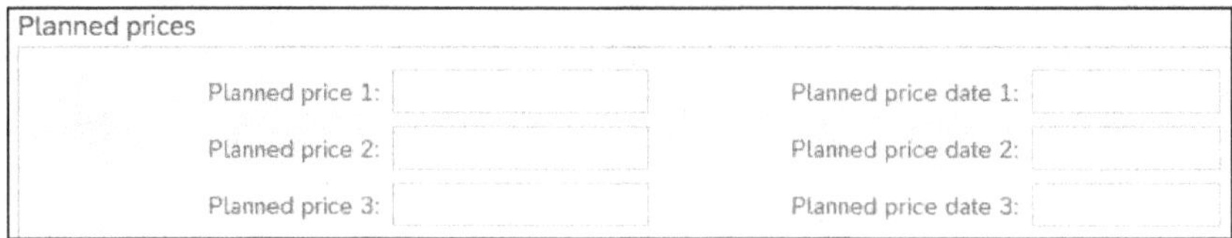

Figure 3.12 Planned Prices Section

You can manually update the **Planned price 1**, **2**, and **3** fields with estimated purchase prices. A standard cost estimate usually retrieves planned prices from these fields if no vendor quotations or purchasing info records exist for purchased items, which is useful when creating cost estimates before you receive vendor quotations early in the lifecycle of a new or modified product. Values entered in the three **Planned price date** fields indicate the date from which values entered in the three **Planned price** fields are valid.

Case Study: Planned Prices and Product Development

A manufacturer wants to retain a history of estimated purchase prices for new materials to gauge estimation accuracy over time. To do this, the manufacturer configures the costing variant so cost estimates search first for a purchasing info record price and then for values in the **Planned price 1**, **Planned price 2**, and **Planned price 3** fields, in that sequence. (We'll discuss costing variant configuration in detail in Chapter 7.)

Early in developing a new product, you create cost estimates to analyze cost viability. At this stage, you might not have received vendor quotations for new components yet, so you enter a rough estimation of the purchase price in the **Planned price 2** field. A cost estimate searching for the component price first searches for a purchasing info record and then the **Planned price 1** field before successfully locating a valid price in the **Planned price 2** field. This iterative process allows product development to develop the new product and accurately estimate its cost continuously.

Product development progresses to the stage when you must manufacture production quantities though you still haven't received vendor quotations for some components. This is a common scenario for many production companies and is even more common when the product development phase is relatively short and dynamic. As companies become more competitive, they need to make new products available with a short and decreasing turnaround time.

Sometime later, purchasing provides an estimated purchase price to accounting to enter in the **Planned price 1** field, and a cost estimate now successfully locates this price without needing to proceed to the **Planned price 2** field. You create, mark, and release cost estimates based on the estimated component price in the **Planned price 1** field, and production of the finished product proceeds.

When purchasing receives vendor quotations, they create or update the vendor purchasing info records directly.

When assembly standard cost estimates are created next, underlying component cost estimates are based on purchasing info record prices without any need to access the **Planned price 1** or **2** fields in the **Costing 2** view. You maintain records of the purchase prices estimated by product development and purchasing in the **Planned Price** fields for future reporting on component price estimation.

The key to the timely release of accurate standard cost estimates is coordination and communication between product development, production, purchasing, and accounting. You can achieve this procedurally, that is, controlled manually by policies and procedures, which many companies achieve with a great deal of success.

The workflow component can be useful in automating some of the communications required between users and departments when researching, developing, and manufacturing new products. For example, when product development indicates that a product is ready to progress from development to production by changing a status, an email can automatically notify purchasing that vendor quotations for components are required.

3.7.3 Valuation Data

Now that we've examined the fields in the **Planned prices** section, we'll consider the **Valuation Data** section fields, as shown in Figure 3.13.

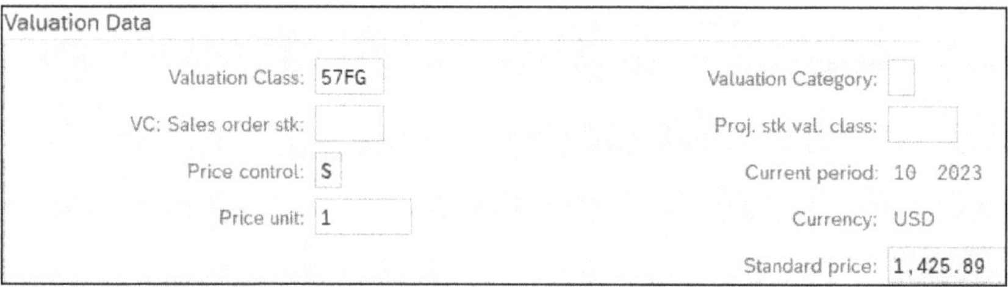

Figure 3.13 Valuation Data Section

The **Valuation Data** section contains fields that define how you valuate a material, including **Valuation Class**, which determines the general ledger accounts posted to automatically during inventory movements, and **Price control**, which determines if inventory is valued by either the standard price or the moving average price.

Valuation Class

The **Valuation Class** field in Figure 3.13 determines which general ledger accounts you update during inventory movement or settlement. The **Valuation Class** field is plant and material-type specific. A material type groups together materials with the same basic attributes such as raw materials, semi-finished products, or finished products. You configure general ledger accounts to receive postings during material movements with Transaction OBYC, which are stored then in database table T030. The quickest way to determine the general ledger accounts in table T030 is with the Data Browser via Transaction SE16H. If you have authorization, run the transaction, and in the selection screen, enter your chart of accounts and valuation class to see a screen like that shown in Figure 3.14.

Chrt/Accts	Trans.	VGCd	AM	ValCl	G/L Acct	G/L Acct
5777	BSX	5777		57FG	200300	200300
5777	GBB	5777	AUA	57FG	300602	300602
5777	GBB	5777	AUF	57FG	300610	300610
5777	GBB	5777	AUI	57FG	400810	400810
5777	GBB	5777	VAY	57FG	400610	400610
5777	GBB	5777	VBR	57FG	400300	400300
5777	PRD	5777		57FG	400920	400920
5777	UMB			57FG	400814	400814

Figure 3.14 Data Browser Listing of General Ledger Accounts in Table T030

From this screen, you can determine, for example, that goods issued to an order or cost center will post to **G/L acct 400300** because it appears on the same row as **Trans.** (Transaction) **GBB** and **AM** (account modifier) **VBR**. (We'll examine account determination in more detail in Chapter 13). The screen in Figure 3.14 provides a quick overview of the general ledger accounts configured for each valuation class (given in the **ValCl** column).

If you don't have authorization, you can ask a colleague with authorization to run Transaction SE16 with the selection screen entries and ask them to send you the results.

Valuation Class for Sales Order Stock and Project Stock

You only need to make entries in the **VC: Sales order stk** (valuation class for sales order stock) and **Proj. stk val. class** (valuation class for project stock) fields shown earlier in Figure 3.13. If you need to assign a different valuation class for these types of stock. This allows you to post inventory movements for these types of stock to specific general ledger accounts. You don't typically need to assign separate valuation classes in these fields. (We'll discuss valuated sales order stock in more detail in Chapter 18.)

Valuation Category

The **Valuation Category** field shown earlier in Figure 3.13 determines which criteria are used to group partial stocks of a material to value them separately. The valuation category is part of the split valuation functionality. Typically, you'll have only one price per material per plant. Split valuation allows you to valuate, for example, batches separately. The moving average price (**V**) is the only **Price control** field setting available if you activate split valuation and enter a valuation category. You assign valuation types to valuation categories and valuation categories to plants with configuration Transaction OMWC or by following the IMG menu path **Materials Management • Valuation and Account Assignment • Split Valuation • Configure Split Valuation**.

To create a material master subject to split valuation, you first need to create a *valuation header record* for the material, which is where you manage the individual stocks of material cumulatively. When creating the material for the first time, enter a value in the **Valuation Category** field and leave the **Valuation Class** field blank. When you save the material, you create the valuation header record.

To create the material for a valuation type, create the material again with Transaction MM01. Because a valuation header record exists, you'll be required to enter the valuation type for the valuation category. Repeat this step for every valuation type planned.

The **Valuation Category** field is also located in the **Accounting 1** view.

Split Valuation Can Require Additional Maintenance

You should always investigate other options before implementing split valuation functionality because you'll have additional material master views for each valuation type.

In some special cases, split valuation is required—for example, when you keep stocks of a material in different companies or countries. In these cases, it might be necessary for group reporting purposes to determine and store the actual cost of manufacture in each separate company or country.

Price Control

The **Price control** field determines whether you valuate inventory at standard (**S**) or moving average (**V**) price, as follows:

- If you assign standard price control (**S**) to a material, you always calculate the value of the material at this price. If goods or invoice receipts contain a price different from the standard price, you post the differences to a price difference account. You don't consider the variance in inventory valuation.
- If you assign moving average price control (**V**) to a material, you automatically adjust the price if goods or invoice receipts are posted with a price that differs from the moving average price. You post the differences to the stock account, and as a result, the moving average price and the value of the stock change.

Price Control for Assemblies and Raw Materials

SAP recommends standard price control for assemblies and finished goods because it creates price stability. It doesn't allow errors or incorrect postings in manufacturing orders to flow through to finished goods inventory valuation. It also allows you to analyze production efficiency via variance postings during finished goods receipt into inventory valued at the standard price.

Many U.S. companies value all materials, including purchased materials, with standard price control. This strategy requires more maintenance because standard cost estimates are required for purchased materials, and you need to analyze purchase price variance (PPV) accounts periodically. Although this strategy involves more work, it does allow you to measure purchasing efficiency by comparing actual prices with planned purchase prices.

Which valuation strategy is the best for your company depends on the level of control required over the purchasing price maintenance process.

Another inventory valuation option is available if more control and analysis options are required. Actual costing automatically calculates and posts all differences from the lowest BOM levels to assemblies and finished goods. Implementation of this functionality requires more analysis and testing. You need to be certain you require this level of control and analysis before proceeding. We'll discuss the Material Ledger and actual costing in more detail in Section 3.8, when we look at the **Accounting 1** view, as well as in Chapter 16.

Current Period

The **Current period** field shown earlier in Figure 3.13 displays the current materials management period, which you set with Transaction MMPV or via the menu path **Logistics • Materials Management • Material Master • Other • Close Period**. This period is independent of the accounting period, which you maintain separately with Transaction OB52.

Material movements can only occur in the current and, if allowed, the previous materials management periods. You have the option of closing the previous materials management period with Transaction MMRV or by following the menu path **Logistics • Production • Master Data • Material Master • Other • Allow Posting to Previous Period**. You typically keep the option to post to the previous period open for several days into the current period to allow period-end analysis and any correction postings to occur. You then disable the option to post to the previous period so the data doesn't change after you run and save accounting and Controlling reports for managers to analyze.

Currency

You determine the **Currency** field during company code configuration, and you cannot change it in the material master screen. You cannot change the company code currency after postings occur. You can view more details about the company code currency with Transaction OB22 or FINSC_LEDGER.

Price Unit

The **Price Unit** field holds the number of units to which the price refers. You can increase the accuracy of the price by increasing the price unit. To determine the unit price, divide the price by the price unit.

You can change the price unit at any time with Transaction MM02. SAP automatically recalculates the prices to match the new price unit. The maximum price unit value you can enter is "10,000."

If you manually enter a price unit greater than the costing lot size, SAP automatically increases the costing lot size to the same value as the price unit. This requirement is because cost estimates calculate costs based on the costing lot size, and rounding issues would arise if you marked and released a standard cost estimate.

Increase Accuracy with Price Unit

Say that the price per unit of a material is $0.215. The problem is that you cannot enter this exact value as a price with a price unit of 1 because the material master **Price** field will only accept a maximum of two values following the decimal point. You enter the exact price by entering a price of $2.15 with a price unit of 10 or a price of $21.50 with a price unit of 100.

Moving Average Price

SAP calculates the moving average price automatically by dividing the material value in the stock account by the quantity of all stocks in the plant. With each valuated goods and invoice receipt that has a different price from the current moving average price, SAP updates the price. If you set the **Price control** at moving average price (**V**), the price in the **Moving price** field determines inventory valuation. If **Price control** is set to **S**, the moving average price is statistical and for information only.

Deactivation of the statistical moving average is recommended in SAP S/4HANA to achieve a significant increase in transactional data throughput for goods movements. For more information, refer to SAP Note 2267835.

Standard Price

The **Standard price** results from releasing a material standard cost estimate and remains constant for at least one period. You post differences between the standard price and actual prices to price difference general ledger accounts.

Typically, companies keep the standard price constant for one year. Companies with rapidly changing prices may decide to release standard cost estimates more frequently, usually every three or six months. This makes variance analysis of the difference between the standard and actual costs easier, although stock valuation changes more frequently.

Now that we've examined the costing fields, we'll look at the fields in the **Accounting** views. If you work with SAP S/4HANA Cloud or use SAP Fiori, you can use the **Manage Material Valuation** app (SAP Fiori ID F2680) to display the same information. You can access the information that controls costing by selecting your chosen material in the selection screen and navigating to the **Costing** tab as shown in Figure 3.15. You can update the planned prices for your material by selecting the **Planned Prices** tab.

To display the standard cost estimate used to set the standard cost for the chosen material, navigate to the **Standard Cost Estimates** tab in Figure 3.16. From here, you can use the **Display Cost Estimate** button to display details of the associated cost estimate.

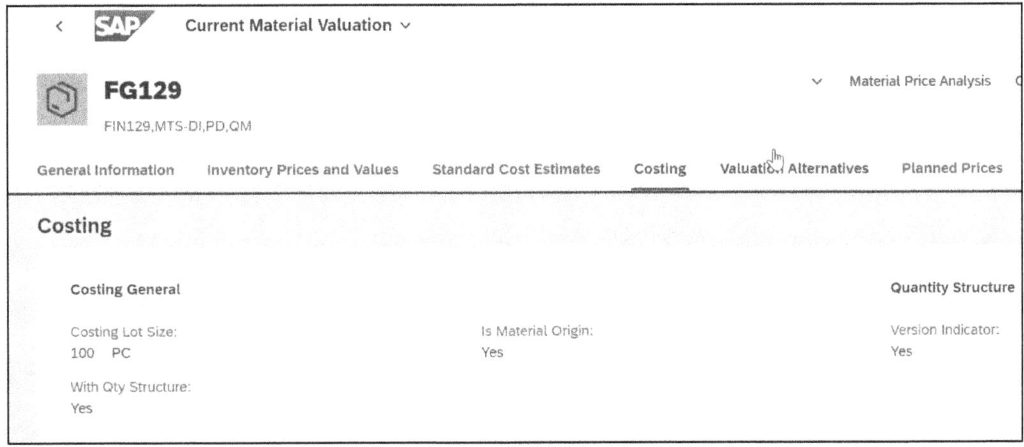

Figure 3.15 Costing Tab in Manage Material Valuations app

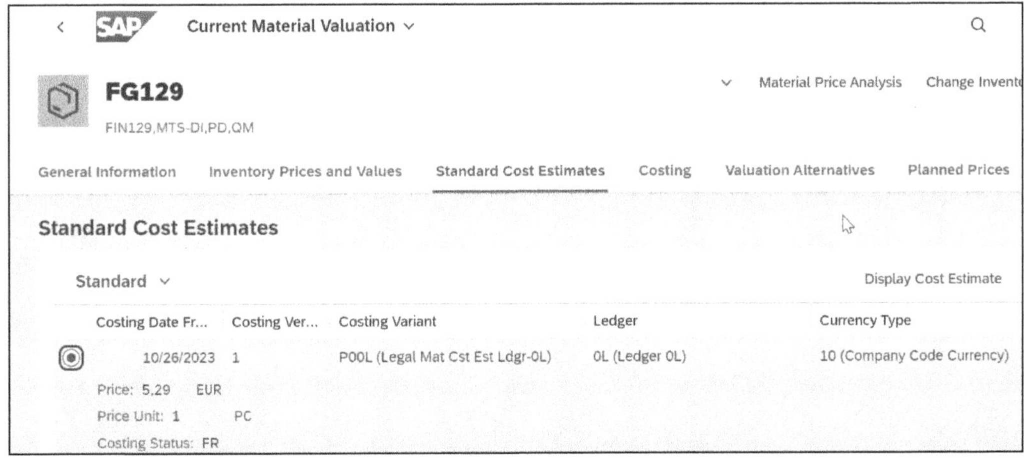

Figure 3.16 Standard Cost Estimates in Manage Material Valuations app

Manage Material Valuations App

Alternatively, you can use the Manage Material Valuations app from SAP Fiori that we showed in the Introduction to view all accounting and costing-related information about a material. The app displays an overview of valuation for materials, including sales order related stocks. A price history is available, and you can see standard cost estimates, change inventory prices, or release planned prices. You can find more information at *http://s-prs.co/v562900*.

3.8 Accounting 1 View

General accounting and management accounting teams typically use the **Accounting** views. Navigate to the **Accounting 1** tab to display the screen shown in Figure 3.17.

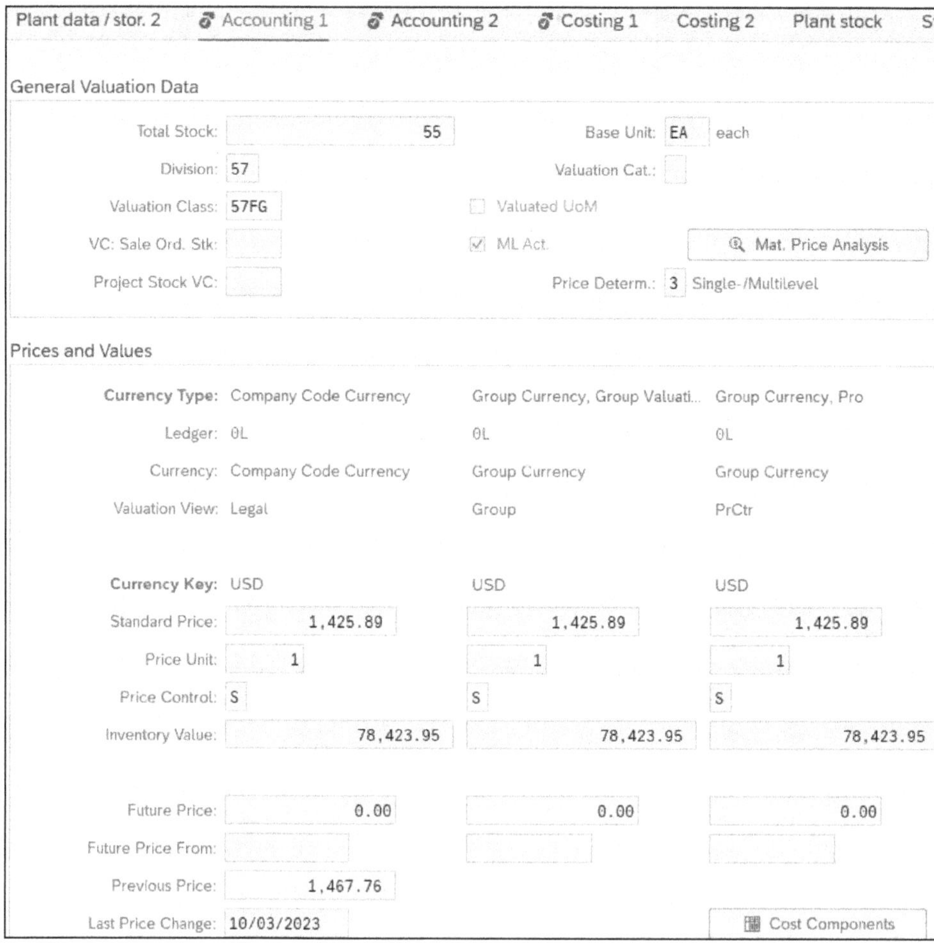

Figure 3.17 Accounting 1 View

The **Accounting 1** view displays fields relevant to actual costing. The **MLAct.** checkbox under **General Valuation Data** is selected, which indicates the Material Ledger is active for this material for this plant. Because the Material Ledger is mandatory for SAP S/4HANA, this field is always selected and cannot be changed.

You record each material-based transaction per period in the Material Ledger, allowing the material inventory to be valued in up to three different currencies or valuations. This information is used as the basis for actual costing, which allows you to roll up and allocate all material and production differences to assemblies and finished products at the end of each period.

The **Accounting 1** view has two sections, **General Valuation Data** and **Prices and Values**. We'll consider the fields in each section, starting with **General Valuation Data**.

3.8.1 General Valuation Data

General Valuation Data contains fields relating to accounting in general that are not included in the more specific sections. Some fields appear in both the **Costing** and **Accounting** views. In the following sections, we'll only consider fields we've not previously discussed in the **Costing** views.

Total Stock

Total Stock indicates the total quantity of all valuated stocks of the material in the plant. This field provides a summary of the total stock of the material inventory without needing to refer to inventory reports or transactions such as Transaction MMBE.

Division

A *division* allows you to group materials based on responsibility for sales or profits from salable materials or services. You always assign a product or service to just one division, which allows you to organize your sales structure around groups of similar products or product lines.

> **Division Example**
>
> If a sales organization sells food and nonfood products through both retail and wholesale distribution channels, each distribution channel could be further split into food and nonfood divisions.

Price Determination

To display a list of possible entries for **Price determ.** (price determination), left-click in the field and press [F4]. You can see the possible price determinations in Figure 3.18.

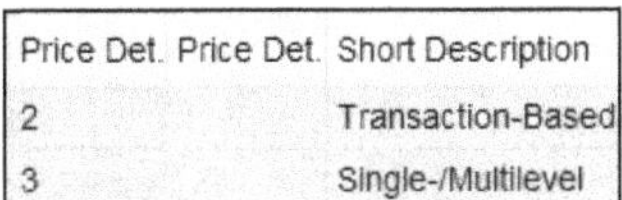

Price Det.	Price Det. Short Description
2	Transaction-Based
3	Single-/Multilevel

Figure 3.18 Material Price Determination Possible Entries

Transaction-Based Price Determination (2)

Transaction-based price determination (**2**) indicates that actual costing is not active. You use this setting if you are interested in carrying standard (S) or moving average (V) prices in multiple currencies or valuations.

You can change price determination following production startup with Transaction CKMM or via the menu path **Accounting • Controlling • Product Cost Controlling • Actual Costing/Material Ledger • Environment • Change Material Price Determination.** You can also determine the material price determination proposed when creating new material masters following production startup with configuration Transaction OMX1 or by following the IMG menu path **Controlling • Product Cost Controlling • Actual Costing/Material Ledger • Activate Valuation Areas for Material Ledger.**

Single Multilevel Price Determination (3)

Single-/Multilevel price determination (**3**) indicates that actual costing is active. You can only use this option if the price control is set at standard price (S). During the month, the periodic unit price (PUP) is effectively a statistical moving average, which changes during goods and invoice receipts. Remember, the price control stays set at the standard price.

Deactivation of the statistical moving average is recommended in SAP S/4HANA to achieve a significant increase in transactional data throughput for goods movements. For more information, refer to SAP Note 2267835.

The PUP is calculated when the period is closed for informational purposes, and you can also use it for material valuation of the closed period if there is a legal requirement to value inventory at actual costs, as is common in countries such as Brazil and Turkey.

Material Price Analysis

Click the **Mat. Price Analysis** button and change the **View** to **Price Determination Structure** to display the screen shown in Figure 3.19.

You can also access this screen with Transaction CKM3 or by following the menu path **Accounting • Controlling • Product Cost Controlling • Actual Costing/Material Ledger • Material Ledger • Material Price Analysis.** In this screen, you can access all material transactions for a period. Transactions are grouped by **Category**, which you can progressively expand to display all transaction documents within a category. You can double-click a transaction document (such as **1000010364**, in this example) to display a Material Ledger document, as shown in Figure 3.20.

This document displays an overview of the inventory change. In this example, the inventory decreased by **15 PC** (pieces) of material **PN57**. You can analyze the transaction further by clicking the hat icon to display the accounting document header information, such as the original transaction code, entry personnel, and time and date of entry.

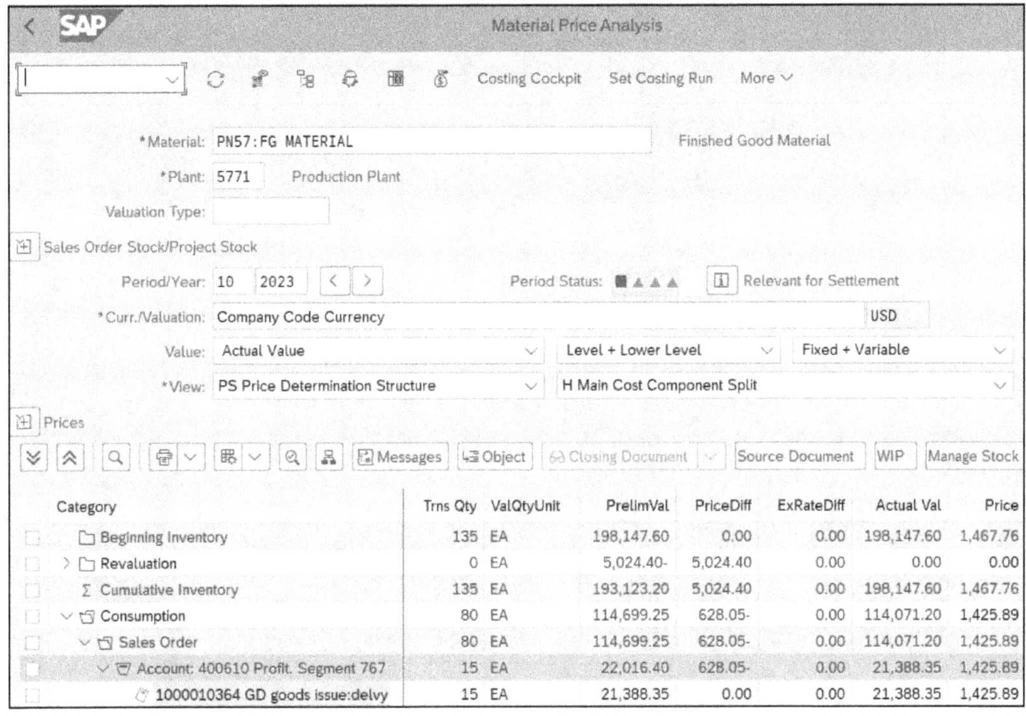

Figure 3.19 Material Price Analysis

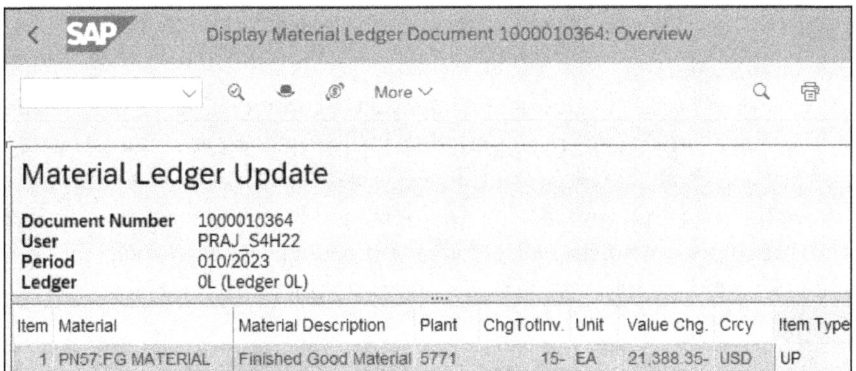

Figure 3.20 Display Material Ledger Document Overview

3.8.2 Prices and Values

Now that we've discussed the fields in the **General Valuation Data** section of the **Accounting 1** view, let's look at the fields in the **Prices and Values** section, as shown in Figure 3.21.

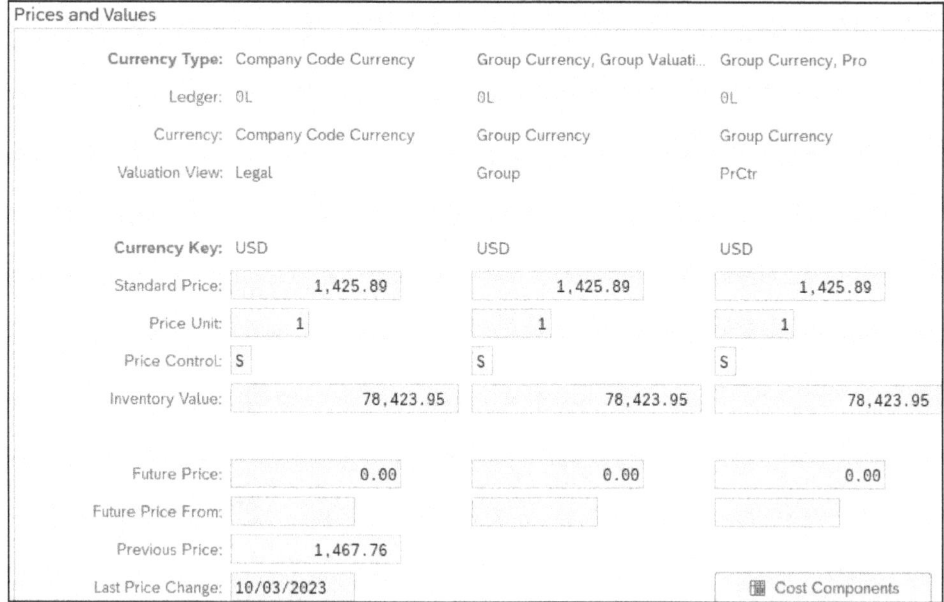

Figure 3.21 Prices and Values Section Accounting 1 View

The **Prices and Values** section contains up to three columns, each representing the three possible currencies and valuation views. We'll look at how this view changes with the introduction of universal parallel accounting in Chapter 19. The combination of currency and valuation view represents a valuation approach for inventory:

- **Company code currency**
  The first column represents values in **Company code currency**, legal valuation, and the leading ledger. When you transfer products between company codes or profit centers, you base legal valuations on transfer prices that include internal profits. You can use this valuation approach by default with or without the Material Ledger activated. The additional functionality of activating actual costing includes the periodic unit price, which represents the actual inventory price for a prior closed period.
- **Group currency**
  The second column represents values in a *group currency* and *group valuation*. A global organization typically uses a group currency in consolidated financial statements and group valuation in management reports, which exclude internal profits when products are transferred between company codes or profit centers.
- **Group currency**
  The third column represents values stored in group currency and profit center valuation.

In the following sections we'll explain the value of the inventory and the impact of the different prices in the material master.

Inventory Value

The **Inventory Value** field holds the value of all valuated stocks of the material in the plant. This value is determined by multiplying the **Total Stock** quantity by the standard or the moving average price, depending on the **Price control** field. In SAP S/4HANA, this value is determined dynamically by reading the data in the `MATDOC` table, rather than being stored on the database.

Future Price and Future Price From

The **Future Price** field provides two mechanisms for updating a material's standard price. The following paragraphs explain each in detail.

You generally only use the first method of updating the standard price with the future price if you have not fully implemented the product cost planning functionality and cannot create standard cost estimates to update standard prices. This method of updating the standard price has limited functionality and yet still gives you some control over the price update because you need to run a separate transaction after entering a price in the **Future price** field to update the **Standard price** field.

To release a future price to become the standard price, follow these steps:

1. Enter the price you intend to become the new standard price for a material in the **Future Price** field in the **Accounting 1** view.
2. Enter a validity date in the **Future Price From** field. SAP will not update the standard price with the future price before this date.
3. Run Transaction CKME or follow the menu path **Logistics • Materials Management • Valuation • Change in Material Price • Release Planned Prices** to display the **Release Planned Price Changes** screen shown in Figure 3.22.

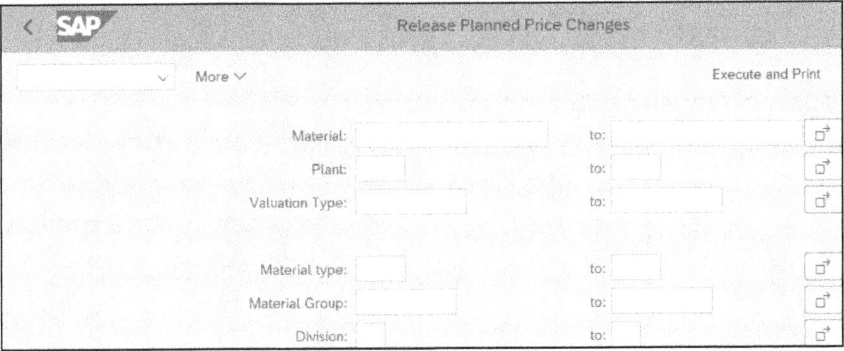

Figure 3.22 Release Planned Price Changes Screen

4. This screen offers many options for selecting a material or range of materials. Enter your data and click the Execute button at the bottom of the screen (not shown), or press F8 to update the standard price in the selected material masters with the future price.

Upon successfully updating the standard price, the **Future Price** and **Future Price From** fields are both blank and ready for the entry of the next future price. SAP will not update the standard price with this procedure if there is either an existing current or future standard cost estimate.

The second method for updating the standard price with the future price involves including the future price in the search strategy sequence in the configuration of the costing variant with Transaction OKKN. (We'll discuss costing variant configuration in detail in Chapter 7.) This field can also provide an extra field in addition to the three **Planned price** fields in the **Costing 2** view to plan prices. If the standard cost estimate successfully updates the standard price using the **Future Price** field, the values in the **Future Price** and **Future Price From** fields remain until you overwrite them.

Note: Maintaining the Future Price

If you cannot maintain the **Future Price** field because it's grayed out, you might need to change a setting in configuration Transaction OMS9 that allows you to maintain field selection for data screens.

Now that we've discussed the **Future price** and **Future Price From** fields, let's look at the **Previous price** and **Last price change** fields.

Previous Price and Last Price Change

These fields are located below the **Future Price** and **Future Price From** fields in the **Accounting 1** view, as shown in Figure 3.23.

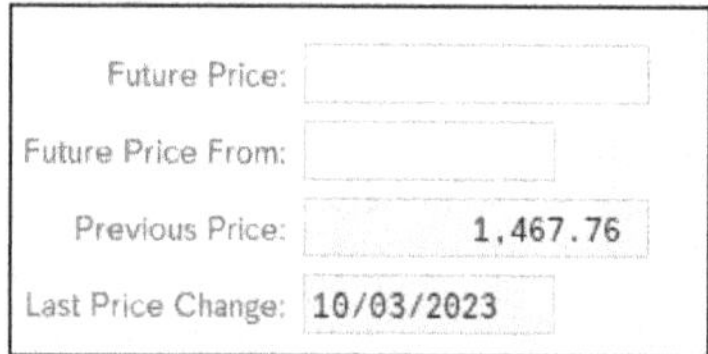

Figure 3.23 Display Previous Price and Last Price Change Fields

If you are using the **Future Price** field to update the standard price without cost estimates, these fields are the only indication you have of the previous standard price and when you released it.

Periodic Unit Price

The **Per. Unit Price** (periodic unit price) is displayed under the **Standard Price** field as shown in Figure 3.24.

Prices and Values	
Currency	MXN
	Company code currency
Standard Price	74.47
Per. Unit Price	21.90
Price Unit	1
Price Control	S
Inventory Value	7,372.53
Val./Per.Unit Prc	2,168.10

Figure 3.24 Periodic Unit Price

During a period, you carry out all transactions with the preliminary valuation price (standard price). The Material Ledger collects data for all activities related to the valuation of materials, and the preliminary valuation price stays constant during the period. You cannot change this price after you have entered data into the Material Ledger for the period.

The differences between the preliminary valuation price and the actual price are recorded per material and posting, and each difference can be displayed during the period as soon as it's entered. The periodic unit price is determined after a period has ended and reflects the actual costs of a material for the closed period. To determine the periodic unit price, the system uses the *cumulative inventory* (all goods received plus beginning inventory) and the *cumulative difference* (all differences between the standard price and the entered price for all goods received and the beginning inventory).

Material price determination calculates the periodic unit price. The **Moving Price** field in the material master is updated with the calculated price for the closed period. The material's price control remains set to standard price (**S**).

Periodic Unit Price Field Name

In a material with price determination **3**, the name of the moving average price field is changed to **Per. Unit Price** (periodic unit price or PUP).

During the revaluation process, the system changes the price control for the material in the previous period from **S** to **V**. The periodic unit price (calculated during material price determination) becomes the valuation price for the previous period.

Previous Period Price

You can only see the previous period price in the **Accounting 1** view because this view contains material valuation per period.

If you decide not to revalue your materials at period end, the periodic unit price of the closed period is for information only, and price control stays as the standard price.

Inventory Value

The **Inventory Value** field indicates the total of all valuated stocks of the material in the plant calculated with the standard price since the **Price Control** is set to **S**. The Material Ledger retains inventory values for each period that you can access by selecting the **Period** tabs.

Value Based on Periodic Unit Price

The **Val./Per.Unit Prc** (value based on periodic unit price) field contains the stock value based on the **Per. Unit Price** (periodic unit price) by multiplying the total stock quantity by the PUP.

Future Price

You can enter a future price for each column corresponding to a different currency or valuation approach. The **Future price** in the first column representing legal valuation in this example includes transfer pricing (internal profits). The **Future price** in the second column representing group valuation excludes transfer pricing. You can create separate cost estimates for each valuation approach that access the future prices for purchased materials.

Future Price From

The standard price will not be updated with the future price before the date specified in the **Future price from** field.

Cost Components

Click the **Cost Components** button shown in Figure 3.21 to display the screen in Figure 3.25.

You can display the actual cost component split for different valuations. To display the actual cost component split, you first need to activate it via the IMG menu path **Controlling • Product Cost Controlling • Actual Costing/Material Ledger • Actual Costing • Activate Actual Cost Component Split**.

If you work with SAP S/4HANA Cloud or SAP Fiori, you can display the same information in the Manage Material Valuations app. Figure 3.26 shows the selection screen of the Manage Material Valuations app with the list of selected materials and their inventory values. Notice that you can see the material prices for each plant per **Ledger**, **Valuation Type**, and **Currency Type** and access the apps to change the material prices using the buttons at the top of the app.

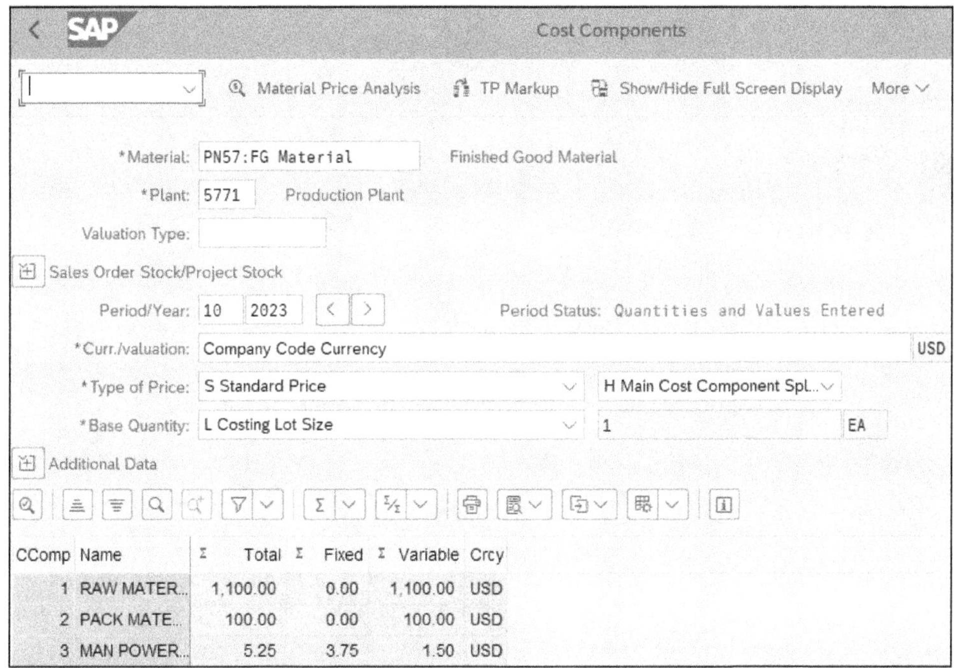

Figure 3.25 Actual Cost Component Split Displaying Valuations

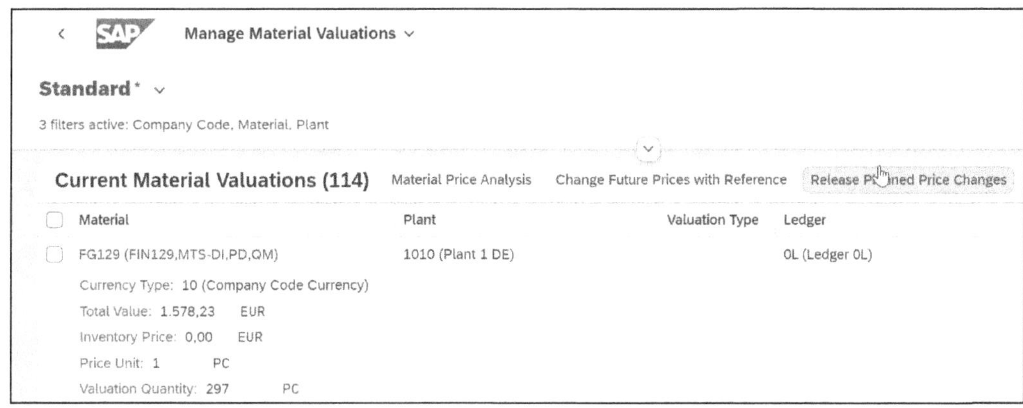

Figure 3.26 Current Material Valuations List in Manage Material Valuations App

To display the valuation details for an individual material, select a material to access the **Current Material Valuation** tab shown in Figure 3.27. Here, we see the link to the **Valuation Class**, the Price **Determinatio**n, and the **Profit Center**, and can use the **Change** button to update price information as required.

Figure 3.27 General Information in Manage Material Valuations App

Now that we've examined fields in the **Accounting 1** view, let's consider fields in the **Accounting 2** view.

3.9 Accounting 2 View

Select the **Accounting 2** tab shown earlier in Figure 3.17 to display the screen shown in Figure 3.28.

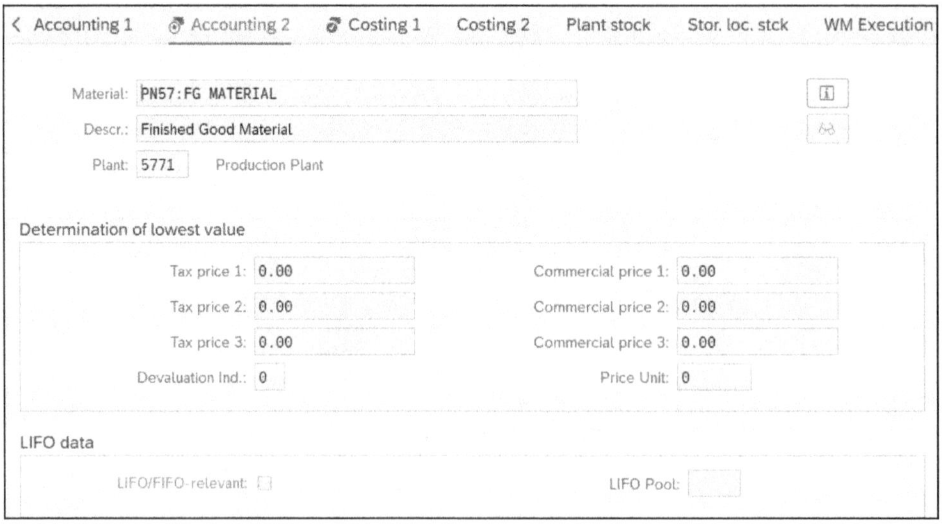

Figure 3.28 Accounting 2 View

The **Tax price** and **Commercial price** fields indicate the tax and commercial values. The distinction between these two valuations isn't commonly observed in the United States or some European countries. You can use these fields in many ways to determine alternate inventory valuations. In the following sections, we'll follow two scenarios for populating the fields with information and then using the information in the fields.

3.9.1 Populate Tax Price and Commercial Price

You can manually enter information in any of the **Tax price** and **Commercial price** fields, or you can automatically populate them. Many options are available for automatically populating these fields. Let's follow one scenario to see how the process works, and then you can apply it to other possible inventory valuation scenarios, as you like.

Run Transaction MRN0 or follow the menu path **Logistics • Materials Management • Valuation • Balance Sheet Valuation • Determination of Lowest Value • Market Prices** to display the screen shown in Figure 3.29.

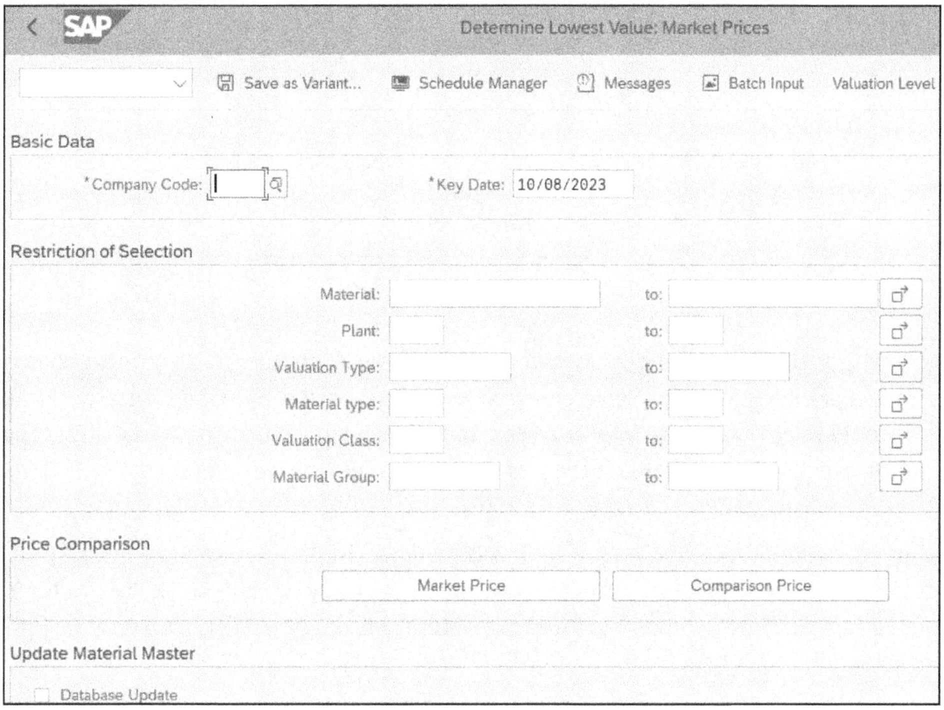

Figure 3.29 Determine Lowest Value - Market Prices

This transaction lets you retrieve prices from receipts, purchase orders, contracts, scheduling agreements, purchasing info records, or standard prices. You can compare these prices with existing prices, such as the current material prices, and you can update any of the **Tax**, **Commercial**, or **Planned price** fields in the **Accounting** and **Costing** views with the results. We'll discuss the fields and buttons in the following sections.

Restriction of Selection

The **Restriction of Selection** section allows you to select which materials determine market prices. To restrict the selection of materials, enter a **Material** range, **Plant**, or any other parameters.

Price Comparison

Click the **Market Price** button to display the screen shown in Figure 3.30.

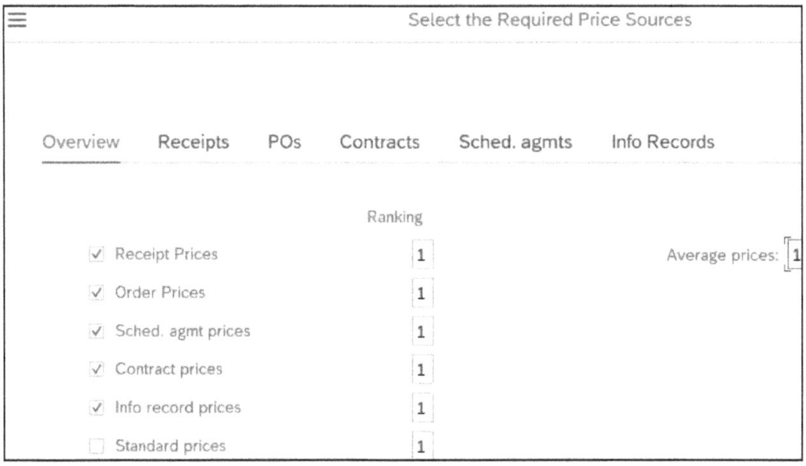

Figure 3.30 Overview Screen: Selecting Market Price Source

Selecting a checkbox on the **Overview** tab causes a corresponding tab to appear on this screen. For example, if you selected the **Standard prices** checkbox, the corresponding tab would appear after the **Info Records** tab. Each tab contains fields for selecting the appropriate data. Let's examine the **POs** (purchase orders) tab as an example. Click the **POs** tab to display the screen shown in Figure 3.31.

This screen indicates that the system will choose the lowest purchase order price during the current fiscal year—in this example, beginning January 2023. The **Analyzed Prices** field shows the number of purchase orders to be included in calculating the lowest price.

Now that we've selected market prices, we can compare them with existing prices. Click the **Comparison Price** button shown earlier in Figure 3.29 to display the screen shown in Figure 3.32.

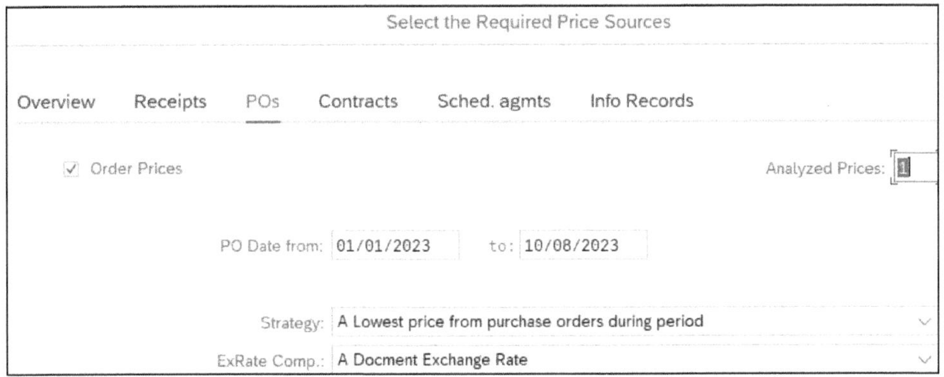

Figure 3.31 Purchase Orders: Selecting Market Price Source

Select a Comparison Price:

Relationship to Market Price

- Lowest Val. Comparisn
- Comparisn Prc as Replcmnt Val.
- No Comparison Price

Price Selection: Lowest of...

Phys. Inventory Prices | Valuation Alternatives

- Current Material Price
- Mat. Price Prev. Month
- Mat. Price Previous Year
- Current Standard Price
- Standard Pr. Prev. Month
- Standard Pr. Prev. Year
- Current MAP
- Tax Price 1
- Tax Price 2
- Tax Price 3
- Commercial Price 1
- Commercial Price 2
- Commercial Price 3

Figure 3.32 Comparison Price Selection

This screen allows you to select which prices to compare with the market price retrieved from purchase orders in our example.

If you'd like to display a report without updating any material master fields, deselect **Database Update**, shown earlier in Figure 3.29 , and click the **Execute** icon or press F8 to display a report such as the one shown in Figure 3.33.

2 Determine Lowest Value: Market Prices 10/17/2023

CoCd	ValA	Material	Val. Type	l Description	Comp.Price	New Price	Price Unit	Change in Percent
0005	0052	1000000531		451 FG GLA - STOCK TRANSFTER P2P	348.00	348.00	1	0.00
0005	0052	1000000533		453 FG GLA - STOCK TRANSFTER /ADD	1,800.00	1,800.00	1	0.00

Figure 3.33 Determine Lowest Value Market Prices Report

This report displays the comparison price (**Comp. Price** column), proposed price (**New Price** column), **Source**, and **Change in Percent**. After reviewing the report, you can do one of two things: run the transaction again with changed parameters or update the prices to fields in the material master. To do the latter, select **Database Update** as shown earlier in Figure 3.29 to display the screen shown in Figure 3.34.

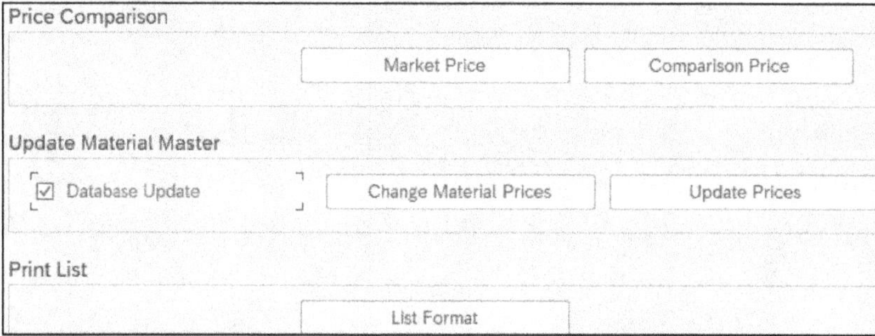

Figure 3.34 Determine Lowest Value Market Prices Database Update

When you select **Database Update**, two new buttons appear in the **Update Material Master** section: **Change Material Prices** and **Update Prices**. Let's review each one.

Click the **Change Material Prices** button to display the screen shown in Figure 3.35.

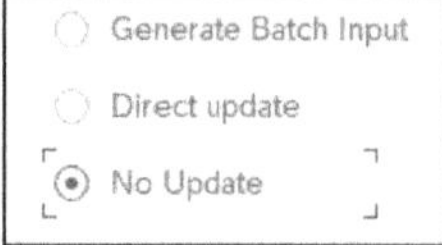

Figure 3.35 Update Material Prices Button Screen

You'll typically leave the default setting of **No Update** selected. With this setting, you'll still be able to update the tax, commercial, and planned prices in the material master. Only select **Direct Update** if you want to update the material standard price with this transaction.

Direct Update

Selecting **Direct update** and executing the transaction might update the standard price, resulting in inventory revaluation. Always test this setting before using it in your production client.

Press Enter in Figure 3.35, and click the **Update Prices** button to display the screen shown in Figure 3.36.

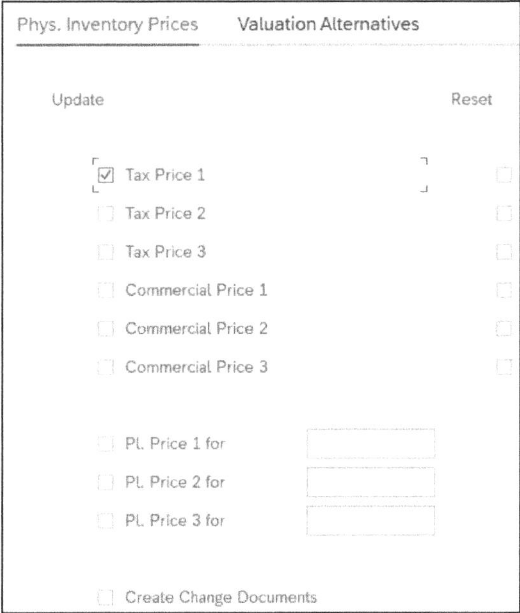

Figure 3.36 Material Master Fields to Update

In this screen, select the material master fields you want to update with the results of the market price valuation. Select **Reset** to initialize the price in the material master field or to set the value to zero as required. Press Enter, and then click the **Execute** icon or press F8.

You'll receive a report similar to the one shown earlier in Figure 3.33, and the system updates the **Tax price 1** field in the **Accounting 2** view as shown in Figure 3.37.

Tax Price Field

You can update the **Tax price 1** field with Transaction MRNO.

In this section, we followed an example of how you can automatically update the **Tax** and **Commercial prices** fields. Now, let's examine how to use these prices after they are populated.

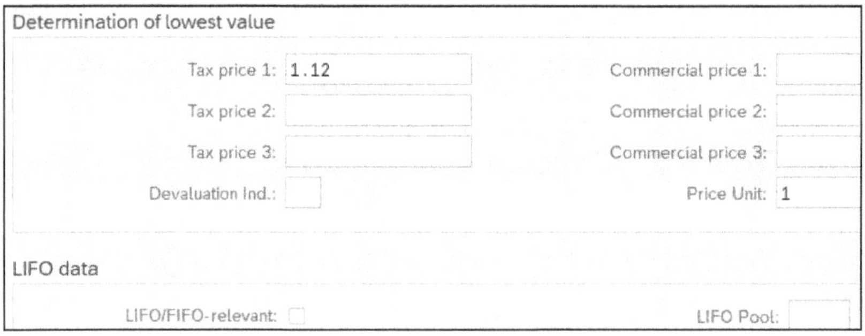

Determination of lowest value

Tax price 1: 1.12 | Commercial price 1:
Tax price 2: | Commercial price 2:
Tax price 3: | Commercial price 3:
Devaluation Ind.: | Price Unit: 1

LIFO data

LIFO/FIFO-relevant: | LIFO Pool:

Figure 3.37 Accounting 2 View, Tax Price 1 Field Updated

3.9.2 Using the Tax and Commercial Price Fields

You can use values in these fields for any purpose. For example, you can create a report that accesses a price in one of these fields, multiply it by stock quantity, and generate an inventory valuation report based on any criteria. Alternatively, you can have a cost estimate access the tax or commercial price and base the standard price on it or populate another field.

We'll discuss costing variant configuration in detail in Chapter 7. But here is a quick preview to demonstrate one of several available methods to use the tax and commercial price fields:

1. Run Transaction OKKN or follow the IMG menu path **Controlling • Product Cost Controlling • Product Cost Planning • Material Cost Estimate with Quantity Structure • Define Costing Variants.**
2. Double-click a costing variant.
3. Click the **Valuation Variant** button and left-click in any field in the **Strategy Sequence** section to display the list of possible entries, as shown in Figure 3.38.

   You can create a valuation variant that searches for the **Tax** or **Commercial** price fields. A cost estimate created with this valuation variant will retrieve a price entered in the corresponding field in the material master.
4. With this cost estimate, you can then update any price field other than the standard costs with Transaction CK24 or by following the menu path **Accounting • Controlling • Product Cost Controlling • Product Cost Planning • Material Costing • Price Update.**
5. Click the **Other Prices** button to display the screen shown in Figure 3.39.
6. Toggle the **Other Prices** and the **Marking** and **Release** buttons as needed, with the other option being used to mark and release standard prices.
7. In this screen, select any material master price field and click the **Execute** icon to release the cost estimate price to the corresponding field in the material master to reflect the new tax valuation.

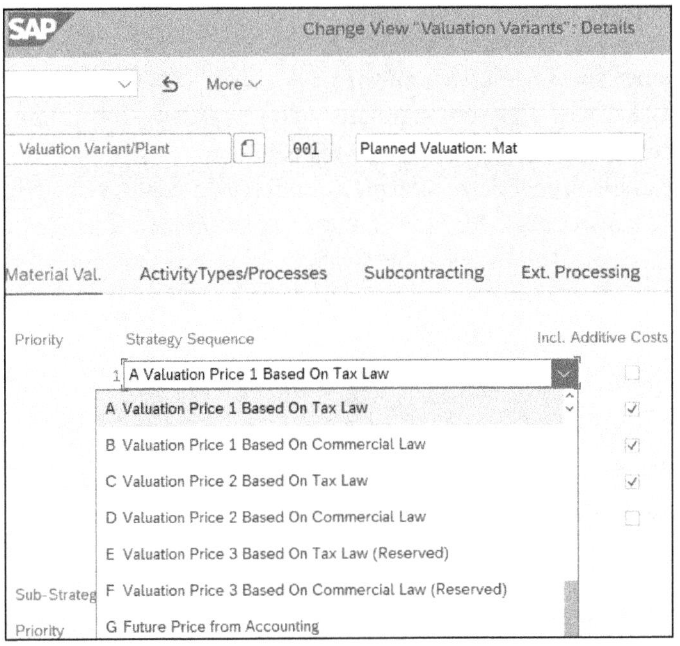

Figure 3.38 Valuation Variant Material Valuation Strategy Sequence

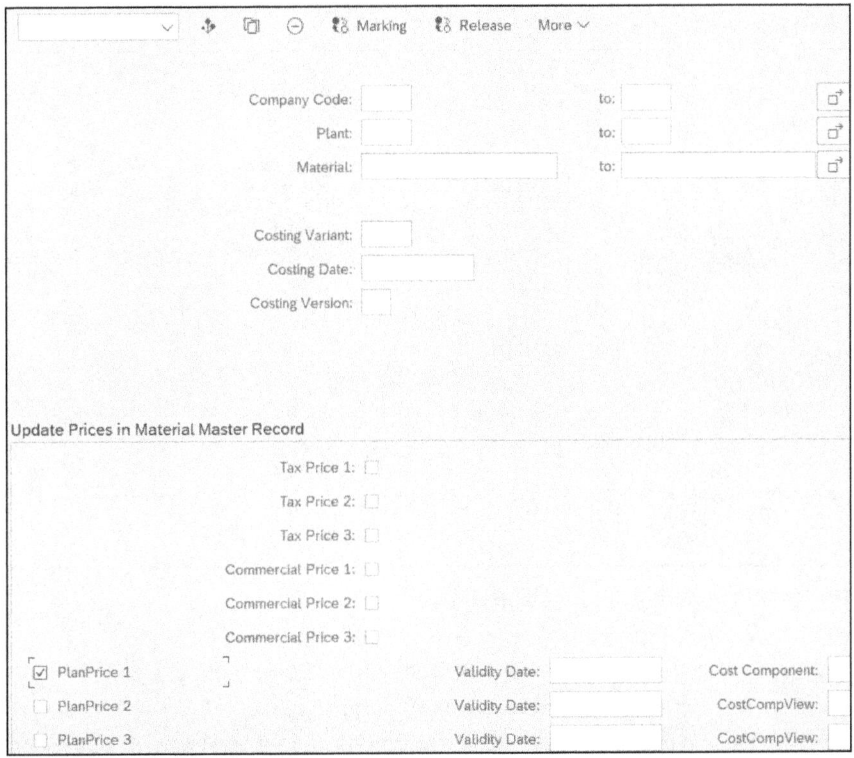

Figure 3.39 Release Other Prices Selection Screen

3.10 Valuation Alternatives

The requirements for balance sheet valuation can vary by country and are driven by tax and trade laws. They can also be driven by company policy to provide comparable results across the organization. If you work with SAP S/4HANA Cloud or use the universal parallel accounting business function, the tax and commercial prices described in the previous section are no longer available, and *valuation alternatives* are used as a basis for balance sheet valuation. The balance sheet valuation alternatives are delivered as part of the scope item BEJ (Inventory Valuation for Year-End Closing). This currently supports the following balance sheet procedures:

- **BSV (Inventory Balance Sheet Value)**
- **LMP (Lowest Value by Market Price)**
- **LMR (Lowest Value by Movement Rate)**
- **ROC (Lowest value by Range of Coverage)**
- **FIFO (First-in, First-Out)**

In the Manage Material Valuations app, you can see the various valuation alternatives used for a given material by selecting it and choosing **Valuation Alternatives**, as shown in Figure 3.40.

Figure 3.40 Valuation Alternatives for Material TG10

Alternatively, you can use the Inventory Balance Sheet Valuation app (SAP Fiori ID F3343) to display the values for all materials as shown in Figure 3.41. You can then use the graphical area at the top of the app to narrow down the selected materials by material type or valuation class.

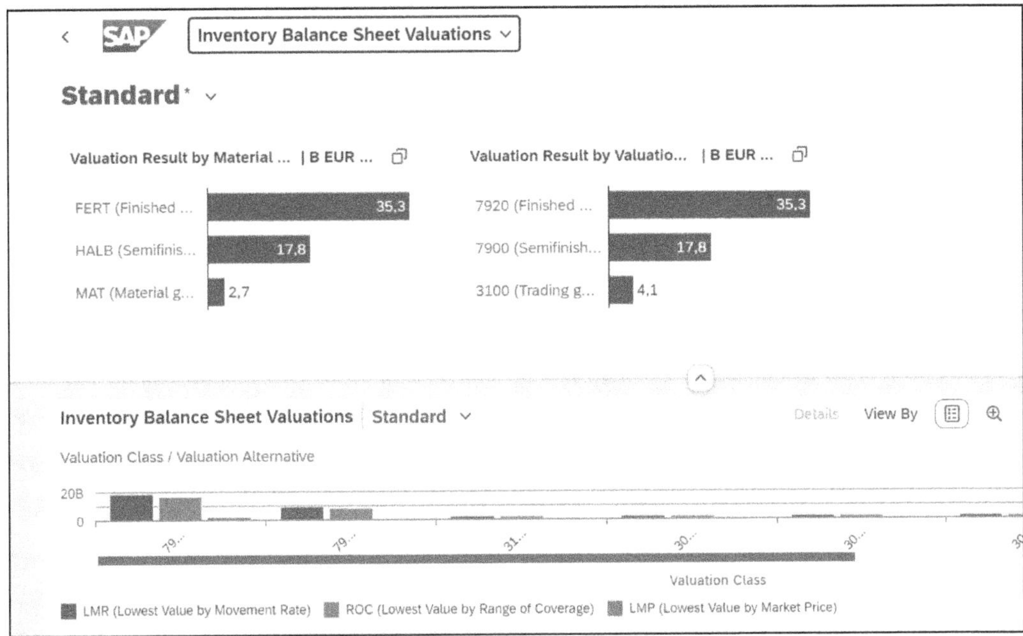

Figure 3.41 Inventory Balance Sheet Valuations App

To access the details of the individual materials, scroll down as shown in Figure 3.42.

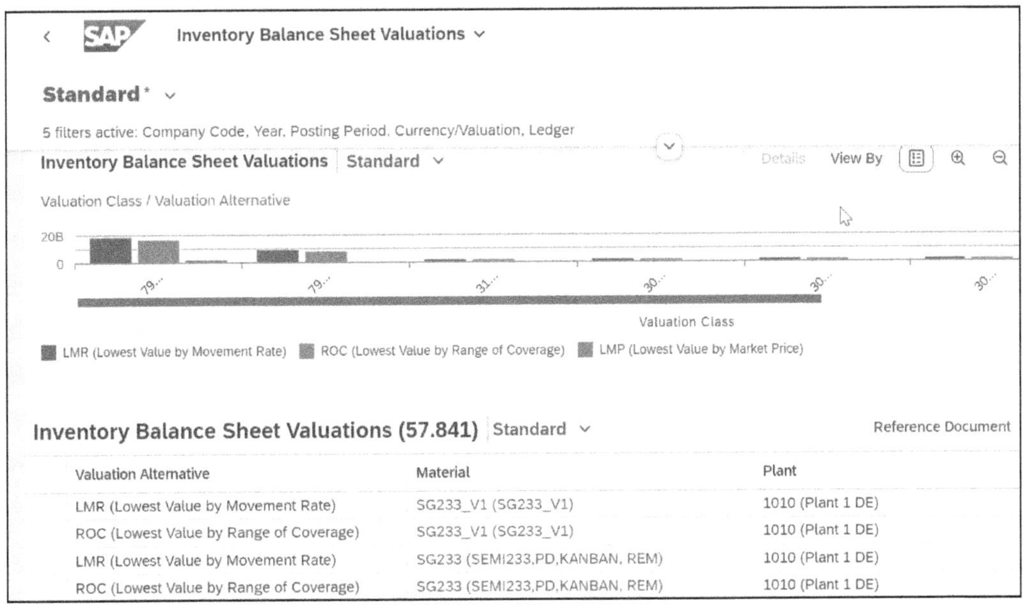

Figure 3.42 Inventory Balance Sheet Valuations

You can select a line from the list of valuation alternatives to understand the method used to calculate the devaluation, as shown in Figure 3.43.

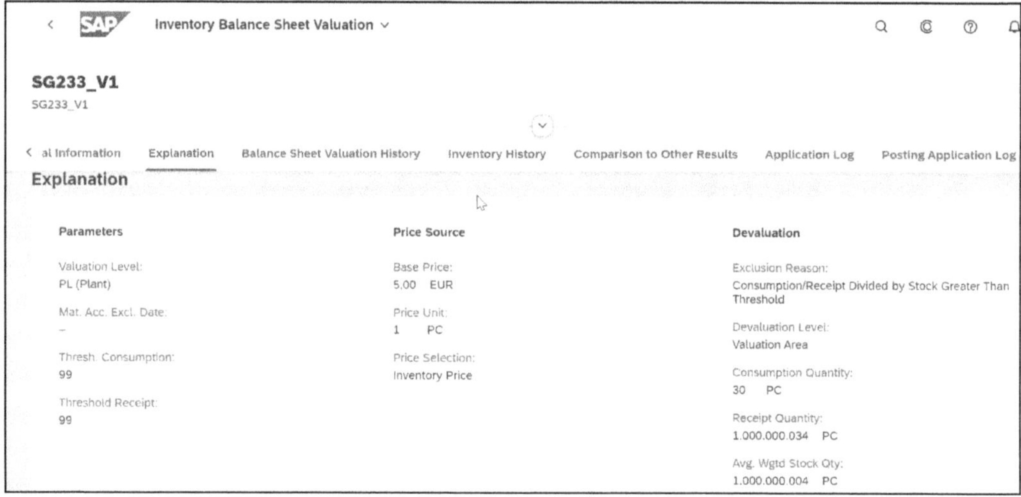

Figure 3.43 Explanation of Inventory Balance Sheet Valuation

3.11 Summary

In this chapter, we discussed the material master fields relevant to SAP S/4HANA product costing. We began with **MRP** fields, which cost estimates refer to when calculating the standard cost. We next examined the **Costing** views, including **Costing 1** and **Costing 2**, which let you access all the fields relevant to management accounting. Finally, we looked at the **Accounting 1** and **Accounting 2** views, which let you access all the fields relevant to financial accounting.

Now that we've considered Controlling and material master data in Chapter 2 and Chapter 3, we'll examine logistics master data in Chapter 4.

Chapter 4
Logistics Master Data

Logistics master data provides information on how you procure and manufacture materials.

4

Now that we've examined Controlling master data in Chapter 2 and material master data in Chapter 3, we'll look at logistics master data in this chapter. This data provides cost estimates with the quantity and price information needed to calculate the standard cost of a product:

- Bills of materials (BOMs) provide component and assembly quantities, while routings provide activity quantities.
- Purchasing information records provide component prices that, together with planned activity prices, provide cost estimates with price information.

A *purchasing information record* (info record) stores all the information relevant to procuring material from a supplier. It contains the purchase price field, which the standard cost estimate uses to determine the purchase price.

We'll examine logistics master data in detail, starting with the BOM.

4.1 Bill of Materials

After all material masters are created and fields are populated correctly, you can use them to create a BOM, as discussed in this section.

4.1.1 Basic Concepts

A BOM is a structured hierarchy of components necessary to build an assembly. BOMs, together with purchasing info records or vendor quotations, provide cost estimates with the information necessary to calculate the material costs of products. You can see an example of a BOM in Figure 4.1.

A cost estimate created for the top-level finished good, **P-100** in this example, selects materials at the lowest level in the BOM first. You cost all materials with material type **ROH** (raw materials) first, then **HALB** (subassemblies), and finally **FERT** (finished goods). You roll up material costs from raw materials through subassemblies to the finished goods as reflected in the **Costing Levels**.

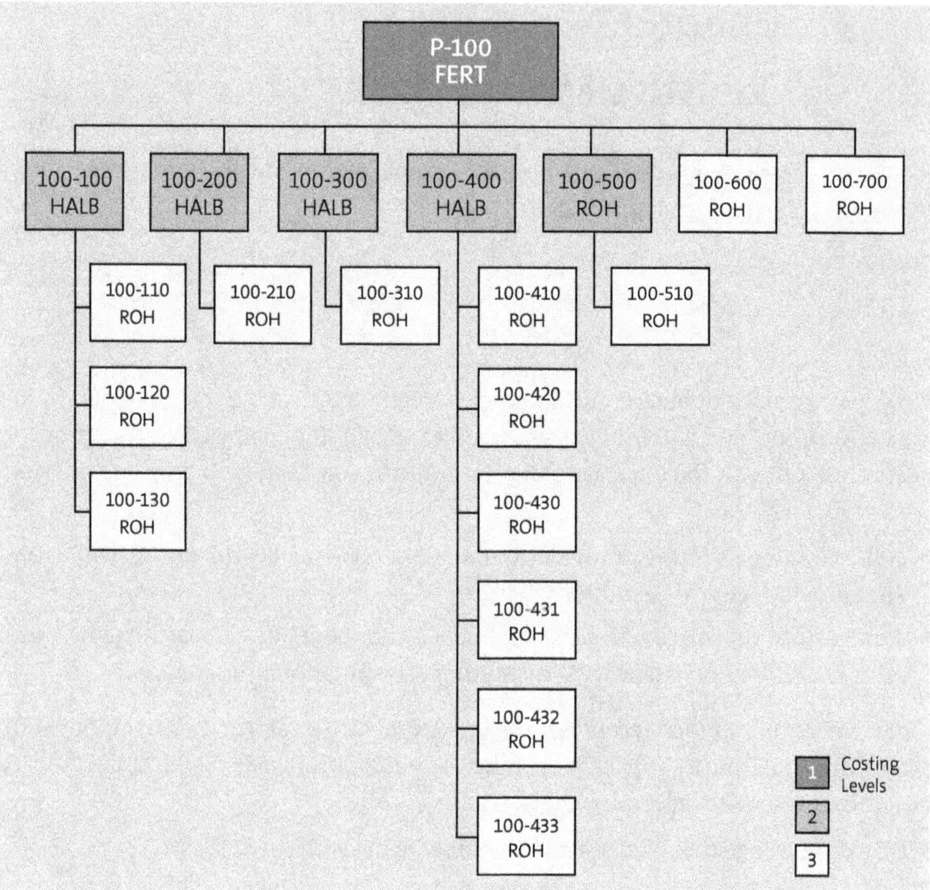

Figure 4.1 Example of a BOM

Each BOM item contains many fields and indicators relevant to product cost controlling, which we'll now examine in detail.

You maintain BOMs with the Manage Bill of Materials app (SAP Fiori ID F2230). You can also maintain BOMs with Transaction CS02 or via the menu path **Logistics • Production • Master Data • Bills of Material • Bill of Material • Material BOM • Change**. The selection screen is shown in Figure 4.2. Type in the **Material**, **Plant**, **BOM Usage**, and **Alternative BOM**. Click in either the **Plant** or **BOM Usage** field and then press F4 to display the list shown in Figure 4.2.

Select the combination of fields for the BOM you want to maintain and press Enter to display the screen shown in Figure 4.3.

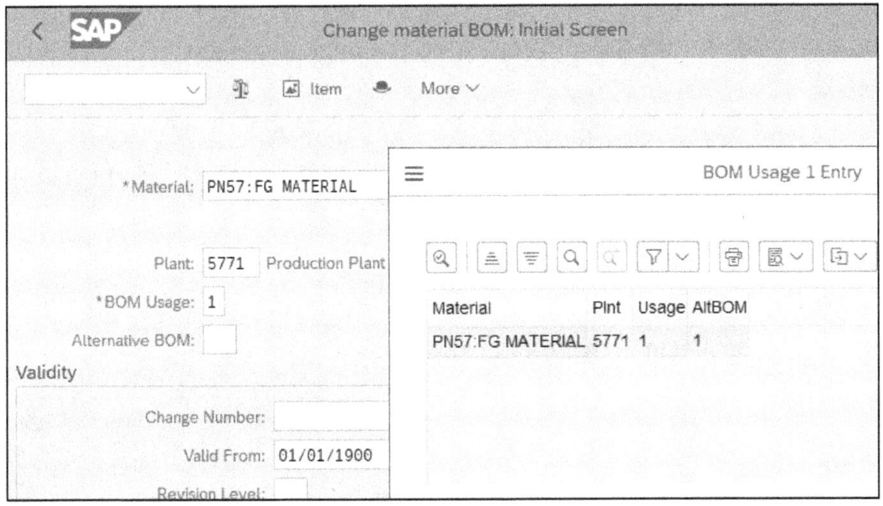

Figure 4.2 Material BOM - Initial Screen

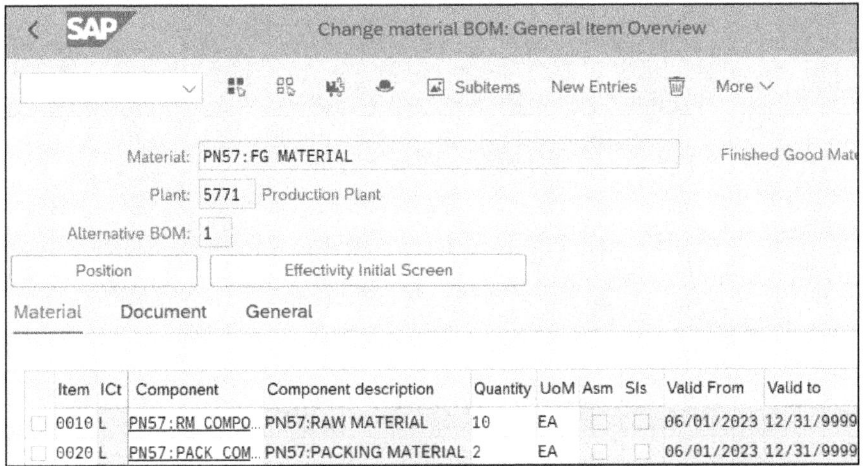

Figure 4.3 Material BOM - General Item Overview

Click the **Position** button to locate a BOM item quickly in the lower part of the screen. Click the **Effectivity Initial Screen** button to display the initial screen shown in Figure 4.4.

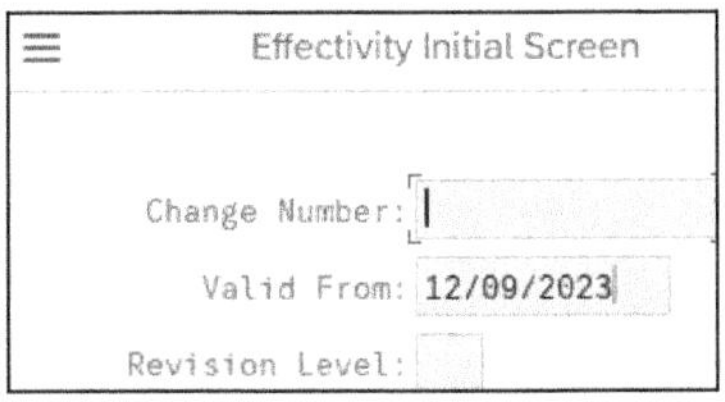

Figure 4.4 Effectivity Initial Screen

You use parameter effectivity to manage multiple configurations of a BOM and routing, making it easy to select the correct items for costing using the same correct BOM and routing.

The **General Item Overview** in Figure 4.3 displays a list of material BOM items required to manufacture the assembly. Let's now examine the fields in the **Basic Data** tab of an individual BOM **Item**.

4.1.2 Basic Data

Double-click any BOM **Item** in Figure 4.3 to display details shown in Figure 4.5.

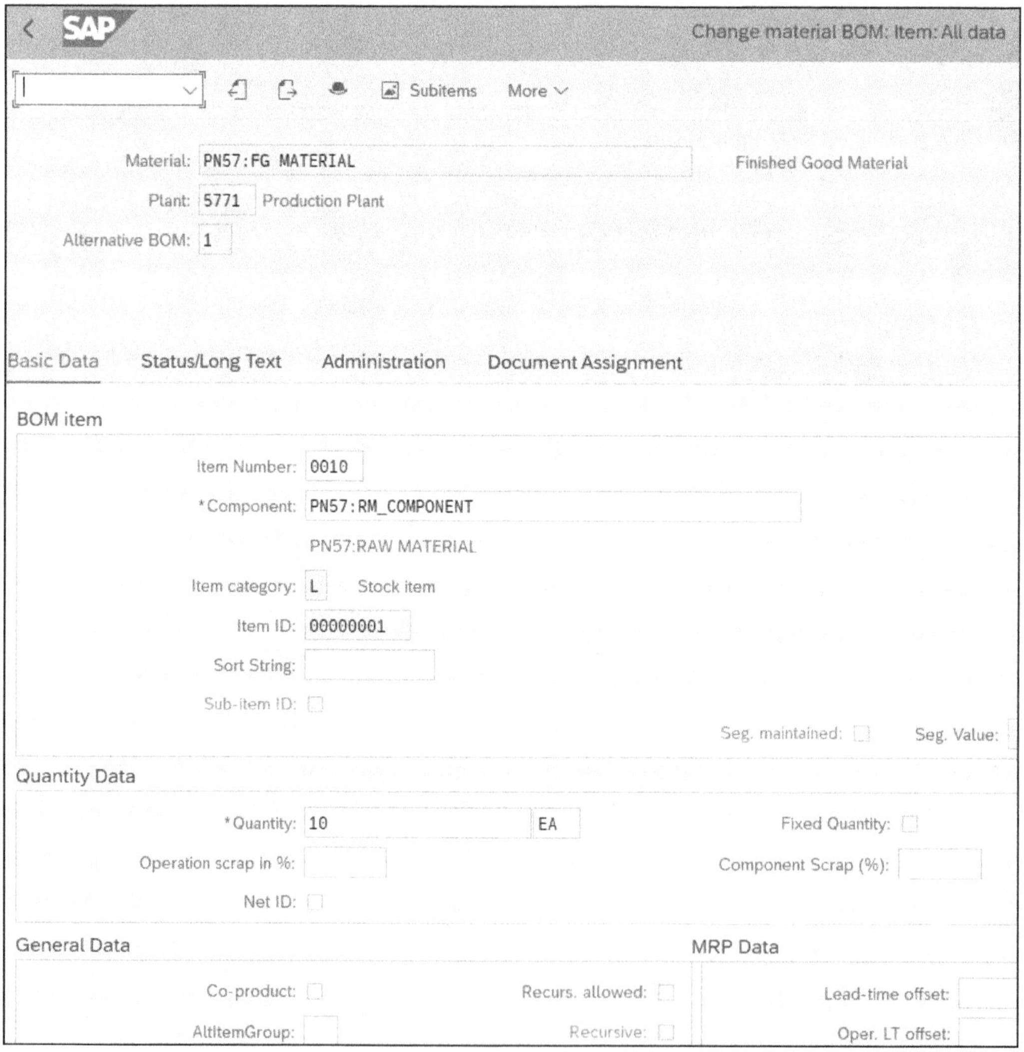

Figure 4.5 Material BOM Item

The **Basic Data** tab of the **BOM Item** includes the **Quantity Data**, **General Data**, and **MRP Data** sections, along with the **Item category**. Let's discuss the fields and checkboxes relevant to product cost controlling:

- **Item Number**
  The **Item Number** (first field) shows the sequence of BOM components.
- **Component**
  The **Component** is the material number of each component.
- **Item Category**
  The **Item Category** groups BOM items into criteria such as stock, nonstock, document, or text items. Some item categories, such as stock and nonstock, are typically relevant to costing, whereas others, such as document or text items, aren't.
- **Quantity**
  The component **Quantity** is related to the product's base quantity. Increasing the base quantity, similar in concept to the price unit, can increase the accuracy of component quantities. You maintain the base quantity in the BOM header, which you access by clicking the header (hat) icon at the top of Figure 4.5.
- **Fixed quantity**
  The **Fixed quantity** checkbox to the right of the **Quantity** indicates that the component quantity is always the same, independent of the assembly or order quantity. For BOM items that only contain text, such as instructions or documentation, the system automatically selects this checkbox.
- **Operation scrap in %**
  Operation scrap in % entered in the BOM item ensures that the input quantity of valuable components inserted in an assembly is reduced. For a particular component, operation scrap allows you to enter a different scrap percentage, less than the assembly scrap percentage.
- **Net ID**
  You select the **Net ID** checkbox to ignore assembly scrap. You must select it to enter **Operation scrap**. Without this checkbox selected, assembly scrap entered in the upper-level assembly increases the quantities of all lower-level components. It is additive to a value entered in the **Component scrap (%)** field.
- **Component scrap (%)**
  Component scrap (%) is the percentage of component quantity that doesn't meet required production quality standards before it is inserted into the production process. An entry in the **Component scrap (%)** field in the BOM item takes priority over an entry in the **MRP 4** view.

BOM Item Component Scrap versus MRP 4 View Priority

A component is used in many assemblies, and the component scrap rate is 10%, which you enter in the **MRP 4** view of the component.

One assembly is manufactured near the inventory store, and only 5% of components are lost or damaged on the way to the production line. A component scrap rate of 5% entered in the component BOM item of this assembly takes priority over the 10% component scrap rate entered in the **MRP 4** view of the component.

- **Co-product**

  A **Co-product** is a valuated product produced simultaneously with one or more other products. When you select the **Co-product** checkbox, you typically enter a negative quantity. Selecting this checkbox allows you to assign the proportion of costs this material will receive in relation to other co-products within an apportionment structure. An apportionment structure defines how you distribute costs to co-products. You can only maintain a BOM item as a co-product if you select the corresponding checkbox in the **MRP 2** view.

  A co-product is of significant financial value, while byproducts are of low financial value. Goods receipts for co-products are the same as normal goods receipts for production orders and use movement type 101.

- **Recurs. allowed**

  A BOM is *recursive* if the product contains a component with the same object number as the superior assembly. Recursiveness is often due to input errors but may be intentional in individual cases. Selecting the **Recurs. allowed** checkbox allows you to maintain a recursive BOM.

Example: Recursiveness Allowed

A cement tile manufacturer has a pile of rejected tile and cement scrap that doesn't have a material number, and it doesn't track the value and quantity in the system. The company grinds the rejected material into a usable material, which you enter in inventory with a value and quantity. Manufacturing the usable material is the activity cost determined by the routing to run the grinding machine.

But a cost estimate requires a BOM and routing to calculate the manufacturing cost, and no material is tracked in the system to enter as a BOM item. This scenario is entirely possible. To solve it, you can enter the output material as a BOM component if you select the **Recurs. allowed** checkbox in the BOM item.

- **Recursive**

  The system sets the **Recursive** checkbox if the exploded BOM for a material contains the material as a BOM item.

Now that we've examined the relevant fields in the **Basic Data** tab, let's examine the fields in the remaining tabs of a BOM item.

4.1.3 Status/Long Text

Click the **Status/Long Text** tab shown in Figure 4.5 to display the details shown in Figure 4.6.

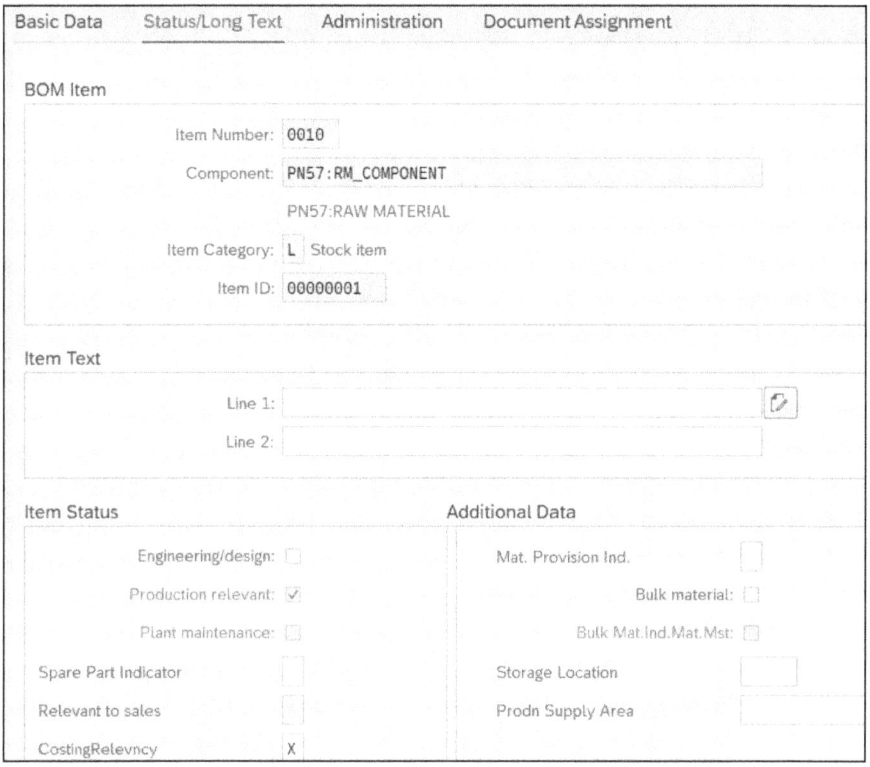

Figure 4.6 BOM Item Status Long Text

The fields in the **Status/Long Text** tab allow you to maintain item text describing each BOM item and to maintain checkboxes in the **Item Status** section to control the extent to which each item can be used in other processes.

In the **Item Text** section, you can enter up to 40 characters in both the **Line 1** and **Line 2** fields. You can enter more characters by clicking the page and pencil icon.

Now, let's consider the items in the **Item Status** section of the **Basic Data** tab of a BOM:

- **Engineering/design**
  Select this checkbox to indicate that this item is relevant to the area responsible for producing a functional design for production.
- **Production relevant**
  The **Production relevant** checkbox determines if the item is relevant to the production process. The system copies items with this checkbox selected to the planned order, and dependent items are calculated. When converted from the planned order, the system automatically copies these items to the production order.

This checkbox isn't changeable since we previously selected the product **BOM Usage** of **1** (**Production**) in the initial screen shown in Figure 4.2.

- **Plant maintenance**
  This checkbox indicates that this item is relevant to plant maintenance procedures. Items that are relevant to plant maintenance are used in maintenance BOMs.

  This checkbox isn't changeable since we previously selected the product **BOM Usage** of **1 Production** in the initial screen shown in Figure 4.2.

- **CostingRelevncy (costing relevancy)**
  The **CostingRelevncy** field determines whether a BOM item is included in costing for a standard cost estimate and the calculation of planned and actual costs for a manufacturing order. When you create a cost estimate for balance sheet purposes, the **CostingRelevncy** field determines how much a BOM item is included in the costing. A blank entry allows you to exclude the cost of some BOM items, such as bulk materials.

  The possible entries are:
  - Blank: not relevant to costing
  - **1**: 50% relevant to costing
  - **2**: 25% relevant to costing
  - **3**: 75% relevant to costing
  - **X**: 100% relevant to costing

Negative BOM Items and Costing Relevancy

For output materials, with the **Co-product** checkbox selected in the **Costing 1** view and a negative BOM item, the **CostingRelevncy** checkbox does not affect the BOM item.

Now, we'll consider the relevant checkboxes in the **Additional Data** section:

- **Bulk material**
  You charge bulk materials—for example, washers and grease—directly to a cost center. Bulk materials aren't relevant for individual costing. The cost of these items is included in the overhead cost components. You cannot select the **CostingRelevncy** checkbox for a bulk material BOM item. If you attempt to select it, you'll receive an error message since the cost of the item would be included in both the material and overhead cost components.

- **Bulk Mat.Ind.Mat.Mst (bulk material indicator in the material master)**
  The **Bulk Mat.Ind.Mat.Mst** checkbox shows whether you defined a bulk material in the **MRP 2** view, which overrides the indicator in the BOM item. If you always use a material as a bulk material, select the checkbox in the material master. You cannot deselect the checkbox in the BOM item if the indicator is set in the material master.

Now that we've studied relevant fields and checkboxes in the **Status/Long Text** tab, let's look at the fields in the **Administration** tab.

4.1.4 Administration

Click the **Administration** tab shown in Figure 4.6 to display the details shown in Figure 4.7.

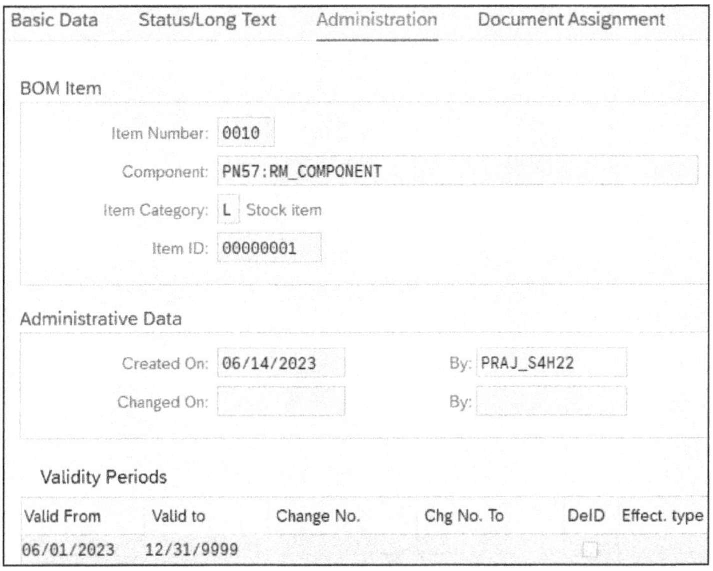

Basic Data | Status/Long Text | Administration | Document Assignment

BOM Item

Item Number: 0010
Component: PN57:RM_COMPONENT
Item Category: L Stock item
Item ID: 00000001

Administrative Data

Created On: 06/14/2023 By: PRAJ_S4H22
Changed On: By:

Validity Periods

Valid From	Valid to	Change No.	Chg No. To	DelD	Effect. type
06/01/2023	12/31/9999			☐	

Figure 4.7 BOM Item Administration Data

This screen shows the following fields:

- **Created On and By**
  These are the BOM item **Created On** and **By** dates. These can be useful if you need to contact the person who created the BOM item for more information.
- **Changed On and By**
  These two fields are the BOM item **Changed On** and **By** dates. When you view a BOM item, you can find more details of its changes by selecting the menu path **Environment • Change Documents.**
- **Validity Periods**
  When you create a standard cost estimate with Transaction CK11N, one of the selection screen fields is **Quantity Structure Date.** This date determines which BOMs and BOM items the system will select for the cost estimate based on **Validity Periods** shown in Figure 4.7.
- **Effect. type**
  Effect. type (effectivity type) provides another option for selecting a BOM item based on **Change No.**

Now, let's examine the fields in the last tab of the **BOM Item** screen.

4.1.5 Document Assignment

Click the **Document Assignment** tab in Figure 4.7 to display the screen shown in Figure 4.8.

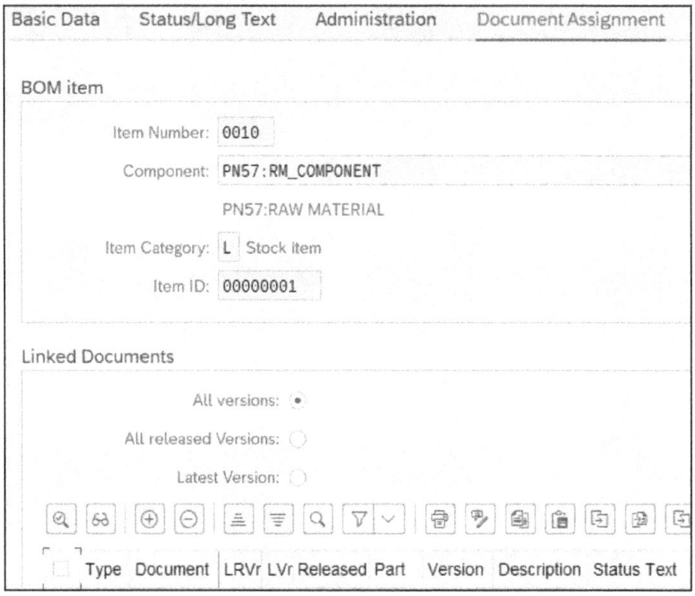

Figure 4.8 BOM Item Document Assignment

The **Linked Documents** section allows you to attach different versions of documents, such as design documents relevant to each BOM item. Now that we've examined BOMs, let's examine work centers. Material masters and BOMs provide cost estimates with material and assembly prices and quantities. To determine labor and activity standard quantities, we need to consider work centers and routings, which list work centers.

4.2 Work Centers

In addition to material masters and BOMs we need to create instructions to manufacture assemblies, as discussed in the following sections.

4.2.1 Basic Concepts

You perform operations at work centers that represent, for example, machines, production lines, or employees, as shown in Figure 4.9.

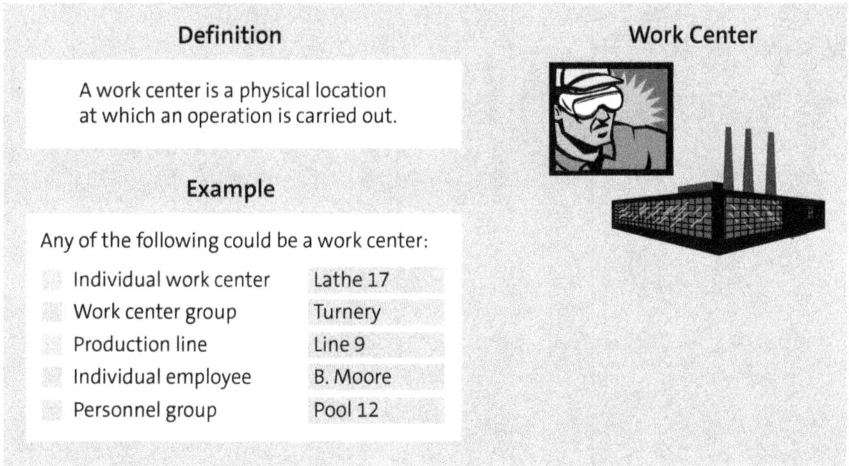

Figure 4.9 Work Center Examples

You define work centers within a plant and assign cost centers. You first create work centers, which you then list in routings with instructions on how to manufacture assemblies, as we'll discuss in Section 4.3.3.

Let's examine the fields in the **Basic Data** tab of a work center.

4.2.2 Basic Data

Each work center contains fields and checkboxes relevant to product cost controlling, which we'll now examine in detail.

You maintain work centers with the Manage Work Centers app (SAP Fiori ID F6175). With this app, you can perform all operations related to work centers. You can navigate to all the relevant apps to view work centers, create and change work centers, and perform mass maintenance of work center capacities.

You can also maintain work centers with Transaction CR02 or via the menu path **Logistics • Production • Master Data • Work Centers • Work Center • Change**. You'll see the selection screen shown in Figure 4.10.

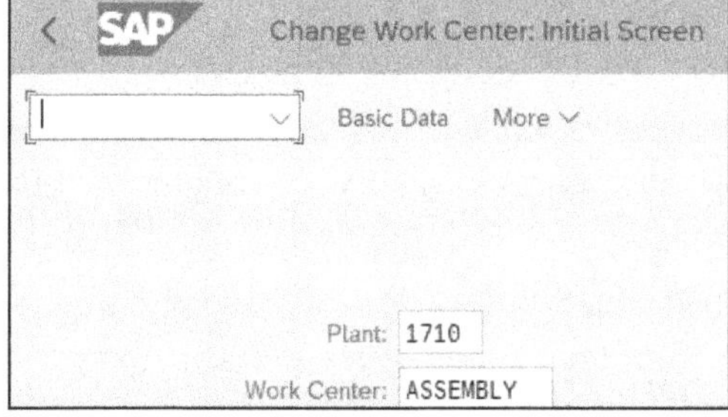

Figure 4.10 Work Center Initial Screen

Type in the **Plant** and **Work Center** and press [Enter] or click the **Basic Data** button to display the screen in Figure 4.11.

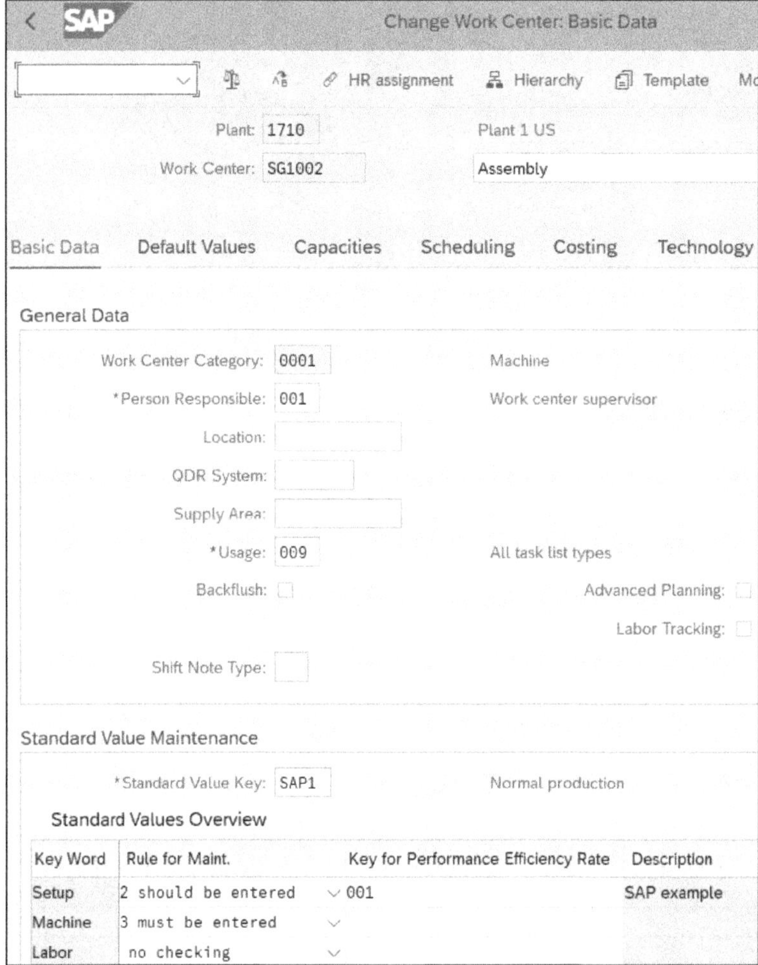

Figure 4.11 Work Center Basic Data

The costing-relevant fields in the **Basic Data** tab determine the work center application area and the **Standard Value Key**, which we'll now explain:

- **Work Center Category**
  You assign a **Work Center Category** to an application area such as production, plant maintenance, or quality management. With Transaction OP40, you configure screens and fields relevant to each work center category.
- **Backflush**
  Select **Backflush** to automatically post component goods issues after their physical issue to an order. You allocate each material component in a BOM to an operation in

the production order. The goods issue occurs automatically during confirmation. Backflushing reduces the amount of manual entries, especially for low-value parts. A consequence of backflushing is that you will have no input quantity variance for components.

SAP considers the **Backflush** checkbox if the **MRP 2** view specifies that a material component will be backflushed at the operation's work center.

- **Standard Value Key**
  The **Standard Value Key** field determines basic data for the up to six standard values available for each work center. Also, it determines how many of the standard values are available at each work center.
- **Key Word**
  The **Key Word** defines the basic use of each standard value—such as **Setup**, **Machine**, and **Labor**—for standard values that you confirm with activity rates for each work center.
- **Rule for Maint. (rule for maintenance)**
  The **Rule for Maint.** determines whether a standard value **should be entered** (optional), **must be entered** (mandatory), or if there is **no checking** in the operation of a routing.
- **Key for the Performance Efficiency Rate**
  The **Key for the Performance Efficiency Rate** is the ratio between an individual's actual output and the planned average output. The system transfers performance efficiency from the work center to the operation, and you cannot change it there. If you don't enter a performance efficiency rate key, the system assumes 100% efficiency. By modifying the performance efficiency rate, you can modify the standard times used by capacity planning and costing.
- **Description**
  The **Description** describes the performance efficiency rate key in more detail.

Performance Efficiency Rate

Entering a performance efficiency rate key of less than 100% increases the planned and actual cost of using an activity. You may use this key if you are uncertain of standard values when using an activity for the first time.

Many companies use an iterative process to increase the accuracy of standard values, which involves using a performance efficiency of 100%. They compare planned with actual standard values and update the planned values as required before subsequent main costing runs. This technique is most effective if you record actual activity quantities during order or operation confirmation.

Now that we've studied the fields and indicators in the **Basic Data** tab, let's examine the fields in the **Default Values** tab.

4.2.3 Default Values

Click the **Default Values** tab shown in Figure 4.11 to display the screen shown in Figure 4.12.

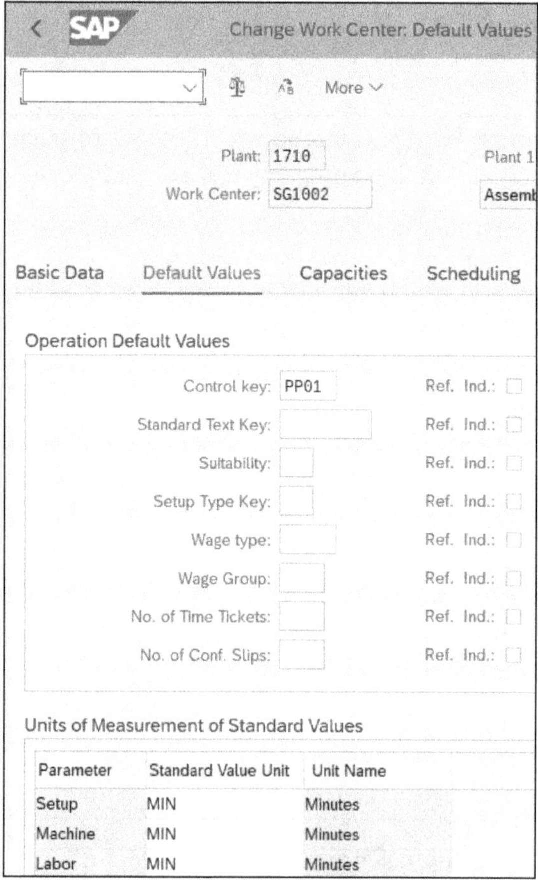

Figure 4.12 Work Center Default Values

The **Default Values** tab contains data that defaults when you add an operation carried out at the work center in a routing. By entering default values, you reduce the effort required when editing operations in routings.

Control Key

A **Control key** specifies how you perform and treat operation activities. Click in the **Control key** field and press F4 to display the dialog box shown in Figure 4.13.

The **Overview** section lists available control keys along with fields and checkboxes. Scroll up or down to see the complete list.

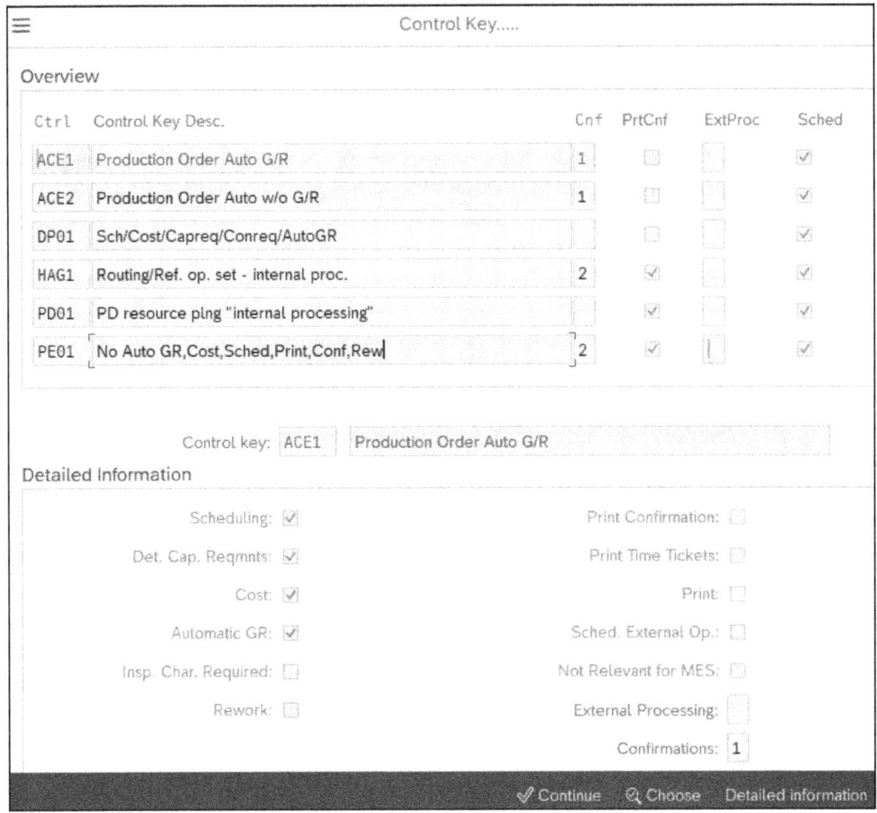

Figure 4.13 Control Key Detailed Information

To display more detailed information, click a control key and the **Detailed Information** button at the bottom. Let's review the fields relevant to product costing:

- **Cost**
  The **Cost** checkbox determines if operations carried out at this work center can be included in planned (cost estimate) and actual (simultaneous) costing.
- **Automatic GR (goods receipt)**
  Automatic GR determines if the system automatically delivers yield from this work center into inventory during final operation confirmation. The manufacturing order automatically receives a credit based on the yield quantity multiplied by the standard price of the material.
- **Rework**
  You don't consider operations carried out at work centers with **Rework** selected when scheduling routings because you don't typically plan for rework since it is not expected.

- **External Processing**
  The control key determines if in-house production or an external supplier performs the operation. This determines if you cost the operation by multiplying the standard value by the planned activity price or by referring to the external processing section of the routing operation details as described in Section 4.3. Here are the three possible entries in the **External Processing** field:
  - **(blank)**
    For internally processed operations
  - **+**
    For externally processed operations
  - **X**
    When both internally and externally processed operations are possible
- **Confirmations**
  An entry in the **Confirmations** field determines if milestone confirmations are carried out at this operation—that is, if all operations up to the preceding milestone operation are confirmed during confirmation. This field also determines if confirmations are mandatory, not possible, or optional at this operation. The following are possible entries in this field:
  - **(blank)**
    Confirmation is possible but not necessary
  - **1**
    Milestone confirmation (not Project System/plant maintenance)
  - **2**
    Confirmation required
  - **3**
    Confirmation not possible

Now that we've examined the control key in detail, let's look at the more fields in the **Default Values** tab shown in Figure 4.12.

Control Key Is Referenced

If you select any of the **Ref. Ind.** (reference indicator) checkboxes in the **Default Values** tab shown in Figure 4.12, the values defaulted into the routing aren't changeable.

> **Deselecting Control Key is Referenced**
>
> When you deselect the **Ref. Ind.** checkbox for a work center used in a routing, the system doesn't copy the default values from the work center to the routing. You must use mass replacement or replace the values manually in the routing.

Unit of Measure of Standard Values

The **Standard Value Unit** contains the work center's unit of measure, as shown previously in Figure 4.12, and determines the activity unit of measure when you add an operation to a routing.

Now, let's look at the **Capacities** section of a work center.

4.2.4 Capacities

Capacity is the ability of a work center to perform a task. It impacts costing to the extent that there can be jumps in the activity usage depending on the requested capacity. Click the **Capacities** tab shown in Figure 4.12 to display the screen shown in Figure 4.14.

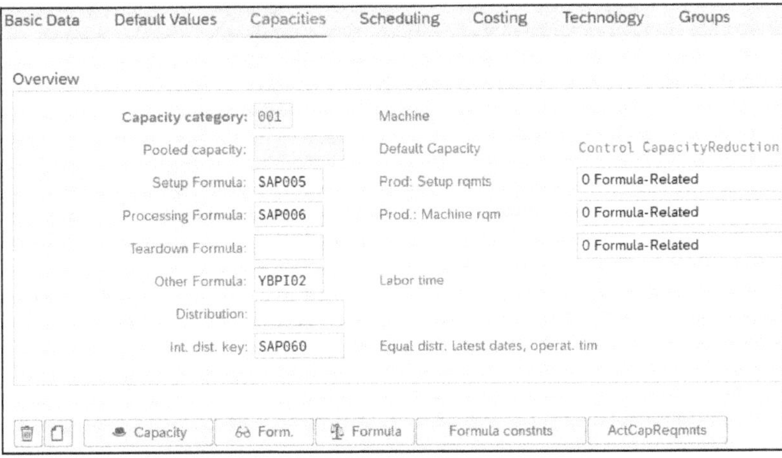

Figure 4.14 Work Center Capacity Overview

Let's start by discussing the **Capacity category** field in the **Capacities** tab. You differentiate capacities according to capacity category. The **Capacity category** allows you to differentiate between machine and labor capacity:

- *Machine capacity* is the availability of a machine based on planned and unplanned outages, and maintenance requirements.
- *Labor capacity* is the number of workers who can operate a machine at the same time.

A capacity category can exist only once per work center. However, capacities at different work centers can have the same capacity category.

To find out more about a capacity category, click the **Capacity** button at the bottom of the **Capacities** tab shown in Figure 4.14. You'll see the capacity header shown in Figure 4.15.

Figure 4.15 Work Center Capacity Header

The following fields are explained as follows:

- **Factory Calendar** defines workdays and nonwork days such as weekends and public holidays
- **Start** and **End Time** for shifts
- **Length of breaks**

Now that we've discussed the **Capacity category**, let's look at the **Pooled capacity** field, shown in Figure 4.16, which identifies an available capacity that several work centers can use.

Pooled Capacity

A *pooled capacity* may be a group of setup personnel who perform setups for many work centers.

If you don't enter a pooled capacity key when assigning capacity to the work center, the system assigns the capacity only to that work center.

Note

With the buttons at the bottom of the **Capacities** tab shown in Figure 4.14, you can perform the following additional functions:

- Delete existing capacity categories
- Add additional capacity categories
- Display or change capacity header information
- Display details of formulas used to determine capacity requirements
- Test formulas with different quantities and standard values
- Set different formula constants
- Make settings for determining the actual capacity requirements

Work center capacity fields are typically set up and tested by production.

The next section of a work center includes scheduling data.

4.2.5 Scheduling

Click the **Scheduling** tab in Figure 4.14 to display work center scheduling shown in Figure 4.16.

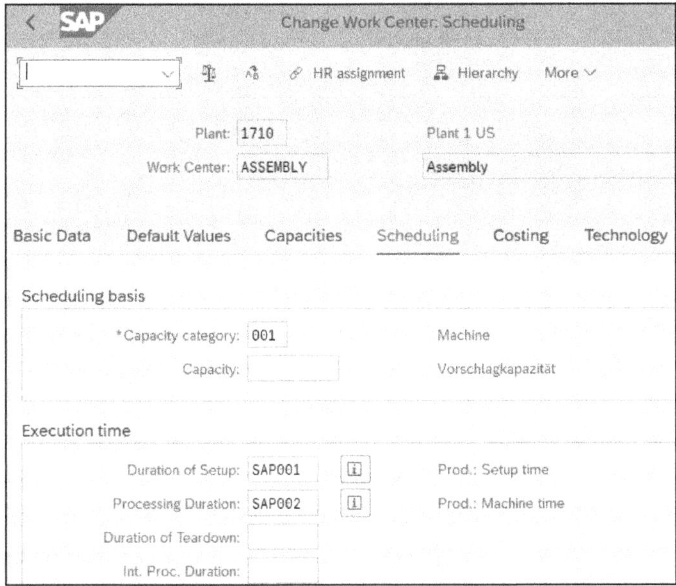

Figure 4.16 Work Center - Scheduling

Scheduling calculates the start and finish dates of manufacturing orders or operations in an order. In the following sections, we'll look at the parameters affecting how these times are calculated.

Capacity Category

You use the operating time of the **Capacity category** to schedule manufacturing orders.

Capacity

The **Capacity** is the name of the capacity.

Setup Formula

In this example, we use the setup formula **SAP001** to calculate the duration of an operation's setup time. If you don't enter a formula, the system uses a scheduling setup time of zero. Click the **Information** icon [i] to the right of the input box for **Duration of Setup** to display details of the formula, as shown in Figure 4.17.

You define the **Formula Key** and **Formula** with configuration Transaction OP54 or by following the IMG menu path **Production • Basic Data • Work Center • Costing • Work Center Formulas • Define Formulas for Work Centers**. You use this transaction to change the fields and indicators displayed in Figure 4.17.

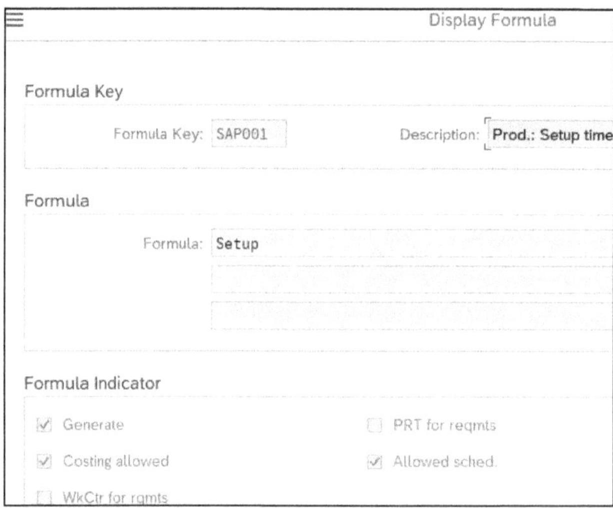

Figure 4.17 Work Center Setup Time Formula

You define formula parameters (**Setup**, in this example) entered in the **Formula** section with configuration Transaction OP51 or via the IMG menu path **Production • Basic Data • Work Center • Costing • Work Center Formulas • Define Formula Parameters for Work Centers**. Double-click the **Setup** parameter to display the details of the parameters shown in Figure 4.18.

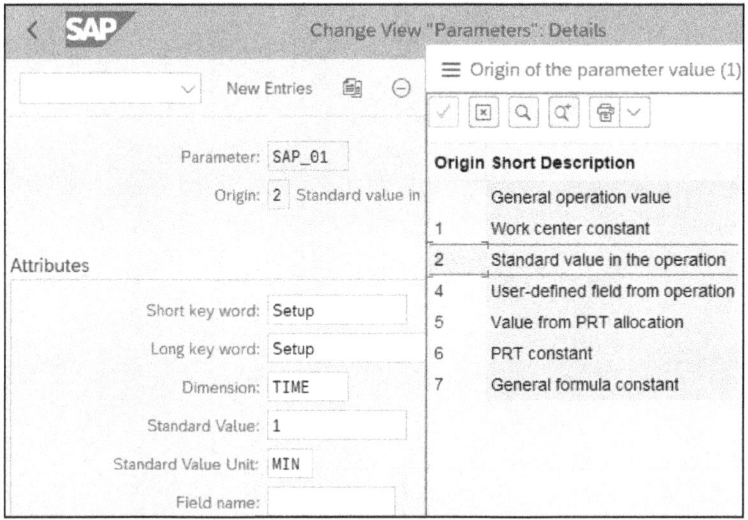

Figure 4.18 Work Center Formula Parameters

Parameter SAP_01 definition indicates that the setup time calculated by the formula in Figure 4.17 is equal to the **Standard value in the operation** of the routing, as we'll discuss in Section 4.3.

Left-click in the **Origin** field and press F4 to display the list of possible parameter value origins, as shown in Figure 4.18. You can derive a parameter from different origins, such as a user-defined field from the operation or a constant.

Process Industries

If you work in a process industry that manufactures the same product on a production line for many process orders sequentially, you won't incur setup time and costs for every process order. To manage this scenario, you can use production campaigns, which you access with Transaction PC02 or by following the menu path **Logistics • Production–Process • Production Campaign**.

Now that we've discussed the setup formula, let's examine the processing formula shown previously in Figure 4.16.

Processing Formula

A *processing formula* is used in scheduling to calculate the **Processing Duration** of an operation. If you don't enter a formula, the system uses a scheduling processing time of zero. Click the **Information** icon [i] to the right of the input box for **Processing Duration** in Figure 4.16 to display details of the formula, as shown in Figure 4.19.

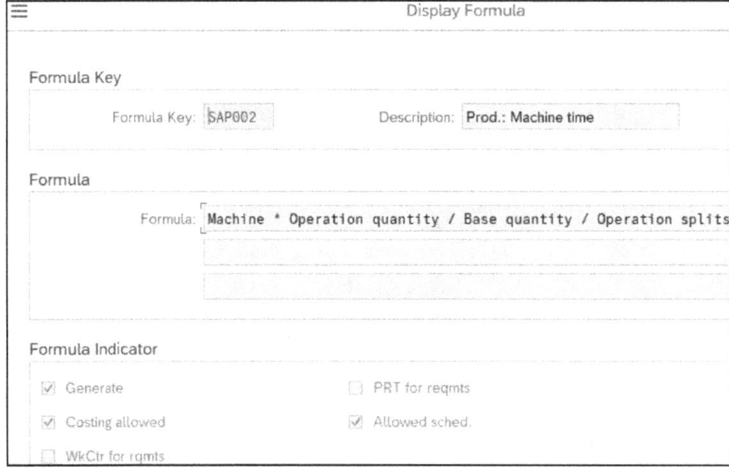

Figure 4.19 Work Center Execution Time Processing Formula

You can figure out how the formula works by determining the details of each formula parameter with the same method used to display Figure 4.19 with configuration Transaction OP51.

SAP Standard Formulas

Production personnel typically set up formula keys and parameters by copying them from SAP-supplied standard formulas. Accounting and Controlling personnel should understand how to analyze formulas when there are any costing issues or messages, especially during initial system implementation.

The **Costing allowed** checkbox in Figure 4.19 controls whether the formula can calculate costs.

Now, we'll consider the next fields in the **Execution time** section of Figure 4.16.

Teardown Formula

Teardown time is needed to restore a work center to its normal state after processing operations. You include the formula key in **Duration of Teardown** in Figure 4.16. Teardown time is similar to setup time because you typically fix both independently of production lot size quantity. If you don't enter a formula, the system uses a teardown time of zero for scheduling.

Equipment Cleaning

Setup and teardown time can also represent equipment cleaning time in food and pharmaceutical industries, for example, between production campaigns, which can be significant to meet regulatory requirements.

Other Types of Internal Processing Formula

You enter an **Int. Proc. Duration** formula for calculating the duration times of other types of internal processing in scheduling (e.g., in networks, in process orders, or in maintenance orders). If you don't enter a formula, the system uses the duration you entered when maintaining an activity or operation in a network or maintenance order.

Now that we've looked at the **Scheduling** section of a work center, the next tab to consider is **Costing**.

4.2.6 Costing

Click the **Costing** tab shown earlier in Figure 4.16 to display the costing fields shown in Figure 4.20.

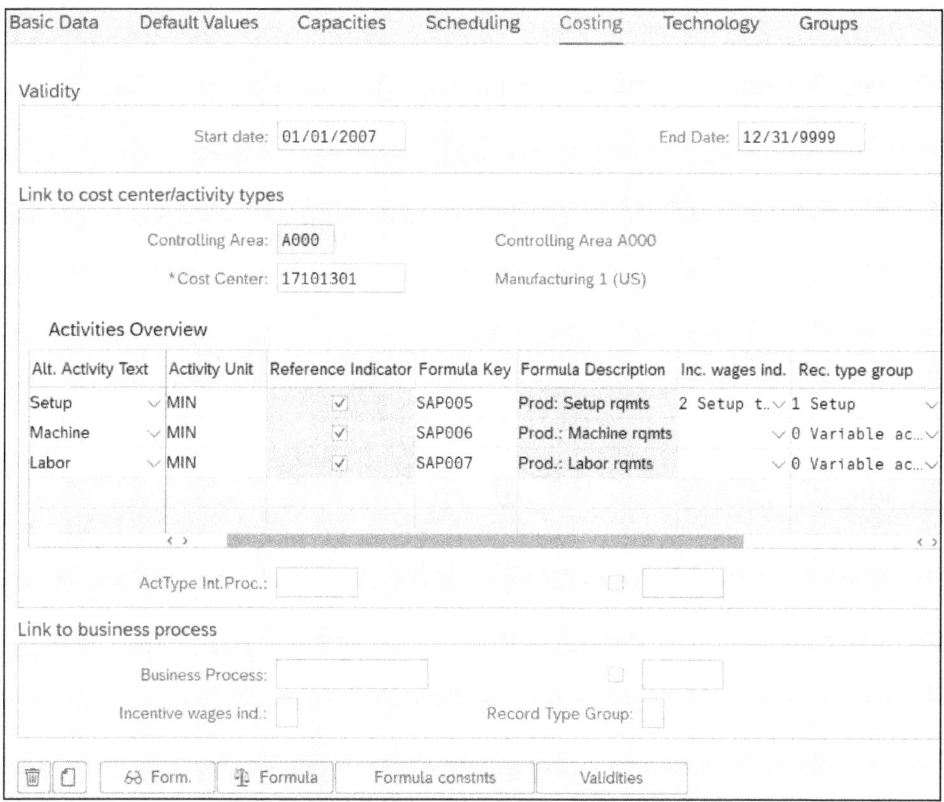

Figure 4.20 Work Center - Cost Center Assignment

A work center uses resources from a cost center to cost the activities performed at the work center. Let's look at the fields in this section:

- **Start date** and **End Date**
  You assign a work center to one cost center for a **Validity** period.

- **Controlling Area** and **Cost Center**
  The cost center that you assign to a work center must exist for the validity period and must belong to the same **Controlling Area** as the **Work Center Plant**. You can assign several work centers in different plants to one cost center. The work centers must belong to the same company code as the cost center.
- **Alt. Activity Text**
  You use **Alt. Activity Text** if you are editing work centers and want to describe an activity on the **Costing** tab.
- **Activity Type**
  You assign an activity type, valid for the cost center, to every standard value in the work center.
- **Activity Unit**
  The **Activity Unit** refers to the time or the quantity unit used to post the consumed activity quantities.
- **Reference Indicator**
  The activity type appears in the routing as a default value. By selecting the **Reference Indicator** checkbox, you can prevent anyone from changing this default.
- **Formula Key and Formula Description**
  The **Formula Key** refers to formulas you use to determine the following:
  - Execution time
  - Capacity requirements
  - Costs of an activity type
  - Total quantity or usage value of a production resource or tool

Formula Test Buttons

The buttons at the bottom of the work center **Costing** tab, shown in Figure 4.20, allow you to test and display formula details.

Now that we've examined BOMs and work centers, let's look at routings. Material masters and BOMs provide cost estimates with material and assembly prices and quantities. Routings are lists of operations performed at work centers that allow you to determine labor and activity standard quantities.

4.3 Routing

A *routing* is a list of tasks containing standard activity times required to perform operations to build an assembly. In the following sections, we'll explain the basic concepts behind the use of the routing for costing purposes and the structure of the routing.

4.3.1 Basic Concepts

Routings, together with planned activity prices, provide cost estimates, the information necessary to calculate the planned activity and overhead costs of products, as well as determine the actual cost during manufacture. You can see an example of a routing in Figure 4.21.

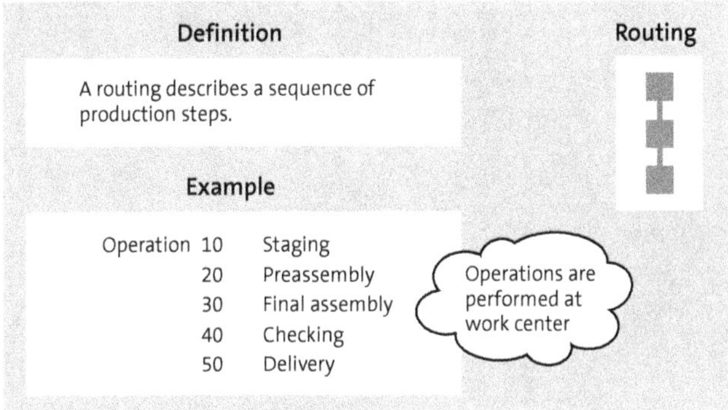

Figure 4.21 Example of a Routing

A routing consists of one or more operations that process and assemble BOM items into subassemblies and final assemblies. You perform operations at work centers, as previously discussed in Section 4.2.

Each routing contains many fields and indicators relevant to product costing, which we'll now examine in detail, including:

- **Routing header**
  Applies to the entire routing
- **Operation overview**
  Lists all operations
- **Operation details**
  Displays details for each operation

You manage routings with the Manage Routings app (SAP Fiori ID F5425). With this app, you can search through existing routings and navigate to the apps to create new routings and change existing routings.

You can also maintain routings with Transaction CA02 or via the menu path **Logistics • Production • Master Data • Routings • Routings • Standard Routings • Change**. You'll see the selection screen shown in Figure 4.22.

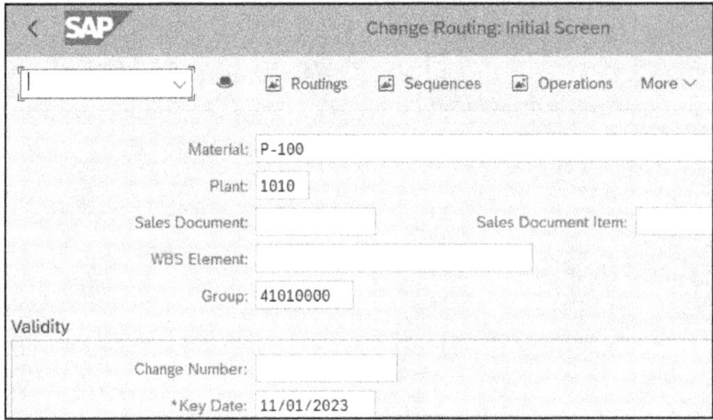

Figure 4.22 Change Routing Initial Screen

Let's examine the routing **Header Details** next.

4.3.2 Header

Type in the **Material** and **Plant** and click the header (hat) icon to display the header shown in Figure 4.23.

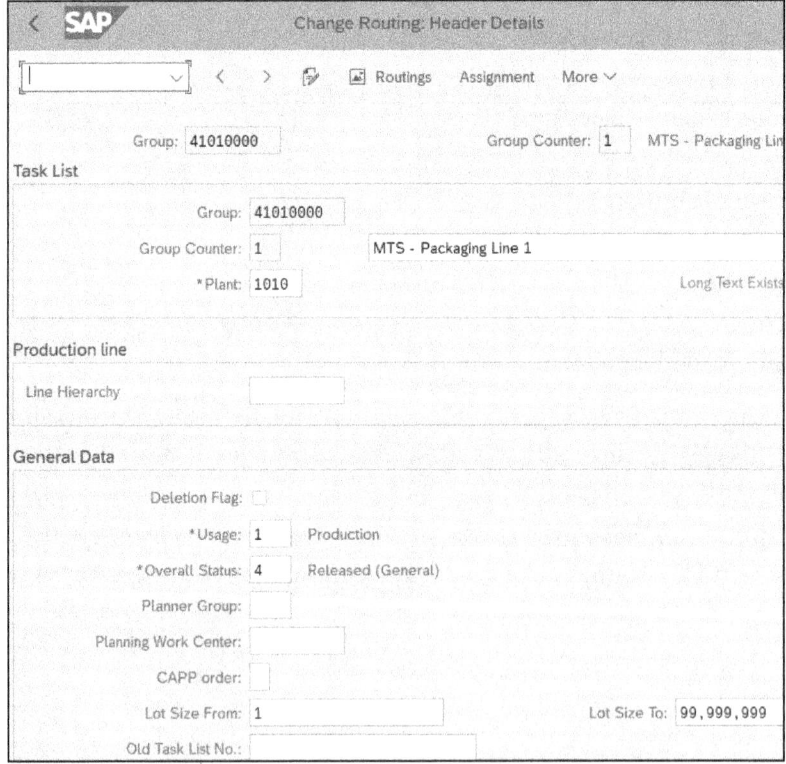

Figure 4.23 Routing Header Details

Let's discuss the relevant fields in this screen:

- **Group**
  In the **Task List** section, the **Group** identifies routings that contain different production steps for the same material.

Group

You have an efficient method of manufacturing a material, but you use a less-efficient method during periods of high demand due to the capacity requirements of in-house production. You create a different task list **Group** for each method of manufacture.

- **Group Counter**
  The **Group Counter** identifies a task list within a task list group. A task list group and group counter uniquely identify a task list.

Group Counter

You use a group counter to identify different lot size ranges to manufacture the same material.

- **Plant**
  This field identifies the **Plant** in which the routing is valid. The system copies the plant to the routing operations as a default value. If you perform the operations in another plant, you can change the default value in the operation. All plants in a routing must belong to the same company code.
- **Usage**
  The task list **Usage** assigns a routing to various work areas, such as production or engineering. This allows you to create several routings to produce one plant material. You can also assign a task list usage to a work order type.

Routing Usage

You can restrict routings of production usage to work orders of a type valid only for production. This avoids the possibility of using development routings in the production process.

  The system uses the task list usage in a production order for automatic task list selection.

- **Overall Status**
  You use **Overall Status** to indicate the processing status of a task list.

Overall Status

You indicate whether the task list is still in the creation or development phase or if it has been released for production with the **Status** field.

- **Assignment**
  Click the **Assignment** (material assignment) button or select the menu path **Routing • Assignment** to display the material assignment as shown in Figure 4.24.

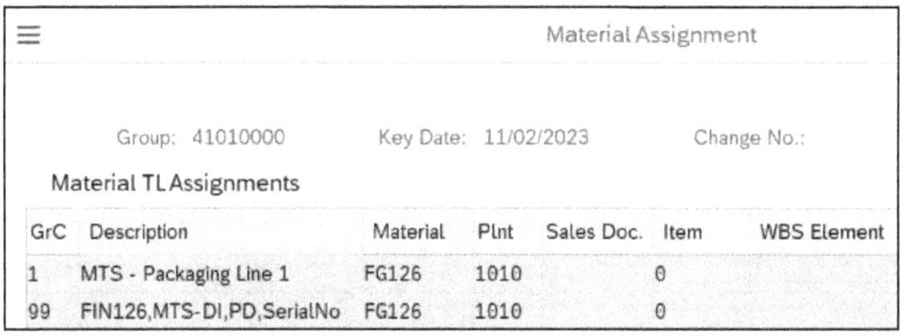

Material Assignment

Group: 41010000 Key Date: 11/02/2023 Change No.:

Material TL Assignments

GrC	Description	Material	Plnt	Sales Doc.	Item	WBS Element
1	MTS - Packaging Line 1	FG126	1010		0	
99	FIN126,MTS-DI,PD,SerialNo	FG126	1010		0	

Figure 4.24 Routing Material Assignment

Material assignment determines which material you produce with a routing. The assignment determines if you can use the routing for material requirements planning (MRP), to create production orders, and for cost estimates for this material.

Routing Material Produced

The material the routing produces determines the manufacturing order credit when you receive the assembly into inventory. The credit is determined by multiplying the goods receipt quantity by the assembly standard price.

Now that we've examined the routing header, let's look at the **Operation Overview**.

4.3.3 Operation Overview

Click the **Operations** button shown earlier in Figure 4.22 or select the menu path **Goto • Operation Overview** to display the operation overview in Figure 4.25.

The **Operation Overview** screen displays a list of operations required to manufacture the assembly. Let's examine the **Operation Details** fields next.

Operation Details

Double-click on an **Operation** to display the operation details in Figure 4.26. The values you see in Figure 4.26 defaulted from the **Work center Assembly**. We've already discussed the relevant fields in the **Operation** section when we looked at work center default values in Section 4.2.3. Now let's examine the **Standard Values** section of Figure 4.26.

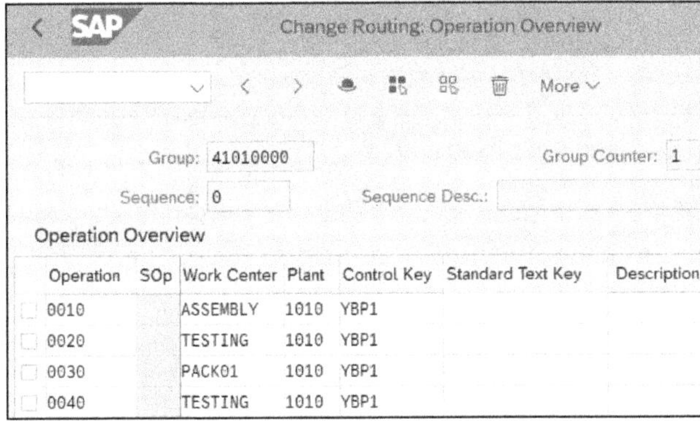

Figure 4.25 Routing Operation Overview

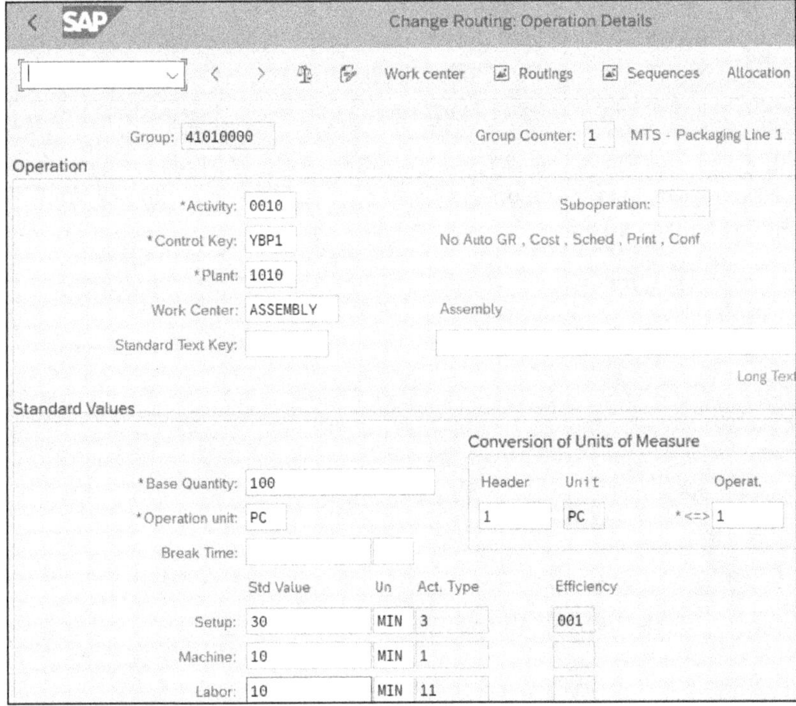

Figure 4.26 Routing Operation Details

Base Quantity

The **Std Value** (standard values) of the operation refer to the **Base Quantity** of the material you are producing. You can increase the accuracy of the standard values by increasing the base quantity. This is similar in concept to increasing the price unit to increase the accuracy of a standard or moving average price.

Operation Unit of Measure

The **Operation unit** of measure used in the operation for the material you are producing is valid for the following:

- Base quantity
- Minimum lot size
- Minimum send-ahead quantity when operations overlap

If you have already maintained a unit of measure in the header, this defaults to the operation unit of measure. You can change the unit of measure in the operation.

Break Time

This is the **Break Time** available to an employee during the operation.

Standard Value

The **Std Value** is how long it typically takes to perform each activity. The standard value multiplied by the planned activity rate of the **Act. Type** (activity type) provides planned labor and overhead costs. You can modify the calculation by a performance efficiency rate and a formula. You roll labor costs up from subassemblies to the finished good.

Now that we've considered the **Operation** and **Standard Values** sections in Figure 4.26, scroll down the **Operation Details** screen until you reach the **General data** section, as shown in Figure 4.27.

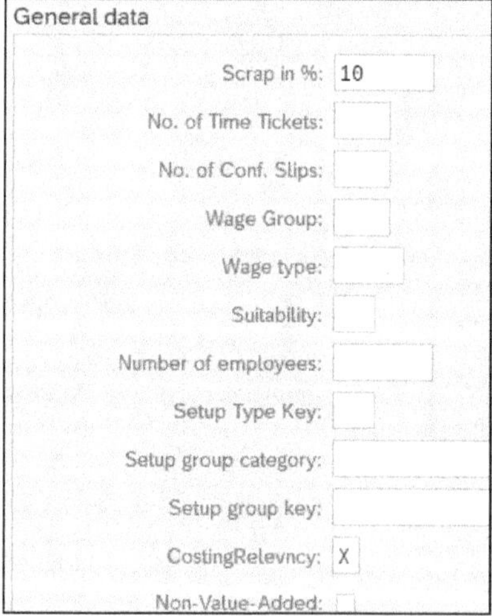

Figure 4.27 General Data Section of Operation Details

Let's discuss the relevant fields in the **General data** section of the operation details.

Operation Scrap

Scrap in % (operation scrap) decreases the quantity in the next operation because operation scrap reduces the quantity to be processed. You take the reduction in quantity into account in scheduling and costing.

Costing Relevancy

CostingRelevncy determines whether the item is included in the standard cost estimate planned costs and actual costs for a production order.

> **Items Not Relevant for Costing**
>
> Operation items are generally relevant for costing, although you can indicate certain items, such as documents and text items, as not relevant for costing.

Non-Value-Added

Having the **Non-Value-Added** checkbox selected flags an operation as not part of the value-added chain. The system then considers the line balance, or the reconciliation between time capacity and time requirements, when evaluating the operations in a production line. For example, you can display the total non-value-added processing time.

Now that we've considered the **General data** section in Figure 4.27, scroll down the **Operation Details** until you reach **External processing**, as shown in Figure 4.28.

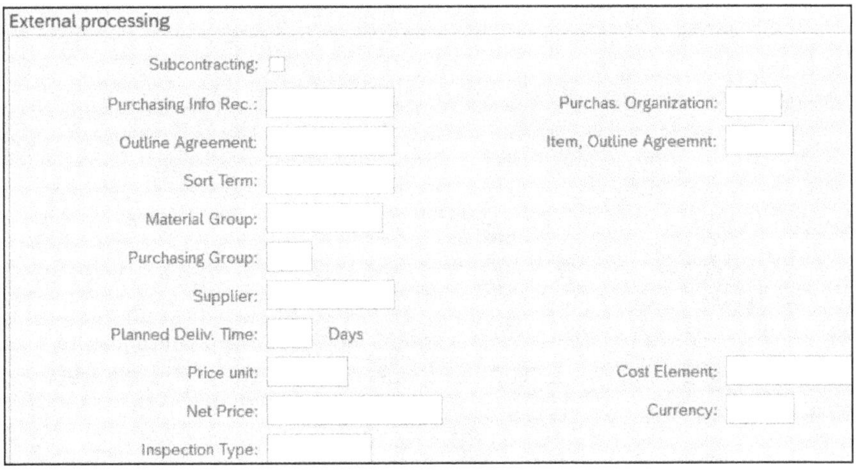

External processing

Subcontracting:
Purchasing Info Rec.:
Purchas. Organization:
Outline Agreement:
Item, Outline Agreemnt:
Sort Term:
Material Group:
Purchasing Group:
Supplier:
Planned Deliv. Time: Days
Price unit:
Cost Element:
Net Price:
Currency:
Inspection Type:

Figure 4.28 External Processing of Operation Details

This is where you enter information about operations performed by an external supplier. The work center control key we discussed in Section 4.2.3 determines the

Operation Details as shown in Figure 4.26. These details determine if an operation is sent to a supplier for external processing. We'll examine the fields in the following sections.

Subcontracting

Selecting the **Subcontracting** checkbox specifies that the operation is subcontracted and that components will be provided to an external supplier. A subcontract purchase order is created for the external supplier.

Purchasing Info Record

You can create a purchasing info record (info record) for the external operation; this contains purchasing information such as supplier and price. If you enter an info record number in the **Purchasing Info Rec.** field and press [Enter], the relevant information defaults from the info record to the **External processing** section of the operation. This defaulted information isn't changeable. If you need to change the information, you first must delete the info record number from the operation.

> **Info Records for External Processing**
>
> Info records for external processing operations aren't associated with a material because you are purchasing a service or activity and not a physical material. To search for external processing info records, left-click in the **Purchasing Info Rec.** field in Figure 4.28, press [F4], and scroll across to **Info Records for External Processing**.

Supplier

Enter the **Supplier** that will carry out the external processing.

Cost Element

Enter the **Cost Element** to identify external processing costs in cost reports.

> **External Processing Cost Elements**
>
> Cost center report S_ALR_87013611—Cost Centers: Actual Plan/Variance identifies costs separately with external processing cost elements that you enter in the **Cost Element** field in Figure 4.28.

Price Unit

The **Price Unit** represents the number of units to which the price refers. You can increase the accuracy of the price by increasing the price unit. To determine the unit price, divide the price by the price unit.

Net Price

In the **Net Price** field, you enter the external processing costs payable to the supplier. You can only maintain the **Net Price** if you have not maintained the purchasing info record conditions. You must then maintain the price on the **Conditions** screen. If you have not maintained the supplier's net price for the material in the purchasing info record, the system proposes the net price of the last purchase order issued to the supplier when you create a new purchase order.

Component Allocation

Click the **Allocation** (Component Allocation) button shown in Figure 4.26 or select the menu path **Goto • Comp. Alloc – Op** to display the component overview in Figure 4.29.

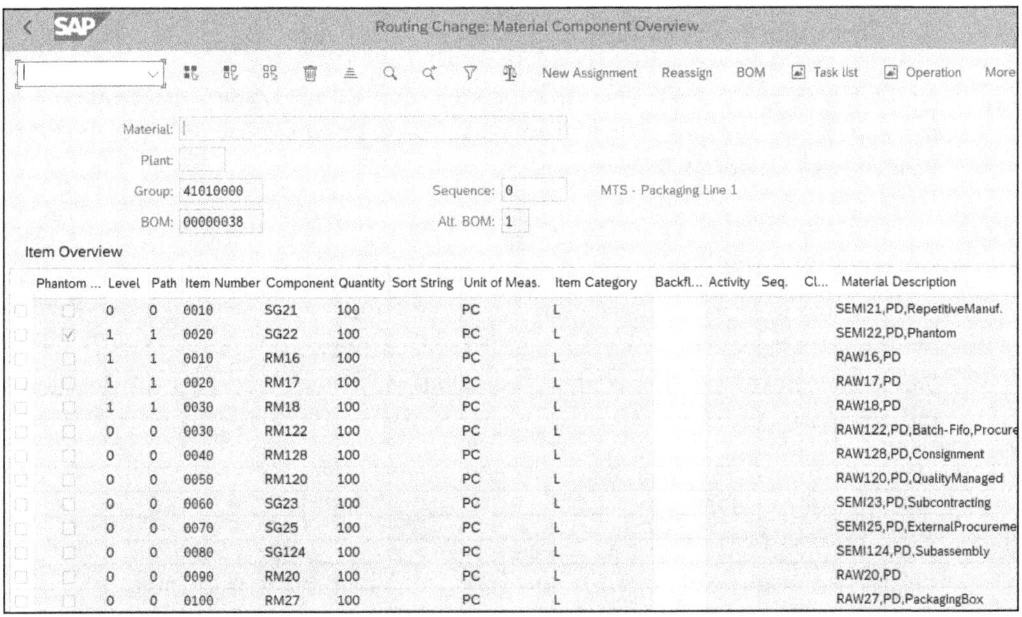

Phantom ...	Level	Path	Item Number	Component	Quantity	Sort String	Unit of Meas.	Item Category	Backfl...	Activity	Seq.	Cl...	Material Description
☐	0	0	0010	SG21	100		PC	L					SEMI21,PD,RepetitiveManuf.
☑	1	1	0020	SG22	100		PC	L					SEMI22,PD,Phantom
☐	1	1	0010	RM16	100		PC	L					RAW16,PD
☐	1	1	0020	RM17	100		PC	L					RAW17,PD
☐	1	1	0030	RM18	100		PC	L					RAW18,PD
☐	0	0	0030	RM122	100		PC	L					RAW122,PD,Batch-Fifo,Procure
☐	0	0	0040	RM128	100		PC	L					RAW128,PD,Consignment
☐	0	0	0050	RM120	100		PC	L					RAW120,PD,QualityManaged
☐	0	0	0060	SG23	100		PC	L					SEMI23,PD,Subcontracting
☐	0	0	0070	SG25	100		PC	L					SEMI25,PD,ExternalProcureme
☐	0	0	0080	SG124	100		PC	L					SEMI124,PD,Subassembly
☐	0	0	0090	RM20	100		PC	L					RAW20,PD
☐	0	0	0100	RM27	100		PC	L					RAW27,PD,PackagingBox

Figure 4.29 Material Component Overview

You assign material components from a BOM to individual operations in the routing in the **Activity** (operation/activity) column. Material components in a BOM not assigned to an operation in the routing are automatically assigned to the first operation when you create a manufacturing order. This is particularly important to ensure that WIP and scrap are calculated correctly if material components are used in the later operations rather than being assigned at the start of manufacturing. It is also important if you use the SAP Fiori app Analyze Costs by Work Center/Operation (SAP Fiori ID F3331) because this shows the material and activity costs per operation and will give an incorrect impression if all material components are shown as belonging to the first operation by default.

You can assign material components from several BOMs or alternative BOMs to a routing. When you create a manufacturing order, you select the BOM and material components assigned to it.

Components Debit the Production Order

When removed from inventory, components debit the manufacturing order. The debit value is determined by multiplying the goods issue quantity by the component price.

Now that we've examined BOMs, work centers, and routings in detail, we'll examine product cost collectors in the next section.

4.4 Product Cost Collector

A *product cost collector (PC)* collects actual costs during the production of an assembly. In the following sections, we'll explain what a product cost collector is and the fields included in the associated master data.

4.4.1 Basic Concepts

Product cost collectors are necessary for repetitive manufacturing and optional for order-related manufacturing. PCCs are the main cost object instead of manufacturing orders when linked for order-related production. You analyze costs per period as shown in Figure 4.30.

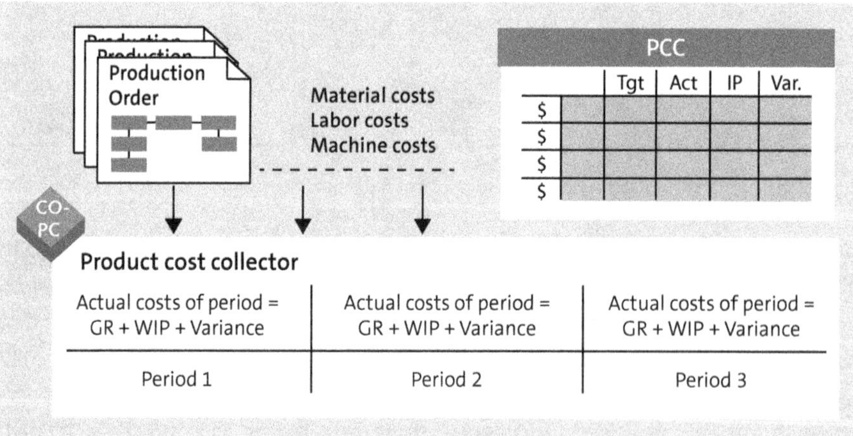

Figure 4.30 Product Cost Collector Analysis per Period

As you post activity confirmations and goods movements to a production order, the system sends the costs automatically to the assigned product cost collector. You analyze product cost collector costs target versus actual with standard report KKBC_PKO.

You perform period-end processing on the PCC. The period costs are divided into the following:

- Actual costs for components issued to the production order and activities confirmed for operations.
- Credits for goods receipts (GR) from the production order into inventory.
- Target costs for work in process (WIP) based on the preliminary cost estimate.
- Variance calculated with the formula *variance = actual period costs – GR – WIP*.

Product cost collectors have several advantages, including:

- Improved period-end closing and reporting performance, since many production orders are assigned to one product cost collector. You view costs for many manufacturing orders collectively per period for a PCC.
- You carry out variance analysis based on a finished good instead of for a manufacturing order.
- You collect costs at the product level independently of the production type. Regardless of whether the production environment is order-related, process manufacturing, or repetitive manufacturing, you collect the production costs for the product on a product cost collector and analyze the costs in each period.
- If production orders remain open for multiple periods, variance reconciliation is easier with product cost collectors.

You can analyze PCC costs with the Product Cost Collector Details Event-Based app (SAP Fiori ID F 6965). This app shows a breakdown of event-based cost postings for repetitive manufacturing orders (order category 05—product cost collectors). It provides a detailed cost comparison of actual and target costs for product cost collectors running with the Event-Based Production Cost Posting solution (3FO).

You maintain PCCs with Transaction KKF6N or via the menu path **Accounting • Controlling • Product Cost Controlling • Cost Object Controlling • Product Cost by Period • Master Data • Product Cost Collector • Edit**. You see a selection screen, as shown in Figure 4.31.

A PCC contains all the information needed to manufacture a product, including fields relevant to cost and variance analysis. To display these fields, follow these steps:

1. Type in the **Material** and **Plant**.
2. Select the **Production Version** checkbox on the left below the plant **SMCO**.
3. Press Enter to display the details of the product cost collector on the right.

We'll now discuss the relevant fields, beginning with the **Data** tab.

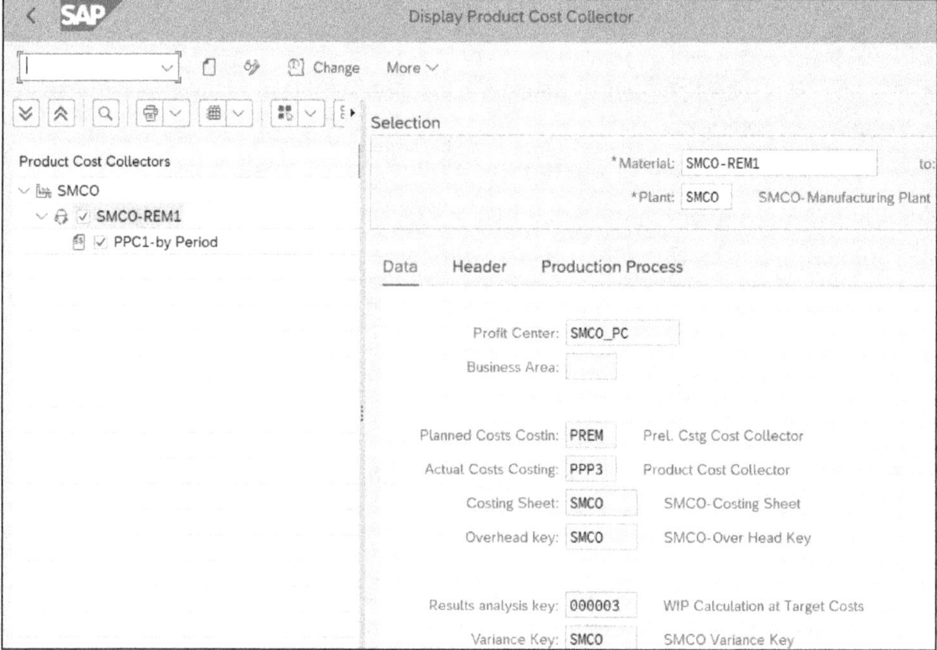

Figure 4.31 Display Product Cost Collector Data

4.4.2 Data

The **Data** tab contains the following fields related to costing, as discussed in the following sections.

Profit Center

A **Profit Center** receives postings in the Universal Journal made in parallel to cost centers and orders. You typically create profit centers based on areas in a company that generate revenue with a responsible manager. You can also access the **Profit Center** field in the **Sales**, **MRP**, and **Plant Data** views.

With profit center accounting active, you'll receive a warning message if you don't specify a profit center during a posting, and all unassigned entries will post to a dummy profit center. You activate profit center accounting with configuration Transaction OKKP, which maintains the controlling area. Profit center accounting using a separate ledger rather than as part of the Universal Journal may no longer be supported from 2025.

Business Area

A **Business Area** is an organizational unit representing a separate area of operations or responsibilities. Financial statements for business areas can be created and used for internal reporting.

> **Profit Centers**
>
> Business areas precede profit centers, which provide more flexibility for organizational changes. Standard and alternate profit center hierarchies aren't available for business areas.

Costing Variant for Planned Costs

A costing variant determines how costs are calculated. The **planned Costs Costin** (costing variant planned) determines the preliminary cost estimate. When compared with actual costs determined by the actual costing variant discussed in the next section, it allows you to carry out production variance analysis.

You define the planned costing variant with configuration Transaction OKKN or via the IMG menu path **Controlling • Product Cost Controlling • Cost Object Controlling • Product Cost by Period • Product Cost Collectors • Check Costing Variants for Product Cost Collectors • Costing Variants to Determine Activity Quantities.** You'll see the screen shown in Figure 4.32.

Costing Variant	Name
PPC1	Standard Cost Est. (Mat.)
PPC2	Mod. Std Cost Est. (Mat.)
PPC3	Current Cost Est. (Mat.)
PREM	Prel. Cstg Cost Collector

Figure 4.32 Planned Costing Variant Overview Screen

A list of available planned costing variants is shown in the **Costing Variant** column. Double-click the standard SAP costing variant **PREM** and then click the **Valuation Variant** button to display the screen shown in Figure 4.33.

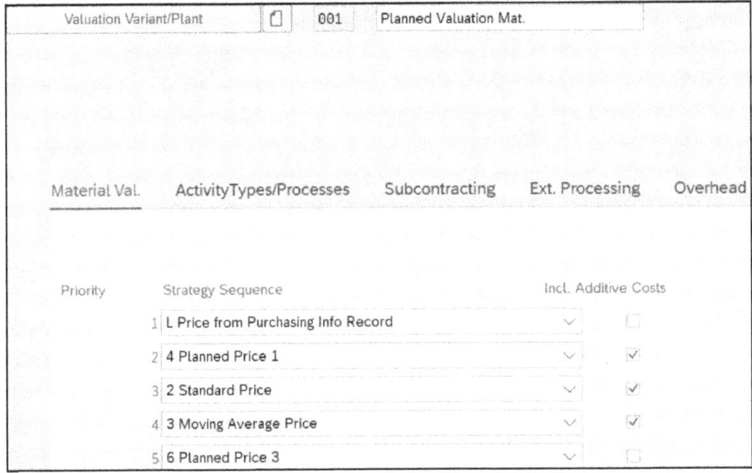

Figure 4.33 Planned Valuation Variant

A planned valuation variant first searches for a component price from a purchasing info record price. You can base the preliminary cost estimate on the most recent vendor quotation, if available, for the material component.

The planned valuation variant also determines which costing sheet you use for preliminary costing in the **Overhead** tab.

Costing Variant for Actual Costs

Actual Costs Costing as shown in Figure 4.31 calculates actual costs by determining the following:

- What activity prices are used to evaluate confirmed internal activities
- Which costing sheet is proposed for calculating overhead costs in period-end closing

You define the actual costing variant with configuration Transaction OPL1 or by following the IMG menu path **Controlling • Product Cost Controlling • Cost Object Controlling • Product Cost by Period • Product Cost Collectors • Check Costing Variants for Product Cost Collectors • Costing Variants for Valuation of Internal Activities**. The screen in Figure 4.34 displays.

Costing Variant	Name
PPP1	Production Order: Planned
PPP2	Production Order: Actual
PPP3	Product Cost Collector

Figure 4.34 Actual Costing Variant Overview Screen

A list of available actual costing variants is displayed in the **Costing Variant** column. Double-click the standard SAP costing variant **PPP3** and then click the **Valuation Variant** button to display the screen shown in Figure 4.35.

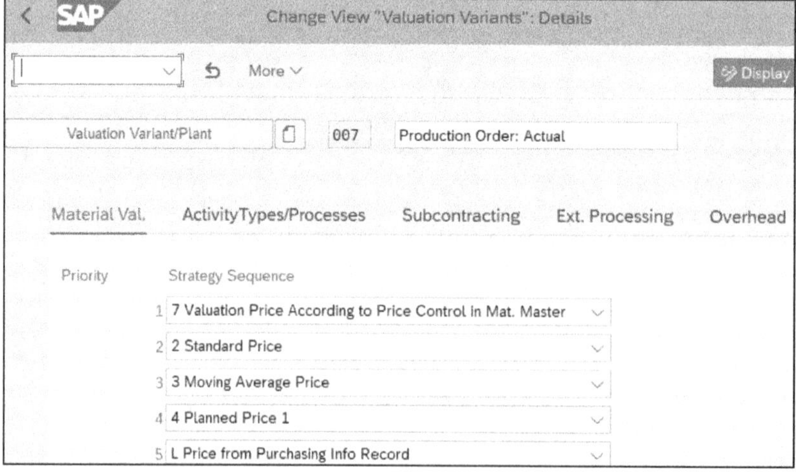

Figure 4.35 Actual Valuation Variant

The actual valuation variant first searches for a material valuation price according to the price control in the **Costing 2** view. If you have the price control set to the moving average price for your material components, this valuation variant will result in a more recent component price than determined by the purchasing info record from the planned valuation variant.

Costing Sheet

The **Costing Sheet** configuration contains the rules for allocating overhead and is defaulted from the valuation variant stored in the costing variant. The costing sheet is contained in the **Overhead** tab of the valuation variant shown in Figure 4.35.

Order Type

The costing variant is stored as a default value according to the order type and plant combination that you define with configuration Transaction OKZ3 or by following the IMG menu path **Controlling • Product Cost Controlling • Cost Object Controlling • Product Cost by Period • Product Cost Collectors • Define Cost-Accounting-Relevant Default Values for Order Types and Plants.** Double-click a plant and order type combination to display the screen shown in Figure 4.36. The values in this screen default into the relevant fields when you create product cost collectors for plant **001** of order type **RM01**, for example.

Now we'll discuss the next fields in the **Data** tab of the product cost collector shown previously in Figure 4.31.

Overhead Key

You use an **Overhead key** to apply different overhead percentages to individual orders or to groups of orders. You assign the overhead key in the overhead rate component of a costing sheet, as we'll discuss in Chapter 5.

Results Analysis Key

Each PCC or order you want to recognize and post a WIP for must contain an **RA Key.** A results analysis key ensures that the PCC or order is included in the WIP calculation during period-end closing.

You can specify the results analysis key (**RA Key**) as a default value for each plant and order type, as shown in Figure 4.36.

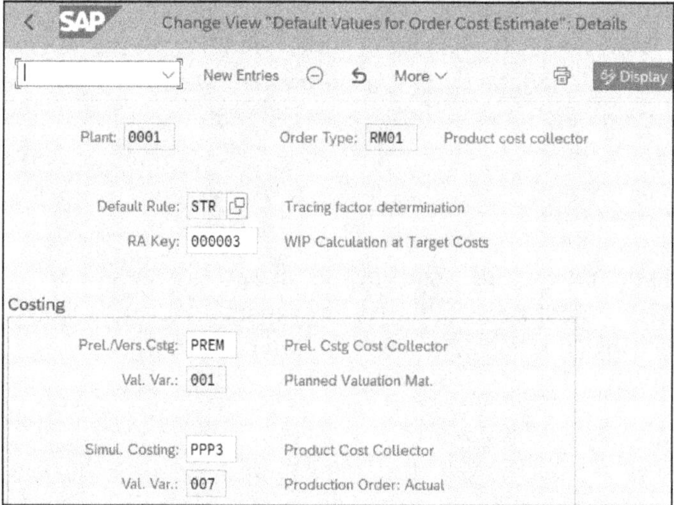

Figure 4.36 Default Values for Order Types and Plants

Variance Key

You only calculate variances on manufacturing orders or product cost collectors containing a **Variance Key**, which is defaulted from the **Costing 1** view when you create manufacturing orders or PCCs. The variance key also determines if you subtract the scrap value from actual costs before determining variances. You can see the variance key in the data tab shown previously in Figure 4.31.

Now that we've discussed the fields in the PCC's **Data** tab, let's examine the fields in the next tab.

4.4.3 Header

Select the **Header** tab shown previously in Figure 4.31 and then click the pencil-and-glasses icon to display the screen shown in Figure 4.37.

The **Header** tab contains several fields relevant to product costing, which we'll now examine.

Description

When you create a product cost collector, the description defaults from the production version text. You can change this description at any time. This can help you identify PCCs in information system reports.

Company Code

You define a PCC by plant, material, and production version. You assign a plant to a company code when initially setting up your organization with configuration Transaction

OX18. The **Company Code** is automatically determined from the plant and isn't changeable.

Material

Because you can enter a range of materials in the **Selection** section of the screen in Figure 4.37, the **Material** field in the **Header** tab represents the material(s) selected in the left side of the screen.

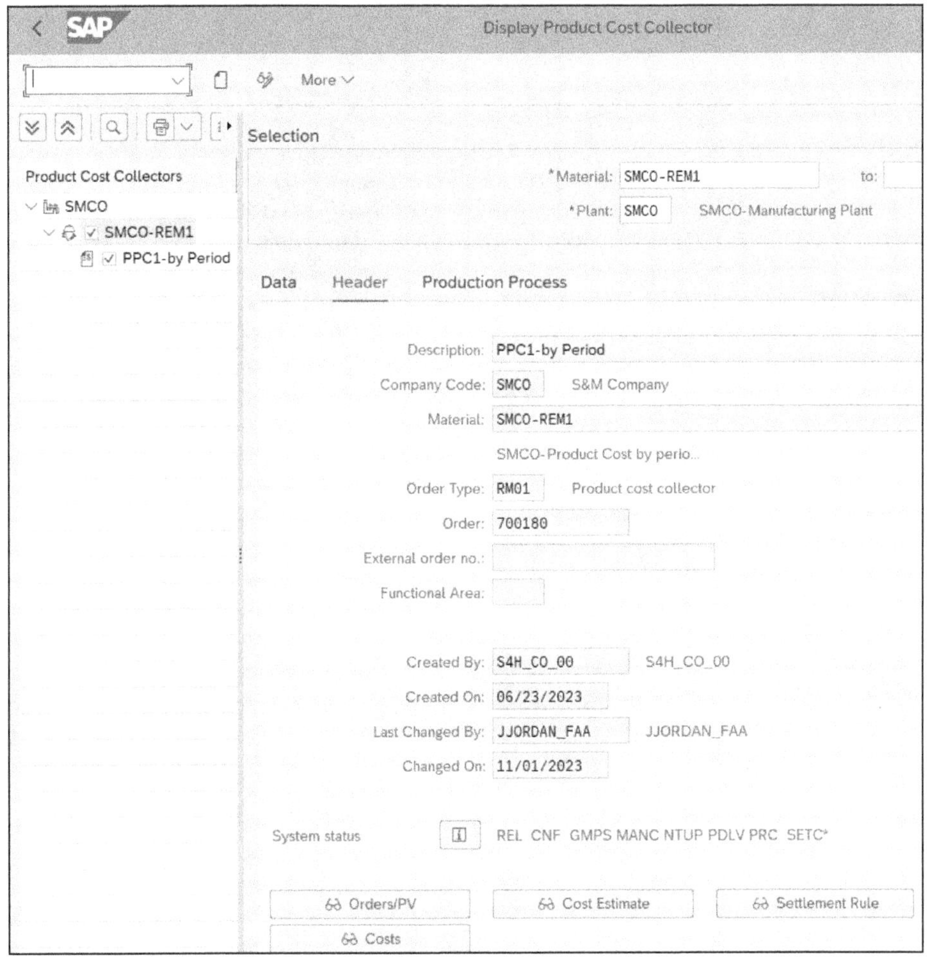

Figure 4.37 Product Cost Collector Header

Order Type

When you initially create a PCC by clicking the **New Page** icon in Figure 4.37, the **Order Type** field is mandatory, as shown by the asterisk next to **Order Type** in Figure 4.38.

Click in the **Order Type** field and press F4 to display a list of possible PCC order types, as shown in Figure 4.38. After you create a PCC, the order type isn't changeable and is grayed out, as shown in Figure 4.37.

Figure 4.38 Create Product Cost Collector

The order type differentiates orders according to number range, costing variants, and settlement profile.

Controlling Level and Production Version

The **Controlling level** for the material determines the characteristics of the production process that you must enter when you create a PCC. In Figure 4.38, the selected **Production Version** radio button in the **Controlling level for material** section means you must enter the production version characteristic when creating the product cost collector. You always use this Controlling level when using production versions or repetitive manufacturing.

If you have already defined a more detailed Controlling level for the plant material, that detailed Controlling level defaults when you create a new product cost collector. Because we've previously created a product cost collector with the most detailed product version Controlling level of the production version, this is the only available Controlling level in Figure 4.38.

External Order Number

You can enter any number in the **External order no.** field shown in Figure 4.37 to easily identify your product cost collectors. You can use this number as a selection criterion during collective processing.

Functional Area

The **Functional Area** field in Figure 4.37 refers to the part of cost of sales (COS) accounting that compares sales revenue with the manufacturing costs of an activity for a given accounting period. You assign expenses posted to the product cost collector to the functional area. The following are typical examples of function areas:

- Research and development
- General and administration
- Sales and distribution
- COS

You report expenses and revenues you cannot assign to functional areas in other profit and loss items sorted according to expense and revenue type. You can enter or change the functional area, provided no postings exist for the product cost collector.

Created By, Created On, Last Changed By, and Changed On

These four fields, which aren't changeable, contain user and date information for the product cost collector creation and last change.

System Status

Click the **Information** icon in Figure 4.37 to the right of **System status** to display detailed status information as shown in Figure 4.39.

The product cost collector **Status** section lists processing steps carried out on the product cost collector. For example, the **REL** checkbox selects automatically when you release the order. The **Syst. Status** section determines which business processes you allow at each stage of the product cost collector processing.

> **Production Order Status PCC**
>
> The status of **PCC** for a manufacturing order identifies that a PCC is used to collect costs. We discuss displaying a list of manufacturing orders connected to a product cost collector in the next section.

Orders/Production Version

Clicking the **Orders/PV** (orders/production version) button in Figure 4.37 will produce one of the following:

- A list of manufacturing orders linked to the product cost collector
- A production version if this is a repetitive manufacturing scenario

Displaying orders or production versions associated with a specific product cost collector can be quicker with this button than with reports such as COOIS (for production orders) or COOISPI (for process orders) because you don't have to complete a selection screen.

Cost Estimate

Clicking the **Cost Estimate** button shown earlier in Figure 4.37 will display a list of preliminary cost estimates created for the product cost collector, as shown in Figure 4.39.

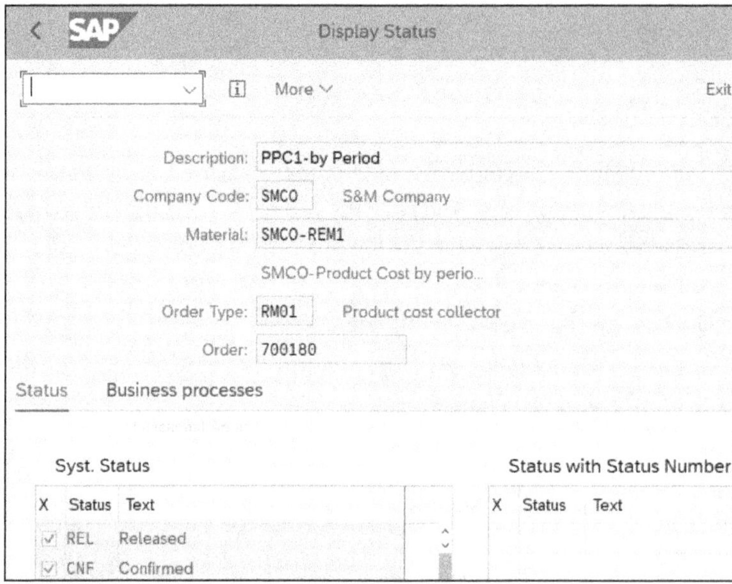

Figure 4.39 Display Status

Double-click a line to display the preliminary cost estimate. You typically want to display the most recent, error-free preliminary cost estimate based on the **Status** and **Costing Date** columns. Click a **Status** and press F4 to display a list of possible entries as shown in Figure 4.40.

Costing status **KA** (**Costed Without Errors**) is required to calculate WIP at target for PCCs. We will discuss WIP in detail in Chapter 15.

To create an individual preliminary cost estimate, click the **Costs** button in Figure 4.37 after ensuring that the product cost collector is in change mode by clicking the glasses-and-pencil icon. This icon toggles between the change and display modes for the product cost collector.

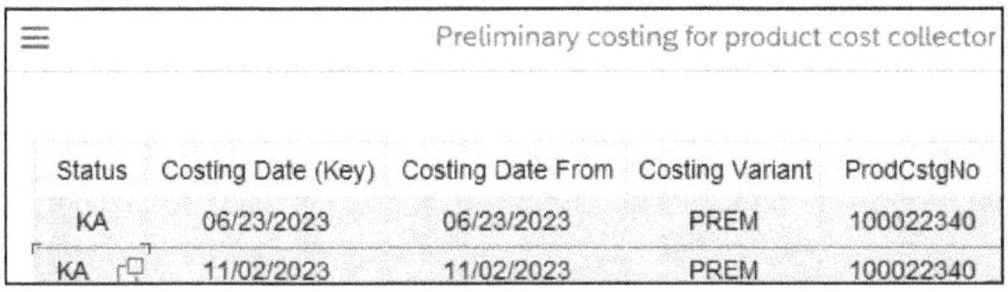

Status	Costing Date (Key)	Costing Date From	Costing Variant	ProdCstgNo
KA	06/23/2023	06/23/2023	PREM	100022340
KA	11/02/2023	11/02/2023	PREM	100022340

Figure 4.40 List of Preliminary Cost Estimates for the Product Cost Collector

You can create preliminary cost estimates collectively with Transaction MF30 or by following the menu path **Accounting • Controlling • Product Cost Controlling • Cost Object Controlling • Product Cost by Period • Planning • Preliminary Costing for Product Cost Collectors**.

Settlement Rule

A **Settlement Rule** determines which portions of product cost collector costs you allocate to receivers. You typically settle a PCC to a material. The valuation class in the **Costing 2** view determines which general ledger accounts are settled to via automatic accounting assignment configuration Transaction OBYC or the IMG menu path **Materials Management • Valuation and Account Assignment • Account Determination • Account Determination Without Wizard • Configure Automatic Postings**.

Costs

Click the **Costs** button in Figure 4.37 to display a detailed report on cumulative costs for a PCC. Detailed reports provide cost element information by row, and target costs, actual costs, and variance by column. You can group the cost element rows by similar business transactions, such as confirmations, goods issues, and goods receipts in the report. You can also access detailed reports in the information system with Transaction KKBC_PKO.

Now that we've discussed the fields and buttons in the **Header** tab, let's examine the fields in the **Production Process** tab.

4.4.4 Production Process

Select the **Production Process** tab shown earlier in Figure 4.37 to display the screen shown in Figure 4.41.

Product cost collectors are created with reference to a production process that describes how a material is produced. The production process has characteristics whose values are unique to that production process. You specify which characteristics to update for the production process via the Controlling level.

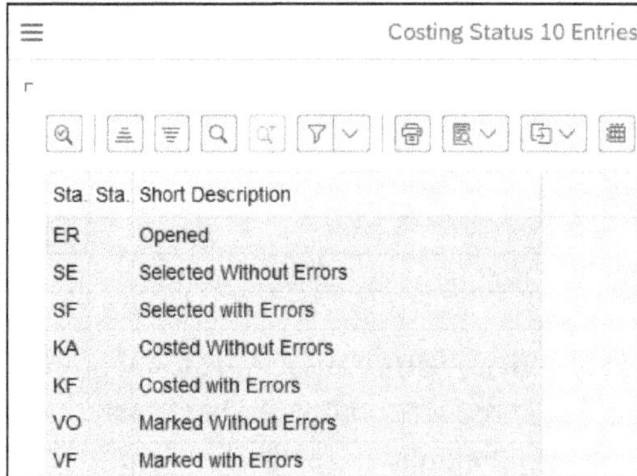

Figure 4.41 Preliminary Cost Estimate Status Possible Entries

The Controlling level you select when you create a product cost collector determines the level at which the costs are collected. You can select the Controlling level by choosing from:

- Production version
- BOM and routing combination
- Material and plant combination

Let's consider the fields in the **Production Process** tab:

- **Description**
  The system creates the name (first field) of a production process automatically from a combination of characteristic attributes defined by the Controlling level. You can manually change the description.
- **Production Process No. (production process number)**
  The system assigns the **Production Process No.** automatically during the creation of the PCC. You cannot change this number.
- **Planning plant**
  A goods receipt takes place for the manufactured material in the **Planning plant**. If the planning and production plants are identical, you don't need to enter the planning plant. The system copies the production plant automatically when you create the product cost collector as shown in Figure 4.38.
- **Production Version**
  The **Production Version** describes the types of production techniques that you can use for a material in a plant and is a unique combination of BOM, routing, and production line. You enter the production version when you first create the PCC.
- **Costing Lot Size**
  The system bases the preliminary cost estimate calculations on the **Costing Lot Size** quantity. If you leave this field blank, the system uses the costing lot size from the **Costing 1** view during the cost estimate calculation.

If you enter a costing lot size in this field, you cannot change the value after you save the PCC. You can, however, change it with Transaction CK91N or by following the menu path **Accounting • Controlling • Product Cost Controlling • Product Cost Planning • Material Costing • Master Data for Mixed Cost Estimate • Edit Procurement Alternatives**. Enter the material and plant and click the glasses-and-pencil icon to display the screen shown in Figure 4.42.

Make an entry in the **Costing Lot Size** field and save the procurement alternative to change the costing lot size entry in the PCC.

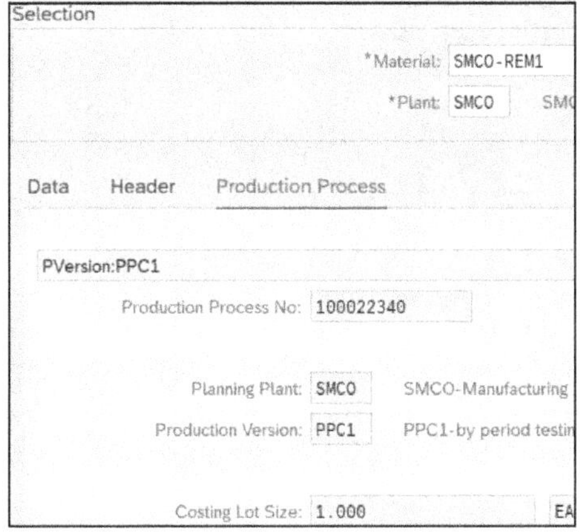

Figure 4.42 Product Cost Collector Production Process

Now that we've looked at the fields in PCCs, let's see how to display a list of PCCs.

4.4.5 List Product Cost Collectors

You display a list of product cost collectors with Transaction S_ALR_87013127 or by following the menu path **Accounting • Controlling • Product Cost Controlling • Cost Object Controlling • Product Cost by Period • Information System • Reports for Product Cost by Period • Object List • Order Selection**. The screen in Figure 4.43 displays.

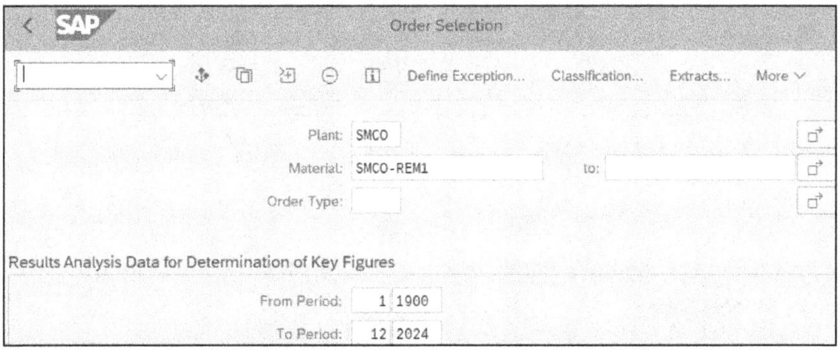

Figure 4.43 Order Selection Initial Screen

Many more selection fields are available than are shown in this initial screen. Click the plus sign icon to display all selection fields, as shown in Figure 4.44.

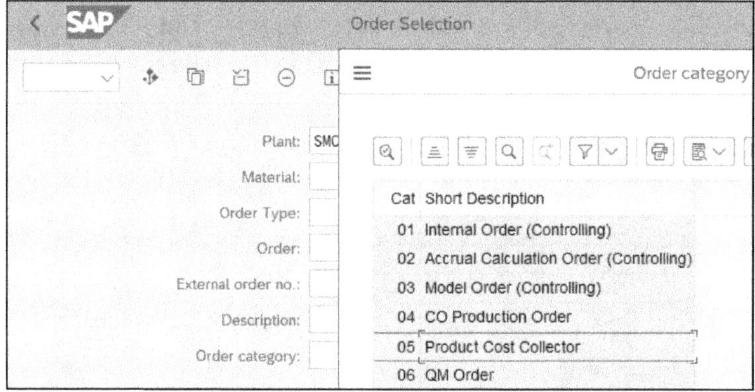

Figure 4.44 Order Selection: All Fields

By selecting orders with **Order category** of **05 Product Cost Collector,** you can restrict the results screen to product cost collectors. You can further restrict the results screen by using the **Order Type** field. After you enter **Order category 05**, the possible entries for the order type on this selection screen are restricted to order types for product cost collectors. You can then enter the order type in the first selection screen shown in Figure 4.43 without expanding the selection list and then save this as a variant.

Limiting the reporting **Period** specifies the periods from which the system reads key figures but doesn't affect the order selection.

Click the **Execute** icon or press F8 to display the results screen shown in Figure 4.45.

Order Selection: Results List

More

Values in Controlling Area Currency USD United States Dollar

Current Data

Order	Material	Plan cost debit	Actual Cost Debit	Crcy	Plan Qty	Act. output qty	Output unit
700180	SMCO-REM1	0.00	3,609.00	USD	0	10	EA
700184	SMCO-REM2	0.00	3,323.97	USD	0	10	EA

Figure 4.45 Order Selection Results Screen

From this report, you can do the following:

- Display a product cost collector directly by clicking an **Order** and selecting the menu path **Extras • Master Data**
- Display a target/actual cost comparison report with cost element detail by double-clicking an order number in the **Order** field

- Continue drilling down through the cost element report to the line item and source document details

You can sort any of the columns in Figure 4.45 to assist with variance analysis.

Now that we've examined BOMs, work centers, routings, and product cost collectors, let's look at purchasing info records in the next section.

4.5 Purchasing Info Record

A purchasing info record stores all the information relevant to procuring material from a supplier. In this section, we'll explain the usage of the purchasing info record for the purposes of product costing and introduce the main fields.

4.5.1 Basic Concepts

A purchasing info record contains a **Purchase Price** field, which the standard cost estimate searches for to determine the purchase price.

You maintain purchasing info records with the Manage Purchasing Info Records app (SAP Fiori ID F 1982). With this app, you can view and manage purchasing info records (info records). You can determine which material can be procured from a supplier at a price for a validity period.

You can also access info records by selecting from a list you display with Transaction ME1M or via the menu path **Logistics • Materials Management • Purchasing • Master Data • Info Record • List Displays • By Material**. The selection screen shown in Figure 4.46 displays.

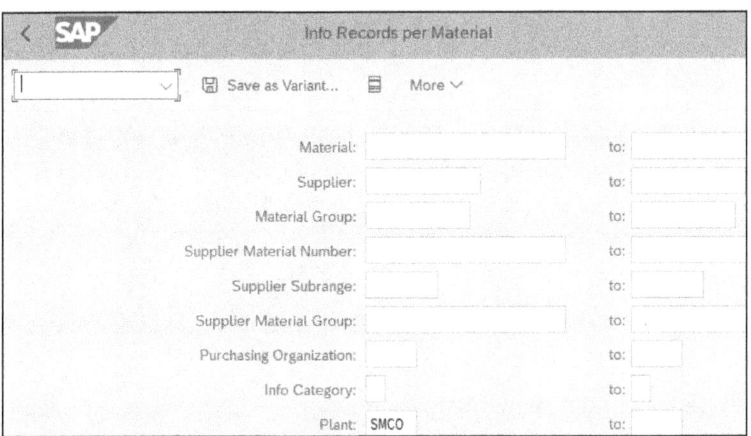

Figure 4.46 Selection Screen for Listing Info Records per Material

Purchasing Info Record Lists

You can list purchasing info records with other criteria, such as by supplier with Transaction ME1L or by material group with Transaction ME1W.

Enter the selection criteria for the purchasing info records you want to display, such as **Material** and **Plant**, and click the **Execute** icon or press F8 to display a list of purchasing info records, as shown in Figure 4.47.

You see purchasing info records listed by **Supplier** for a **Material** in this screen. Purchasing info records contain several sections. The first section of an individual purchasing info record accessed from the list in Figure 4.47 is **Purch. Organization Data 1**.

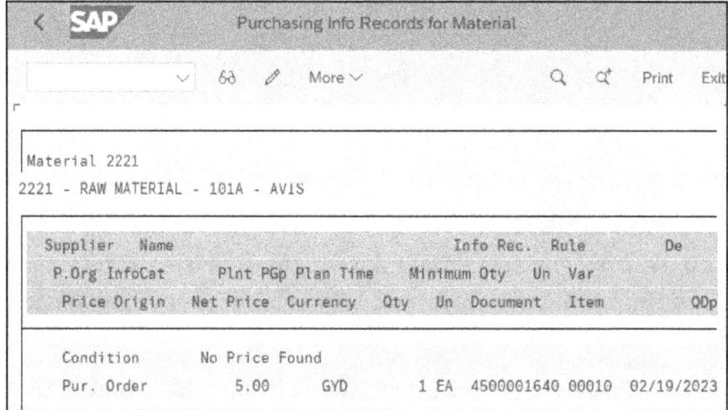

Figure 4.47 List of Info Records for Materials

4.5.2 Purchasing Organization Data 1

Select a purchasing info record in Figure 4.47 and click the pencil icon to display the screen shown in Figure 4.48.

You see purchasing data relevant to a particular purchasing organization for a **Supplier** and **Material** combination. A purchasing organization procures materials and services and negotiates the **Conditions** of purchase with suppliers. Let's discuss the two price fields in detail:

- **Net Price**
  The system calculates the **Net Price** after taking all discounts and surcharges into account. You do one of the following:
  - Calculate and enter the net price manually
  - Enter the gross price, discount, and surcharge conditions, as we'll discuss in Section 4.5.3, and allow the system to calculate the net price

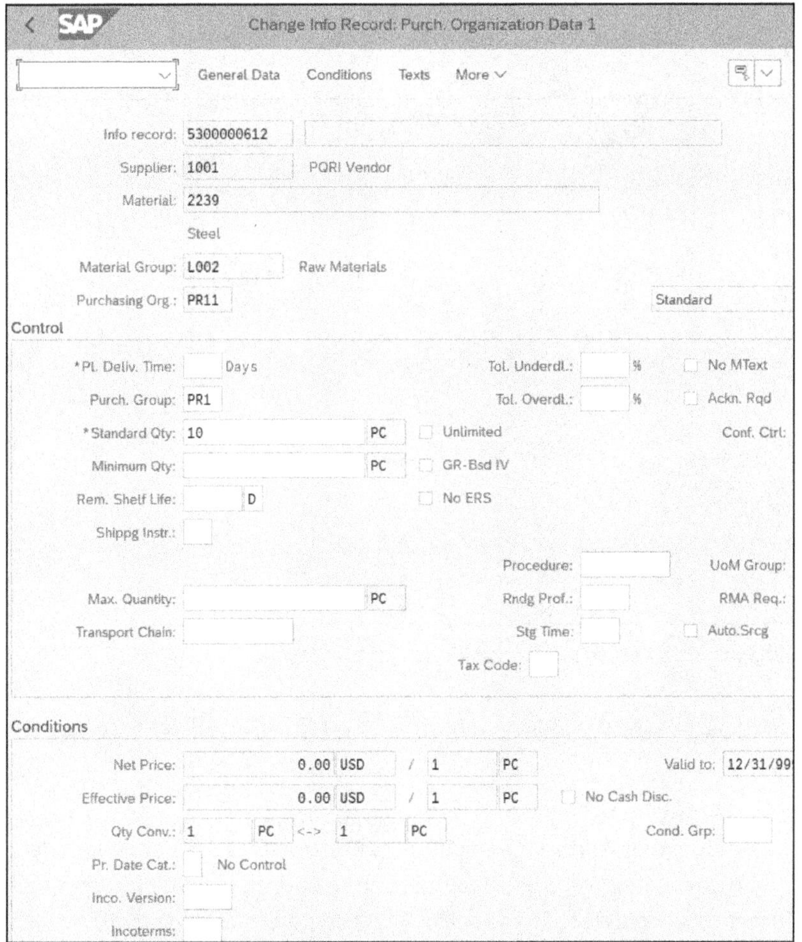

Figure 4.48 Purch. Info Record Organization Data 1

Info Record Net Price

You can only maintain the net price if you have not yet maintained the info record conditions, as we'll discuss in Section 4.5.3. You must then maintain the price on the **Change Gross Price Condition** screen.

- **Effective Price**
  The **Effective Price** is the end price after taking all conditions, such as cash discounts and delivery costs, into account. The system determines this price automatically.

4.5.3 Conditions

Purchasing conditions allow you to record multiple supplier quotations for materials and services and discounts, surcharges, and other supplements in the system. Click the

Conditions button shown at the top of Figure 4.48 to display the conditions screen shown in Figure 4.49.

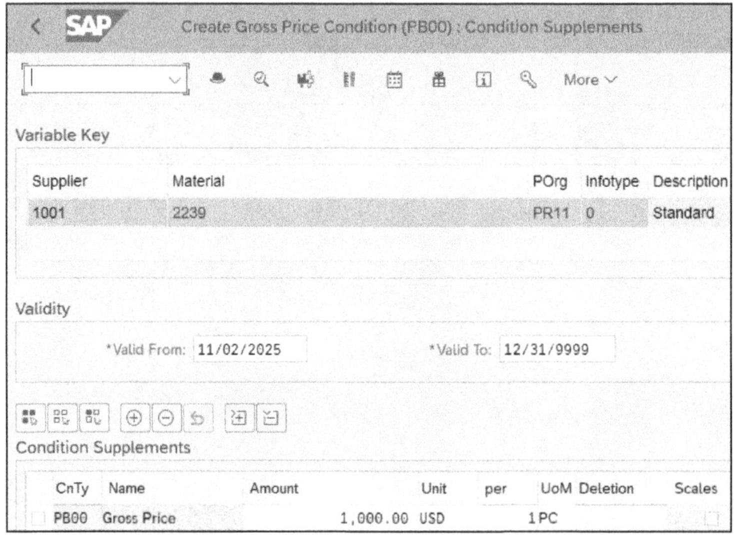

Figure 4.49 Purchasing Info Record Price Conditions

Let's examine each part of the **Gross Price Condition** screen in the following sections.

Variable Key

This section contains **Supplier**, **Material**, **POrg** (purchasing organization), and plant data. The **Infotype** (info record category) column contains one of the following four possible purchasing info record categories that you select when creating the info record:

- **Standard**
  Specifies that the info record is used only for standard purchase orders.
- **Subcontracting**
  Specifies that you use the info record with subcontracting orders. In subcontracting, you supply material parts to an external vendor who manufactures the complete assembly.
- **Pipeline**
  This specifies that the info record is used for pipeline withdrawals. Pipeline materials such as oil or water flow directly into the production process, and stock quantities don't change during withdrawal.
- **Consignment**
  Specifies that the info record is used for consignment withdrawals. The supplier maintains a stock of materials at a customer site. The supplier retains ownership of the materials until they are withdrawn from consignment stores.

Now let's consider the next sections of the **Price Condition** screen.

Validity

Validity dates are info record price conditions that allow you to record previous, present, and future quotation prices. A cost estimate will access a price condition with a validity period corresponding to the valuation date of the cost estimate.

Condition Supplements

The **Gross Price** of **1,000.00 USD per 1 PC** (piece) is the supplier quotation, excluding discounts or surcharges. You can enter discounts and surcharges as additional condition supplements in the lines following the gross price condition. The system will automatically consider any supplements when calculating the net price as discussed previously in Section 4.5.2.

You can also enter **Scales** representing supplier quotations with reduced prices for greater purchase quantities. When the **Scales** checkbox is selected, more than one vendor price is entered based on the quantity ordered. Double-click on the condition line to display details of the scales. The screen shown in Figure 4.50 displays.

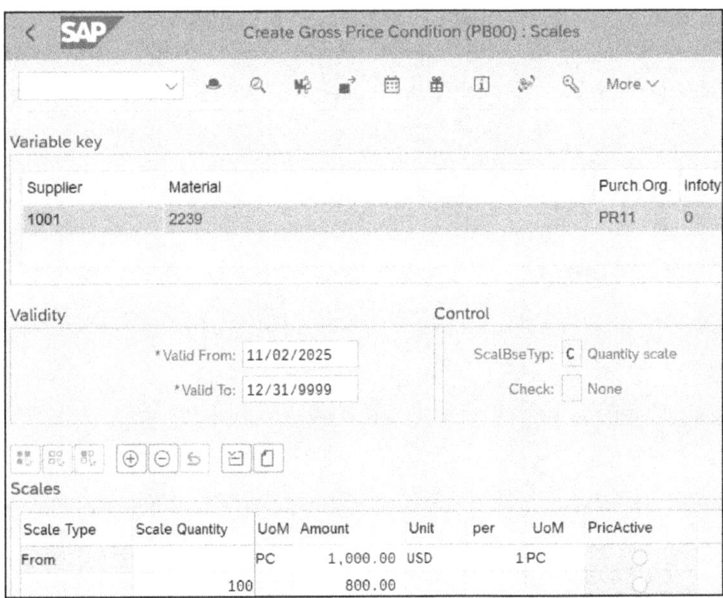

Figure 4.50 Purchasing Info Record Scales

The scale in Figure 4.50 indicates that the purchase price for a quantity between 1 and 99 is 1,000.00 USD, which discounts to 800.00 USD for a **Quantity** of 100 or more. A cost estimate determines which scale price to access based on the costing lot size. During a costing run, the cost estimate costing lot size is determined from the **Costing 1** view.

Costing Lot Size

The costing lot size should be set as close as possible to typical purchase order quantities of components to minimize purchase price variance (PPV) postings because the standard cost estimate calculation is based on the costing lot size.

You can view details of changes to the purchasing info record price resulting from new supplier quotations by clicking the **Validity Periods** (calendar) icon at the top of the screen, in the middle.

4.5.4 Plant-Specific Purchasing Info Records

You can create purchasing info records automatically when you create a purchasing document. With some simple settings, you can specify if the automatically created information records are plant-specific. If you don't, a cost estimate may instead choose the lowest-cost general information record.

You can display scheduling agreements, which arc long-standing purchasing documents, with Transaction ME2M or by following the menu path **Logistics • Materials Management • Purchasing • Purchase Order • List Displays • By Material**. Type in material, plant, the scope of the list, and document type LP, and then execute to display a list of agreements. Double-click a line to display a purchasing schedule. Click in the **Info-Update** field and display a list of possible entries, as shown in Figure 4.51.

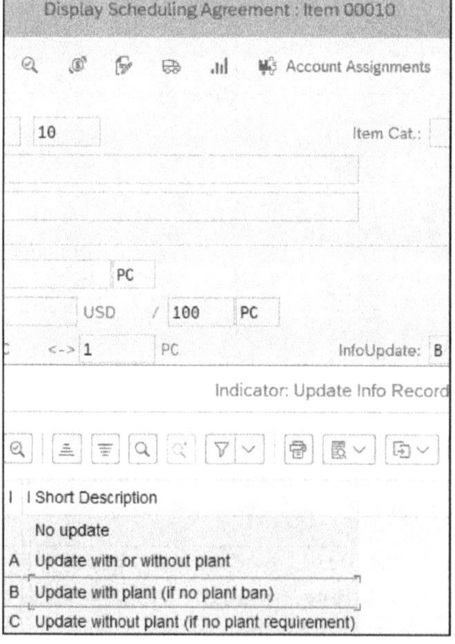

Figure 4.51 Purchasing Schedule Agreement

Select option **B Update with plant** to default to this field and ensure you automatically create plant-specific purchasing information records. To determine if the related purchasing information record is plant-specific, select the menu path **Environment • Info Record** and click the **Purch. Org. Data 1** button to display the screen in Figure 4.52.

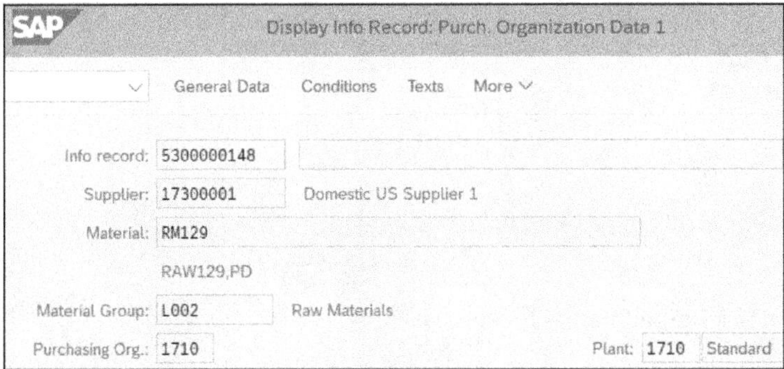

Figure 4.52 Plant-specific Purchasing Info Record

An entry in the **Plant** field means the info record is plant-specific, and a valuation variant will only search within the plant if all purchasing info records are plant-specific.

Follow these two steps to ensure purchasing documents automatically create plant-specific info records:

1. The first step is configuring default values for buyers with Transaction OMFI or via the IMG menu path **Materials Management • Purchasing • Environment Data • Define Default Values for Buyers**. Double-click the text **Settings for Default Values** and either change an existing default values setup or create your own. In the **Info record update** section, click in the **Purchase Order** field and display a list of possible entries, as shown in Figure 4.53.

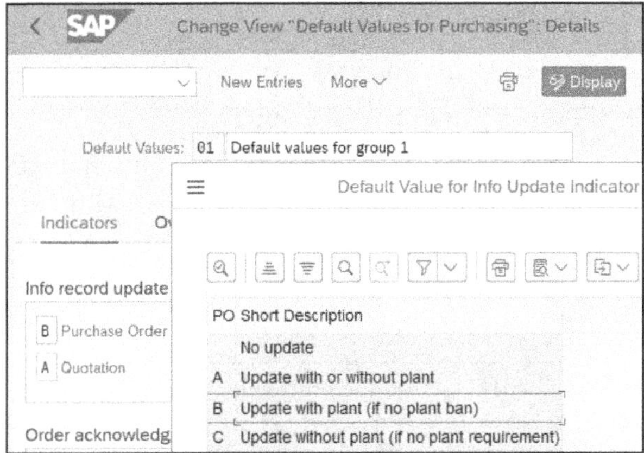

Figure 4.53 Default Value for Buyers: Purchasing Info Record Update

Choose **B** for **Update with plant** and save your entry.

2. The next step is to assign this default value to buyers. You carry out this assignment either as a collective update with Transaction SU10 or by assigning the **Default Values for Purchasing** to your own profile by selecting the menu path **System • User Profile • Own Data** on any screen. Click the **Parameters** tab and enter the **Default Values for Purchasing** profile you created for **Parameter ID EVO**, as shown in Figure 4.54.

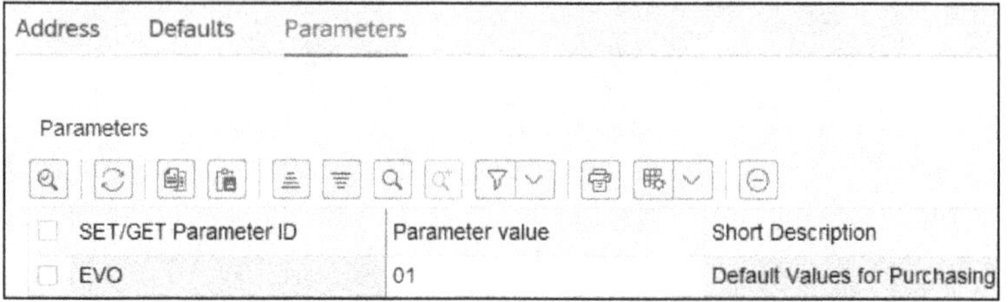

Figure 4.54 Parameter ID for Default Values for Purchasing

While you can set a specific info record update for each scheduling agreement, you can also select the purchase order **InfoUpdate** (info record update) checkbox in the **Material Data** tab, as shown in Figure 4.55.

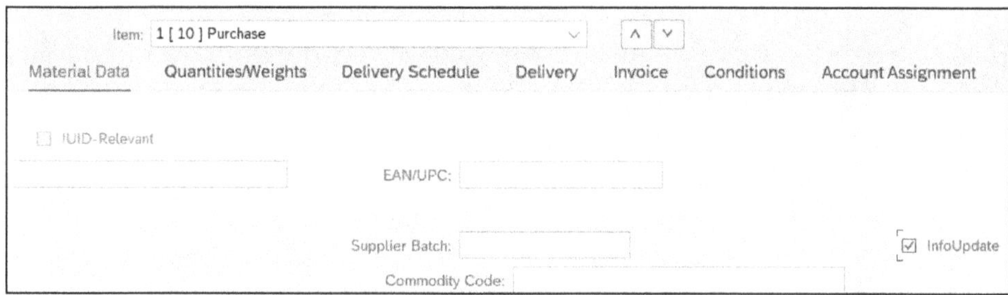

Figure 4.55 Purchasing Info Record Update Indicator in Purchase Order

Selecting the **InfoUpdate** checkbox causes the following:

- If two info records exist, the system updates the info record with a plant.
- If one info record exists, with or without a plant, the system updates it.
- If no info record exists, and you specify **B** as a buyer default, the system creates an info record with a plant. Otherwise, the system creates an info record without a plant.

SAP Note: Info Record Update

SAP Note 569885 contains more details about the purchase order **InfoUpdate** checkbox (EKPO-SPINF) shown in Figure 4.55.

Now that we've examined purchasing info records in detail, let's examine source lists in the next section.

4.6 Source List

A *source list* specifies the allowed material suppliers per plant within a validity period. It shows when a material may be ordered from a supplier or under a long-term purchase agreement. You can specify that you must maintain a source list for a material by selecting the **Source list** checkbox in the **Purchasing** view.

You access a source list by selecting from a list with Transaction MEOM or by following the menu path **Logistics • Materials Management • Purchasing • Master Data • Source List • List Displays • By Material**. You will see the selection screen shown in Figure 4.56.

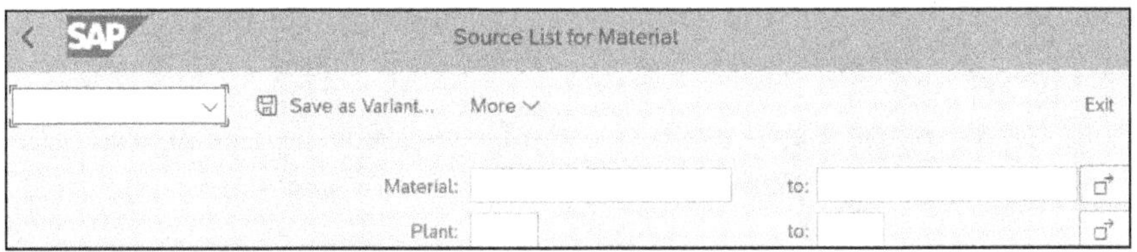

Figure 4.56 Source List for Material Selection Screen

The fields in this screen aren't mandatory. If you aren't sure which materials have source lists, type in a **Plant** and execute to display all source lists for the plant, as shown in Figure 4.57.

Source List for Material

More Print

Material 1000000554 Plnt 1710

Valid from	to	Supplier	Plnt	POrg	OUn	Agmt.	Item	Dis
10/12/2023	10/12/2099	10605		1710				
10/12/2023	10/12/2099	10606		1710				

Figure 4.57 Source List for Material Results List

The allowed sources of supply for **Material 1000000554** in **Plant 1710** are listed. Click the **Edit** (pencil) icon to maintain the source list, as shown in the example in Figure 4.58.

Material: 1000000593 JB,1150KG,WHITE

Plant: 1710 Plant 1 US

Source List Records

Valid from	Valid to	Supplier	POrg	PPl	OUn	Agmt	Item	Fix	Blk
10/19/2023	10/12/2099	17258001	1710				0		
10/19/2023	10/12/2099	17258002	1710				0		
10/19/2023	10/12/2099	17258003	1710				0		

Figure 4.58 Maintain Source List: Overview Screen

Each **Supplier** allowed to supply **Material 1000000593** in **Plant 1710** is listed. The last two columns influence cost estimate results:

- **Fixed source of supply**
  Select the **Fix** checkbox to specify the preferred supplier within a validity period. A cost estimate will first search for a valid info record for a **Supplier** with this checkbox selected, even if there are valid info records available with a lower price. If you don't select any checkboxes, a cost estimate will search for the info record with the lowest price.
- **Blocked source of supply**
  You select the **Blk** checkbox to block a source of supply. A cost estimate will not search for info records associated with a blocked source of supply.

If a cost estimate doesn't retrieve the info record price you expect, checking source list settings can often uncover the reason.

4.7 Summary

In this chapter, we discussed logistics master data in detail. We looked first at BOMs, which contain component and assembly quantities, and then at routings, which contain activity quantities. We considered work centers, where you carry out operations and activities, and how you list work centers in routings. We also looked at purchasing info records, which contain the purchase price field and other purchasing information.

Now that we've considered master data in this and the previous two chapters, we'll examine costing configuration, beginning with the costing sheet in Chapter 5.

Chapter 5
Costing Sheets

Costing sheet configuration determines how cost estimates calculate overhead costs. A base cost is multiplied by a rate to debit a production order and credit a cost center.

In the previous three chapters, we discussed Controlling, material, and logistics master data, which provide cost estimates with the quantity and price information necessary to calculate the standard cost to procure or manufacture a material. In this chapter, we discuss *costing sheet configuration*, which determines how cost estimates calculate overhead costs. We will explore overhead costs, the base to which overhead costs are applied, the overhead rate, and the credit key.

Period-Based Costing

At period-end overhead, work in process (WIP) and variances are calculated and settled. This moves costs from the sending cost center to the receiving order and ensures that the overhead rates are correctly reflected in the WIP and variances.

Event-Based Costing

As of SAP S/4HANA release 2022, event-based processing is available with the universal parallel accounting business function. Goods movements and confirmations represent events that trigger the overhead calculation according to the costing sheet. Depending on the order's status, this triggers either the posting of a journal entry for the WIP or the cancellation of any existing WIP and the calculation of production variances.

5.1 Overhead

In addition to direct material and labor costs, overhead costs are typically included as a separate component of the finished product standard price. Overhead costs may include building lease, insurance, and general office staff not directly involved in the production process.

You can allocate overhead with any of these three methods, which we discuss in the following sections:

- Activities
- Templates
- Costing sheets

5.1.1 Activities

You can increase the planned activity price to include the overhead. An alternative is to create separate overhead activity types with Transaction KL01 to represent activities other than the manufacturing steps. The advantages of allocating overhead with activity types include the following:

- Real-time posting during activity confirmation
- No configuration requirement

Disadvantages of dedicated overhead activity types include the increased production data setup required in work centers and routings and possible increased maintenance during activity confirmations since the extra activities also need to be confirmed.

Activity types describe activities provided by cost centers allocated to receiving objects, such as manufacturing orders.

We discuss a detailed example of confirming an activity in Chapter 13, Section 13.2.3 .

5.1.2 Templates

You allocate overhead costs with templates via drivers and formulas. Templates use functions to access data to determine the allocation proportions dynamically. They also use formulas to perform calculations on the data accessed by functions. Because these functions differ depending on which environment you call them from, it's important to pick the right environment for your template. Environment 001 (cost estimate/production orders) is the default environment for product costing.

For example, you can use the number of employees in a department to calculate personnel planned overhead costs.

You create a template with Transaction CPT1 or by following the menu path **Accounting • Controlling • Product Cost Controlling • Product Cost Planning • Product Cost by Order • Template Allocation • Individual Processing • Extras • Template • Create**. The screen shown in Figure 5.1 is displayed.

Name your **Template** and select an **Environment.** *Environments* determine the columns available in a template. You define an environment with Transaction CTU6 or via the IMG menu path **Controlling • Product Cost Controlling • Product Cost Planning • Basic Settings for Material Costing • Templates.**

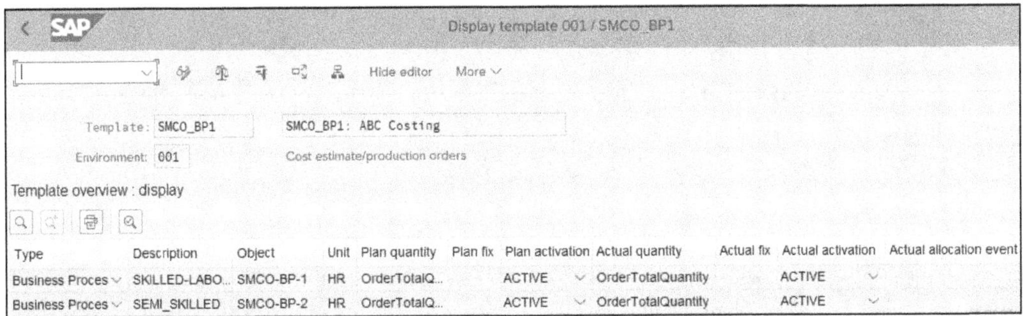

Figure 5.1 Create Template

At period close, you run template allocation for production orders with Transaction CPTA or by following the menu path **Accounting • Controlling • Product Cost Controlling • Period-End Closing • Product Cost by Order • Template Allocation • Individual Processing**. The template allocation is performed automatically when you create a standard cost estimate.

5.1.3 Costing Sheets

Costing sheets are the most popular way to allocate overhead. The advantages of using costing sheets include the following:

- Flexibility in allocating overhead across products or product groups
- Less manufacturing master data maintenance

Configuration is required, however, as explained in the following sections. If you work with best-practice content or in SAP S/4HANA Cloud, these settings are delivered via scope items. We'll look at the delivered settings and how to adjust them at the end of this chapter. In a greenfield implementation, you can choose whether you work with the delivered best-practice content or whether you create the settings for the costing sheet from scratch.

5.2 Costing Sheets

Let's inspect the configuration of a costing sheet to see how it works. To view the settings, use configuration Transaction KZS2 or follow the IMG menu path **Controlling • Product Cost Controlling • Product Cost Planning • Basic Settings for Material Costing • Overhead • Define Costing Sheets**. The screen shown in Figure 5.2 is displayed.

Available **Costing Sheets** are listed on the right of this overview screen. If you are working with universal parallel accounting, activate a costing sheet for event-based overhead calculation by selecting the checkbox in the **Evt. based** column.

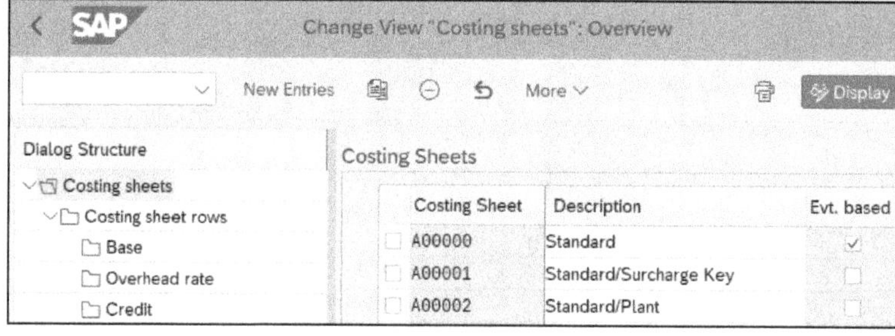

Costing Sheet	Description	Evt. based
A00000	Standard	✓
A00001	Standard/Surcharge Key	
A00002	Standard/Plant	

Figure 5.2 Costing Sheet Overview

You can use existing costing sheets or copy one and create your own. Let's choose an example costing sheet and examine the components. Select the first costing sheet (**A00000**, in this example) and double-click **Costing sheet rows** on the left. The screen shown in Figure 5.3 is displayed.

The three costing sheet components—**Base**, **Overhead rate**, and **Credit**—are listed on the left of the screen, while the **Costing sheet rows** are displayed on the right.

When overhead is calculated during period-end processing, two things happen:

- The manufacturing order or product cost collector receives a debit.
- A cost center receives a credit with the calculated overhead value.

Let's now examine each costing sheet component in detail.

SAP
Change View "Costing sheet rows": Overview
New Entries Variable List More
Dialog Structure
Costing sheets
Costing sheet rows
Base
Overhead rate
Credit
Procedure: A00000 Standard
Costing sheet rows

Row	Base	Overhead Rate	Description	From	To Row	Credit
10	B000		Material			
20		C000	Material OH	10		E01
30			Material usage......			
40	B001		Wages			
45	B002		Salaries			
50		C001	Manufacturing OH	40	45	E02
60			Manufacturing costs...	40	50	
70			Cost of goods manufactured...			
80		C002	Administration OH			E03
90		C003	Sales OH	70		E04
100			Cost of goods sold...			

Figure 5.3 Costing Sheet Rows Overview

5.3 Calculation Base

A *base* is a group of accounts to which overhead is applied. Each account identifies unique cost types within a cost estimate, such as raw material or machining labor costs. These costs, identified by the base, are multiplied by an overhead rate to determine the overhead value in the cost estimate.

To see how accounts are entered in a base, select any row with an entry in the **Base** column in the screen shown in Figure 5.3 and double-click **Base** at the left of the screen. The screen shown in Figure 5.4 is displayed.

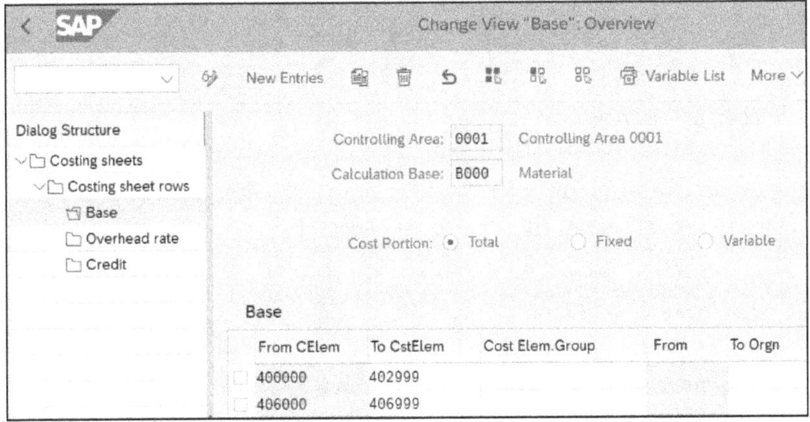

Figure 5.4 Calculation Base Overview

You can enter individual accounts or ranges in the **From CElem** and **To CstElem** columns. You also can enter a cost element group in the **Cost Elem.Group** column. You can subdivide within cost elements by entering origin groups in the **To Orgn** column and in the **Costing 1** view. You can also divide the calculation base into fixed and variable costs if necessary, by selecting either the **Fixed** or **Variable** radio button. Now that we've discussed how bases work, let's examine the next cost sheet component: the overhead rate.

5.4 Overhead Rate

The *overhead rate* is a percentage factor applied to the value of the calculation base (group of accounts). To see how percentage rates are entered in a calculation rate, select any row with an entry in the **Overhead rate** column shown earlier in Figure 5.3 and double-click **Overhead rate** at the left of the screen. The screen shown in Figure 5.5 is displayed.

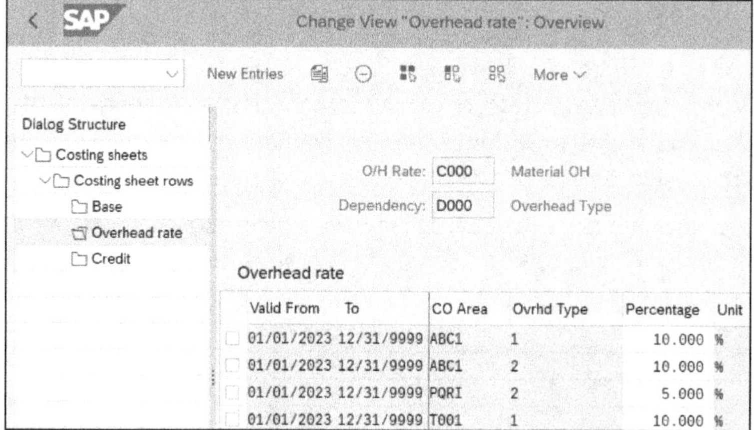

Figure 5.5 Overhead Rate Overview

Overhead rates are date-dependent, with **Valid From** and **To** dates, allowing different rates to be entered per fiscal year or fiscal period if necessary.

Rate per Period

Before using this functionality at its most detailed level, be sure the increased accuracy of overhead allocation offsets the maintenance effort required. If you enter different rates per period instead of per year, do you benefit from the perceived increased accuracy?

The **Dependency** field allows the same overhead rate to be applied to all materials within a plant or company code. Other dependencies allow different rates to be applied per order type or overhead key. Overhead keys can be entered per individual manufacturing order or product cost collector. Dependency provides a high level of control and flexibility but also increases setup and maintenance requirements.

5.4.1 Percentage Overhead

You can maintain overhead dependencies with Transaction KZZ2 or by following the IMG menu path **Controlling • Product Cost Controlling • Product Cost Planning • Basic Settings for Material Costing • Overhead • Costing Sheet: Components • Define Percentage Overhead Rates.** The screen shown in Figure 5.6 is displayed.

Select a **Percentage overhead** on the right and double-click **Details** on the left of the screen to maintain percentage overhead rates like the screen shown earlier in Figure 5.5. The advantage of the screen in Figure 5.6 is that you can also create and maintain overhead dependencies. Click the **New Entries** button.

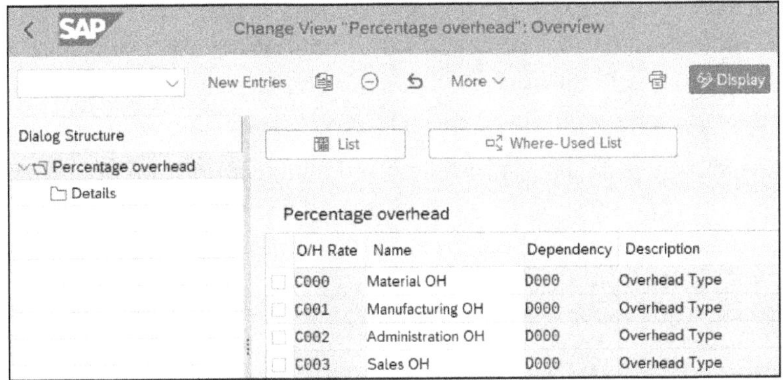

Figure 5.6 Maintain Percentage Overhead Rates

Click in the **Dependency** field and press F4 to display a list of possible entries. An explanation for some of the commonly used **Dependency** keys follows:

- **D010: Overhead Type/OH Key**
  Create different overhead rates for each material in a plant.
- **D020: Overhead Type/Plant**
  Create the same overhead rates for every material in a plant.
- **D030: Overhead Type/Company Code**
  Create the same overhead rates for every material in all plants in a company code.

To use a dependency based on an overhead key, such as **D010** in the first item in the preceding list, you must first create overhead keys and link them to overhead groups as described in the following sections.

Overhead Keys

You can maintain overhead keys with Transaction OKOG or via the IMG menu path **Controlling • Product Cost Controlling • Product Cost Planning • Basic Settings for Material Costing • Overhead • Define Overhead Keys**. The screen shown in Figure 5.7 is displayed.

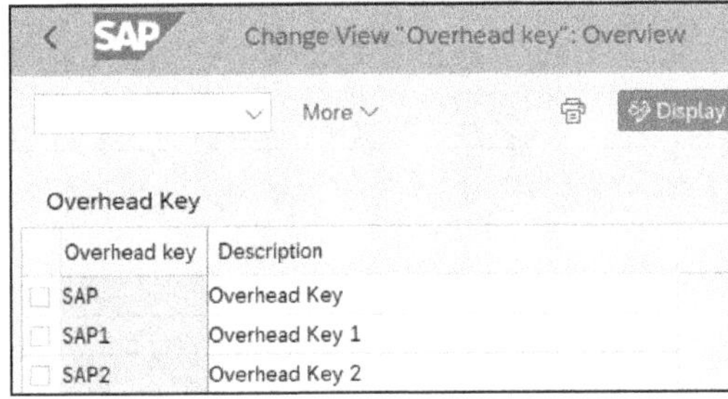

Figure 5.7 Define Overhead Keys

You can either use an SAP-delivered overhead key or create your own key with any alphanumeric combination of up to six characters. The overhead key is assigned to a product cost collector in the **Data** tab as shown in Figure 4.30 in Chapter 4, Section 4.4. Manufacturing orders contain the overhead key in the **Control data** tab.

The overhead *key* defaults to product cost collectors and manufacturing orders, which are created by an overhead *group* specified in the **Costing 1** view of the material to be manufactured. In the following section, we'll examine how to assign overhead keys to overhead groups.

Overhead Groups

You maintain overhead groups with Transaction OKZ2 or by following the IMG menu path **Controlling • Product Cost Controlling • Product Cost Planning • Basic Settings for Material Costing • Overhead • Define Overhead Groups**. The screen shown in Figure 5.8 is displayed.

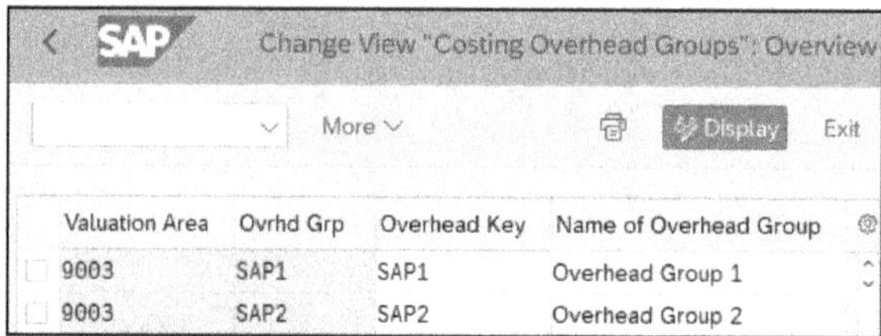

Figure 5.8 Define Overhead Groups

As you create overhead groups, you assign an overhead key, which links the overhead group in the material master with the overhead key in product cost collectors and manufacturing orders.

Overhead Groups in the Costing 1 View and Overhead Keys in Production Orders

Suppose you want to apply an overhead rate of 10% to one group of materials and 20% to another. To do this, you create two overhead groups and two overhead keys:

You enter overhead **groups** as follows:

- **SAP10** in the **Costing 1** view of all materials in the first group
- **SAP20** in the **Costing 1** view of all materials in the second group

You create two rows in the costing sheet:

- Row **10**: Link percentage overhead 10% to overhead key SAP10
- Row **20**: Link percentage overhead 20% to overhead key SAP20

Now that we've created and linked overhead keys and groups, let's look at how the costing sheet is determined.

Default Costing Sheet

Product cost collectors and manufacturing orders contain a **Costing Sheet** entry that determines overhead costs based on overhead keys and overhead groups. You maintain the overhead-related fields for a production order with Transaction CO02 or by following the menu path **Logistics • Production • Shop Floor Control • Order • Change**. Click the **Control** tab to display the screen shown in Figure 5.9.

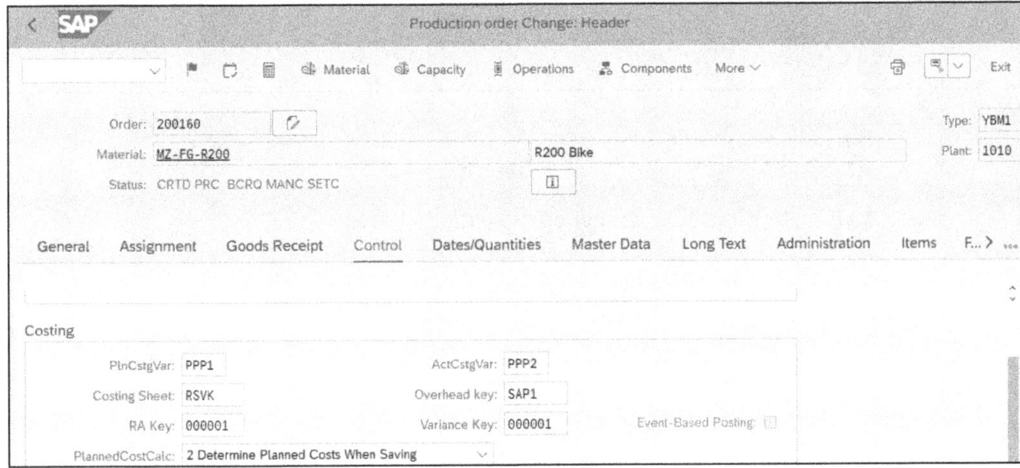

Figure 5.9 Production Order Control Tab

In this example, **Costing Sheet RSVK** defaults from the valuation variant contained in **PlnCstgVar** (plan costing variant) **PPP1**. You maintain costing variant PPP1 with Transaction OPL1 or by following the IMG menu path **Controlling • Product Cost Controlling • Cost Object Controlling • Product Cost by Period • Product Cost Collectors • Check Costing Variants for Product Cost Collectors • Costing Variants for Valuation of Internal Activities.**

Follow these steps to display the costing sheet assignment:

1. Run Transaction OPL1 and double-click Costing Variant **PPP1**.
2. Click the **Valuation Variant** button.
3. Click the **Overhead** tab.

The screen shown in Figure 5.10 is displayed.

Costing Sheet RSVK in the **Valuation Variant** corresponds to the costing sheet in Figure 5.9. The costing variant in Figure 5.9 is determined by the production order **Type,** which in this example is **YBM1**, in the upper-right corner. You maintain order types with Transaction OKZ3 or by following the IMG menu path **Controlling • Product Cost Controlling • Cost Object Controlling • Product Cost by Order • Manufacturing Orders • Define Cost-Accounting-Relevant Default Values for Order Types and Plants.** Double-click a plant and order type combination to display the screen shown in Figure 5.11.

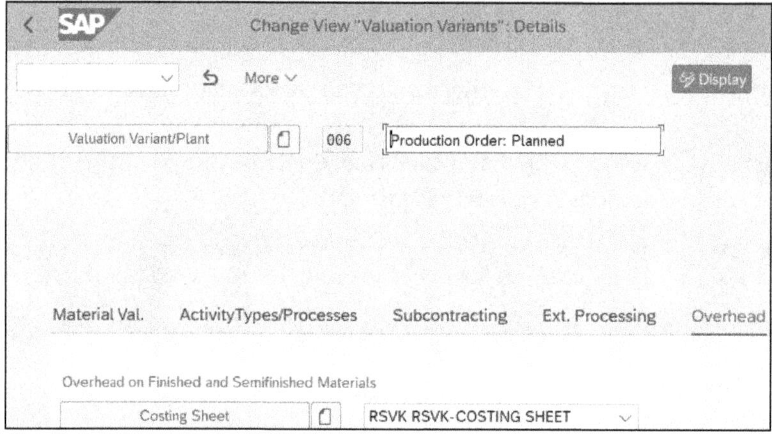

Figure 5.10 Costing Sheet Assignment to Valuation Variant

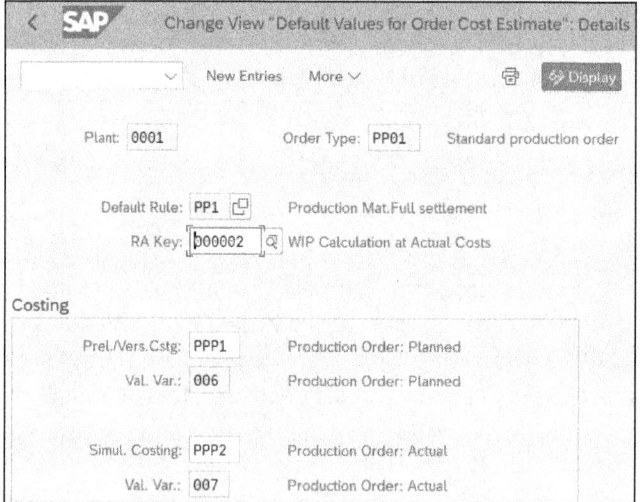

Figure 5.11 Default Values for Plant and Order Type

The two costing variants—**PPP1** and **PPP2**, in this example—correspond to the two costing variants shown earlier in Figure 5.9.

Sales Order Costing Sheets

The requirements class in the **Procurement** tab of a sales order line item proposes the default costing sheet for a sales order. We discuss sales order controlling in detail in Chapter 17.

If the **Copy Costing Sheet** indicator in the requirements class is selected, this costing sheet can be transferred to the assigned manufacturing orders.

If the **Pass on Costing Sheet** checkbox in the costing type is selected, the costing sheet is transferred from the sales order to subordinate materials. You maintain the sales order costing type with Transaction OKY9. Double-click a costing variant, click the **Costing Type** button, and click the **Misc.** tab to display the screen shown Figure 5.12.

We discuss costing types in detail in Chapter 7.

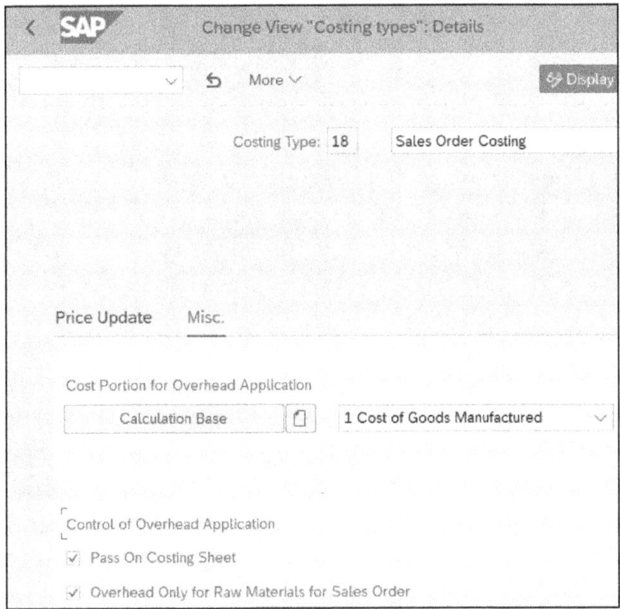

Figure 5.12 Sales Order Costing Type: Pass On Costing Sheet Checkbox

The **Pass On Costing Sheet** checkbox specifies that, in sales order costing, the calculation of overhead for semifinished products (individual requirements) that go into the finished product listed in the sales order item is carried out in the costing sheet specified in the sales order item. This ensures that the conditions for overhead calculation are specific to the sales order. If you don't select this checkbox, the system uses the costing sheet from the valuation variant entered in the costing variant as described earlier in this section.

Now that we've examined percentage overhead rates, let's look at quantity-based overhead rates.

5.4.2 Quantity-Based Overhead

Quantity-based overhead allows you to determine the amount charged depending on the production quantity based on a given rate. It is a more accurate method of distributing overhead costs than percentage-based overhead but is less complicated to set up and maintain than using the activity-based costing (ABC) module.

Example: Quantity-Based Overhead Rates

You can define quantity-based overhead rates, such as $100 per piece or percentage-based rates, as described in the previous section.

Percentage overhead rates are recommended if you want to allocate overhead to cost elements for a cost center independent of the quantity. A quantity-based rate allocates overhead per unit of measure of a cost element. You can combine percentage and quantity-based rates.

Example: Combined Quantity and Percentage-Based Overhead Rates

You could define a rate of 10% on a cost element (percentage-based) plus an additional $300 for every 100 hours (quantity-based).

For quantity-based overhead, you must set the **Record Qty** indicator in the account cost element master data as shown in Chapter 2, Section 2.1.2.

You maintain quantity-based overhead rates with Transaction KZM2 or via the IMG menu path **Controlling • Product Cost Controlling • Product Cost Planning • Basic Settings for Material Costing • Overhead • Costing Sheet: Components • Define Quantity-Based Overhead Rates**. The screen shown in Figure 5.13 is displayed.

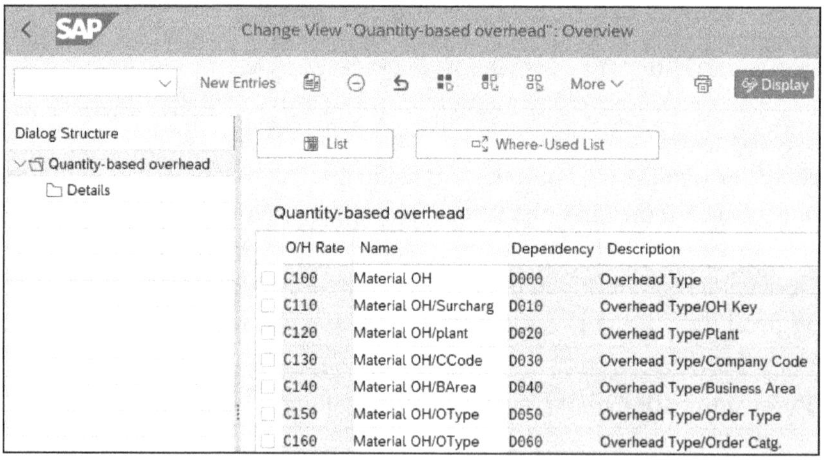

Change View "Quantity-based overhead": Overview

New Entries More Display

Dialog Structure
Quantity-based overhead
Details

List Where-Used List

Quantity-based overhead

O/H Rate	Name	Dependency	Description
C100	Material OH	D000	Overhead Type
C110	Material OH/Surcharg	D010	Overhead Type/OH Key
C120	Material OH/plant	D020	Overhead Type/Plant
C130	Material OH/CCode	D030	Overhead Type/Company Code
C140	Material OH/BArea	D040	Overhead Type/Business Area
C150	Material OH/OType	D050	Overhead Type/Order Type
C160	Material OH/OType	D060	Overhead Type/Order Catg.

Figure 5.13 Maintain Quantity-Based Overhead Overview

Select a **Quantity-based overhead** on the right and double-click **Details** on the left of the screen to maintain quantity-based overhead rates, as shown in Figure 5.14.

In this screen, you enter the **Amount** per unit of measure (**Uni**). The rates are date-dependent, which allows you to keep a history of rates as you maintain the rates per fiscal period or year. Overhead type (**Ovrhd Type**) allows you to maintain separate rates for plan, actual, and commitment costs. Click in the **Ovrhd Type** field and press F4 to display the list of possible entries as shown in Figure 5.15.

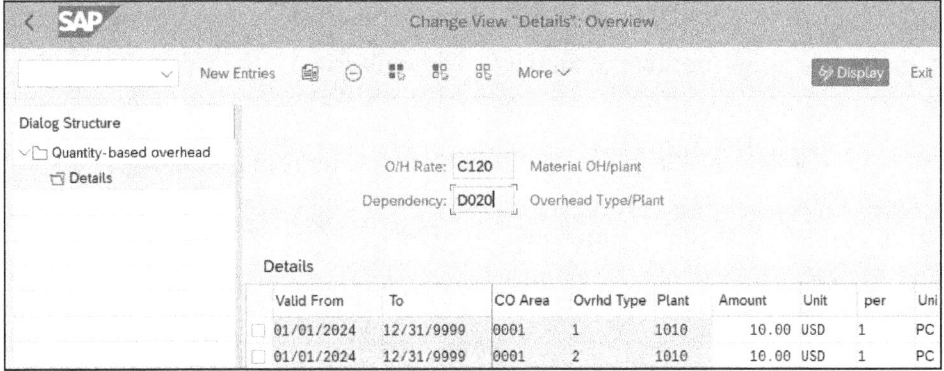

Valid From	To	CO Area	Ovrhd Type	Plant	Amount	Unit	per	Uni
01/01/2024	12/31/9999	0001	1	1010	10.00	USD	1	PC
01/01/2024	12/31/9999	0001	2	1010	10.00	USD	1	PC

Figure 5.14 Maintain Quantity-Based Overhead Rates

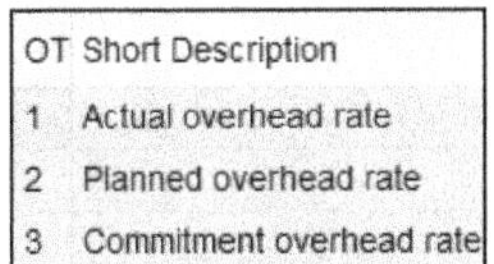

OT	Short Description
1	Actual overhead rate
2	Planned overhead rate
3	Commitment overhead rate

Figure 5.15 Overhead Type Possible Entries

Typically, you maintain the same rates for actual and planned overhead type. We discussed dependency, overhead keys, groups, and default costing sheets in the previous subsections of Section 5.4.1.

Tip: Maintaining Percentage-Based and Quantity-Based Overhead Rates

You can also maintain percentage-based overhead rates with Transaction S_ALR_87008275 and quantity-based overhead rates with Transaction S_ALR_87008272 or by following the menu path **Accounting • Controlling • Product Cost Controlling • Cost Object Controlling • Product Cost by Order • Period-End Closing • Current Settings.**

SAP: Current Settings for Costing Sheets

Refer to SAP Note 310768 for further information on current settings for costing sheets. Current settings allow you to perform configuration transactions via the standard user menu path. Because you might need to update costing sheets routinely, you can allow users to easily make these changes.

Now that we've examined bases and overhead rates, let's examine the final costing sheet component: the credit key.

5.5 Credit Key

You assign a credit key in the **Credit** column shown earlier in Figure 5.3 to each row with an entry in the **Overhead rate** column. During overhead allocation, two things happen:

- A manufacturing order or product cost collector is debited
- A cost center is credited

The *credit key* defines which cost center receives the credit. To display how a cost center is entered in a credit key, select any row with an entry in the **Credit** column shown earlier in Figure 5.3 and double-click **Credit** at the left of the screen. The screen shown in Figure 5.16 is displayed.

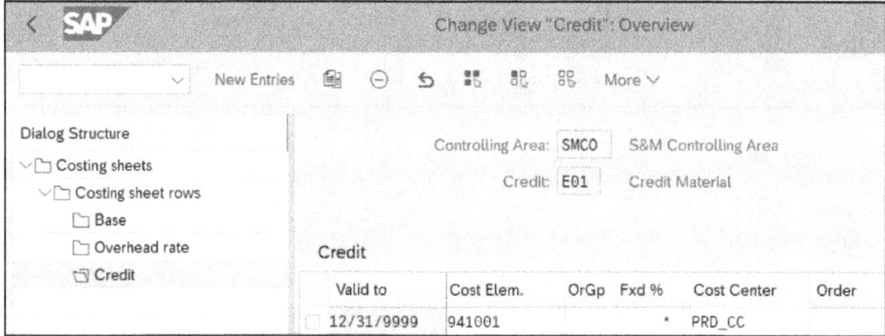

Figure 5.16 Credit Overview

Enter the cost center to receive the overhead credit in the **Cost Center** column. A secondary cost element is a required entry in the **Cost Elem.** column. The secondary cost element identifies the planned overhead cost in the cost estimate and the actual overhead debit in manufacturing order and product cost collector cost reports.

The **Credit Overview** screen also identifies the overhead credit to the cost center in cost center reports such as the standard report S_ALR_87013611—Cost Centers: Actual Plan/ Variance.

An entry in the **FXD %** field of Figure 5.16 will ensure that the fixed and variable costs are correctly assigned in the contribution margin scheme. The default entry is an asterisk (*), as shown in Figure 5.16, which means that the fixed and variable portions of the surcharge are determined in the same way as the fixed and variable costs in the calculation **Base**, as shown previously in Figure 5.4.

5.6 SAP S/4HANA Cloud Configuration (Scope Item BEG)

The best-practice settings for overhead calculation are delivered with scope item BEG. In SAP S/4HANA Cloud, this is a default scope item, so the settings will automatically be available for you. If you work with SAP S/4HANA Private Cloud or on-premise SAP

S/4HANA, you can choose to use these settings during your initial implementation. Since the chart of accounts is also delivered as best-practice content, you may not need to adjust the calculation base, but you will need to adjust the overhead rate to include the percentage rates appropriate to your organization and the credit key to reference the correct cost center.

To make these changes in SAP S/4HANA Cloud, you must be assigned to the role SAP Business Process Expert. To access the user interface for configuration, choose **Manage Your Solution • Configure Your Solution** and select **Finance • Product Costing**, as shown in Figure 5.17. In SAP S/4HANA Private Cloud or on-premise SAP S/4HANA, you can access the configuration using the menu paths described in the previous sections.

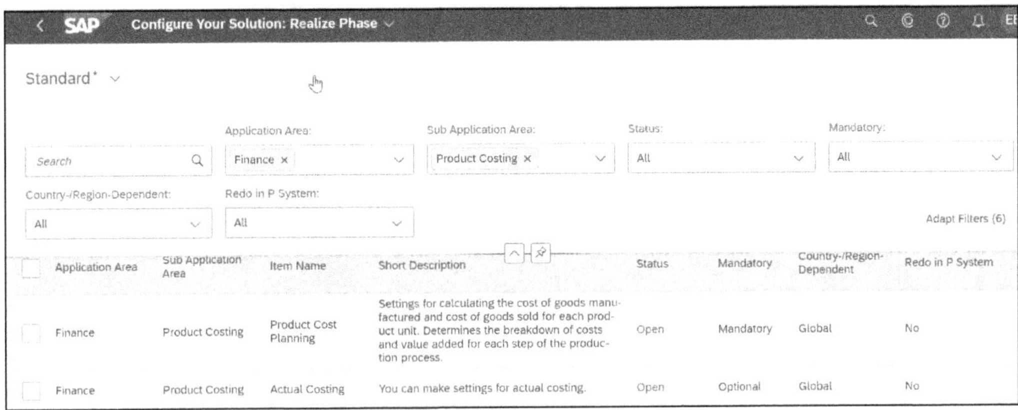

Figure 5.17 Configuration Settings in SAP S/4HANA Cloud

To display the detailed settings for Product Cost Planning, select **Product Costing • Product Cost Planning** and choose **Define Costing Sheets**, as shown in Figure 5.18, or enter the ID **102922**.

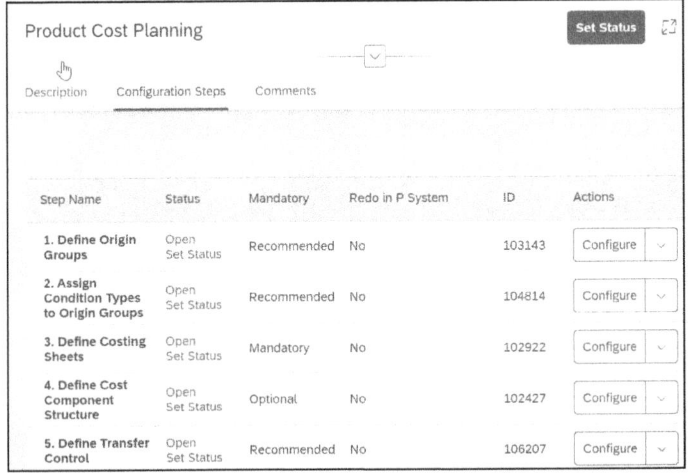

Figure 5.18 Cloud Settings for Product Cost Planning

5.6.1 Delivered Costing Sheets

Sample costing sheets are delivered for each country for use in production and in material requirements planning, where the first four digits of the costing sheet represent the company code of the relevant country and the next two digits production (PC) or planning (PP), as shown in Figure 5.19. There is an additional costing sheet for event-based overhead calculation for those countries that have released scope item 3FO (Event-Based Production Cost Posting). You will recognize the screens in the **Define Costing Sheets** dialogs, but you'll notice some restrictions. For example, you cannot create quantity-based overhead rates in SAP S/4HANA Cloud.

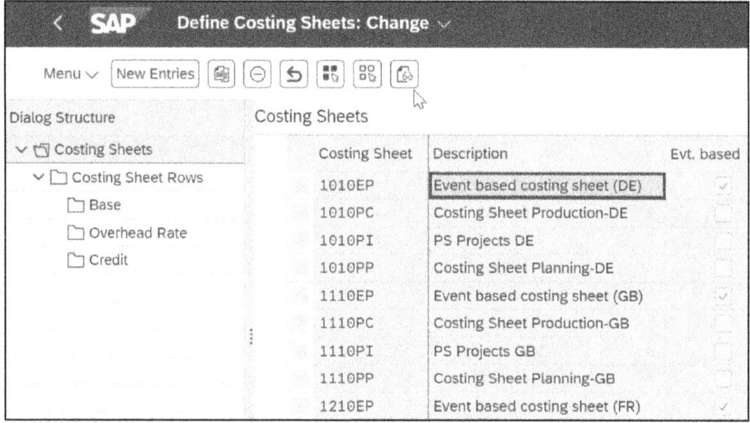

Define Costing Sheets: Change

Costing Sheet	Description	Evt. based
1010EP	Event based costing sheet (DE)	✓
1010PC	Costing Sheet Production-DE	
1010PI	PS Projects DE	
1010PP	Costing Sheet Planning-DE	
1110EP	Event based costing sheet (GB)	✓
1110PC	Costing Sheet Production-GB	
1110PI	PS Projects GB	
1110PP	Costing Sheet Planning-GB	
1210EP	Event based costing sheet (FR)	✓

Figure 5.19 Sample Costing Sheets from a Best-practice System

The costing sheet rows delivered for production have two calculation bases—**Material** and **Production**—and allow you to define overhead rates for **Material Overhead** and **Production Overhead** that are plant-dependent, as shown in Figure 5.20. At a minimum, you will want to adjust the overhead rates and the credit key to reflect your organization's needs, but you can also create a costing sheet from scratch, as we explained in Section 5.2.

Change View "Define Costing Sheets": Overview

Procedure: 1010PC Costing Sheet Production-DE

Row	Base	Overhea...	Description	From	To Row	Credit
100	Y001		Material			
109		YOH1	Material overhead	100		Y10
200	Y002		Production			
209		YOH2	Production overhead	200		Y20
300			Production overhead			

Figure 5.20 Costing Sheet Rows for a Sample Costing Sheet

The costing sheet rows delivered for use in planning include additional rows for administration and sales, as shown in Figure 5.21. These take the figure calculated in **Row 300 (Cost of goods manufactured)** and apply additional overhead rates to this figure using the conditions in **Row 305 (Administration ovh)** and **Row 306 (Sales overhead).**

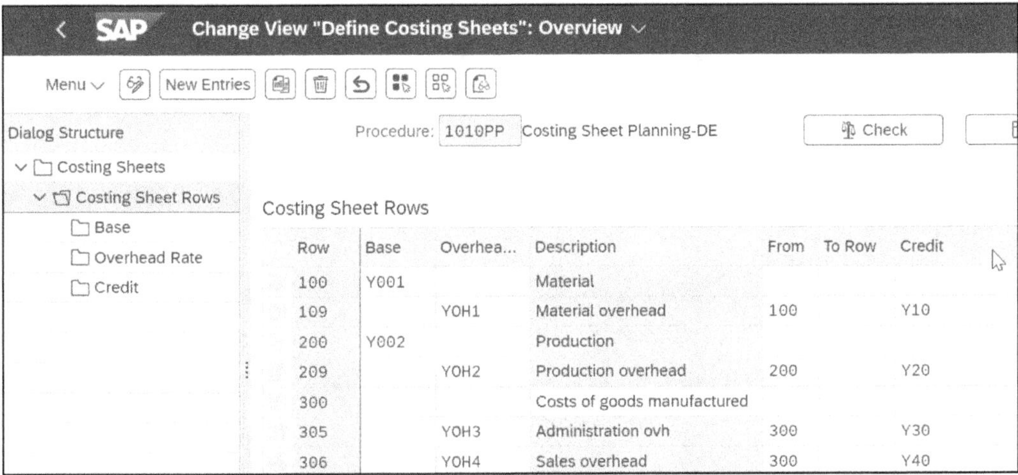

Figure 5.21 Costing Sheet Rows for a Sample Costing Sheet

5.6.2 Calculation Base

To check the material accounts to be used as a basis for the material overhead, select **Row 100** and choose **Base**, as shown in Figure 5.22. Here, we see that the material accounts are those in **Cost Elem.Group 1200_CE**. If you work with the standard chart of accounts, you shouldn't need to make changes here unless you define a costing sheet from scratch.

Figure 5.22 Calculation Base for Material Costs

5.6.3 Overhead Rate

To adjust the material overhead for each plant, select **Row 109** and choose **Overhead Rate**, as shown in Figure 5.23. Scroll through the list to find the relevant plants and replace **7.000%** with the required percentages.

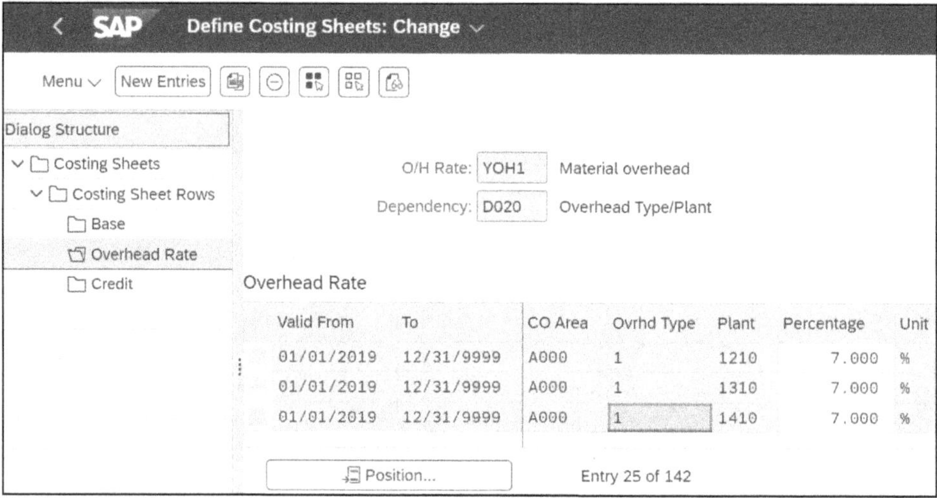

Figure 5.23 Overhead Rate for Material Costs by Plant

5.6.4 Credit Key

To change the cost center to be credited for the material overhead, select **Row 109** and choose **Credit**, as shown in Figure 5.24. If you work with the standard chart of accounts, you won't need to adjust the account/cost element, but you will want to enter the appropriate **Cost Center** for your organization.

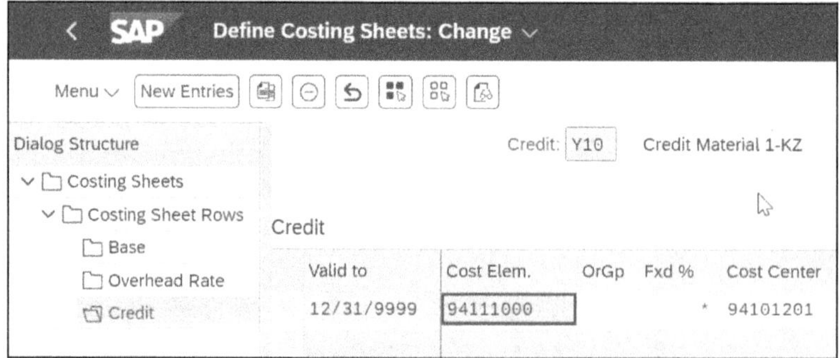

Figure 5.24 Cost Center to be Credited for Material Overhead

5.7 Summary

In this chapter, we discussed how costing sheet configuration allows cost estimates to calculate overhead costs. We looked in detail at the three costing sheet components: the calculation base, overhead rates, and credit keys. We also examined overhead calculation settings delivered with scope item BEG in SAP S/4HANA Cloud.

Now that we've examined master data in previous chapters and costing sheets in this chapter, structures are in place for the cost estimate to determine material, labor, and overhead costs. The next step is to instruct the cost estimate on grouping costs together for the cost component view in the cost estimate. Grouping costs complements the basic itemization view, which is a simple listing of items in the cost estimate. The most common cost components are materials, labor, and overhead. We'll examine cost components in further detail in Chapter 6.

Chapter 6
Cost Components

Cost components group similar costs such as material, labor, and overhead to carry them consistently through the cost structure. You assign accounts/cost elements to cost components and cost components to cost component views in the cost component structure and use these for product cost reporting and profitability analysis.

In previous chapters, we looked at material masters, BOMs, routings, and costing sheets in preparation for creating cost estimates. In this chapter, we'll examine cost components as another requirement to create cost estimates.

Cost components are fundamental to transparency in management reporting to determine the cost of manufacturing assemblies and it's important to establish a clear structure for your cost components from the outset.

6.1 Basic Concepts

Cost components group costs of similar type by account and cost element. The three basic cost components are:

- Material
- Labor
- Overhead

You can break down these basic cost components into a maximum of 120 components as needed. The origin group in the **Costing 1** view allows you to report at a more detailed level than cost elements, so you might use this to separate different types of material costs. You also have the option of reporting on fixed and variable costs as separate cost components, so you might use this to create separate components for fixed labor costs and variable labor costs. We'll look at cost component configuration in other sections of this chapter. In this section, we'll focus on the basic concepts of cost components.

Note: Actual vs Plan Cost Components

Actual and plan cost components are based on the same configuration.

6.1.1 Cost Estimates

You view plan cost components in a cost estimate. You can find cost estimates to display with Transaction S_P99_41000111 or by following the menu path **Accounting • Controlling • Product Cost Controlling • Product Cost Planning • Information System • Object List • For Material • Analyze/Compare Material Cost Estimates.**

A cost estimate defaults to the **Itemization** view, as shown in Figure 6.1.

Cost Component View	Total Costs	Fixed Costs	Variable Costs	Currency
Cost of Goods Manufactured	7,078.49	6,925.34	153.15	USD

Itemization in Controlling Area Currency

ItmNo	Item Cat.	Resource	Cost Element	Total Value
1	E	4010040001 4010_WC1 CLE	994303	1,585.11
2	E	4010040001 4010_WC1 MACH	994301	847.95
3	E	4010040001 4010_WC1 LAB	994302	3,862.70
4	M	4010 4045RM10	400005	60.53
5	M	4010 4045RM20	400005	48.43
6	M	4010 4045RM30	400005	30.27
7	G	4010040001 994101	994101	13.92
8	G	4010040001 994102	994102	629.58
				= 7,078.49

Figure 6.1 Cost Estimate Itemization View

Itemization displays a list of costs by **Resource, Item Cat.** (item category) and **Cost Element.** Item category defines cost origin, internal activity (**E**), material (**M**), and overhead (**G**) in this example. Each item category is linked with a cost component; for example, we could choose to define a material cost component by assigning cost element **400005** when we configure cost components in the next section of this chapter.

You can change the **Itemization** view to a **Cost Component** view by either double-clicking **Cost of Goods Manufactured** at the top of Figure 6.1, or by using the menu path **Costs • Display Cost Components.** You'll see the screen displayed in Figure 6.2.

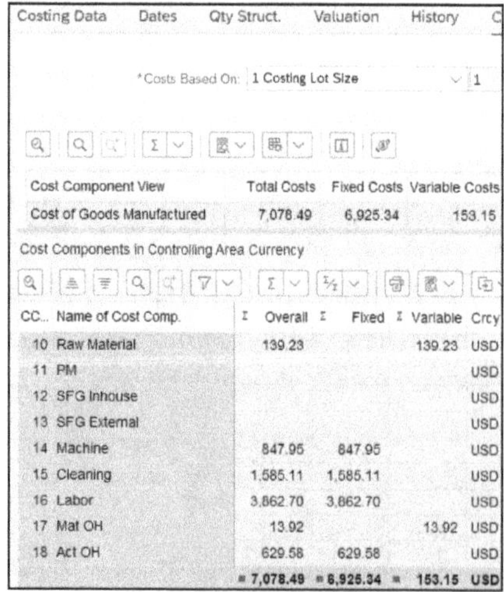

Costing Data | Dates | Qty Struct. | Valuation | History

*Costs Based On: 1 Costing Lot Size

Cost Component View	Total Costs	Fixed Costs	Variable Costs
Cost of Goods Manufactured	7,078.49	6,925.34	153.15

Cost Components in Controlling Area Currency

CC...	Name of Cost Comp.	Overall	Fixed	Variable	Crcy
10	Raw Material	139.23		139.23	USD
11	PM				USD
12	SFG Inhouse				USD
13	SFG External				USD
14	Machine	847.95	847.95		USD
15	Cleaning	1,585.11	1,585.11		USD
16	Labor	3,862.70	3,862.70		USD
17	Mat OH	13.92		13.92	USD
18	Act OH	629.58	629.58		USD
		= 7,078.49	= 6,925.34	= 153.15	USD

Figure 6.2 Cost Component View

The **Raw Material** cost of **139.23** in Figure 6.2 is the sum of the three material items in the **Itemization** view in Figure 6.1.

6.1.2 Cost Component Rollup

In multilevel costing structures, the cost component split provides information about the cost of the original components rolled up to assemblies.

A cost estimate created for a finished good or subassembly first navigates to the lowest-level BOM components via the procurement type in the **MRP 2** view, which we discussed in Chapter 3. It then rolls up the individual material cost components through subassemblies to the finished product, keeping the structure of the cost components consistent during the process. This provides you with transparency to the material cost of multilevel structures and differs from the account assignment that you see in financial accounting, where the general ledger account used to record the goods movement changes with the valuation class of the associated material. To allow you to navigate through the costs of different levels, the **Costing Structure** displays by default on the left of a cost estimate, as shown in Figure 6.3.

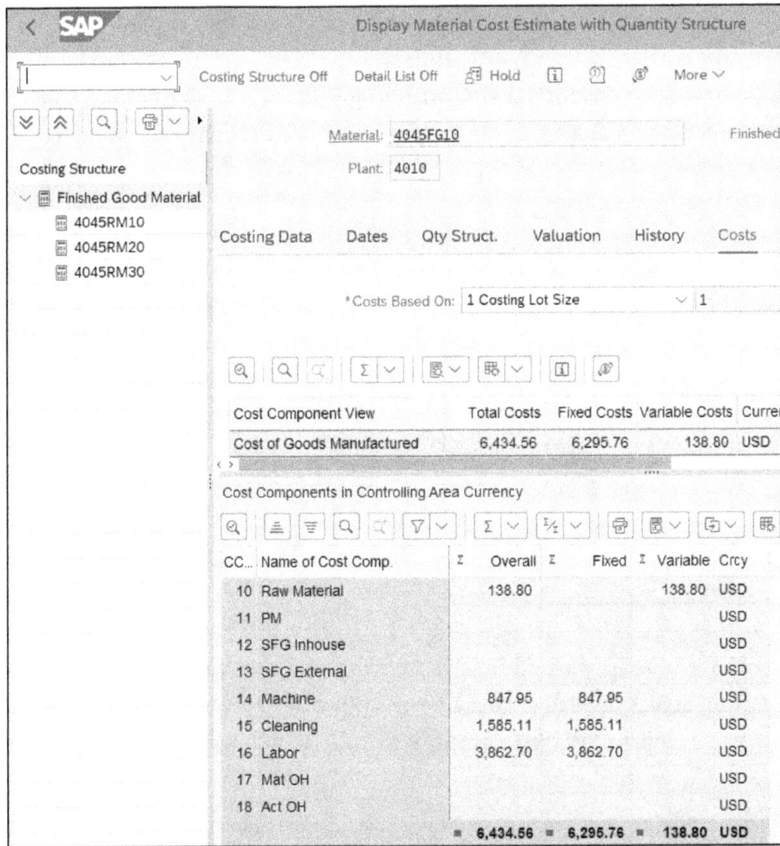

Figure 6.3 Cost Estimate with Cost Component Split

Double-click on a calculator icon in the **Costing Structure** on the left to display the cost estimate details on the right.

There are two fundamentally different types of cost component split:

- **Cost of goods manufactured**
  The *costs of goods manufactured* shows the raw materials, machine, labor, and overhead costs needed to produce the material and is built up in accordance with the cost elements for the various cost postings.
- **Primary cost component split**
  The *primary cost component split* is a different way of looking at the machine and labor costs and shows the primary costs used to supply these activities, such as energy, depreciation, steam, water, and so on, alongside raw material costs and overhead.

Now that we've discussed what cost components are and their purpose, let's look at the configuration.

6.2 Cost Components with Attributes

You access cost component structure configuration with Transaction OKTZ or via the IMG menu path **Controlling • Product Cost Controlling • Product Cost Planning • Basic Settings for Material Costing • Define Cost Component Structure**. The screen in Figure 6.4 displays.

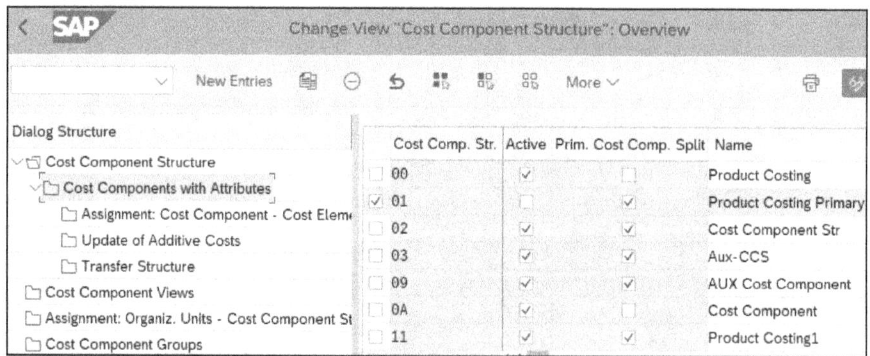

Figure 6.4 Cost Component Structure Overview

You'll see **Cost Comp. Str.** (cost component structures) listed in the first column on the right side. You can select the **Prim. Cost Comp. Split** checkbox to nominate a cost component structure as a primary cost component split, which we discuss further in

Section 6.2.8. The **Active** checkbox must be selected to create cost estimates or to calculate activity prices with a cost component structure. You can use an existing cost component structure or create your own. We'll choose cost component structure **01** in Figure 6.4 and examine the components in the following sections.

> **Note: Active Cost Component Structures**
>
> You need to deselect the **Active** checkbox before you can save changes to a cost component structure. After completing the changes, you will need to return to this overview screen and reselect the **Active** checkbox, or you'll receive error messages when attempting to create cost estimates based on this cost component structure.

6.2.1 Cost Component Structure

Select cost component structure **01** (shown selected) and double-click **Cost Components with Attributes** on the left to display the screen shown in Figure 6.5.

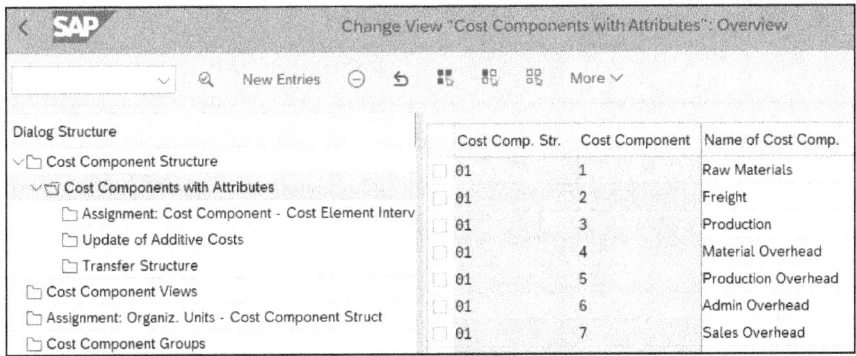

Figure 6.5 Cost Components with Attributes Overview

You'll see **Cost Components** listed on the right. You can use existing cost components or copy and create your own. Each cost component structure can contain up to 120 cost components that contain only variable costs or up to 60 cost components that contain both fixed and variable costs.

Either select a cost component and click the magnifying glass icon or double-click the cost component to display the cost component details in Figure 6.6.

We'll examine each section of this screen in the following sections.

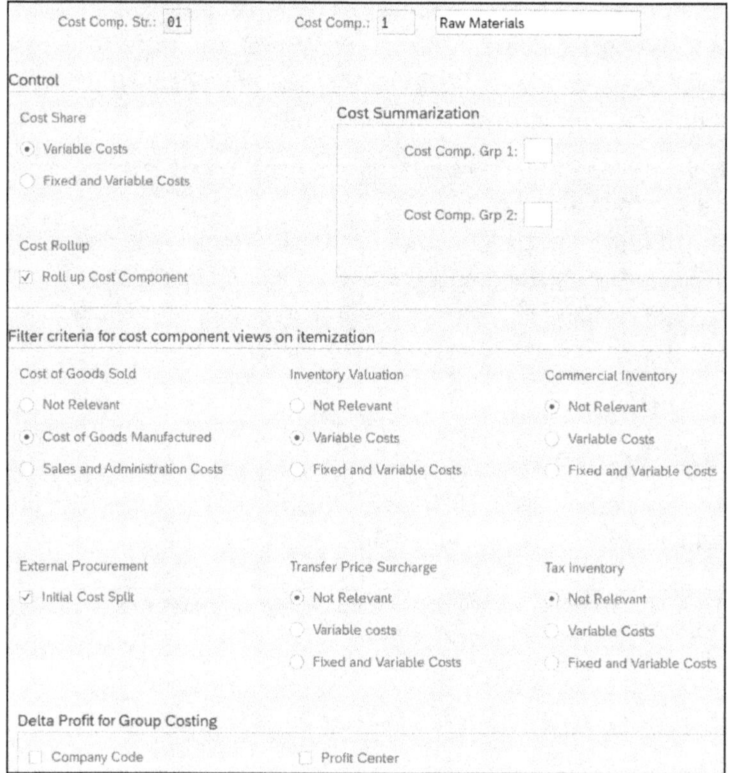

Figure 6.6 Cost Components with Attributes Details

6.2.2 Control

We'll first examine the **Control** section shown in Figure 6.7.

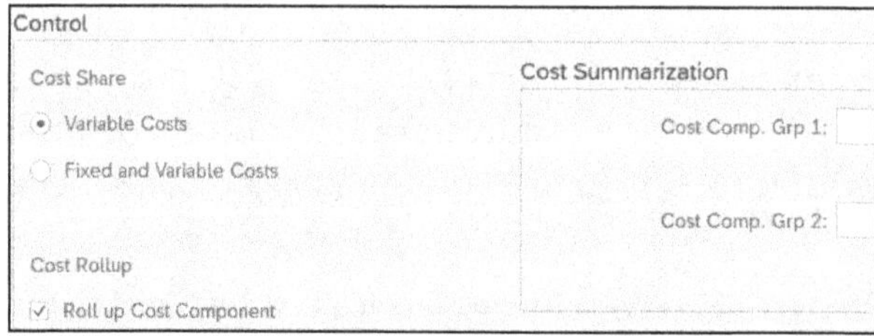

Figure 6.7 Control Section of Cost Component Details

Cost Share controls whether you show just variable costs or both fixed and variable costs in separate columns in cost estimates. You typically select **Fixed and Variable Costs** unless you don't work with fixed costs.

Roll up Cost Component determines whether the costing results of a cost component roll up into the next-highest costing level. You must roll up costs that are relevant to cost of goods manufactured (COGM) or inventory valuation.

You can see if the costs associated with this cost component are relevant to COGM or inventory valuation by inspecting the fields below the **Roll up Cost Component**, as shown in Figure 6.8.

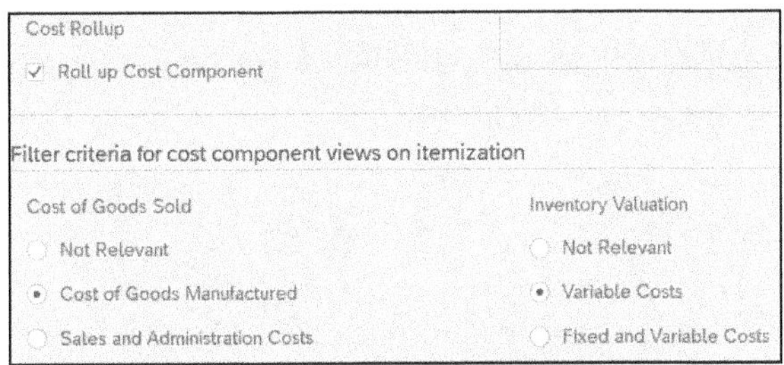

Figure 6.8 Inventory Valuation Fields Require Cost Rollup

You'll receive an error message if you deselect **Roll up Cost Component** with these fields selected:

- **Cost of Goods Manufactured**
- **Inventory Valuation** relevant for either **Variable Costs** or **Fixed and Variable Costs**

This guarantees that you roll up all COGM and ensures that all manufacturing costs are rolled up by cost components for standard cost estimates.

6.2.3 Filter for Cost Component Views on Itemization

Now that we've looked at the **Control** section, let's examine the remaining indicators in **Filter criteria for cost component views on itemization** in Figure 6.9.

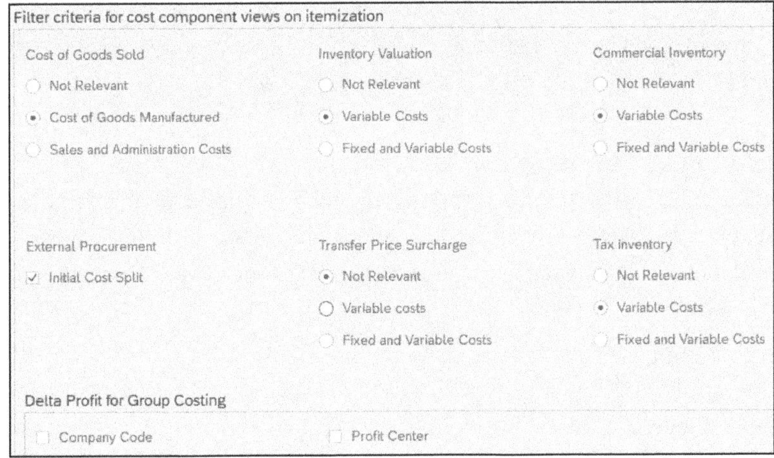

Figure 6.9 Filter Criteria for Cost Component Views on Itemization

You associate a cost component with a cost component view on itemization. This allows you to filter costs displayed in a cost estimate. Let's examine each section in detail:

- **Cost of Goods Sold**
  You define the *cost of goods sold (COGS)* for the cost component. The typical setting is to have **Cost of Goods Manufactured** selected as shown.

Example: Cost of Goods Manufactured

You select **Cost of Goods Manufactured** and then create a cost component view with **Cost of Goods Manufactured** selected in the filter. You display all cost components with **Cost of Goods Manufactured** selected in a cost estimate with this cost component view.

We'll look at the cost estimate display in more detail as we discuss the cost component view in Section 6.3.

- **Inventory Valuation**
  Inventory Valuation allows you to define which cost components are relevant for inventory valuation. You typically select **Fixed and Variable Costs** to allow you flexibility.

Note: Sales and Administration Costs

Sales and administration costs are typically not relevant for inventory valuation and don't appear in a cost component view based on inventory valuation. They are flagged as **Sales and Administration Costs** to distinguish them from **Cost of Goods Manufactured**. You can create a second cost component view to display sales and administration costs and a third to display both COGM and sales and administration costs. This is just one example, and you can create many cost component views.

- **Commercial Inventory and Tax Inventory**
  When you need to report on commercial and tax inventory values, select these fields as relevant. We discussed these fields in the **Accounting 2** view section of Chapter 3. They form the basis for cost estimates to valuate inventory for tax purposes at year end.
- **Transfer Price Surcharge**
  The following example presents a typical situation where you may select **Transfer Price Surcharge** in Figure 6.9.

Example: Additive Costs and Transfer Price Surcharge

Some companies create additive cost estimates to represent legal profits when transferring materials between company codes and profit centers.

You associate these additive costs with **Transfer Price Surcharge** and then create cost component views that include or exclude cost components with this selected. This provides you with flexibility when viewing inventory valuation with and without markup due to transfer price.

- **External Procurement**
  Select **Initial Cost Split** under **External Procurement** in Figure 6.9 to assign a cost component to an initial cost split. You have the option to group together all externally procured materials and associated costs in a cost component view. You can create an additive cost estimate for all procurement costs, such as purchase price, freight charges, insurance contributions, and administrative costs, and include these costs in the initial cost split. An additive cost estimate allows you to enter costs manually in a unit cost estimate that you can then associate with a standard cost estimate.

6

Now that we've discussed the **Filter criteria for cost components views on itemization** section, let's discuss the **Delta Profit for Group Costing** section of Figure 6.9.

6.2.4 Delta Profit for Group Costing

You select the **Delta Profit for Group Costing** checkboxes to display internal profits between company codes and profit centers as a cost component. This will only occur if you have activated the Material Ledger with multiple valuations. You can only select the checkbox for one cost component per cost component structure. Delta profits cannot be relevant for inventory valuation, and you cannot assign cost elements to cost components with either of these selected.

Let's discuss the **Cost Summarization** section of the **Cost Components with Attributes Details** screen next.

6.2.5 Cost Summarization

You can see the **Cost Summarization** in Figure 6.10.

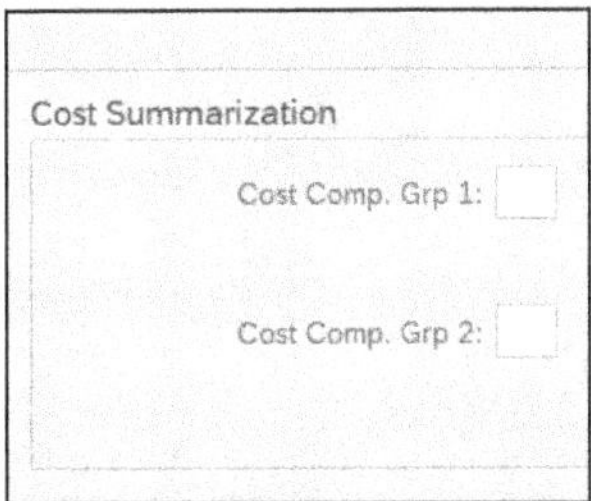

Figure 6.10 Cost Summarization Section

Cost components aren't available to add as columns in standard cost estimate collective reports or in costed multilevel bills of materials (BOMs). If you create a cost component group for each cost component, however, and enter the cost component group in the **Cost Comp. Grp 1** field of the corresponding cost component, you can then display the cost component groups as columns in standard cost estimate list reports. This report is important when comparing cost components across products or a range of products. Many companies have a requirement for such a report, which you can set up with a standard configuration.

The addition of a second cost component group in the **Cost Comp. Grp 2** field enables different levels of summarization. A cost accountant may need to carry out cost analysis of each cost component, such as three types of labor, across a range of products. An operations manager, on the other hand, may only be interested in analyzing total labor. In this case, you can create summary cost component groups and assign them to the corresponding cost components.

Tip: Viewing Cost Component Group 2 Reports

When viewing cost component groups as columns in cost estimate list reports, you need to select **Extras • Activate Cost Comp. Group 2** from the menu bar to view the summarized cost component values in the corresponding columns.

Now that we've discussed all the fields in the **Cost Components with Attributes Details** screen, we'll examine how you assign cost elements to cost components in the next section.

6.2.6 Assign Cost Elements to Cost Components

Let's take a raw material cost component as an example and examine its configuration. Select the **Raw Materials** cost component shown previously in Figure 6.5 and double-click **Assignment: Cost Component – Cost Element**. The screen shown in Figure 6.11 displays.

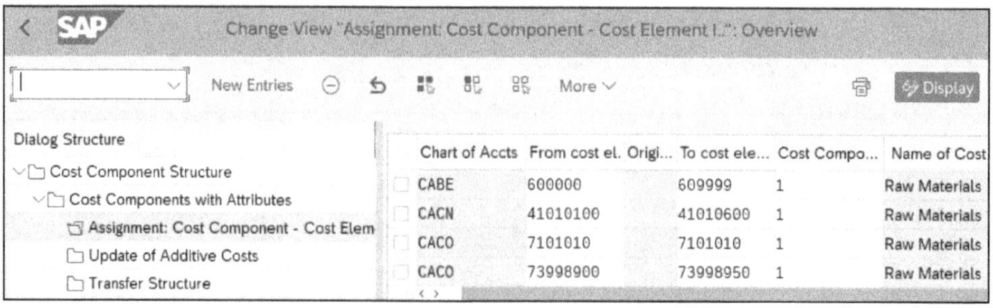

Chart of Accts	From cost el.	Origi...	To cost ele...	Cost Compo...	Name of Cost
CABE	600000		609999	1	Raw Materials
CACN	41010100		41010600	1	Raw Materials
CACO	7101010		7101010	1	Raw Materials
CACO	73998900		73998950	1	Raw Materials

Figure 6.11 Cost Element Assignment Overview

Individual cost elements or cost element ranges are assigned to cost components in the **From cost el.** (from cost element) and **To cost elem.** (to cost element) columns.

For detailed reporting within cost elements, you can make an entry in the corresponding **Origi** (origin group) column. If you enter an origin group without entering cost elements, the cost component will include the cost of all materials with the origin group regardless of the cost element.

> **Note: Cost Elements and Origin Group Are Not Mandatory**
>
> The **From cost el.**, Origin group, and **To cost ele.** fields aren't mandatory. If you leave these fields blank, the system will assign all costs in the cost estimate not assigned to a cost component to the cost component with a blank entry. You could make this a cost component for "other costs."

To use origin groups when setting up component groups, use the following procedure.

You create origin groups with Transaction OKZ1 or via the IMG menu path **Controlling • Product Cost Controlling • Product Cost Planning • Basic Settings for Material Costing • Define Origin Groups**. The origin group key is a four-digit alphanumeric field, so you can create as many origin groups as necessary.

You can enter an origin group key in the origin group column in Figure 6.11 immediately after you create it. You'll also enter the origin group in the **Costing 1** view of all relevant material masters and create cost estimates for costs to appear in the cost component containing the origin group.

Now that we've discussed how to assign cost elements to cost components, let's examine how to assign cost elements for cost estimates without quantity structures.

6.2.7 Update of Additive Costs

Double-click **Update of Additive Costs** shown on the left in Figure 6.11 to display the screen shown in Figure 6.12.

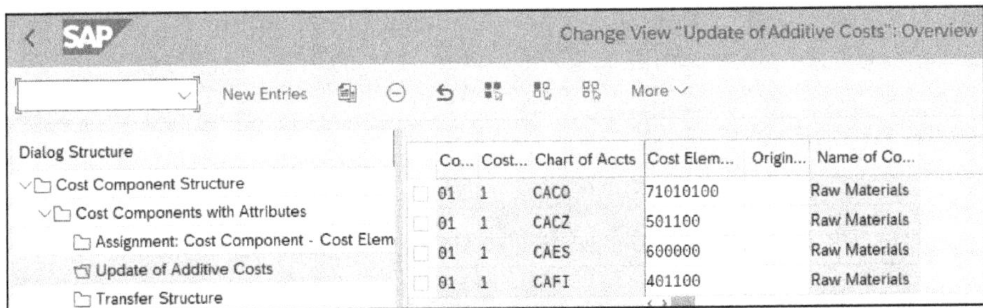

Figure 6.12 Update of Additive Costs Overview

You assign a cost element for additive costs and for a cost component in cost estimates without quantity structure—that is, without a BOM or routing. We'll discuss cost estimates without quantity structures in more detail in Chapter 11.

Next, we'll set up a transfer structure for the primary cost component split.

6.2.8 Transfer Structure

Double-click **Transfer Structure** on the left in Figure 6.12 to display the screen shown in Figure 6.13.

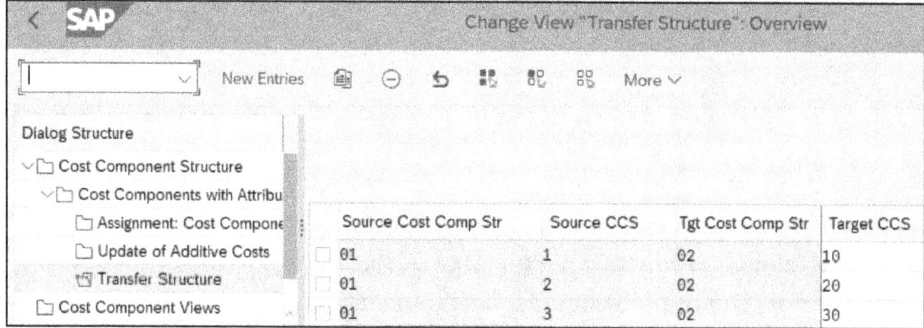

Figure 6.13 Cost Component Transfer Structure Overview

A *transfer structure* transfers the costs from the cost components of one cost component structure to the cost components of another cost component structure. For example, you can specify a source cost component from the primary cost component split determined during activity price calculation to a target cost component assigned to the COGM cost component split.

A *primary cost component split* allows you to view primary costs such as energy, wages, and depreciation. You can see the impact of changes on your primary costs—such as increases in energy prices, higher wage costs, or changes in depreciation rules—on product profitability provided that you have the system calculate the planned costs for the activity prices.

Without the primary split, these costs are not separately visible within the activity rate and are grouped as manufacturing costs in product costing and margin analysis.

Example: Transfer Structure Mapping

In Figure 6.13, source cost components from the primary cost component split (activity rate) map to the target COGM cost component split (production costs) as detailed by the following list with cost component descriptions:

- Source cost component 1 (materials) maps to target 10 (components)
- Source cost component 2 (wages) maps to target 20 (labor)
- Source cost component 3 (salaries) maps to target 30 (labor)

When you have a significant primary cost, such as energy, you may decide to create an energy cost component in both source and target cost component structures and map them in a one-to-one relationship. Generally, you'll create more primary cost components than COGM cost components.

You can also create a component in both the source and target cost component structures for miscellaneous costs posted to the cost center.

You must assign all components from the source cost component structure into the transfer structure, or you'll receive an error message when attempting to activate the source cost component structure.

In addition to setting up a transfer structure, you need to ensure that the reference information is in place to deliver the details of the primary costs. To do this, you must calculate the plan activity price automatically with Transaction KSPI and set up an additional cost component split to separate out the various parts of the activity price (depreciation, energy, wages, labor, and so on). This structure is used to break out the activity price into its primary cost components during activity price calculation. You then map the primary cost components from the activity price to the primary costs in the cost estimate using the transfer structure.

You assign a cost component split, usually a primary cost component split, to the planning version so the activity price can be analyzed by primary cost components. You can do this with Transaction OKEQ or by following the IMG menu path **Controlling • General Controlling • Organization • Maintain Versions**. Select a **Version**, double-click **Settings for Each Fiscal Year**, double-click a fiscal year, and then click the **Price calculation** tab to display the screen shown in Figure 6.14.

You assign a cost component structure in the **Cost Comp. Str.** field. This field isn't mandatory, and you don't have to assign a primary cost component split when automatically calculating the activity price. You need to enter a cost component split and then calculate the activity price to display a cost estimate with the primary cost component split. You also need to have defined a transfer structure and assigned an auxiliary cost component structure to organizational units, as we'll discuss in Section 6.4.

Now that we've discussed cost components with attributes, let's examine the cost component view.

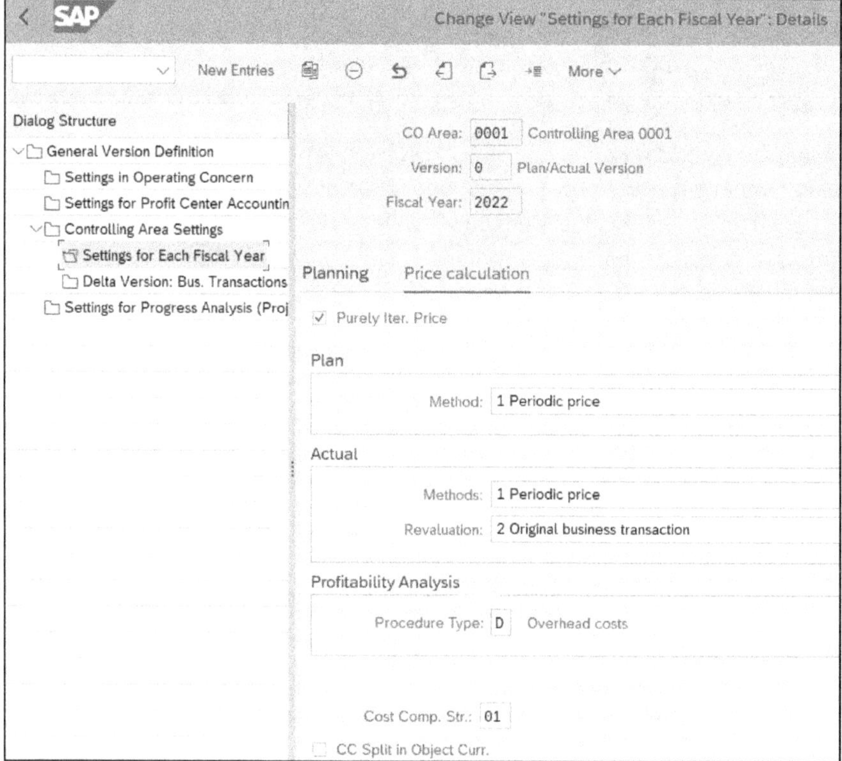

Figure 6.14 Assign Primary Cost Component Structure to Price Calculation

6.3 Cost Component View

You assign each cost component to a cost component view. When you display a cost estimate, you choose a cost component view that filters the cost components appearing in the cost estimate. In the simplest case, you assign all cost components in a cost component structure to one COGM or COGS view, and this is the only view you would use when displaying a cost estimate.

In this section, we'll explain how to set up cost components and view them during costing. We'll also explain how to include other elements, such as overhead, in the structure and how to flag the costs for further analysis.

6.3.1 Setting Up Cost Components

Assigning a cost component to a cost component view is a two-step process. First, you enter filter criteria when defining the details of each cost component, as shown previously in Figure 6.6.

You must meet certain relationships and conditions when setting the filter criteria, as we discussed in Section 6.2. For example, all **Cost of Goods Manufactured** entries must also be relevant for inventory valuation.

The second step in assigning a cost component to a cost component view is to define each cost component view. To do this, double-click **Cost Component Views** in the cost component structure as shown previously at the bottom left of Figure 6.13 to display the screen shown in Figure 6.15.

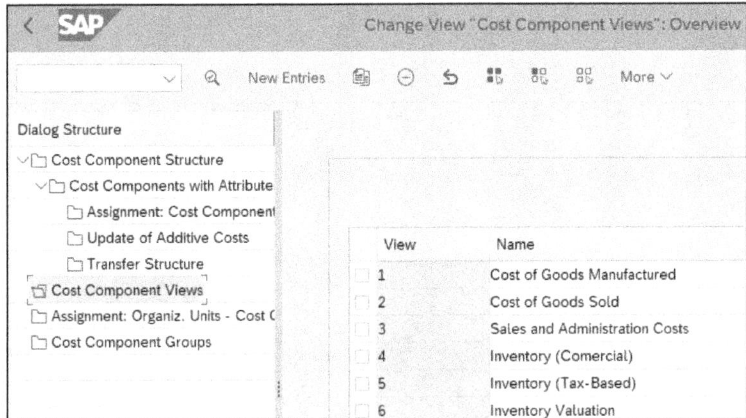

Figure 6.15 Cost Component Views

Available cost component views are listed on the right. You can use existing cost components or create your own. The cost component **View** key is a two-digit numeric field that allows you to create up to 99 cost component views.

The first five views are standard and will appear in cost estimates if you don't change the standard settings. You can change the number and order of cost component views listed in a cost estimate by selecting **Settings • Cost Display** from the menu bar when displaying a cost estimate.

Let's look at the detail behind a cost component view. Double-click cost component **View 1 Cost of Goods Manufactured** to display the cost component view details in Figure 6.16.

The description of the cost component view typically corresponds to the cost components filtered.

Tip: Compare Cost Component Detail with View Detail

Compare the filter criteria in the cost component details previously shown in Figure 6.6 with the filter list in the cost component view shown in Figure 6.16. There are nine corresponding filter criteria on each screen, although it's not immediately obvious because they aren't in the same order. For example:

- **Cost of Goods Manufactured** in Figure 6.6 has a corresponding entry as the third filter criteria (shown selected) in Figure 6.16.
- **Sales and Administration Costs** in Figure 6.6 has a corresponding entry as the second filter criteria in Figure 6.16.
- **Initial Cost Split** in Figure 6.6 has a corresponding entry as the first filter criteria in Figure 6.16.

You can match the nine filter criteria in the same way.

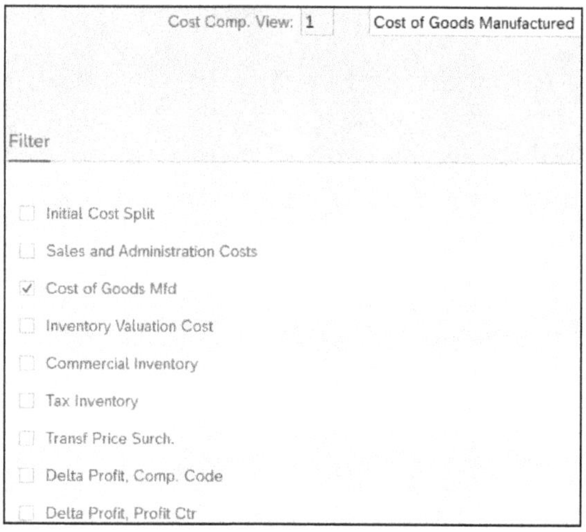

Cost Comp. View: 1 Cost of Goods Manufactured

Filter

☐ Initial Cost Split
☐ Sales and Administration Costs
☑ Cost of Goods Mfd
☐ Inventory Valuation Cost
☐ Commercial Inventory
☐ Tax Inventory
☐ Transf Price Surch.
☐ Delta Profit, Comp. Code
☐ Delta Profit, Profit Ctr

Figure 6.16 Cost Component View Details

You can create your own cost component views by matching the cost component and cost component view filter criteria. You may need to create your own cost component views in addition to the five supplied standard. You have flexibility in setting up cost component views to meet your reporting requirements.

Now that we've examined how to define and set up cost component views, let's look at how you put them to use. First, we'll look at the cost estimate screen.

6.3.2 Cost Estimate Display

You display a cost estimate the Manage Costing Runs–Estimated Costs app (SAP Fiori ID F1865), with Transaction CK13N or by following the menu path **Accounting • Controlling • Product Cost Controlling • Product Cost Planning • Material Costing • Cost Estimate with Quantity Structure • Display**. You see an example of a cost estimate display in Figure 6.17.

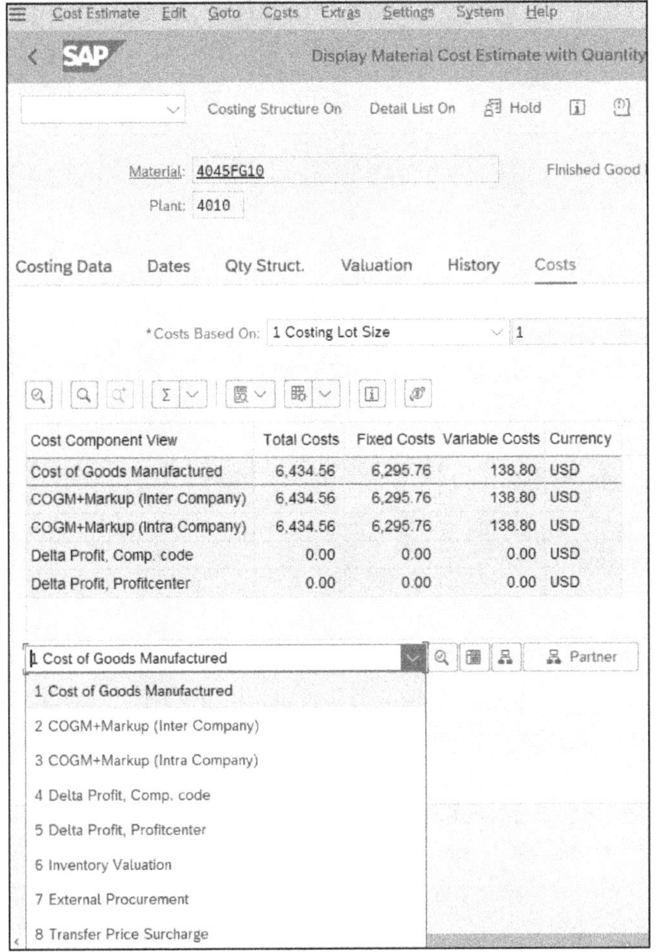

Figure 6.17 Cost Estimate Displaying Cost Component View Dropdown

Tip: Find Cost Estimates Quickly

There can be many cost estimates per material. Quickly display the current released cost estimate by clicking the **Current** button in the **Costing 2** view of the material.

If you don't know the material, you can list cost estimates in a plant with Transaction S_P99_000111.

You have two ways to display details of the cost components included in each cost component view:

- Double-click a view listed in the **Cost Component View** column.
- Click **Cost of Goods Manufactured** to display a dropdown list of all cost component views, and then click a cost component.

You can adjust which cost components appear in a cost estimate by selecting **Settings • Cost Display** from the menu bar shown at the top of Figure 6.17. The screen in Figure 6.18 displays.

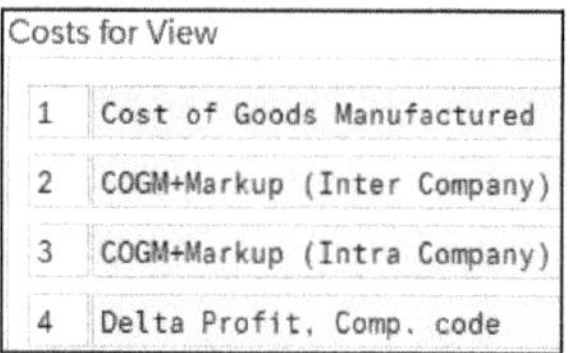

Costs for View	
1	Cost of Goods Manufactured
2	COGM+Markup (Inter Company)
3	COGM+Markup (Intra Company)
4	Delta Profit, Comp. code

Figure 6.18 Costs for View in Cost Estimate

Entries you make in the first column determine the number and order of cost component views listed in the cost estimate screen. If you leave a field blank, the cost component view will not appear in the cost estimate.

Now that we've examined cost component views in cost estimates, let's look at how they are used to calculate overhead.

6.3.3 Overhead Calculation

You can use cost component views in the calculation of overhead for semifinished goods in finished goods. In the costing type, you specify a cost component view as the calculation base for overhead. You maintain a costing type with Transaction OKKI or via the IMG menu path **Controlling • Product Cost Controlling • Product Cost Planning • Material Cost Estimate with Quantity Structure • Costing Variant: Components • Define Costing Types**. Double-click a **Costing Type** (**01** in this example) and then select the **Misc.** tab to display the screen shown in Figure 6.19.

Entering a cost component view as a calculation base allows you to determine overhead for semifinished goods on a calculation base independent of total cost estimate costs. Click the **Calculation Base** button in Figure 6.19 to display a list of possible entries for the cost component view.

Note

The system enters semifinished products in the cost estimate with costs, including COGM, sales and distribution costs, and administration costs. You can calculate material overhead based on only COGM for the semifinished product by entering the corresponding cost component view as the calculation base.

Now we'll look at how other components use cost component views.

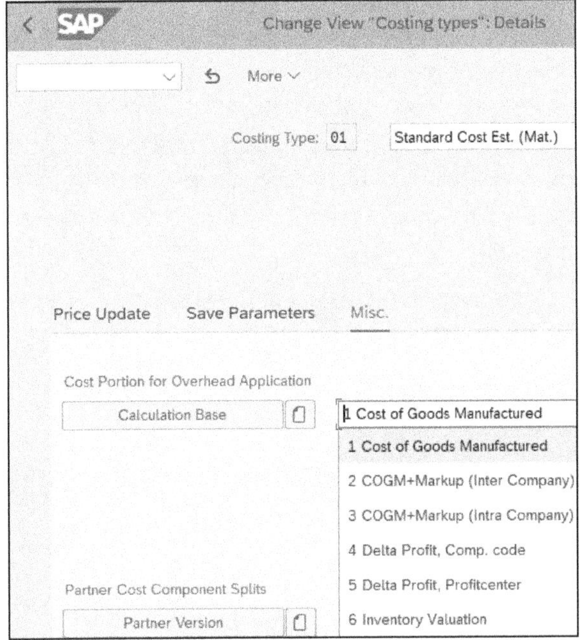

Figure 6.19 Cost Component View as Calculation Base for Overhead

6.3.4 Other Components

Other components use cost component views to categorize costs, as described here:

- **Sales and distribution**
  The COGS cost component view determines which costs to use as a basis for pricing to determine a net value for the sales order item.
- **Profitability analysis**
  The COGS cost component view determines which costs are compared to sales revenue to calculate the contribution margin for each product. In the case of costing-based profitability analysis, the cost components are mapped to value fields for the purposes of profitability reporting. In the case of margin analysis, the cost components are mapped to COGS accounts in order to split the total cost of goods sold to reflect the values for the underlying cost components.
- **Materials management**
  The inventory valuation cost component view determines which costs go into the standard price of the material and also determines the tax and commercial prices.

Now that we've looked at cost component views, let's examine how you assign cost components to organizational units.

6.4 Organizational Units of Cost Components

The company code determines which cost component structure the standard cost estimate uses, which ensures that the same cost component structure is used for all plants and costing variants in a company code. If you use different cost component structures in different plants, the standard cost estimate in one plant cannot access the results of standard cost estimates in another plant. You cannot transfer costing data for materials transferred from one plant to another.

For other cost estimates, the cost component structure is determined through the combination of company code, plant, and costing variant.

To assign organizational units, double-click **Assignment: Organiz. Units** shown previously in Figure 6.15. The screen in Figure 6.20 displays.

Change View "Assignment: Organiz. Units - Cost Component..": Overview

More ∨ Display

Company Code	Plant	Costing Variant	Valid from	Cost Comp Structure (Main CCS)	Name	Cost Comp Structure (Aux. CCS)
++++	++++	++++	01.01.1900	01	Product Costing	ZK
++++	++++	PS06	01.01.2000	01	Product Costing	ZK

Figure 6.20 Assignment of Organizational Units

You can assign cost component structures to organizational units such as **Company Code**, **Plant**, and **Costing Variant**. Some entries in this example employ masking. In the **Plant** column, all rows have the entry **++++**, which is a shorthand method of assigning cost component structures to all plants. Specific entries always take priority over masked entries. If you plan to perform group costing, this has its own costing variant, and you should ensure that the cost component structure to be used for this purpose applies to all plants.

Now that we've discussed assigning organizational structures in general, let's look first at assigning the main cost component structure and then at the auxiliary cost component structure.

6.4.1 Main Cost Component Structure

The *main cost component split* is the principal cost component split used by a standard cost estimate to update a standard price. The main cost component split can be for COGM or for a primary cost component split. You assign the main cost component structure in Figure 6.20. You may need to widen the columns so you can see the entire text in the column headings as shown in Figure 6.21.

Cost Comp Structure (Main CCS)	Name	Cost Comp Structure (Aux. CCS)
01	Product Costing	ZK
01	Product Costing	ZK

Figure 6.21 Main and Auxiliary Column Heading Text

You assign the main cost component structure in the first column and the auxiliary cost component structure in the last column. You don't have to assign an auxiliary cost component structure, but you do need to assign a main cost component structure to each row you define.

Now we'll discuss the auxiliary cost component structure in more detail.

6.4.2 Auxiliary Cost Component Structure

When viewing a cost estimate with an activated auxiliary cost component structure, you can switch between the main and auxiliary cost component structures.

You assign an auxiliary cost component structure to an organizational unit in the last column in Figure 6.21. Follow these additional steps to activate the auxiliary cost component structure:

1. Create an auxiliary cost component structure.
2. Create a transfer structure mapping the auxiliary cost components to the main cost components, as shown earlier in Figure 6.13.
3. Assign the auxiliary cost component structure to the plan version per fiscal year in the **Price calculation** tab shown previously in Figure 6.14.
4. Assign the auxiliary cost component structure to organizational units, as shown in Figure 6.20.
5. Automatically calculate the plan activity price with Transaction KSPI.

You don't have to enter an auxiliary cost component structure. In most cases, the COGM cost component structure alone provides sufficient cost component reporting. The auxiliary cost component structure is available if you need additional reporting on primary costs.

Next, we'll look at a way to check assignment to organizational units.

6.4.3 Check Assignment to Organizational Units

You can check the assignment of cost component structures to organizational units by inspecting the costing variant. Take costing variant PPC1, for example. Display costing variant PPC1 with Transaction OKKN or double-click the costing variant in the **Costing Data** tab of a cost estimate. Select the **Assignments** tab and click the **Cost Component Structure** button to display the assignments in Figure 6.22.

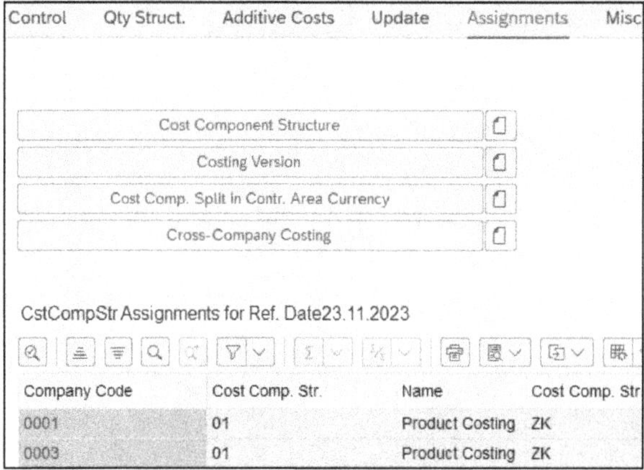

Figure 6.22 Cost Component Structure Assignments for Costing Variant PPC1

The main cost component structure **01** is assigned to company codes **0001** and **0003**, and the auxiliary cost component structure assigned is **ZK**.

You can also assign a cost component structure to a plant. To add **Plant** as a column in Figure 6.22, click the down-pointing arrow to the right of the grid icon at the far right and select **Change Layout** to see the **Change Layout** screen shown in Figure 6.23.

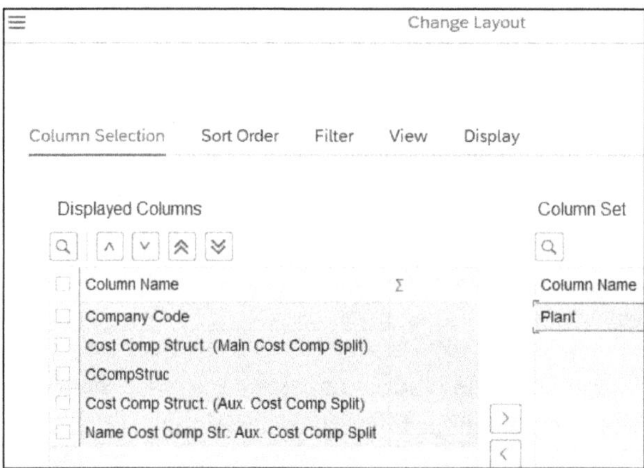

Figure 6.23 Change Layout to Include the Plant Column

Double-click **Plant** to move it across to **Displayed Columns** and press [Enter] to display the screen shown in Figure 6.24.

CstCompStr Assignments for Ref. Date23.11.2023

Company Code	Cost Comp. Str.	Name	Cost Comp. Str.	Name	Plant
0001	01	Product Costing	ZK	ZK Cost Component Str	0001
0003	01	Product Costing	ZK	ZK Cost Component Str	0003

Figure 6.24 Cost Component Structure Assignments, Including Plant

You can now see the **Plant** column in the cost component structure assignments for the costing variant.

Now that we've examined assigning organizational units and cost component structures and how to check the assignment, let's look at cost component groups.

6.5 Cost Component Groups

You can include cost component groups as columns in reports that display cost components, such as in the following transactions:

- **Transaction S_P99_41000111**
  Analyze/compare material cost estimates.
- **Transaction CK86_99**
  Costed multilevel BOM.
- **Transaction CK13N**
  Costed multilevel BOM on left side of the cost estimate display.

A costed multilevel BOM is a hierarchical overview of the values of all items of a costed material according to the costed quantity structure (BOM and routing).

You cannot include cost components as columns in these reports. You can, however, assign cost components to cost component groups that you can add to the reports. The procedure to do this is detailed in the following sections.

6.5.1 Create Cost Component Groups

To maintain cost component groups, double-click **Cost Component Groups** shown earlier in Figure 6.15. The screen in Figure 6.25 displays.

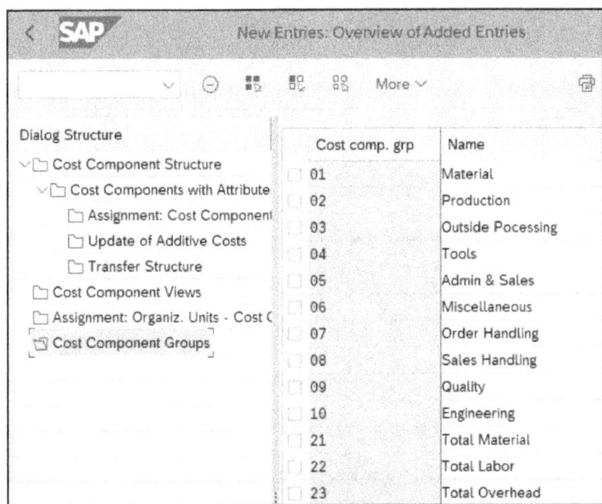

Figure 6.25 Cost Component Groups

Cost component groups are listed on the right. You can use existing cost component groups or create your own. The **Cost comp. grp** (cost component group) key is a two-digit numeric field.

Creating all of your cost component groups with two digits (i.e., "02" instead of "2") makes it easier to follow the order of cost component groups listed. For example, cost component group 2 would be listed between cost component groups 19 and 20 on this screen, not between 1 and 3 as you might initially expect.

In Figure 6.25, cost component groups **01** to **10** correspond to unique individual cost components. Cost component groups **21**, **22**, and **23** correspond to summary groups of cost components.

Now that we've looked at how to create and number cost component groups, let's see how to assign them to cost components.

6.5.2 Assign Groups to Cost Components

We discussed assigning cost component groups to cost components in Section 6.2.5. Let's review how to make this assignment. Display cost component attribute details as shown in Figure 6.26.

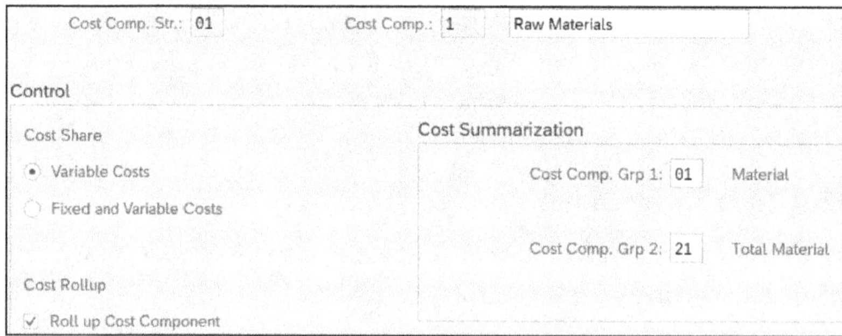

Figure 6.26 Cost Component Attribute Details - Cost Summarization

In the **Cost Summarization** section, you can assign two cost component groups to a cost component. You have flexibility in how to assign these groups. Let's discuss an example of how you can assign each group.

Example: You Can Assign Two Cost Component Groups to Each Cost Component

A cost accountant requires reporting on cost components in cost estimate lists and in costed multilevel BOMs. She created and assigned one cost component group to each cost component in the **Cost Comp. Grp 1** (cost component group 1) field shown in Figure 6.26.

The operations manager requires a summarized cost component view. He created three additional cost component groups, combining or subtotaling cost components into total material, labor, and overhead. He assigned these summary cost component groups to corresponding cost components in the **Cost Comp. Grp 2** (cost component group 2) field.

Now that we've discussed assigning cost component groups to cost components, let's see how you assign them as columns to reports.

6.5.3 Assign Groups to Report Columns

Cost component groups can be added as columns in many standard reports, such as Transaction S_P99_41000111 (List Cost Estimates). Run the transaction, click the **Change Layout** icon, and move the required cost component columns on the right side of the screen to the left, as shown in Figure 6.27.

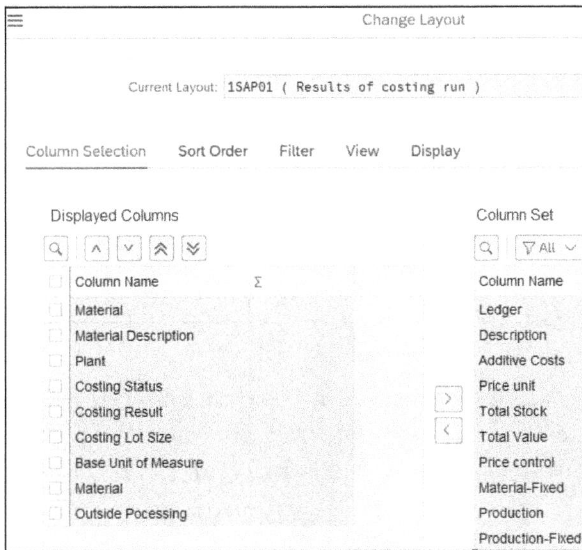

Figure 6.27 Add Cost Component Groups to Report Columns

The cost component groups become available as report columns, and you see them listed in the **Column Set** column in Figure 6.27. You add the cost component groups to report columns by selecting them and clicking the left-pointing arrow icon. In this example, we've moved **Material** and **Outside Processing** cost component groups to **Displayed Columns**.

Note

Cost component groups ending in **Fixed** in Figure 6.27 represent the fixed portion of costs. These cost component groups are available if you select **Fixed and Variable Costs** when defining the cost component attributes as you saw in Figure 6.26.

Press [Enter], and the required cost components will appear in the cost estimate report as shown in Figure 6.28.

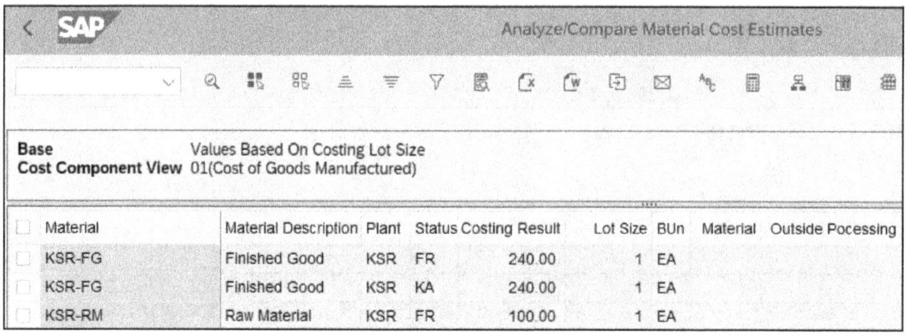

Analyze/Compare Material Cost Estimates

Base Values Based On Costing Lot Size
Cost Component View 01(Cost of Goods Manufactured)

Material	Material Description	Plant	Status	Costing Result	Lot Size	BUn	Material	Outside Pocessing
KSR-FG	Finished Good	KSR	FR	240.00	1	EA		
KSR-FG	Finished Good	KSR	KA	240.00	1	EA		
KSR-RM	Raw Material	KSR	FR	100.00	1	EA		

Figure 6.28 Cost Component Groups Displayed as Columns

The material and outside processing cost components now appear as the **Material** and **Outside Processing** columns in the cost estimate report. Material and outside processing cost components don't add up to the total cost shown in the **Costing Result** column because not all cost components are included in this example. Simply use the same procedure to add other cost components as required.

You can use a similar technique to add cost component group columns to standard costed multilevel BOM reports. You can view these reports with Transactions CK86_99 and CK13N.

Tip: How to View Cost Component Group 2 Columns

When viewing cost component groups as columns in cost estimate list reports, you need to select **Extras • Activate Cost Comp. Group 2** from the menu bar of the initial selection screen to view the summarized cost component values in the corresponding columns in the results screen.

Several active cost component structures may be available in Transaction OKTZ (shown previously in Figure 6.4). You'll need to determine to which cost component structure you'll assign your newly created cost component groups. To do this, display a standard cost estimate with Transaction CK13N and click the **Costing Data** tab to display the screen shown in Figure 6.29.

Click the **Costing Variant** text to display the costing variant. Select the **Assignments** tab and then click the **Cost Component Structure** button to display the cost component structure assignment for each company code, as shown in Figure 6.30.

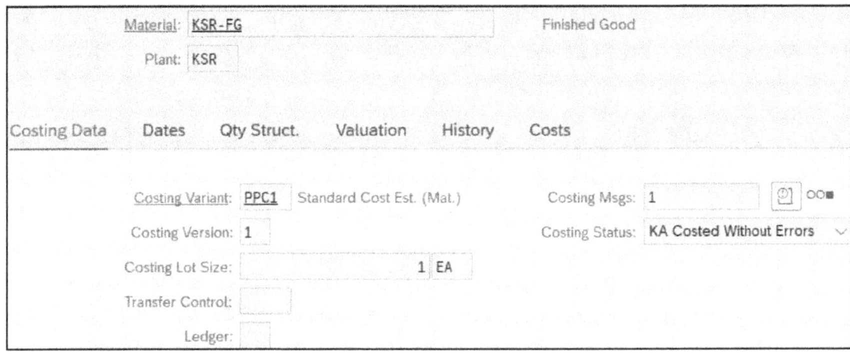

Figure 6.29 Standard Cost Estimate Costing Variant

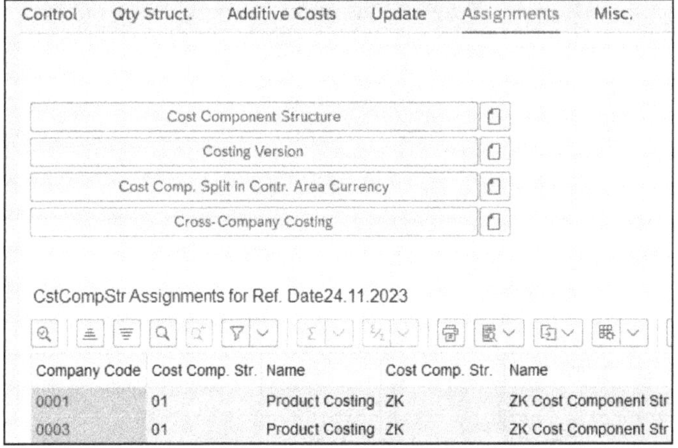

Figure 6.30 Costing Variant Cost Component Structure Assignments

The main cost component structure assigned to company codes **0001** and **0003** is **01: Product Costing**, and the auxiliary cost component structure assigned is **ZK**. You'll only see the last column if you assigned an auxiliary cost component structure as discussed in Section 6.4.

In this example, if you are setting up cost component groups for company code **0001**, you should deactivate cost component structure **01**, assign the cost component groups to the cost components, and then reactivate cost component structure **01**.

6.6 SAP S/4HANA Cloud Configuration (Scope Item BEG)

The best-practice settings for the cost component structure are delivered with scope item BEG (standard cost calculation). In SAP S/4HANA Cloud, this is a default scope item, so the settings will automatically be available for you. If you work with SAP S/4HANA Private Cloud or on-premise SAP S/4HANA, you can choose to use these settings to accelerate your initial implementation.

In this section, we'll explain how to configure cost components in SAP S/4HANA Cloud and what is delivered as standard business content.

6.6.1 Configuration Prerequisites

To make changes to the configuration in SAP S/4HANA Cloud, you'll need to be assigned to the role SAP Business Process Expert. To access the user interface for configuration, choose **Manage Your Solution • Configure Your Solution** and select **Finance • Product Costing** as shown in Chapter 5, Figure 5.17. In SAP S/4HANA Private Cloud or on-premise SAP S/4HANA, you can access the configuration using the menu paths described in the previous sections and make whatever changes you need.

To display the detailed settings for Product Cost Planning, select **Product Costing • Product Cost Planning** as shown in Chapter 5, Figure 5.17 and choose **Define Cost Component Structure** or enter the ID **102427**, as shown in Figure 6.31.

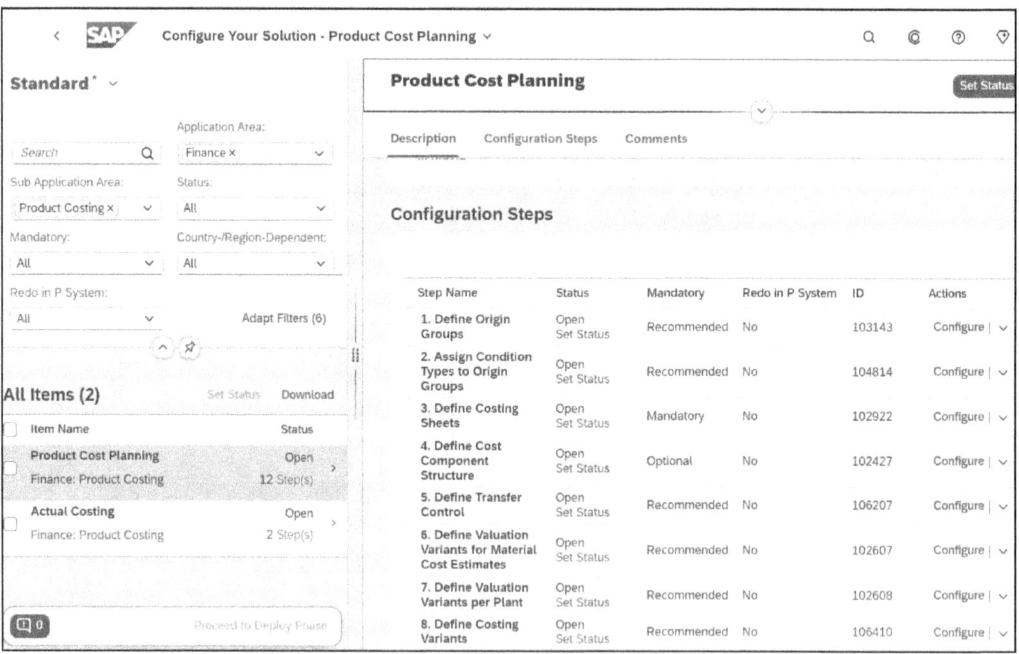

Figure 6.31 Configuration Settings in SAP S/4HANA Cloud

Because some functions, such as the primary cost component split, are not supported in the public cloud, the **Dialog Structure** shown in Figure 6.32 is reduced compared to the **Dialog Structure** shown earlier in Figure 6.4. Other functions, such as group valuation, are activated using an additional scope item (5W2), so you will only see these settings if the appropriate scope item has been activated in your system.

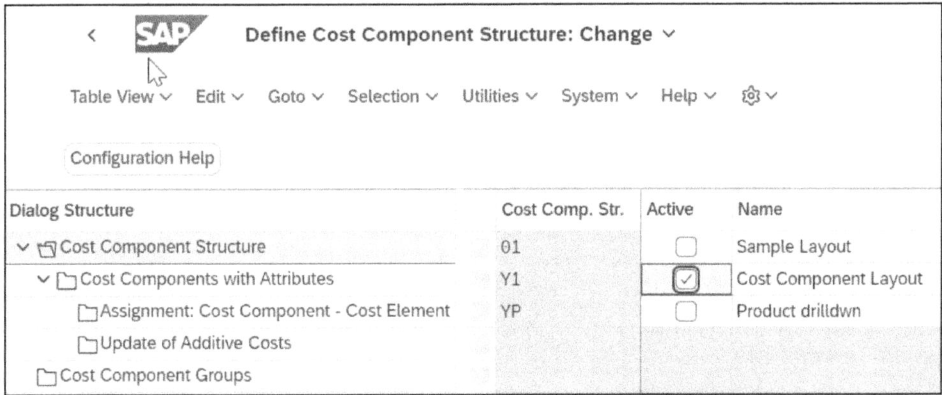

Figure 6.32 Dialog Structure for Cost Component Structure in SAP S/4HANA Public Cloud

6.6.2 Delivered Cost Component Structures

In addition to the sample cost component structure that we saw in Figure 6.4, cloud systems and systems using best-practice content include the cost component structures shown in Figure 6.32 with the naming convention **Y***, where **Y1** is used to set the standard costs for inventory valuation and **Y2** for drilldown reporting. **Y1** is active by default and is active in all plants. Notice that you can change this layout, but you cannot create a new one or assign it to a specific plant. The cost component structure **Y1** contains 11 cost components, as shown in Figure 6.33, plus an additional cost component **999** for group valuation that was activated using scope item 5W2 (Group Valuation). The numbering convention separates the material-related costs (direct materials, co- and byproducts, third party, freight and landed costs, and material overhead) from the production-related costs (personnel time, machine time, setup time, production overhead) with additional cost components for sales and administration and any additional costs. Cost component structure **YP** contains the same cost components, with the addition of a cost component for WIP (work in process) and minus the component for group valuation.

The settings to assign the general ledger accounts to the cost components are provided as part of the content delivery, with the relevant general ledger accounts delivered with the default chart of accounts being assigned to the appropriate cost components. There is also a default general ledger account for the additive costs. When you set up a cloud system, there are special tools to change the general ledger accounts if you have your own chart of accounts.

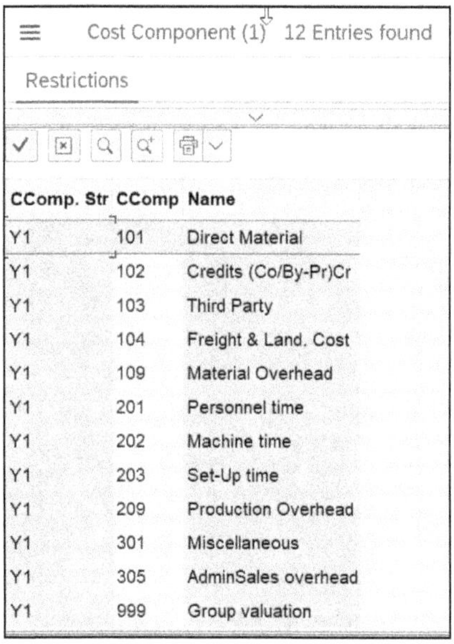

Cost Component (1) 12 Entries found

Restrictions

CComp. Str	CComp	Name
Y1	101	Direct Material
Y1	102	Credits (Co/By-Pr)Cr
Y1	103	Third Party
Y1	104	Freight & Land. Cost
Y1	109	Material Overhead
Y1	201	Personnel time
Y1	202	Machine time
Y1	203	Set-Up time
Y1	209	Production Overhead
Y1	301	Miscellaneous
Y1	305	AdminSales overhead
Y1	999	Group valuation

Figure 6.33 Standard Cost Components in Public Cloud and Best Practices

Figure 6.34 shows the assignment of various material accounts to cost component **101** for direct materials and an expense account to cost component **103** for third party.

If you work with freight and landed costs, you'll also need to check the condition types in the purchase info record and link them with origin groups. One **Origin Group** (**YFC**) is delivered by default, and this is assigned to all condition types, but you might want to refine the assignment by adding further origin groups to separate the associated purchase conditions with different cost components. To add origin groups, choose **Define Origin Groups** or enter the ID **103143** in Figure 6.31. You'll then need to link the new origin groups with the conditions in the purchase info record by choosing **Assign Condition Types to Origin Groups** or entering ID **104814** in Figure 6.31. Replace origin group YFC with your own origin groups to separate the costs for the different condition types, as shown in Figure 6.34.

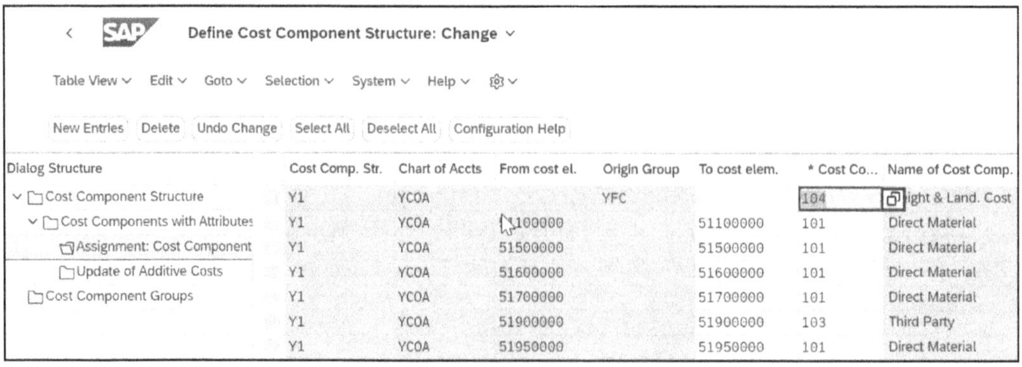

Define Cost Component Structure: Change

Table View Edit Goto Selection System Help

New Entries Delete Undo Change Select All Deselect All Configuration Help

Dialog Structure
- Cost Component Structure
  - Cost Components with Attributes
    - Assignment: Cost Component
    - Update of Additive Costs
  - Cost Component Groups

Cost Comp. Str.	Chart of Accts	From cost el.	Origin Group	To cost elem.	* Cost Co...	Name of Cost Comp.
Y1	YCOA		YFC		104	Freight & Land. Cost
Y1	YCOA	5100000		51100000	101	Direct Material
Y1	YCOA	51500000		51500000	101	Direct Material
Y1	YCOA	51600000		51600000	101	Direct Material
Y1	YCOA	51700000		51700000	101	Direct Material
Y1	YCOA	51900000		51900000	103	Third Party
Y1	YCOA	51950000		51950000	101	Direct Material

Figure 6.34 Assignment of Accounts and Origin Groups to Cost Components

6.6.3 Update of Additive Costs

There are two ways to capture freight costs in SAP S/4HANA Cloud—either via the conditions in the purchasing info record as described in the previous section or as additive costs, adding costing items manually as described in Section 6.2.7. If you create additive costs manually, then these must be assigned to a cost component according to the configuration shown in Figure 6.35.

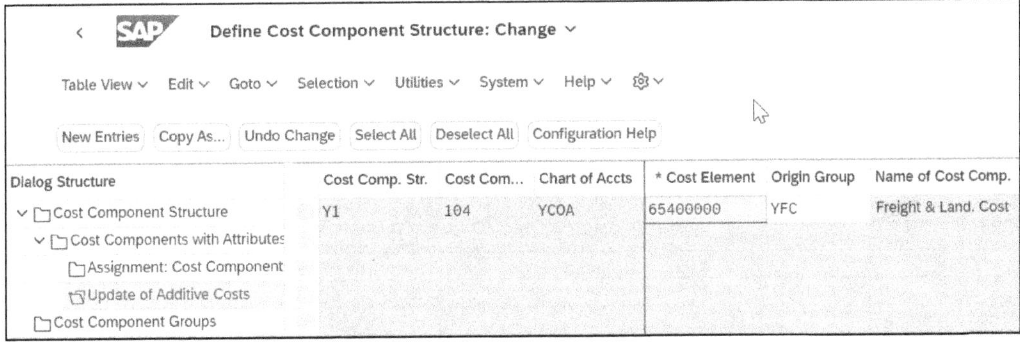

Figure 6.35 Update of Additive Costs

6.6.4 Supported Cost Component Views

SAP S/4HANA Cloud offers only two cost component views. For each cost component, you can specify that it is to be considered either as cost of goods manufactured (in order words, relevant for inventory valuation) or sales and administration (in other words, not part of inventory valuation). Figure 6.36 shows the assignment of the **Direct Material** cost component to the **Cost of Goods Manufactured** view. All costs are considered fixed and variable, and all costs are rolled up, so these settings are hidden.

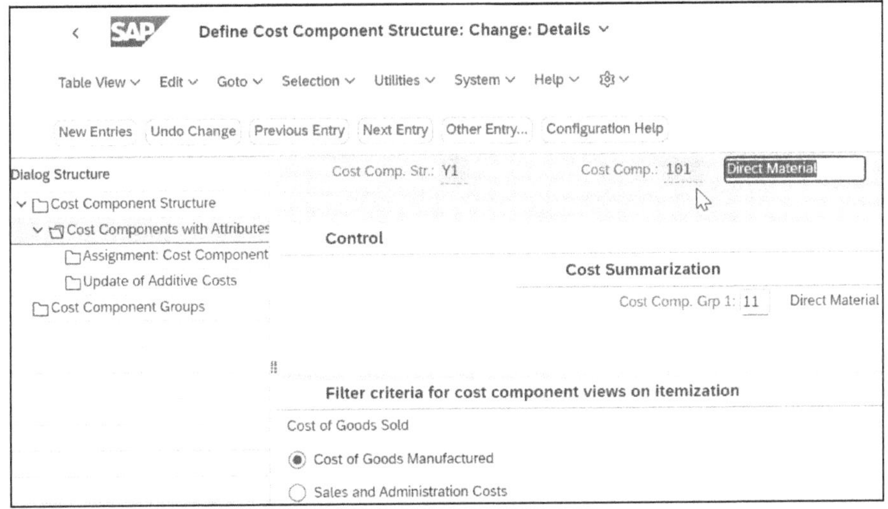

Figure 6.36 Details of Direct Material Cost Component, Showing Cost Component Views

6.6.5 Defining Cost Components

In SAP S/4HANA Cloud, you cannot create a new cost component structure, but you can add cost components to the structures shown earlier in Figure 6.32. Notice that in Figure 6.36, once you are within a cost component structure, you can use the **New Entries** button to add further cost components to meet your individual needs as shown in Figure 6.37. You'll then have to assign the underlying general ledger accounts to this cost component as described in Section 6.2.6.

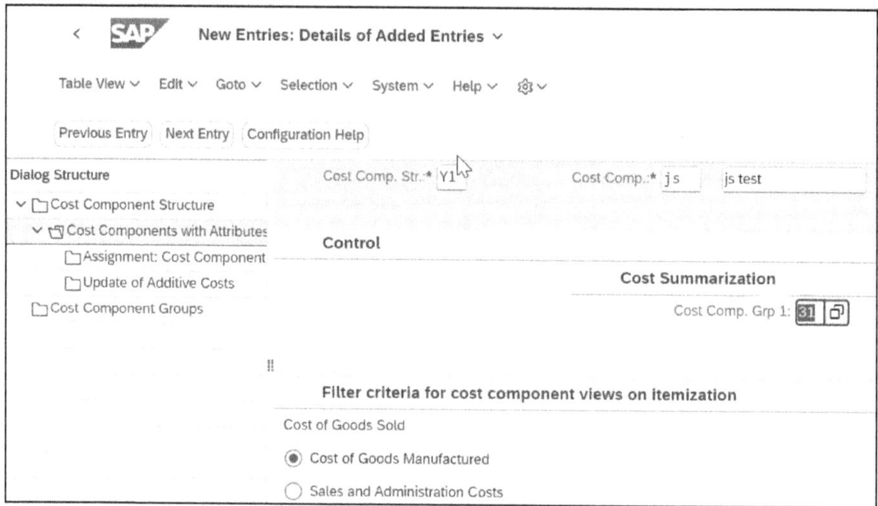

Figure 6.37 Adding a Cost Component

6.7 Summary

In this chapter, we discussed cost components and how they relate to product costing. We looked at how cost components group costs of a similar type by cost element. We defined cost component views and examined how to assign auxiliary cost component structures, transfer structures, and organizational units. We also looked at how to create and assign cost component groups.

Now that we've examined cost components in this chapter, we'll continue the preparation for creating cost estimates by examining costing variant configuration in Chapter 7 and Chapter 8.

Chapter 7
Costing Variant Components

A costing variant contains all the control parameters and settings for creating a cost estimate, such as determining prices when costing materials and activities.

7

Now that we've examined master data, costing sheets, and cost components in previous chapters, let's look at the costing variant. You base every cost estimate on a costing variant. The costing variant contains all the control parameters and settings for costing, such as selecting prices when costing materials, activities, and business processes.

A costing variant contains six control components:

- Costing type
- Valuation variant
- Date control
- Quantity structure control
- Transfer control (optional)
- Reference variant (optional)

We'll examine each of these components in detail in the following sections. First, we'll look at how you set up and configure the costing variant, and we'll then examine each component. In Chapter 8, we'll examine the costing variant tabs in detail.

7.1 Define Costing Variant

You access costing variant configuration with Transaction OKKN or via the IMG menu path **Controlling • Product Cost Controlling • Product Cost Planning • Material Cost Estimate with Quantity Structure • Define Costing Variants**. The screen shown in Figure 7.1 is displayed.

Costing Variants

Costing Variant	Name
PPC1	Standard Cost Est. (Mat.)
PPC2	Mod. Std Cost Est. (Mat.)
PPC3	Current Cost Est. (Mat.)
PREM	Prel. Cstg Cost Collector

Figure 7.1 Define Costing Variants

Standard costing variants are listed in Figure 7.1. You can use these or create your own.

Creating Costing Variants

If you need settings different from the system/cloud-supplied best-practice costing variants, you can create your own beginning with the letters X, Y, or Z without changing the standard ones so you can reference them.

Costing variant PPC1 is system-supplied for standard cost estimates. You can create your own costing variant—for example, ZPC1—and make changes while referring to PPC1. You also know from the naming logic that you copied ZPC1 from PPC1 settings.

Cloud systems and systems using best practice content include additional costing variants shown later in Figure 7.65 with the naming convention **PY***.

If your changes involve settings in either the costing type or valuation variant discussed in Section 7.2 and Section 7.3, you should first copy the system-supplied components and then make the changes to your own costing variant components. You can then create your own costing variant that includes your costing type and valuation variant. Note that you cannot change the system-supplied components in the cloud.

You cannot change the costing type or valuation variant if you copy a standard costing variant; you must create your own costing type and valuation variant first. When you create and save a new costing variant, the costing type and valuation variant are no longer changeable.

Standard costing variants for group and profit center valuation are not system-supplied, so you need to create your own and link them to a costing type to support group and profit center valuation. If you work with the best-practice settings, the costing variant for group valuation is delivered with scope item 5W2 and for profit center valuation with scope item 6VQ.

Double-click a costing variant in Figure 7.1 to display the screen shown in Figure 7.2.

The **Control** tab lists the six costing variant control parameters as buttons. The **Costing Type** and **Valuation Variant** texts are grayed out and not changeable because they form the key structure of costing results.

Although it's technically possible to have two costing variants with the same costing type and valuation variant, you should never configure this to prevent the possibility of overwriting data. The costing variant is not a key in the KEKO table.

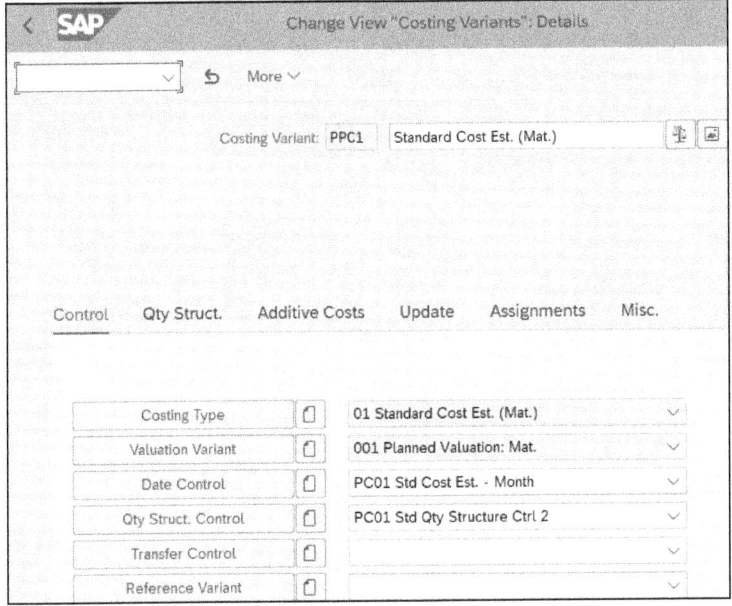

Figure 7.2 Costing Variant Details Control Tab

Overview Icon

Click the **Overview** icon at the right to display an overview of all the costing variant component settings.

We'll examine the six components here and then the costing variant tabs in Chapter 8.

7.2 Costing Type

To display all costing types, click the new page icon next to the **Costing Type** button. The screen in Figure 7.3 is displayed.

Costing Type	Name
01	Standard Cost Est. (Mat.)
0L	Leading Costing Type
10	Inv. Costing: Tax Law
11	Inv. Costing: Comm. Law
12	Mod. Std Cost Est. (Mat.)
13	Ad Hoc Cost Estimates
19	Product Cost Collector
NL	Non-Leading Costing Type

Figure 7.3 Costing Type Overview

Available standard costing types are listed in the overview screen. You can either use a standard costing type or create your own. We'll return to this topic in Chapter 19 to look at the additional costing types delivered to use with universal parallel accounting, where the costing type is linked with the ledger for the purposes of legal valuation.

Costing Type Key

When you create your own costing type, you must use an alphanumeric key because the system reserves numeric keys for the standard keys. A common use case is to add separate costing types for group and profit center valuation if you aren't using the best-practice settings.

Double-click a costing type to display the details shown in Figure 7.4.

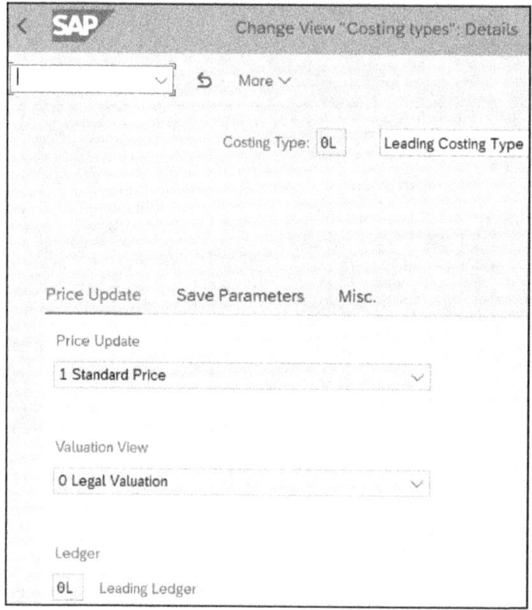

Figure 7.4 Costing Type Price Update Tab

Let's examine the fields in the **Price Update** tab displayed when you initially view the costing type.

7.2.1 Price Update

The **Price Update** tab contains three fields: **Price Update**, **Valuation View**, and **Ledger**. The **Leading Ledger** contains journal entries for all business transactions for legal purposes, while a non-leading ledger contains management accounting journal entries only.

The **Ledger** is only visible in the **Costing Type** with the universal parallel accounting business function activated, which we'll discuss in detail in Chapter 19. **Price Update** controls whether you can update the material master and which fields, as shown in Figure 7.5.

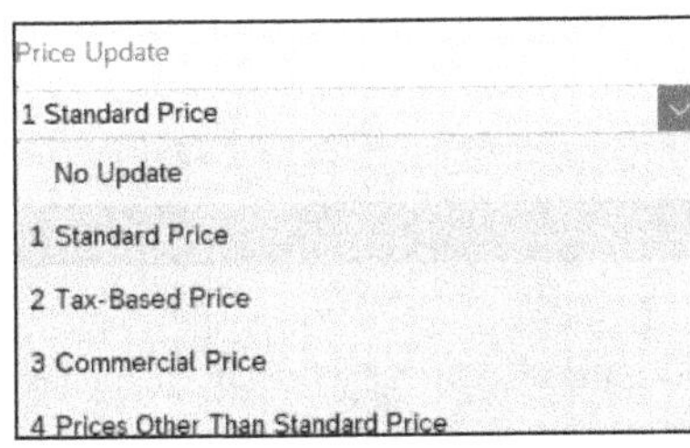

Figure 7.5 Price Update Possible Entries

The system only allows one costing type to update the standard price for each valuation view. Inventory is typically valued based on the **Legal Valuation** selection, as shown previously in Figure 7.2. When a costing type contains **1: Standard Price** in the **Price Update** field, this field isn't changeable.

Because you normally assign each costing type to a unique costing variant, only one costing variant, such as PPC1, can update the standard price. This allows you to control who can update the standard price by restricting access to the costing variant containing this costing type. As we'll see in Chapter 19, universal parallel accounting allows the update of a different standard price for each ledger, so you'll find that multiple costing variants and costing types are delivered that reference the various ledgers.

You can change the entry in the **Price Update** field for costing types that don't contain standard price in the **Price Update** field because updating other material master fields, such as **2: Tax-Based** and **3: Commercial Price**, isn't as critical because those fields don't directly affect inventory valuation, but only the balance sheet valuation performed at year-end.

You typically use the **4: Prices Other Than Standard Price** entry to update the **Planned Prices** fields in the **Costing 2** view.

Product Development Team

Product development needs to create cost estimates to determine the cost viability of proposed new products and when modifying existing products. To allow the product development team the flexibility to create their own cost estimates, you create a costing variant containing a costing type with **No Update** in the **Price Update** field for their exclusive use. Product development team members can create as many cost estimates as needed with their costing variant without any concern that they could influence the standard price.

Now let's discuss the **Valuation View** field in the **Price Update** tab as shown in Figure 7.6.

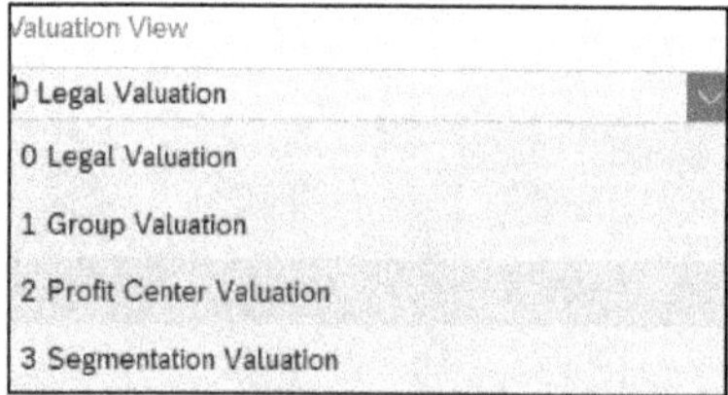

Figure 7.6 Price Update Valuation View Possible Entries

If you are working with one valuation view, you'll only need to create a costing type for **Legal Valuation**. If you are working with parallel valuation, you'll create costing types for each valuation for which you need to create cost estimates. You can activate parallel valuation per controlling area by assigning a currency and valuation profile to the controlling area with Transaction 8KEM, as shown in Figure 7.7.

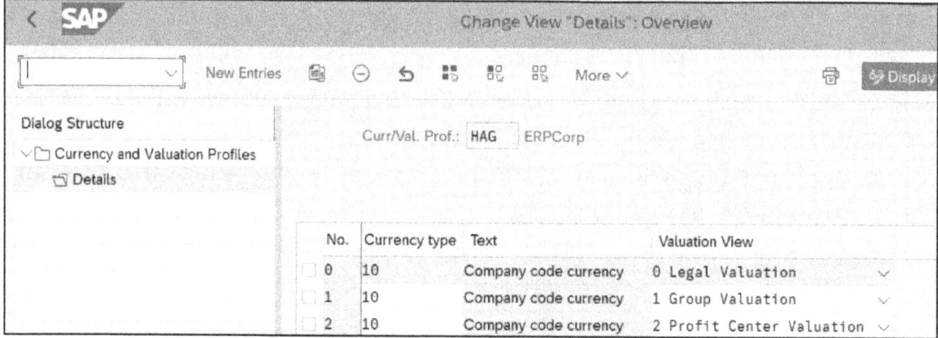

Figure 7.7 Currency and Valuation Profile

The currency and valuation profile approach is different with universal parallel accounting activated, as described in detail in Chapter 18.

When running reports with parallel valuation, you need to specify which valuation view to base the report on. Let's review each valuation view:

- **Legal Valuation**
  This is the only valuation view you can have if you aren't using parallel valuation. You base currency on the company code currency, and internal profits for transactions between internal trading partners are included. Transfer pricing determines the markup, representing internal profits added to materials transferred between internal trading partners.
- **Group Valuation**
  You typically base group-consolidated reports on group valuation. Here are two aspects you should consider:

- Group currency
- Profit elimination between internal trading partners

You must convert all currencies into the group currency, and you must eliminate internal profits between internal trading partners.

- **Profit Center Valuation**
  You reproduce business transactions between profit centers for management approaches based on internal prices. These can differ from the transfer prices agreed between company codes.
- **Parallel Cost of Goods Manufactured**
  This view is for international groups that are required to valuate their inventories in accordance with multiple accounting principles, for example, in accordance with International Financial Reporting Standards (IFRS) following group guidelines as well as in accordance with local accounting principles. You can valuate cost of goods manufactured (COGM) using multiple accounting principles in parallel.

You typically create parallel valuation for a controlling area with the material ledger. In Chapter 3, we discussed the extra columns available for each valuation view in the **Accounting 1** view with the Material Ledger.

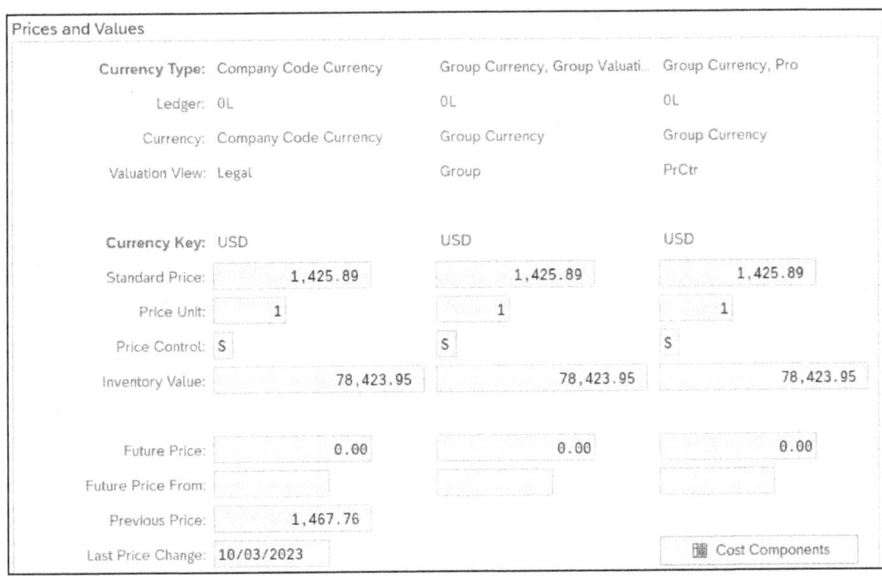

Figure 7.8 Accounting 1 View

Let's review the extra columns as shown in Figure 7.8:

- The first column contains prices based on the **Legal Valuation** view. You update the **Standard Price** by a standard cost estimate based on a costing variant with legal valuation in the **Valuation View** field of the costing type.

- The second column contains a **Standard Price** based on the **Group Valuation** view. You can update the standard price in this view with a cost estimate based on a costing variant by entering a group valuation in the **Valuation View** field of the costing type. In this case, you can retrieve a component group price from the **Future price** field in the **Group currency** column shown in Figure 7.8.

The **Accounting 1** view will look different with UPA activated, as you can have all currency/valuation combinations, which we'll discuss in more detail in Chapter 19.

Now that we've discussed the fields in the **Price Update** tab shown previously in Figure 7.4, let's discuss the next tab.

7.2.2 Save Parameters

Click the **Save Parameters** tab shown previously in Figure 7.4 to display the screen shown in Figure 7.9.

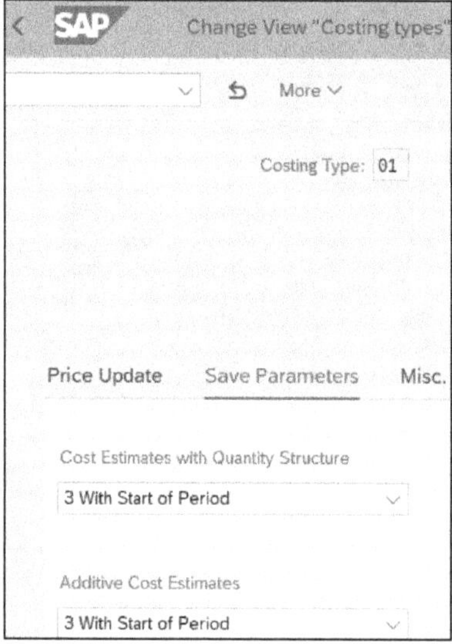

Figure 7.9 Costing Type Save Parameters Tab

We'll examine the fields in the **Save Parameters** tab in the following sections.

Cost Estimates with Quantity Structure

The **Cost Estimates with Quantity Structure** value determines which date is included in the key when you save a cost estimate, as shown in Figure 7.10.

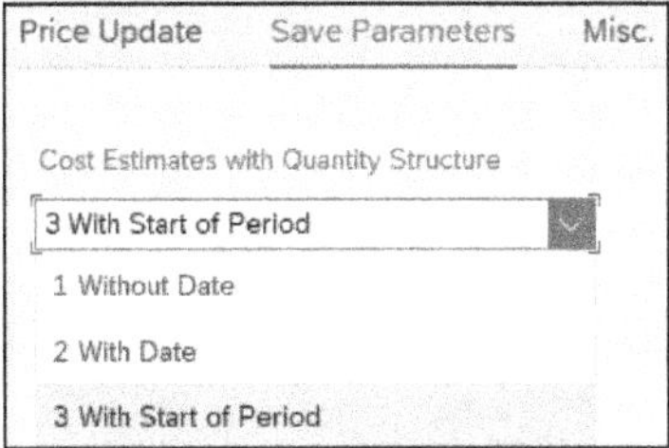

Figure 7.10 Save Parameters Cost Estimates Date

Here is a description of each of the three possible entries:

- **1: Without Date**
  With this option, you don't save the cost estimate with a date. You may use this for ad hoc cost estimates created on a one-off basis for reporting only.
- **2: With Date**
  The start date of the cost estimate will be applied as the key date. You may use this for repetitive manufacturing, or in simulation cost estimates.
- **3: With Start of Period**
  The costing type used to create standard cost estimates must contain **3: With Start of Period**. If you attempt to enter another option, you'll receive a message stating you can only save a standard cost estimate with the start of the period as the key date. This is because the standard cost estimate is tied to the inventory valuation in the material master, and this always references a period. For this reason, you can only create a standard cost estimate with reference to the accounting period, and the first day of the period is a key date.

Now we'll discuss the next field in the **Save Parameters** tab.

Additive Cost Estimates

The **Additive Cost Estimates** field determines the date included in the key when you save additive cost estimates, as shown in Figure 7.11.

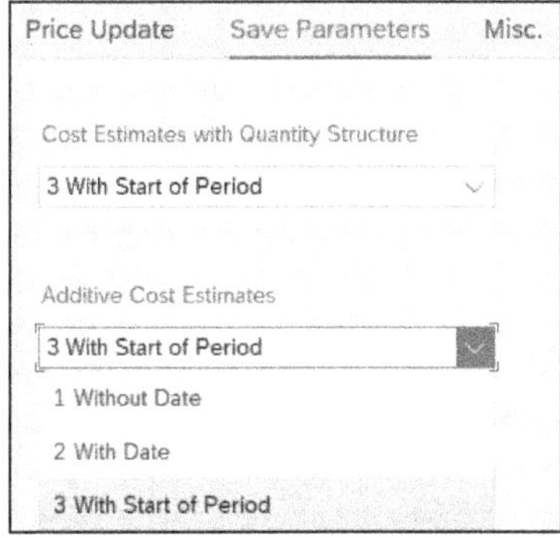

Figure 7.11 Save Parameters Additive Cost Estimates Date

We discussed these dates in the previous section. If you don't intend to use additive cost estimates, you can enter **1 Without Date**. Additive cost estimates are used in conjunction with the cost estimate based on the BOM and the routing and need the same save parameters to connect the cost estimates.

Now that we've discussed the **Save Parameters** tab, let's discuss the next tab.

7.2.3 Miscellaneous

Click the **Misc.** (miscellaneous) tab shown in Figure 7.11 to display the screen shown in Figure 7.12.

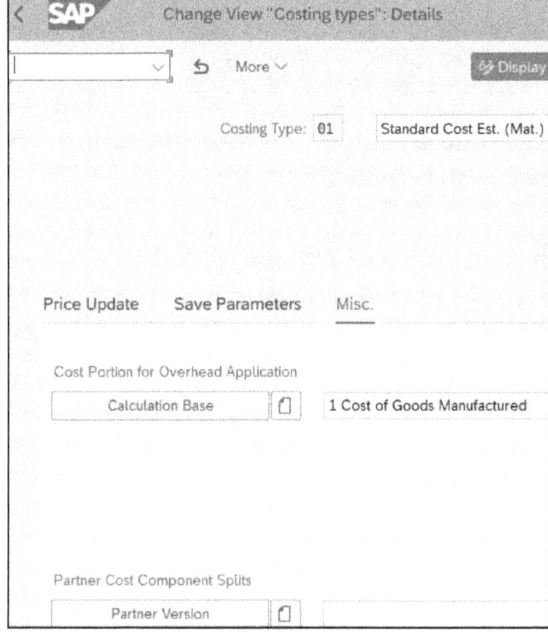

Figure 7.12 Costing Type Misc. Tab

The **Misc.** tab contains fields to calculate overhead in alternative ways, such as cost component views and partner versions.

Cost Portion for Overhead Application

You can use cost component views to calculate overhead for semifinished goods in finished goods. Cost component views filter cost components based on cost components and cost component filters. Click **1 Cost of Goods Manufactured** to display a list of possible entries as shown in Figure 7.13.

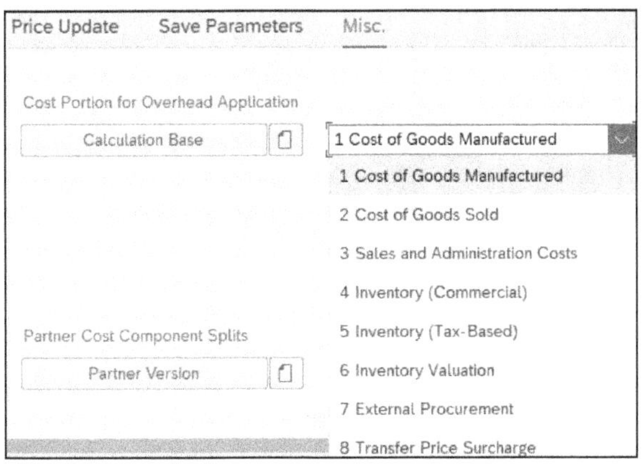

Figure 7.13 Misc. Tab Cost Component Views

Entering a cost component view as a calculation base allows you to determine overhead for semifinished goods on a calculation base independent of total cost estimate costs. Typically, you'll use a calculation base of the **1 Cost of goods manufactured** to calculate overhead for semifinished goods. This means you exclude **3 Sales and administration costs** from the calculation base.

Click the **Calculation Base** button to see details of the cost component view. To branch out to the cost component views configuration screen, click the new page icon.

Now we'll examine the next field in the costing type **Misc.** tab.

Partner Cost Component Splits

Click the **Partner Version** new page icon to see a list of partner versions. Double click on a partner version to view details, as shown in Figure 7.14.

If you are using a group costing parallel valuation, you can specify a partner version of the costing type. (We discussed group costing and group cost estimates in Section 7.2.1.) The **Partner Version** contains a key that uniquely defines a combination of a partner and a direct partner. A *partner* is a business unit (such as a plant or company code) involved in a value-added process, which you assign by making selections in Figure 7.14.

A direct partner transfers its product or service to another partner directly. You can generate a cost component split for each partner by the group costing function, showing all value-added segments. To hide the portion of the value added by the direct partner procuring the product or service on transfer to the receiving partner, you can subsume the data under the value added by the direct partner. The choices available when you select the **Direct Partner** tab are the same as the list shown for the **Partner** tab in Figure 7.14.

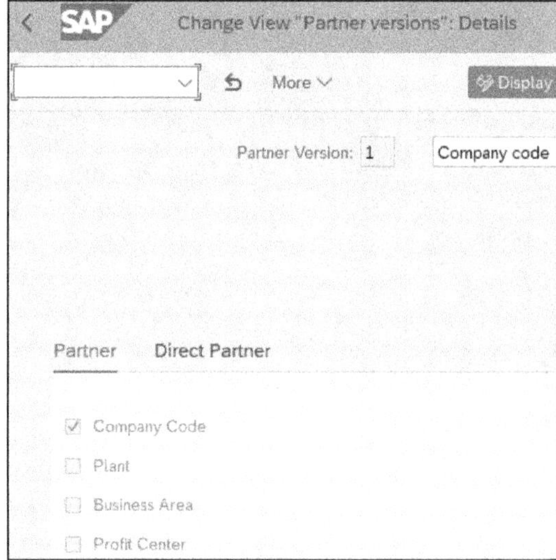

Figure 7.14 Partner Version Details

When you create a cost estimate with a costing type with an assigned partner version, the **Partner** cost component split button becomes available, as shown in Figure 7.15.

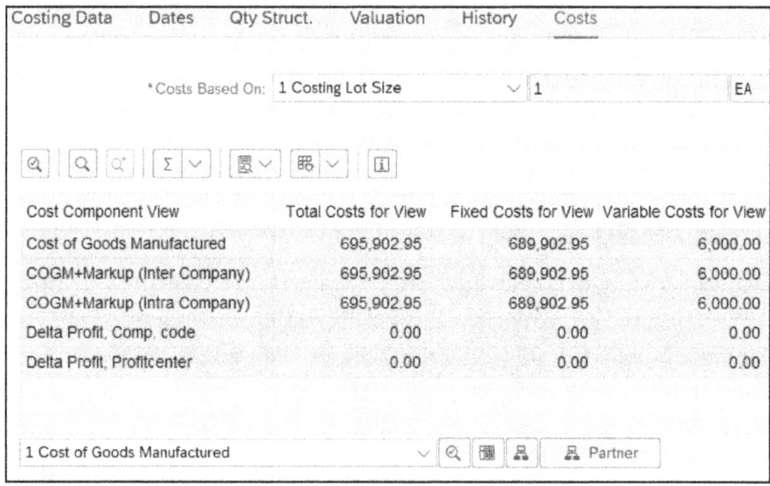

Figure 7.15 Partner Cost Component Split Button

Click the **Partner** button to display the split shown in Figure 7.16.

This is the cost component split for the cost estimate shown earlier in Figure 7.15. To view a partner cost component split, click the two-person icon to display a list of partner views as shown in Figure 7.17. Double-click a partner view to display the associated costs.

Now that we've discussed the costing type, let's look at the valuation variant.

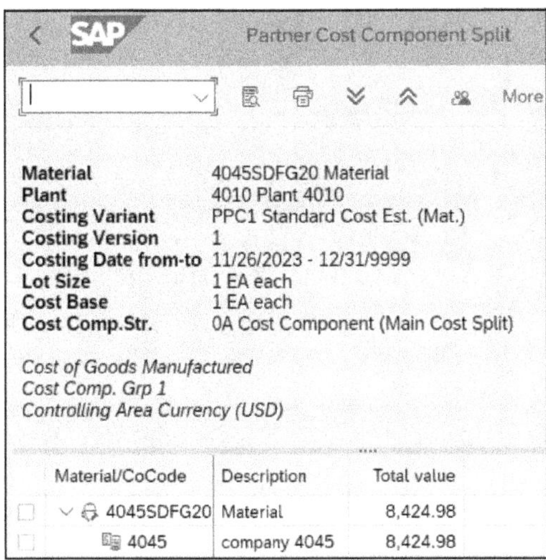

Figure 7.16 Partner Cost Component Split

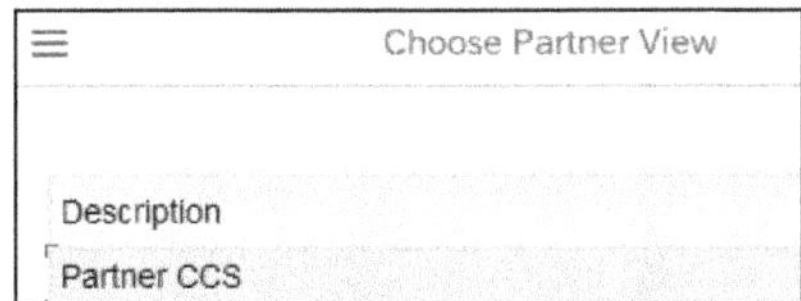

Figure 7.17 Choose Partner View

7.3 Valuation Variant

To display all valuation variants, click the new page icon next to the **Valuation Variant** button shown earlier in Figure 7.2 to display the list shown in Figure 7.18.

Valuation Variant	Name
001	Planned Valuation Mat.
002	Current Valuation: Mat.
003	Tax-Based Valuation
004	Commercial Valuation
005	MRP Valuation
006	Production Order: Planned
007	Production Order: Actual
008	Planned Valuation: Orders
009	Target Valuation: Mat.

Figure 7.18 Valuation Variant Overview

You can use a standard valuation variant or create your own. The valuation variant contains the parameters required for the valuation of a cost estimate. Let's discuss the details contained in a valuation variant in the following sections. We'll return to this topic in Chapter 19 when we look at the new valuation variants introduced with universal parallel accounting.

Access Valuation Variants

You can also access valuation variants with Transaction OKK4 or via the IMG menu path **Controlling • Product Cost Controlling • Product Cost Planning • Material Cost Estimate with Quantity Structure • Costing Variant: Components • Define Valuation Variants**.

7.3.1 Material Valuation

Double-click a valuation variant to display the screen shown in Figure 7.19.

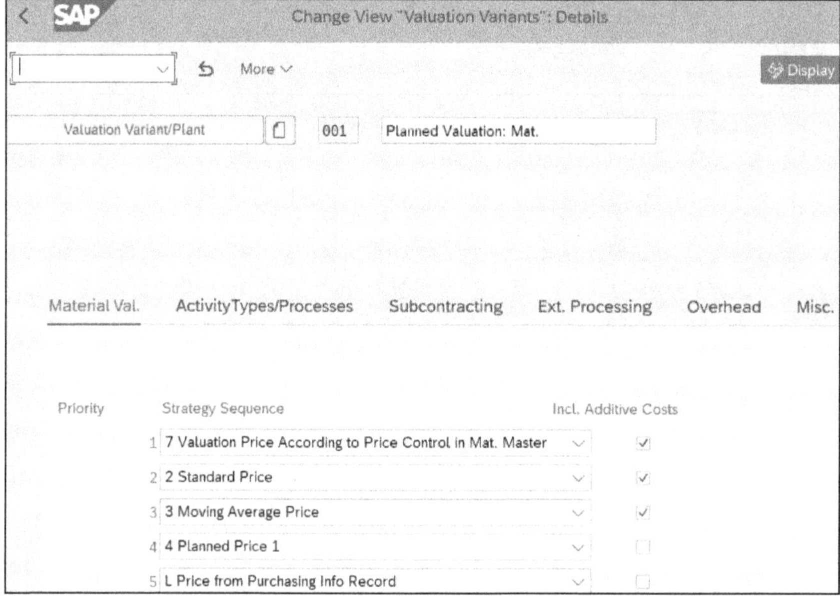

Figure 7.19 Valuation Variant Details

The **Material Val.** (material valuation) tab displays when you initially view a valuation variant.

Note: Valuation Variant/Plant

You can use different valuation strategies or different overhead rates in plants that belong to the same company code. Click the **Valuation Variant/Plant** button to assign a valuation variant to a plant. Otherwise, the same valuation variant applies to all your plants.

In material valuation, you define the **Search Sequence** for prices for component materials. In the following sections, we'll explore the strategy sequence to determine the material prices and the sub-sequence to select purchasing prices and discuss the use of additive costs.

Strategy Sequence

The **Strategy Sequence** directs the search for entries in **Price** fields in the material master, purchasing info record, and condition types. The search continues for the fields in a defined priority (in the **Priority** column) until a valid price is located. This price is taken as the component price, which is multiplied by the component quantity from the BOM to determine the material value, which then rolls up to the highest-level material cost component. If a price isn't located, you'll receive an error message.

You cannot release a standard cost estimate to update the standard price until you eliminate all error messages. After you eliminate the errors, for example, by entering a valid purchasing info record price, you can create the standard cost estimate again and mark and release it.

You can enter up to five strategies prioritized, as shown in the **Priority** column in Figure 7.19.

Example: Strategy Sequence

Enter the following strategy sequence in the **Material Val.** tab:

- **4: Planned Price 1**
- **2: Standard Price**
- **3: Moving Average Price**

If the system finds a valid entry in **Planned Price 1** in the **Costing 2** view, it will take this as the material price. If it doesn't find a value in **Planned Price 1**, a price in the **Standard Price** field is used. If there is no standard price, the **Moving Average Price** is used.

We also presented a case study involving valuation strategies in Chapter 3, Section 3.5.2.

Left-click in a field to display a list of possible entries. The screen shown in Figure 7.20 is displayed.

Scroll down the list to see all possible entries. A **Strategy Sequence** involving Planned Price fields may involve manually updating **Planned Price 1** in the material master. You can also automatically update the **Planned Price 1** field with cost estimates with a costing variant that retrieves prices from other fields. This is typically the valuation strategy you use when purchasing info records aren't maintained. A purchasing info record stores all the information relevant to the procurement of a material from a supplier, including the latest quoted purchase prices.

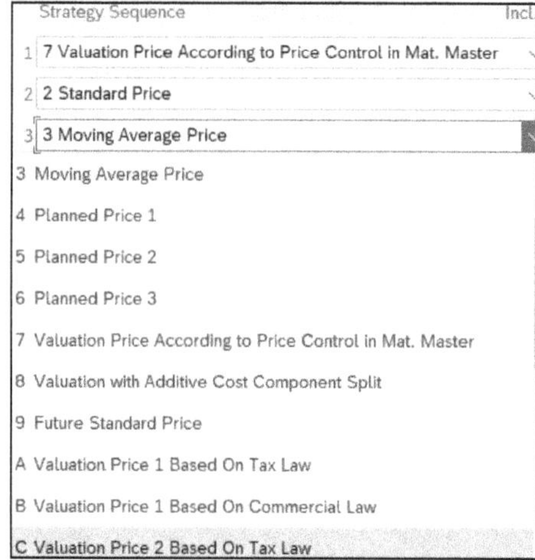

Figure 7.20 Material Valuation Strategy Possible Entries

A more streamlined approach involves adding purchasing info records as part of a strategy sequence. Because the purchasing department maintains purchasing info records to store supplier quotation prices, you don't need to maintain entries in material master **Planned Price** fields.

> **Note: Info Record Advantages**
>
> Another advantage of Purchasing maintaining purchasing info records is that the latest valid supplier price defaults into the **Purchase Price** field when you create a purchase requisition or a purchase order.

Sub-Strategy Sequence with Purchasing Info Record Valuation

When you choose **Price from Purchasing Info Record** as a valuation strategy, additional fields become available in the **Material Val.** Tab, as shown in Figure 7.21.

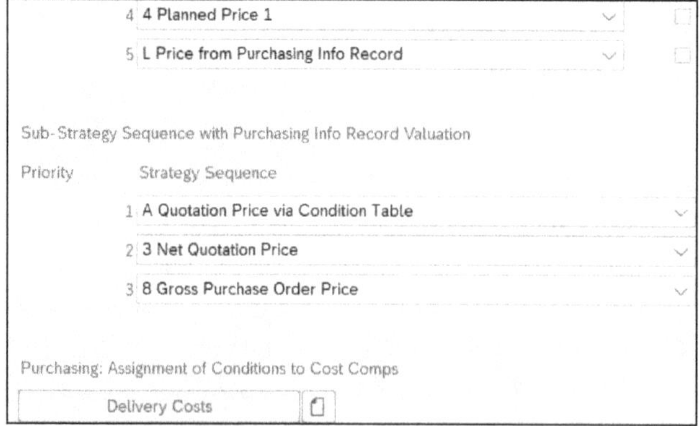

Figure 7.21 Sub-Strategy Sequence with Purchasing Info Record

Additional valuation strategies are required because multiple prices are available within purchasing info records. To display a list of possible entries in the purchasing info record valuation strategy fields, left-click in a field. The screen shown in Figure 7.22 displays.

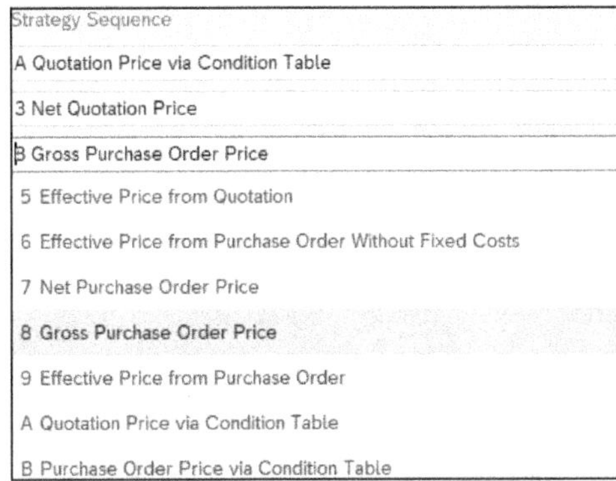

Figure 7.22 Purchasing Info Record Valuation Strategies

There are five types of purchasing prices:

- **Gross price (strategies 4 and 8)**
  Gross price doesn't take any discounts or surcharges into account. The system-supplied gross price condition is PB00. *Conditions* are stipulations agreed with vendors for prices, discounts, and surcharges. You can maintain conditions in quotations, purchasing info records, outline agreements, and purchase orders. An *outline agreement* is a longer-term purchase arrangement with a vendor for the supply of materials or the performance of services.
- **Net price (strategies 3 and 7)**
  Net price is calculated from the gross price by taking discounts and surcharges into account. You can enter either the purchasing info record net price or gross price. You can only maintain the net price, however, if you have not maintained any purchasing info record conditions. After you create the conditions, you must maintain the price in the **Conditions** screen that contains the gross price.

  If you don't maintain the vendor's net price in the purchasing info record, the system proposes the net price of the last purchase order issued to the vendor when a new purchase order is created.
- **Effective price (strategies 5 and 9)**
  An effective price is the end price after taking *all* conditions, including cash discounts and delivery costs, into account. You first subtract vendor discounts and rebates to calculate the net price, and then you add on other costs such as freight charges and customs and duty. The system automatically determines the price. Display the purchasing info record or purchase order conditions to determine how the price was calculated.

- **Effective prices without fixed costs (strategies 2 and 6)**
  With these strategies, the effective price is calculated from all of the conditions that are also contained in the assigned calculation schema, with the exception of those with a condition type assigned to the *fixed amount* calculation type (data element KRECH = B). See SAP Note 351835 for more details.
- **Price via condition table (strategies A and B)**
  You can assign the purchasing condition type to an origin group. You can maintain this table using the **Delivery Costs** button, which we'll discuss further in the next section.

You retrieve the prices discussed in the preceding list from either the purchasing info record or the last purchase order, as discussed in the following points:

- **Prices from a quotation (strategies 2 to 5)**
  The net and effective quotation prices are maintained in the **Purchasing Organization Data 1** data view, and the gross quotation price and other conditions are maintained in the **Conditions** data of a purchasing info record.

  The system updates purchasing info record conditions if the **InfoUpdate** checkbox in the **Material Data** tab of a purchase order item is selected. SAP Note 569885 contains more details about the **InfoUpdate** checkbox.
- **Prices from a purchase order (strategies 6 to 9)**
  A record of the last purchase order is maintained by the system in the **Purchasing Organization Data 2** data view of a purchasing info record. You can access this screen by selecting **Goto • Purch. Org. Data 2** from the menu bar when viewing a purchasing info record, or you can display the last purchase order directly by selecting **Environment • Last Document.** You can enter the last purchase order prices and conditions as valuation strategies in Figure 7.22.

  Note that you can only retrieve the price from purchase info records and not purchasing contracts.

Note: Purchase Order Info Records

You can display the info record that a purchase order referenced by selecting **Environment • Info Record** from the purchase order menu bar. You can also display a list of purchasing info records for a material with Transaction ME1M or for a supplier with Transaction ME1L. Double-click a purchasing info record in the lists to display details.

We discussed purchasing info records in detail in Chapter 4.

Purchasing: Assignment of Conditions to Cost Components

Now that we've reviewed the strategy sequence sections of the **Material Val.** tab, we'll discuss **Delivery Costs** as shown in Figure 7.23.

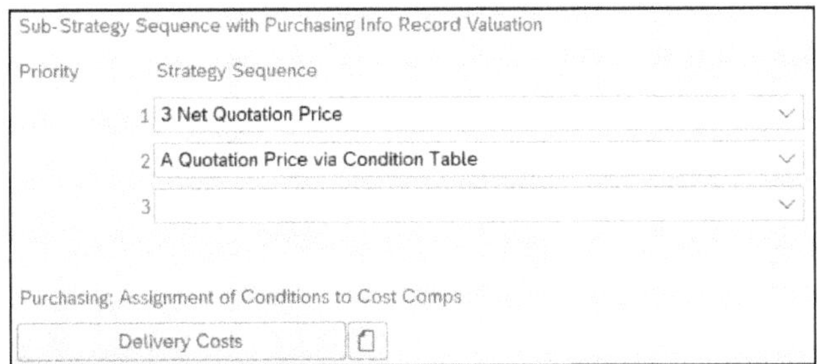

Figure 7.23 Delivery Costs Section of the Material Val. Tab

You can assign purchasing info record condition types to origin groups via the **Delivery Costs** button. A *condition type* is a key that identifies a condition that can provide a detailed breakdown of material costs such as gross price, duty, freight, and insurance. This assignment provides a row in the cost estimate itemization under the same cost element for each assigned origin group. You can also assign the origin groups to separate cost components to provide more detailed reporting on material costs in cost estimates.

You can assign condition types to origin groups by either clicking the **Delivery Cost** button, with Transaction OKYO, or via the IMG menu path **Controlling • Product Cost Controlling • Product Cost Planning • Selected Functions in Material Costing • Raw Material Cost Estimate • Assign Condition Types to Origin Groups**. You'll see a screen like the one shown in Figure 7.24.

Condition Type	Valuation View	Subtype of Valuation View	Valuation Variant	Controlling Area	Company...	Valuation area	Origin Group
FGW1	0		0Y1	A000			YFC
FQU1	0		001	A000			YFC
FQU1	0		00L	A000			YFC
FQU1	0		02L	A000			YFC
FQU1	0		0Y1	A000			YFC
FRA2	0		001	C005			100
FVA1	0		001	A000			YFC
FVA1	0		00L	A000			YFC
FVA1	0		02L	A000			YFC
FVA1	0		0Y1	A000			YFC
PB00	0		001	C005			Z1

Figure 7.24 Assignment of Condition Types to Origin Group

You assign each **Condition Type** to an **Origin Group** for each **Valuation area** (plant) in a **Company** (company code). Examples of condition types you may assign to origin groups include duty, freight, and insurance. You can see these costs separately in the cost component split of a raw material cost estimate if you assign these origin groups to separate cost components. See Chapter 6 for more details on assigning origin groups to cost components.

Note: Purchasing and Condition Types

Assigning condition types to origin groups is only useful if purchasing assigns costs with condition types in purchasing info records and purchase orders. When considering using condition types in material costing, you should review the requirements with purchasing.

Now that we've discussed assigning condition types to origin groups, we'll examine how to include additive costs in cost estimates.

Include Additive Costs

In the **Material Val.** tab shown earlier in Figure 7.21, you can select checkboxes that allow the inclusion of additive costs to raw material cost estimates depending on which valuation strategy is successful in locating a valid price, as shown in Figure 7.25.

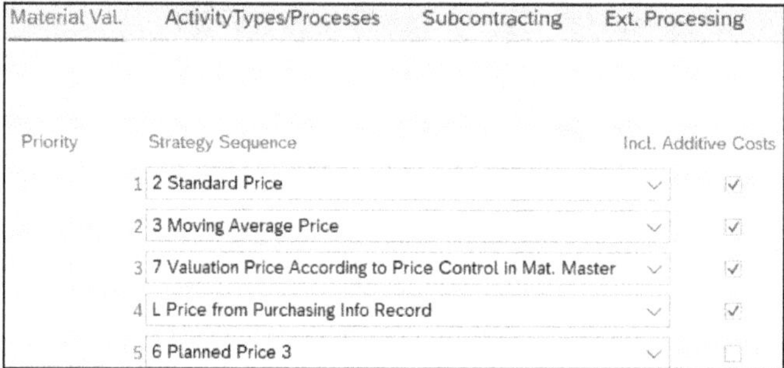

Figure 7.25 Include Additive Costs Indicators

Additive costs allow you to enter costs manually in a unit cost estimate (spreadsheet in SAP) that you can add to an automatic cost estimate with quantity structure. By selecting a checkbox, you allow manual costs, if they exist, to add to an automatic cost estimate depending on the successful valuation strategy. We'll follow some examples of when to include additive costs in raw material cost estimates.

Examples: Additive Costs

Here are some examples of where additive costs may be used:

- A pharmaceutical company needs visibility of the price markup, which represents internal profits when they transfer a material between company codes in the United States. Additive costs are created that represent the price markup, which is then included in the standard cost estimate if you use a legal cost estimate rather than an additional group cost estimate.
- A tile manufacturer adds freight costs when shipping a material internally from one plant to another plant using additive costs. You can add freight costs to materials purchased from external vendors via purchasing info record condition types (as

explained in the previous section), but you cannot use purchasing info records for stock transfers between plants.

Now that we've examined the **Material Val.** tab in Figure 7.25, we'll move on to the next tab.

7.3.2 Activity Types and Processes

Here you define the search sequence for internal activities and processes. The **Activity-Types/Processes** tab divides into the **Strategy Seq.** and **Cost Center Accounting** sections, which we'll discuss next.

Strategy Sequence

The *strategy sequence* directs the system to search **Price** fields in activity types and processes. The system continues to search the specified fields in the priority you define until a valid price is located. The system takes this price as the activity price, which it multiplies by the quantity to determine the activity value, which it then rolls up to the highest-level activity cost component. If the system doesn't successfully locate a price, it issues an error message that must be resolved before you can release the cost estimate to update the standard price.

Select the **ActivityTypes/Processes** tab in Figure 7.25 and click in any **Strategy Seq.** (strategy sequence) field to display a list of possible entries as shown in Figure 7.26.

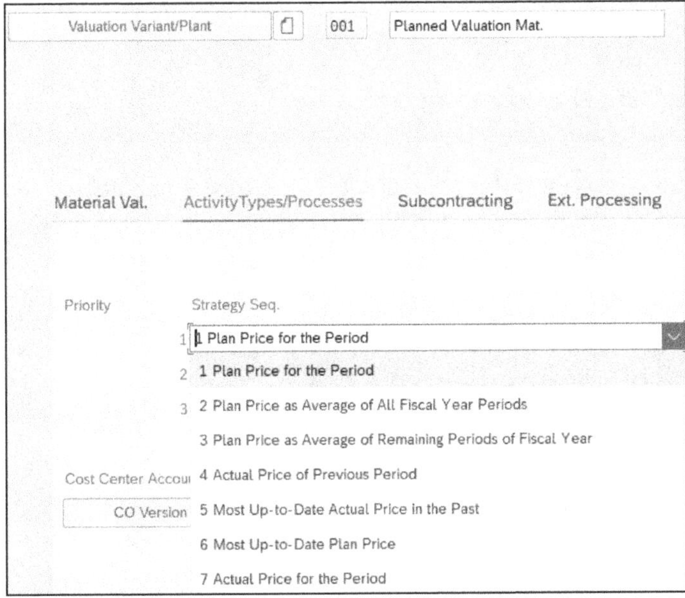

Figure 7.26 Activity Type Valuation Strategy Possible Entries

You can choose from three types of activity prices:

- **Plan price (strategies 1, 2, and 3)**
  With these strategies, you plan activity prices per fiscal period. You have a choice of the following:
  - **1: Plan Price for the Period**
  - **2: Plan Price as Average of All Fiscal Year Periods**
  - **3: Plan Price as Average of Remaining Periods of Fiscal Year**

  An advantage of using an average plan price is that it may decrease variance during the year. The standard price is calculated using one activity price. If you base the cost estimate calculation on the plan activity price of one period, the activity price may vary in the remaining periods, resulting in a variance. To make variance analysis easier, one option is to base the cost estimate on an average plan price, which should then reduce the variance occurring due to activity price variance. Some companies may need to analyze this variance, however, so it depends on your management reporting requirements and also on how often you carry out main costing runs.
- **Actual price (strategies 4 and 7)**
  You have a choice of using the following:
  - **4: Actual Price of Previous Period**
  - **7: Actual Price for the Period**

  You can calculate the actual price with Transaction KSII, which allocates all remaining cost center debits to manufacturing orders or product cost collectors, which is useful if you carry out a main costing run every period.
- **Most up-to-date price (strategies 5 and 6)**
  You have a choice of using the following:
  - **5: Most Up-to-Date Actual Price in the Past**
  - **6: Most Up-to-Date Plan Price**

  These two strategies are especially useful if you don't get a chance to update planned activity prices for the following fiscal year before activities attempt to access them. The following case study provides an example of how using the most up-to-date price strategy can be helpful when placed as a priority 2 or 3 in the strategy sequence.

Case Study: Most Up-to-Date Plan Price

Manufacturing orders typically use costing variant PPP1 with valuation variant 006, which calculates order-planned costs. This system-supplied valuation variant has **Plan Price for the Period** as the only activity price strategy.

Production may need to create manufacturing orders for the next fiscal year before reaching the end of the current fiscal year. If you have not yet calculated and entered the plan activity rate for the next fiscal year, then you generate error messages when you attempt to calculate the manufacturing order plan costs.

However, if you include **Most Up-to-Date Plan Price** as a priority 2 strategy, the system will successfully access the plan activity price for the current fiscal year, even though there is no plan price in the period in the next fiscal year in which you create the manufacturing order.

Now that we've looked at activity type and processes strategy sequences, we'll examine the next section of the **ActivityTypes/Processes** tab.

Cost Center Accounting

When entering plan activity rates, you also enter a **Version**. Version 0 is typically set up for both plan and actual costs and is the version you enter in the **Cost Center Accounting** section of the **ActivityTypes/Processes** tab, as shown in Figure 7.27.

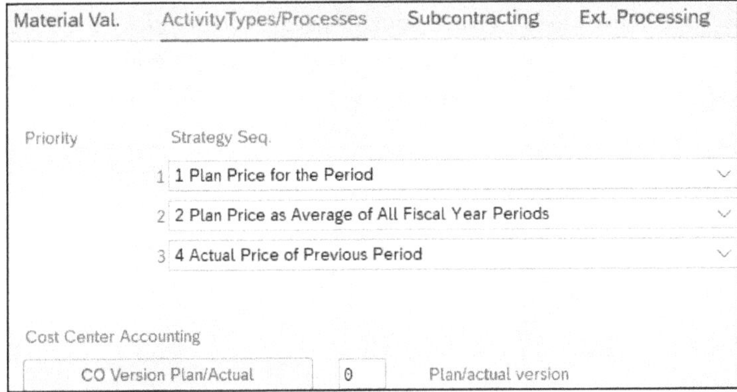

Figure 7.27 Activity Type CO Version Plan/Actual

You can create up to a thousand (999) versions, if necessary, to configure alternate plan scenarios because the **Version** key is a three-digit alphanumeric field. You typically carry out scenario analysis in other versions and then copy the most likely scenario to version 0. You base cost estimates on the activity type rates of version 0 if entered in the screen shown in Figure 7.27. Because version 0 also contains actual data, you can carry out variance analysis to explain the difference between plan and actual costs. This process changes with universal parallel accounting because activity prices can be ledger specific. You can continue to use the same price for all ledgers. This means that you won't be able to use Transaction KP26 to enter activity rates but must instead log on with the Overhead Accountant role and use the SAP Fiori app Manage Cost Rates – Plan to enter the activity rates manually or to show the result of the data transfer from SAP Analytics Cloud for planning.

You can see details of all versions by either clicking the **CO Version Plan/Actual** button shown in Figure 7.27, with Transaction OKEQ, or following the IMG menu path **Controlling • General Controlling • Organization • Maintain Versions.** You'll see a screen similar to the one shown in Figure 7.28.

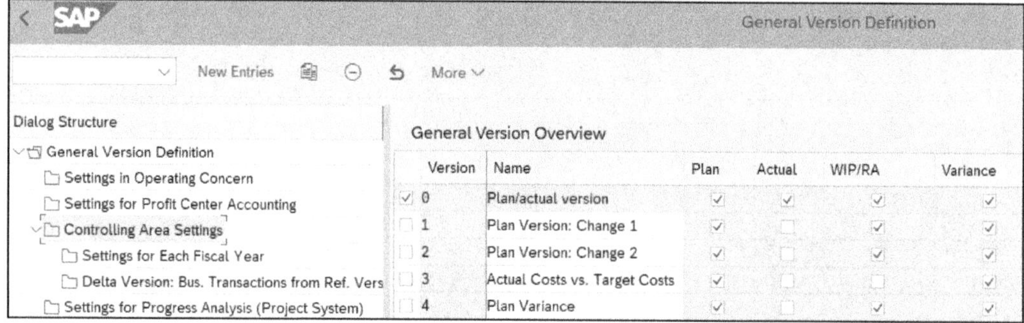

Figure 7.28 General Version Definition

In the **Version 0** row, both the **Plan** and **Actual** are selected, along with **WIP/RA** and **Variance**. To see further details about version 0, select the first row and double-click **Controlling Area Settings** to see the screen shown in Figure 7.29.

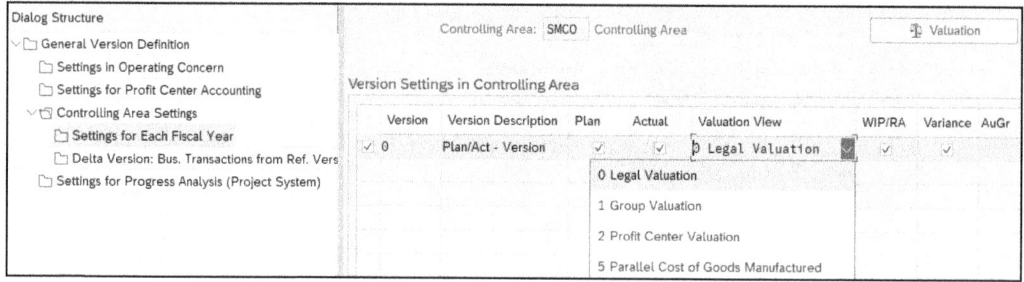

Figure 7.29 Controlling Area Settings

Controlling Area 1000 contains **Version 0** based on **Legal Valuation** as shown in the **Valuation View** column. If you use parallel valuation and transfer prices, you define parallel actual versions in addition to the operational version 0. Click the **Valuation View** field to see a list of possible entries as shown in Figure 7.29. If you are only using one valuation, you'll see **Legal Valuation** in this field. If you are using multiple valuations, you can check the settings by clicking the **Valuation** button. Select the **Version 0** row and double-click **Settings for Each Fiscal Year** to display a list of fiscal years as shown in Figure 7.30.

Double-click one of the years in the **Year** column to display the detailed **Version Settings**.

Now that we've looked at **ActivityTypes/Processes**, let's examine the fields in the valuation variant **Subcontracting** tab.

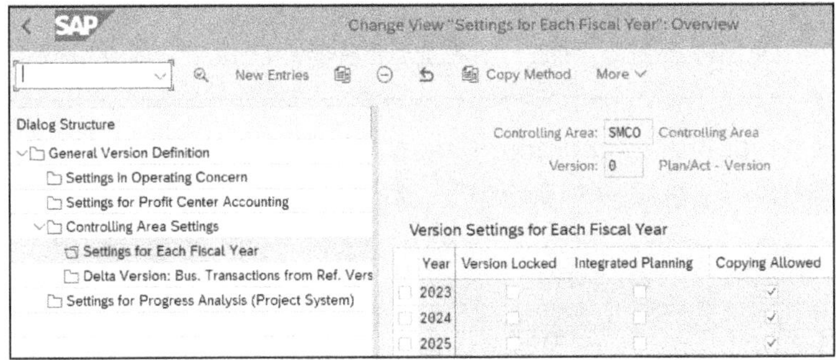

Figure 7.30 Settings for Each Fiscal Year

7.3.3 Subcontracting

In subcontracting, you send components to an external supplier who manufactures the complete assembly and returns the goods to you. The supplier provides a quotation that you enter in a subcontracting purchasing info record. In this section we'll look at the strategy sequences to select the price of the subcontracting, purchase costs, and external processing.

Strategy Sequence

You enter a strategy sequence for determining a subcontracting price by selecting the **Subcontracting** tab of the valuation variant, as shown in Figure 7.31.

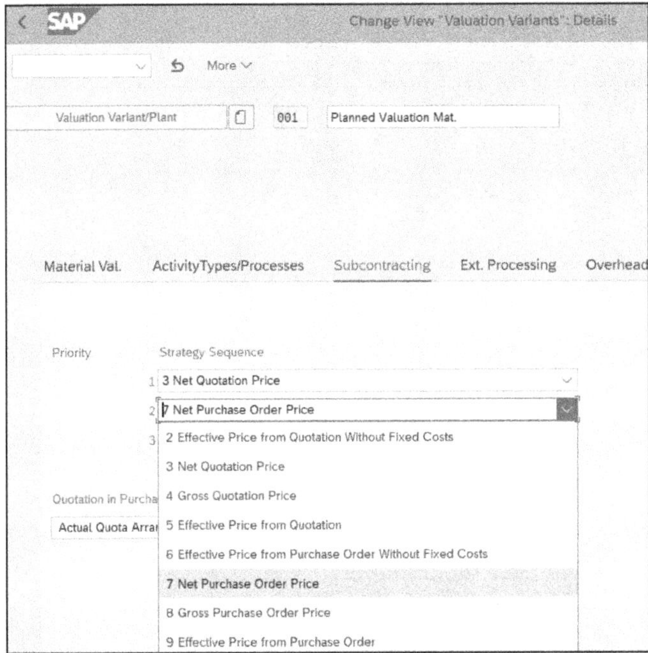

Figure 7.31 Subcontracting Strategy Sequence Possible Entries

You enter the **Strategy Sequence** to search for prices in the purchasing info record and last purchase order. Click any field to display the list of possible entries as shown in Figure 7.31.

A special procurement type of subcontracting in the **MRP 2** view of an assembly indicates that this material is subject to subcontracting.

Note: Purchasing Info Record Categories

There are four purchasing info records categories:

- Standard
- Subcontracting
- Pipeline
- Consignment

For example, if the special procurement type indicates that this is a subcontract material, the cost estimate will only search for subcontracting category purchasing info records.

Now that we've considered the **Strategy Sequence** section of **Subcontracting**, let's look at the next section.

Quotation in Purchasing

Click the **Quotation in Purchasing** field to display a possible entry list as shown in Figure 7.32.

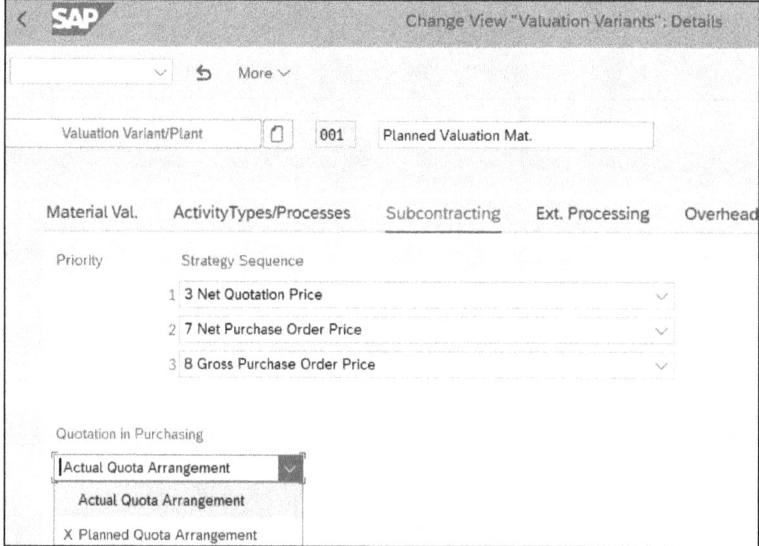

Figure 7.32 Quotation in Purchasing Possible Entries

You use purchasing **Planned Quota Arrangements** and **Actual Quota Arrangements** to create a mixed price for materials that are manufactured at different external suppliers with parts provided by the customer. You can specify whether the quotas of the individual suppliers entered in the source list for the material to be processed should be determined through the **Planned Quota Arrangement** or the **Actual Quota Arrangement**. The source list specifies allowed sources (purchasing info records) of a material for a plant within a validity period. If you have entered quota arrangements for multiple suppliers for a subcontracting item, the system uses the supplier with the highest planned quota in this example.

Next, let's look at the fields in the **Ext. Processing** (external processing) tab of the valuation variant.

7.3.4 External Processing

Select the **Ext. Processing** tab shown in Figure 7.32 to display the screen shown in Figure 7.33.

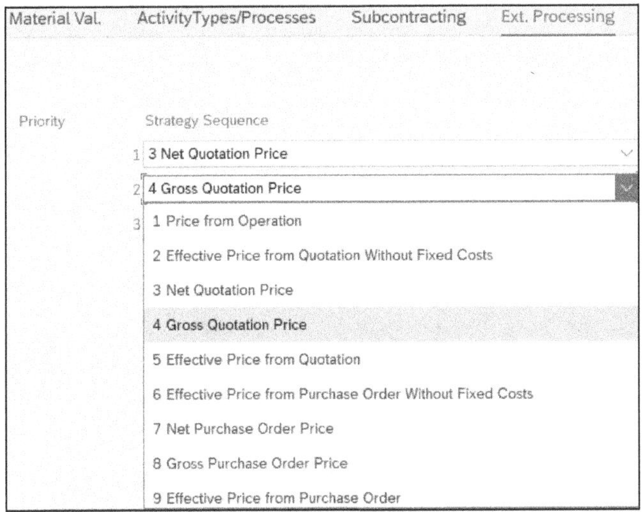

Figure 7.33 External Processing Strategy Sequence Possible Entries

We discussed purchasing info records and purchase order prices in Section 7.3.1. But we've not yet discussed the first possible entry, **Price from Operation**, shown in Figure 7.33. We need to differentiate clearly between subcontracting and external processing:

- **Subcontracting**
  You send components to the external supplier and then receive the completed assembly from the supplier into inventory.
- **External processing**
  You assign operations in a routing to an external supplier for processing. You send

the unfinished assembly to the supplier. The external supplier carries out the operation, for example, degreasing, and then the supplier returns the unfinished assembly to you for further processing in the remaining operations in the routing.

When a supplier carries out an operation externally, you maintain the data relevant to costing directly in the operation. You access this data with Transaction CA02 or via the menu path **Logistics • Production • Master Data • Routings • Routings • Standard Routings • Change**. Enter a routing with an external operation in the selection screen, click the **Operations** button, and double-click an operation with external processing (usually PP02 control key) to display the screen shown in Figure 7.34.

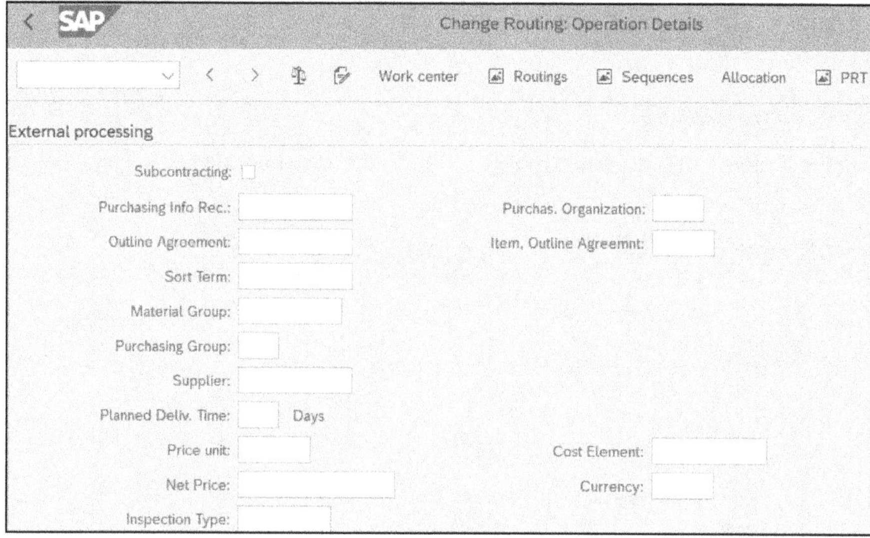

Figure 7.34 External Processing Section of Operation Details

You have two options for entering external processing price details:

- **Direct entry**
  You can enter the price and associated information directly in the operation fields.
- **Purchasing info record**
  You can store external processing cost information in a purchasing info record. When you enter the purchasing info record number and press Enter, the system retrieves information from the purchasing info record and automatically populates the relevant fields of the operation. Information retrieved from the purchasing info record isn't changeable in the operation. If you need to change the operation details fields, delete the purchasing info record number, and press Enter. The information remains, and all fields become changeable.

Select **Subcontracting** in Figure 7.34 if you intend to send components with an unfinished assembly to an external supplier. You can only use a subcontracting category purchasing info record in this case unless you enter the costing information directly in the operation fields.

Let's look at the next valuation variant tab.

7.3.5 Overhead

Select the **Overhead** tab to the right of the **Ext. Processing** tab shown previously in Figure 7.32 to display the screen shown in Figure 7.35.

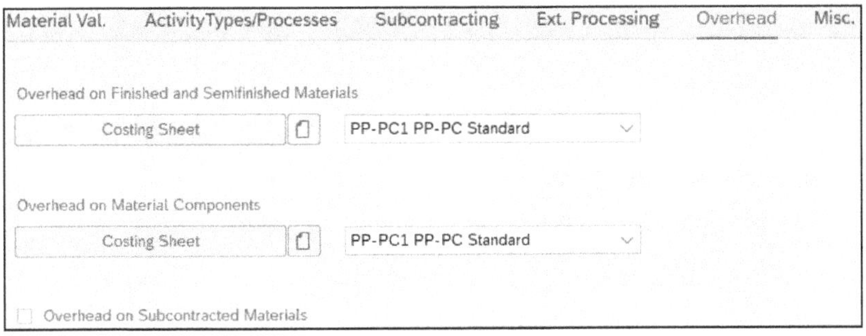

Figure 7.35 Valuation Variant Overhead Tab

Let's examine the three sections of the **Overhead** tab next.

Overhead on Finished and Semifinished Materials

In this section, you associate a costing sheet with the valuation variant for assemblies for the calculation of overhead. Let's discuss the options in this section:

- **Costing Sheet**
  Click this button to maintain the costing sheet shown in the field to the right of the button. Refer to Chapter 5 for details of how to maintain a costing sheet.
- **New page**
  Click this icon to create a new costing sheet or to maintain details of another costing sheet.
- **Costing Sheet**
  Click this field to display a list of possible entries for costing sheets. The costing sheet entered in this field determines the calculation of overhead on finished and semifinished materials.

Now let's look at the next section in the **Overhead** tab.

Overhead on Material Components

You can calculate overhead on raw materials based on your entry in this section. The **Costing Sheet** button, new page icon, and **Costing Sheet** field work in the same way as in the previous section.

Note: Overhead for Non-Stock Related Components

These overhead costs are usually for cost components that are nonstock-related because you don't typically use an overhead costing sheet with the purchasing process. You cannot allocate overhead when posting goods receipts and invoices. You would always post a purchase price variance (PPV) because the standard price would be greater than the goods and invoice receipt cost by the amount of overhead calculated in the standard cost estimate.

With specific additional purchasing costs such as freight and duty, you can apply these as purchasing conditions that you can configure to post to specific general ledger accounts other than a PPV account.

Overhead on Subcontracted Materials

Select this checkbox to apply overhead to subcontracted materials.

Let's now look at the last tab in the valuation variant.

7.3.6 Miscellaneous

Select the **Misc.** tab in Figure 7.35 to display the screen shown in Figure 7.36.

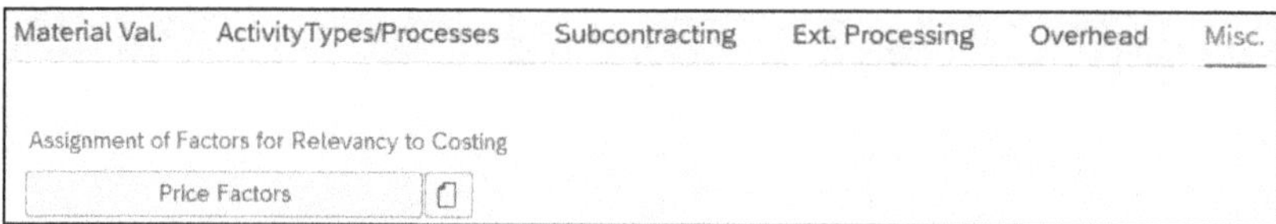

Figure 7.36 Valuation Variant Miscellaneous Tab

With the **Price Factors** button, you can control the extent to which a BOM item, operation, or suboperation in a routing is included in costing. In most cases, you set the item as either fully relevant to costing (selected) or not relevant to costing (blank). These system-defined price factors are always available. First, we'll look at how you set the **Relevancy to Costing** checkbox in a BOM item, and then we'll examine how you define price factors.

You can maintain a BOM item with Transaction CS02 or by following the menu path **Logistics • Production • Master Data • Bills of Material • Bill of Material • Material BOM • Change.** Double-click a BOM item and select the **Status/Lng Text** tab to display the screen shown in Figure 7.37.

Click in the **CostingRelevncy** field and press F4 to display a list of possible entries as shown in Figure 7.37. The **Not relevant to costing** and **X: Relevant to Costing** entries are system defined and always available. You can define your own indicators with Transaction OKK9 or by following the IMG menu path **Controlling • Product Cost Controlling •**

Product Cost Planning • Price Update • Parameters for Inventory Cost Estimate • Define Relevancy to Costing, as shown in Figure 7.38.

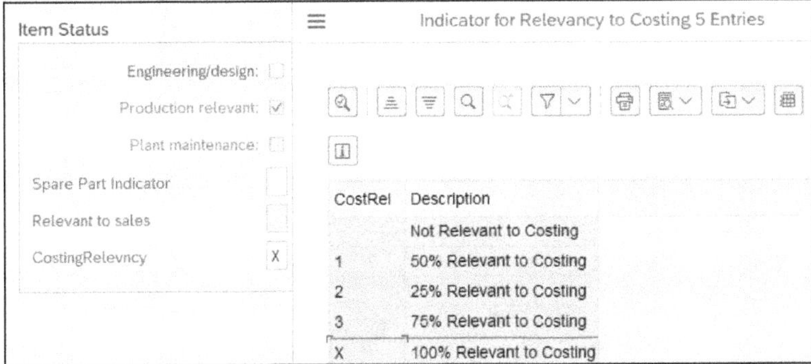

Figure 7.37 BOM Item Relevancy to Costing Indicator

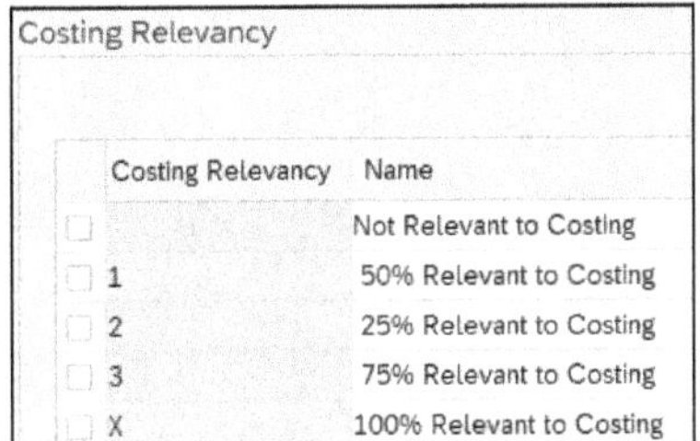

Figure 7.38 Maintain Costing Relevancy Indicators

Costing Relevancy indicators **1: 50% Relevant to Costing**, **2: 25% Relevant to Costing** and **3: 75% Relevant to Costing** are user defined in this example. The **Costing Relevancy** key is a one-character alphanumeric entry, allowing you to create up to 34 user-defined indicators.

Note: System Supplied Indicators

You cannot delete the system-supplied **Costing Relevancy** indicators.

After you have defined your own costing relevancy indicators, you can assign price factors by either clicking the **Price Factors** button in the **Misc.** tab of the valuation variant as shown earlier in Figure 7.36, using Transaction OKK7, or via the IMG menu path **Controlling • Product Cost Controlling • Product Cost Planning • Price Update • Parameters for Inventory Cost Estimate • Define Price Factors**. The screen in Figure 7.39 displays.

You assign **Price Factors** to **Costing Relevancy** indicators to devalue items in an inventory cost estimate. You can assign factors separately for fixed costs in the **Fxd Prc. Factor** column and variable costs in the **Var. Price Factor** column.

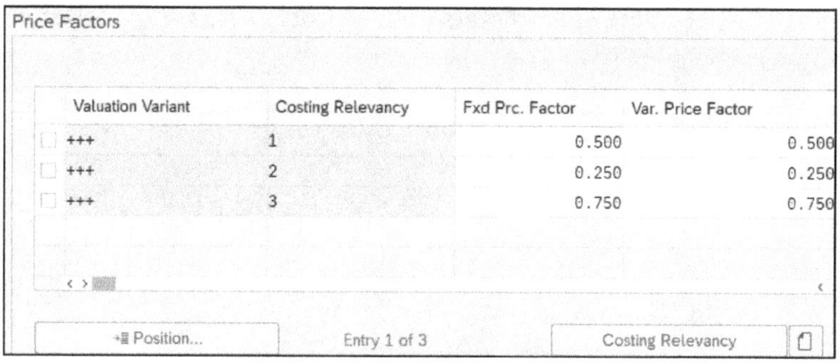

Price Factors

	Valuation Variant	Costing Relevancy	Fxd Prc. Factor	Var. Price Factor
☐	+++	1	0.500	0.500
☐	+++	2	0.250	0.250
☐	+++	3	0.750	0.750

Position... Entry 1 of 3 Costing Relevancy

Figure 7.39 Factors for Relevancy to Costing

The **Valuation Variant** entries of three plus signs (**+++**) indicate that the price factors are valid for all valuation variants that don't have specific entries. In this example, you would assign fixed and variable price factors of **0.600** for **Valuation Variant 004** and **Costing Relevancy 1**, whereas all other valuation variants you assign a price factor of **0.500** for **Costing Relevancy** indicator 1.

Example: Costing Relevancy Indicator

You allocate packaging materials in part to sales and administration costs that you shouldn't capitalize. In this case, you can mark all the BOM items, or mark the operations in a routing, directly related to packaging with a **Costing Relevancy** indicator you define for this purpose. Together with the valuation variant, this indicator determines the price factor used in an inventory cost estimate. An *inventory cost estimate* assesses the tax-based and commercial prices for purchased parts and populates the same fields of the finished and semifinished products with the calculated value.

In this example, if you assign **Costing Relevancy** indicator **3** to all packaging-related BOM items, these items will devalue by a factor of 0.750 in an inventory cost estimate.

You can click the **Costing Relevancy** button at the bottom of the screen in Figure 7.39 to maintain **Costing Relevancy** indicators shown in Figure 7.39.

Now that we've finished discussing the fields contained in the valuation variant, we'll look at the next costing variant component.

7.4 Date Control

Date control determines the proposed dates when you create a cost estimate and whether the user can change these dates. You maintain date control either by clicking the **Date Control** button in the **Control** tab of a costing variant, with Transaction OKK6, or via the IMG menu path **Controlling • Product Cost Controlling • Product Cost**

Planning • Material Cost Estimate with Quantity Structure • Costing Variant: Components • Define Date Control. The screen in Figure 7.40 displays.

Date control	Name
PC01	Std Cost Est. - Month
PC02	Std Cost Est. - Subseq.FY
PC03	Std Cost Est. - Period
PC04	Current Cost Est.- Dates
PC05	Mod. Std Cost Est. Dates
PC06	Std Cost Estimate - Old
PC07	Sales Order Cost Estimate

Figure 7.40 Date Control Overview

The standard system contains predefined date controls. You can use these without making any changes. Double-click a date control to display the screen shown in Figure 7.41.

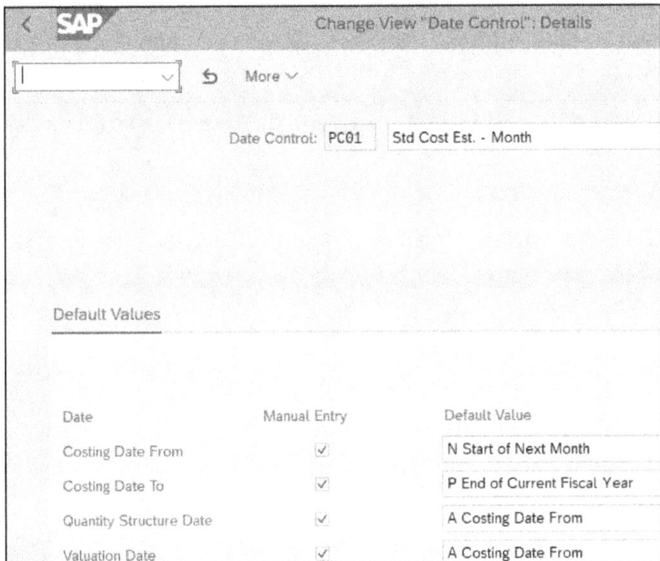

Figure 7.41 Date Control Details

When you create a cost estimate, the system proposes four dates corresponding to those assigned in the **Default Values** tab. If you select the **Manual Entry** checkbox, the user can change the proposed dates when creating a cost estimate. Let's examine the default value for each date:

- **Costing Date From**
  Costing Date From determines the validity start date of the cost estimate. The cost estimate cannot be marked and released until you reach the start date. You can change the start date to a previous date and create the cost estimate. However, you cannot save, mark, or release a standard cost estimate with a start date in the past.

You typically create standard cost estimates for release at the start of next month as shown in the **Default Value** column in Figure 7.41, which provides you with an opportunity to correct problems indicated by cost estimate messages. You can also review the costs calculated during the current month and prior to the next month.

You base modified (**Mod.**) standard cost estimates on the current quantity structure and planned prices, and you use these estimates to cost materials during the fiscal year to analyze cost developments. The system-supplied default **Costing Date From** for the modified standard cost estimate is today's date.

You base **Current** cost estimates on the current quantity structure and prices, and you use them to cost materials during the fiscal year to analyze development costs. The system-supplied default **Costing Date From** for the current standard cost estimate is the start of the current month.

- **Costing Date To**
  Costing Date To determines the validity finish date of the cost estimate. Variance calculation requires a standard cost estimate that is valid for the entire fiscal year. This date is typically set to the maximum possible date or the end of the current fiscal year for a standard cost estimate.

  The current and modified standard cost estimates' default **Costing Date To** is usually set to the end of the current month.

- **Quantity Structure Date**
  The **Quantity Structure Date** determines which BOM and routing you select for the cost estimate. Because these can change over time, it's useful to be able to select a particular BOM or routing by date.

  You usually set the default date for *standard* cost estimates to be the same as the **Costing Date From** entry. The default date for the current cost estimate date is usually set to today's date, and the default date for the modified standard cost estimate is usually set to the start of the current month.

- **Valuation Date**
  The **Valuation Date** determines which material and activity prices you select for the cost estimate. Purchasing info records can contain different supplier-quoted prices valid for different dates. Likewise, you can plan different activity prices per period within a fiscal year, which can be useful during product development, for instance, to hold the valuation date constant while changing the quantity structure date to isolate the cost effect of changing the structure of a BOM.

  You usually set the default date for standard cost estimates to be the same as the **Costing Date From** entry. The default date for the current cost estimate is usually set to today's date, and the default date for the modified standard cost estimate is usually set to the start of the current month.

Now that we've examined the fields contained in date control, let's look at the next costing variant component.

7.5 Quantity Structure Control

You can use quantity structure control in cost estimates with quantity structure to specify how the system searches for alternative BOMs and routings. You can maintain quantity structure control either by clicking the **Quantity Structure** button in the **Control** tab of a costing variant, with Transaction OKK5, or via the IMG menu path **Controlling • Product Cost Controlling • Product Cost Planning • Material Cost Estimate with Quantity Structure • Costing Variant: Components • Define Quantity Structure Control**. The screen in Figure 7.42 displays.

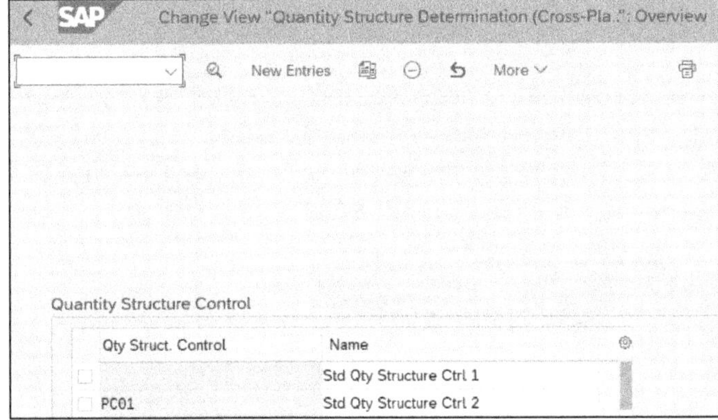

Figure 7.42 Quantity Structure Determination Overview

The standard system contains predefined quantity structure controls. You can use these without making any changes, or you can create your own. Select a quantity structure control and click the details icon or double-click a quantity structure control (**PC01** in this example) to display the screen shown in Figure 7.43.

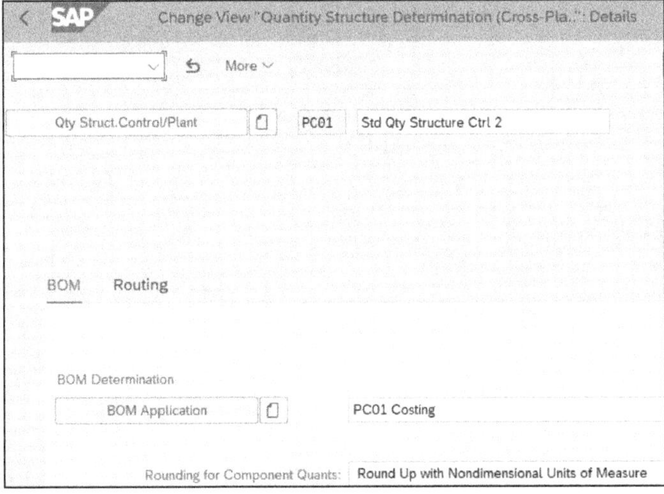

Figure 7.43 Quantity Structure Determination Details

Let's examine the details of how a BOM is selected by quantity structure determination by examining the fields in the **BOM** tab in the following sections.

> **Note: Quantity Structure Control Per Plant**
>
> You can use different quantity structure controls in plants that belong to the same company code if necessary. Click the **Qty Struct. Control/Plant** button to assign a quantity structure control to a plant. Otherwise, the same quantity structure controls apply to all plants in a company code.

7.5.1 Bill of Materials

BOM Application represents a process to automatically determine alternative BOMs in the different organizational areas within a company. PC01 is the system-supplied BOM application for use with costing variant PPC1 for standard cost estimates. Click the **BOM Application** button to display the screen shown in Figure 7.44.

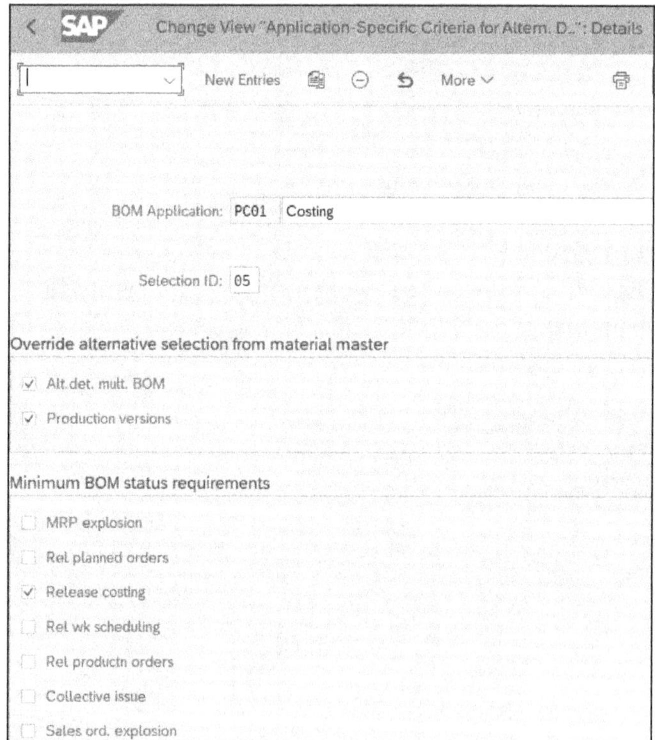

Figure 7.44 BOM Application Details

We'll analyze each section of the **BOM Application** screen in the following sections.

Selection ID

Selection ID determines the order in which the system searches for alternative BOMs or based on BOM usage that identify a specific section of your company such as production, engineering, or costing. You can maintain details of a selection ID with Transaction OPJI or via the IMG menu path **Controlling • Product Cost Controlling • Product Cost Planning • Material Cost Estimate with Quantity Structure • Settings for Quantity Structure Control • BOM Selection • Check BOM Selection**. The screen in Figure 7.45 displays.

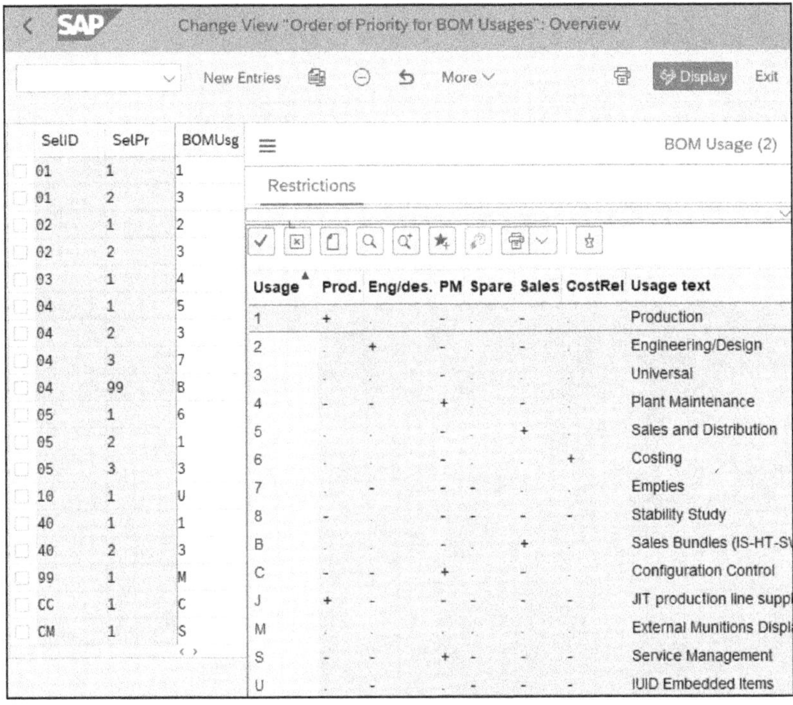

Figure 7.45 Selection ID Order of Priority for BOM Usages Overview

Let's examine **SelID** (selection ID) **05** because this is contained in BOM application PC01, which is used by standard cost estimates with system-supplied settings. The **SelPr** (selection priority) column contains entries that assign priority to the **BOMUsg** (BOM usage) entries in the last column for each selection ID.

Click a **BOMUsg** field and press the F4 key to display a list of possible entries as shown in Figure 7.45. These settings relate to entries that you can make in BOM item status, which controls the areas of a company relevant to the BOM item.

First, we'll discuss the meanings of the settings in the possible entries list and then see how this affects BOM item status:

- **Plus sign (+)**
  This BOM item status indicator must be selected.

- **Minus sign (-)**
  This BOM item status indicator cannot be selected.
- **Period (.)**
  This BOM item status indicator may be selected.

Now let's see how the BOM **Usage** of **1** for production affects the BOM item status indicators as shown in Figure 7.46.

Figure 7.46 BOM Item Status Indicators for Production BOM

The **Production relevant** checkbox in the **Item Status** section is selected and cannot be changed for this BOM item. This corresponds to the plus sign in the BOM **Usage** setting for production BOMs in Figure 7.45. The system copies BOM items with this indicator selected to the planned order and calculates their dependent requirements. When you convert the planned order to a production order, the system automatically copies these items to the production order.

The **Plant maintenance** checkbox isn't selected, and there is no entry in the **Relevant to sales** field. You cannot change these checkbox and field settings. This corresponds to the minus sign in the BOM **Usage** setting for production BOMs in Figure 7.45. You use items relevant to plant maintenance in maintenance BOMs. In sales order processing for variant products, you see items that are relevant to sales as order items for the header material.

You can change the **Engineering/design** checkbox and the **Spare part indicator** and **CostingRelevncy**. This corresponds to the period signs in the BOM **Usage** setting for production BOMs in Figure 7.44.

Engineering and design items are relevant to the area responsible for producing a functional design for production. The **Spare part indicator** defines the item as a spare part. The **CostingRelevncy** field controls whether a BOM item, operation, or suboperation in the routing is included in costing.

Now that we've discussed how BOM usage controls BOM item status indicators, let's examine each of the three entries for **Selection ID 05** in Figure 7.44 to see how this controls BOM selection:

- **Selection ID 05, priority 1**
  BOM usage 6 (**Costing**) is assigned the first priority. You usually copy costing BOMs from BOMs with a usage of production at the start of each fiscal year before the main costing run. You can adjust costing BOMs that aren't reflected in production BOMs. This isn't generally recommended because standard cost estimates should represent production costs.
- **Selection ID 05, priority 2**
  BOM usage 1 (**Production**) is assigned the second priority. If no costing BOM exists, a standard cost estimate will search for a production BOM.
- **Selection ID 05, priority 3**
  BOM usage 3 (**Universal**) is assigned the third priority. If there are no costing or production BOMs, standard cost estimates will search for a universal BOM.

Override Alternative Selection from Material Master

Now that we've discussed how selection ID automatically selects BOMs, we'll discuss the next section of the BOM application screen (shown earlier in Figure 7.45) as shown in Figure 7.47.

In the **MRP 4** view, you can enter a method for selecting alternative BOMs, which we discussed in detail in Chapter 4. If you select **Alt.det. mult. BOM** (alternative determination for multiple BOMs), the system will search for a specific alternative BOM for a certain date. You can define alternative BOMs for a date with Transaction OPPP or via the IMG menu path **Controlling • Product Cost Controlling • Product Cost Planning • Material Cost Estimate with Quantity Structure • Settings for Quantity Structure Control •**

BOM Selection • Check Alternative Selection for Multiple BOM. The screen in Figure 7.48 displays.

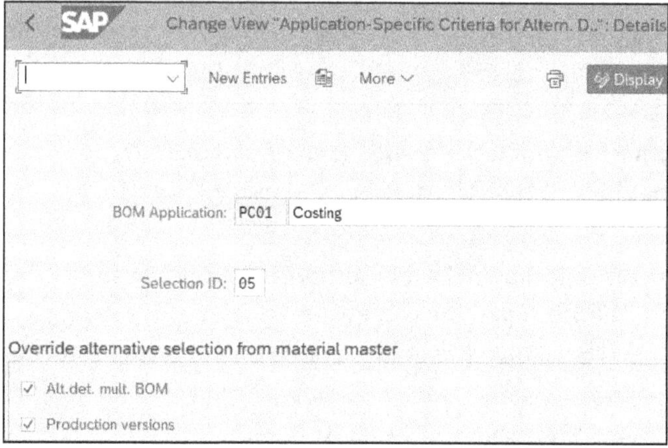

Figure 7.47 Override Alternative Selection from Material Master

Material	Plnt	BOM Usg	Valid From	AltBOM
PN57:FG MATERIAL	5771	1	11/27/2023	1
PN57:FG MATERIAL	5771	1	11/27/2023	3

Figure 7.48 Alternative BOM Determination Overview

You enter the date from which the alternative BOM is available. After this date, a cost estimate will access this BOM, which overrides the method set for selecting alternative BOMs set in the **MRP 4** field.

In the same section of the BOM **Application** screen shown in Figure 7.47 is the **Production versions** checkbox. This checkbox shows that when you select alternatives automatically according to application, you use the production versions maintained in the material master as selection criteria.

Minimum Bill of Materials Status Requirements

Now let's discuss the checkboxes in the remaining section of the BOM **Application** screen shown earlier in Figure 7.44. Before the system selects an alternative BOM with the selection ID in the BOM application, the alternative BOM must meet minimum BOM status requirements. Selections in Figure 7.44 indicate these requirements.

You display a BOM status with Transaction CS02 or via the menu path **Logistics • Production • Master Data • Bills of Material • Bill of Material • Material BOM • Change**. Enter the BOM in the selection fields, press Enter, and then click the **Header** (hat) icon to display the BOM header as shown in Figure 7.49.

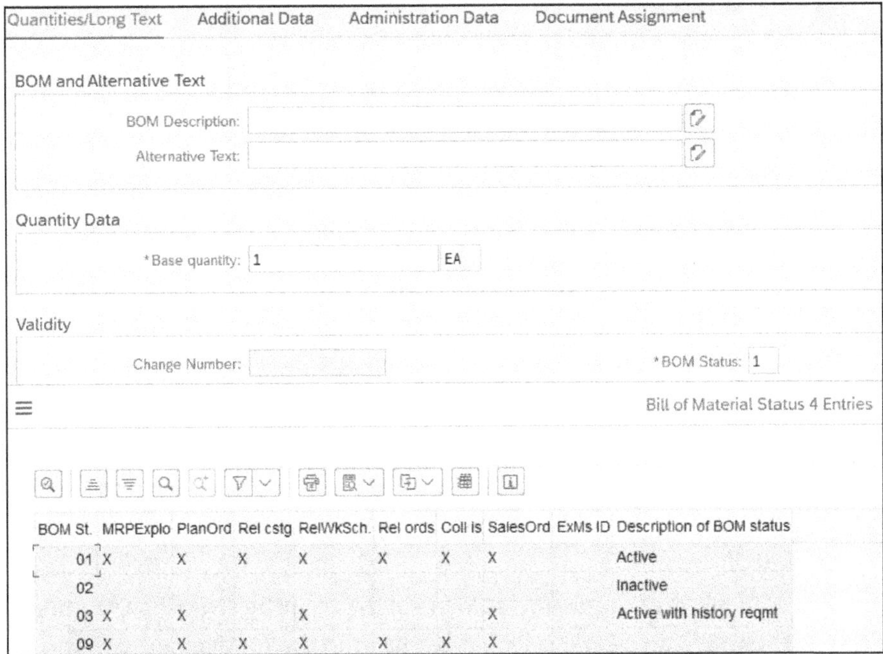

Figure 7.49 BOM Header Overview with BOM Status Possible Entries

Click the **BOM status** field and then press F4 to display the list of possible entries for **BOM status** shown in Figure 7.50. The BOM status checkboxes and descriptions in Figure 7.49 correspond to the seven BOM status columns in Figure 7.50.

Note: BOM Item vs Header Level Status

Here is a review of both types of BOM status utilized by BOM applications:

- **BOM item status**
  This applies to each BOM item. BOM usage controls the BOM item status checkboxes and fields, which you use during BOM selection with selection ID.
- **BOM header level status**
  This applies to all BOM items. Checkboxes in the BOM application set the minimum requirements for BOM status for selecting a BOM.

There is one remaining field we need to examine in the **BOM** tab of the **Quantity Structure Determination** screen shown earlier in Figure 7.42.

Rounding for Component Quants

You can see **Rounding for Component Quants** in the **BOM** tab in Figure 7.50.

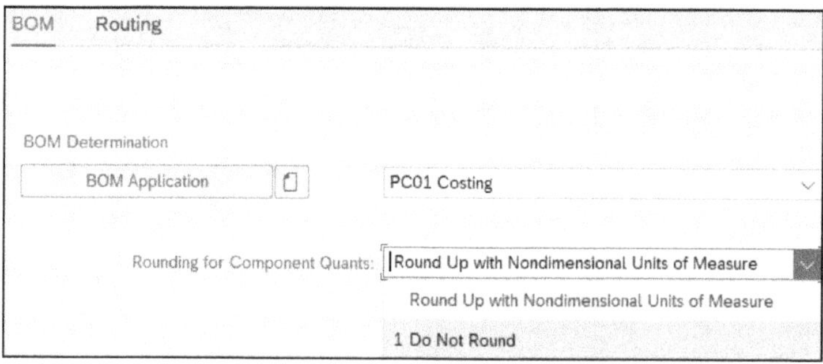

Figure 7.50 Rounding Indicator in Cost Estimate with Quantity Structure

Click the field to display a list of possible entries, as shown in Figure 7.50. When costing a material, this entry specifies whether the required quantity of a BOM component with a unit of measure that can only assume whole number values (such as "piece") rounds up when the calculation of scrap-adjusted quantities produces a required quantity that isn't a whole number.

Now that we've studied the fields in the **BOM** tab of the **Quantity Structure Determination** screen as shown Figure 7.42, we'll look at the fields in the **Routing** tab.

7.5.2 Routing

Select the **Routing** tab shown in Figure 7.50 to display the screen in Figure 7.51.

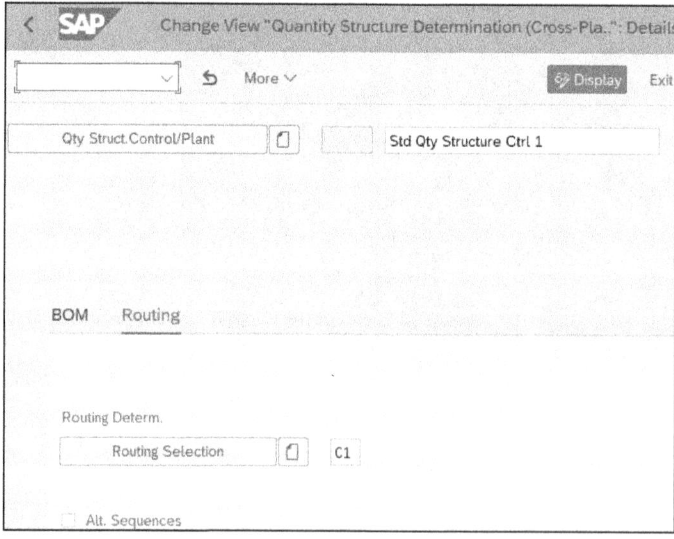

Figure 7.51 Routing Selection Details

Routing selection determines how cost estimates automatically select routings. Click the **Routing Selection** button to display the details shown in Figure 7.52.

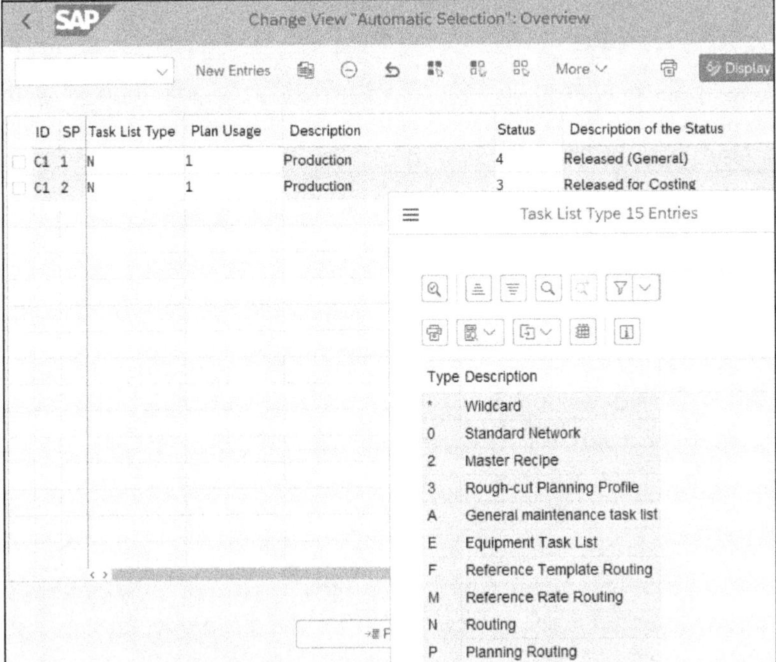

Figure 7.52 Routing Selection ID Details

Let's examine the columns and details of the first row on the priority list to understand how a routing is automatically selected:

- **ID**
  This column specifies the selection ID, **C1** in this case.
- **SP** (selection priority)
  This column specifies the sequence in which the system searches for each routing.
- **Task List Type**
  The entries in this column specify the type of routing. Click a **Task List Type** field and press F4 to display the list of possible entries shown in Figure 7.42. From this list, you can see that the selection ID first searches for a **Routing** of **N** and, if unsuccessful, then searches for an **Reference Operation Set** of **S**. A *reference operation set* is a routing type that defines a sequence of operations that repeats regularly, which simplifies entering data in a routing.
- **Plan Usage/Description**
  These columns refer to a specific section of your company such as production, engineering, or plant maintenance. This concept is like BOM usage.
- **Stat (status)/Description of the status**
  These columns indicate the processing status of a task list. You may assign a different status when the part is in research and development and when production releases it.

Selection ID 01 will select a released production routing first if one is available when creating a standard cost estimate.

Now let's examine the **Alt. Sequences** (alternative sequences) checkbox in the **Routing** tab as shown in Figure 7.53.

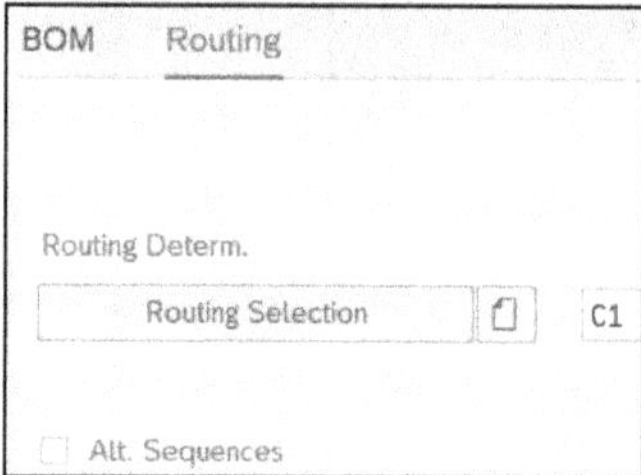

Figure 7.53 Alternative Sequences Transferred from Routing to Order

Select this checkbox to specify that you can transfer alternative sequences from the routing into the order. An alternative sequence is a sequence of operations that you can use as an alternative to several consecutive operations from the standard sequence. You can use alternative sequences, for example, if the production process differs according to lot size.

Now that we've looked at the quantity structure control component of the costing variant, let's examine transfer control.

7.6 Transfer Control

When you create a cost estimate for a finished good, the system first creates cost estimates for all the lowest-level components and then progressively rolls up the costs through the BOM hierarchy, level by level.

During a six-month (semiannual) or annual costing run, you typically create new cost estimates for all components and assemblies. During the year, if a new product contains all new components, a new standard price is needed for all components. But some of the components might already exist and be shared with other existing products. Changing the existing standard price for these components could lead to a planned variance on production orders for existing products.

Transfer control allows the transfer of existing component and subassembly cost estimates to new product cost estimates. This is also known as *partial costing*.

You maintain transfer control by clicking the **Transfer Control** button in the **Control** tab of a costing variant, with Transaction OKKM, or via the IMG menu path **Controlling • Product Cost Controlling • Product Cost Planning • Material Cost Estimate with Quantity Structure • Costing Variant: Components • Define Transfer Control**. The screen in Figure 7.54 displays.

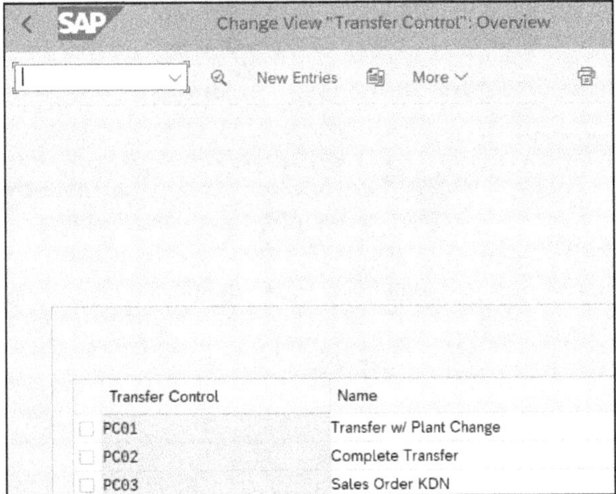

Figure 7.54 Transfer Control: Overview

The standard system contains the predefined transfer controls displayed in Figure 7.54. You can use these or copy one and create your own. Let's examine **Transfer Control PC02** as an example. Double-click **PC02** to display the screen shown in Figure 7.55. We'll discuss the fields in the **Single-Plant** and the **Cross-Plant** tabs in the following sections.

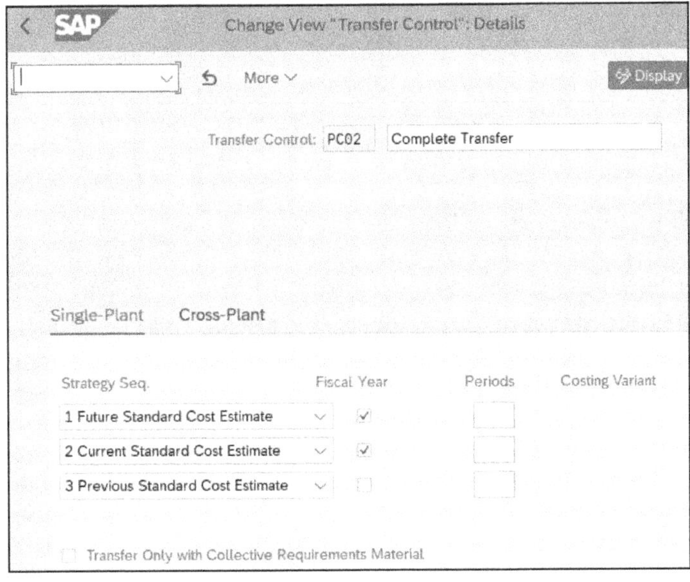

Figure 7.55 Single Plant Transfer Control Details

7.6.1 Single-Plant

The **Single-Plant** tab indicates that you transfer existing cost estimates for subassemblies and components within a plant to a new product cost estimate. This avoids re-costing existing materials between main costing runs. In this section, we'll discuss how to apply transfer control within a single plant. Then, in the following section, we'll discuss how to apply transfer control across plants.

Note

Transfer control defined in the **Single-Plant** tab doesn't apply to materials transferred across plants using the special procurement key in the **MRP 2** view. You define transfer control separately for materials transferred across plants in the **Cross-Plant** tab shown in Figure 7.55.

Strategy Sequence

The strategy sequence (**Strategy Seq.**) determines the order in which the system searches for existing cost estimates. If the system cannot find an existing cost estimate that meets the requirements of any of the strategies, it creates a new cost estimate. Click a strategy field to display a list of possible entries, as shown in Figure 7.56.

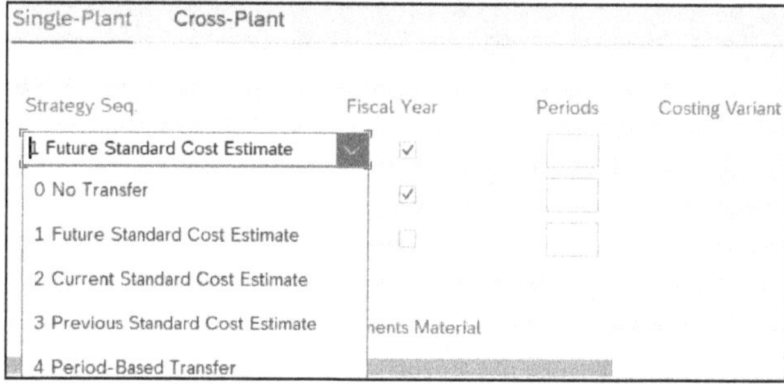

Figure 7.56 Transfer Control Strategy Sequence Possible Entries

If the **Strategy Seq.** field contains **2 Current Standard Cost Estimate, 3 Previous Standard Cost Estimate**, or **1 Future Standard Cost Estimate,** the system will only select a cost estimate with a start date within the current fiscal year or within specified periods.

Fiscal Year

If you select **Fiscal Year**, the system will search for a cost estimate with a start date within the current fiscal year.

Periods

For current, previous, and other cost estimates, the **Periods** field refers to the number of periods in the past (i.e., before the start date of the cost estimate).

For future cost estimates, the **Periods** field refers to the number of periods in the future (i.e., after the start date of the cost estimate) and in the past (before the start date).

If you select **4 Period-Based Transfer** (period-based cost estimates), a **Periods** entry has no effect because, in this case, the system searches for cost estimates that have exactly the same keys (i.e., costing type, valuation variant, costing version, and, if applicable, period).

A **Periods** entry of zero means the current period. If you don't make any entries, the system interprets this as zero and searches for existing cost estimates in the current period only.

Note: Other Cost Estimates

If you enter **Other Cost Estimates** as a **Strategy**, two additional columns become available in Figure 7.56. The additional **Costing Variant** and **Costing Version** columns provide you with additional selection capability for selecting cost estimates.

Now let's examine the remaining checkbox in the **Single-Plant** tab shown previously in Figure 7.55.

Transfer Only with Collective Requirements Material

This checkbox is dependent on another setting made in the material master. We'll discuss this dependent setting first and then discuss this transfer control checkbox.

The **Dependent Requirements** checkbox for individual and collective requirements in the **MRP 4** view determines whether the dependent (lower-level) requirements of an assembly can be grouped together (collective) or must be treated separately (individual), as described here:

- A **Collective Requirements** setting means the material is produced or procured for various requirements.
- An **Individual Requirements** setting means the material is specially manufactured or procured for a sales order.

If you select **Transfer Only with Collective Requirements Material** in Figure 7.55, individual requirements materials, especially for sales orders, are costed again even when a cost estimate already exists, which is useful if your costs are dependent on lot size.

Don't select this checkbox if existing costing data is also to be transferred for materials controlled by individual requirements. The system then searches for a cost estimate in the sequence of the specified strategies.

Now that we've investigated the fields and settings in the **Single-Plant** tab in Figure 7.55, let's look at the **Cross-Plant** tab.

7.6.2 Cross-Plant

Select the **Cross-Plant** tab shown earlier in Figure 7.55 to display the screen shown in Figure 7.57.

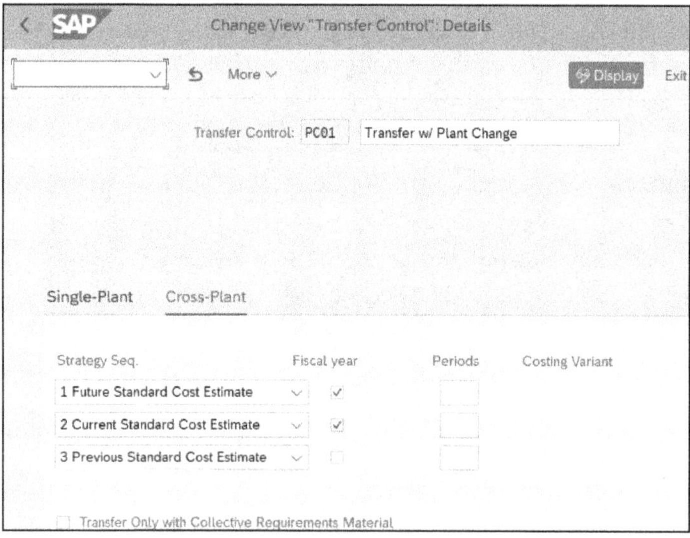

Figure 7.57 Cross-Plant Transfer Control Details

The fields and checkboxes in this tab are the same as those in the **Single-Plant** tab, and they work in the same way. Let's follow an example of how you can set up the checkboxes in these two tabs.

If your company has several plants and an individual cost accountant is responsible for each plant, you may decide to use the standard SAP transfer control PC01. The **Single-Plant** tab settings are shown in Figure 7.58.

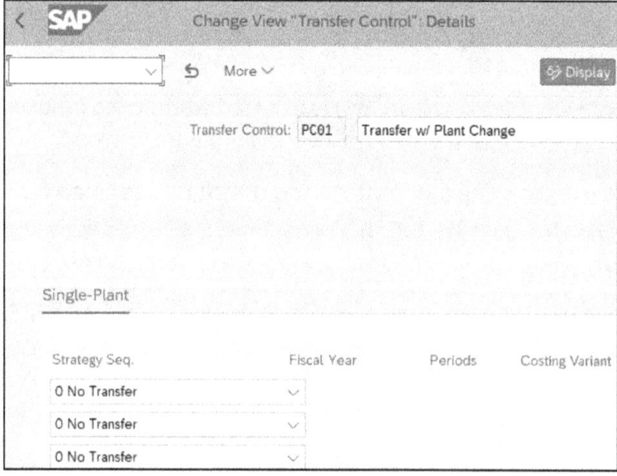

Figure 7.58 Single-Plant Transfer Control PC01

Costing runs using transfer control PC01 will create and release new cost estimates within a single plant without using any existing cost estimates. The costing run, however, will use existing cost estimates from other plants if available because the cross-plant transfer control strategy is the same as shown in Figure 7.57. This strategy allows cost accountants to create and release new cost estimates only in their own plants.

Now that we've looked at the transfer control component of the costing variant, let's examine the final costing variant component: the reference variant.

7.7 Reference Variant

A *reference variant* is a costing variant component that allows you to create material cost estimates or costing runs based on the same quantity structure to improve system performance or make comparisons between cost estimates.

Whereas transfer control allows you to transfer an entire existing cost estimate, you can choose which parts of a reference variant to transfer, such as material components or internal activities.

You maintain a reference variant either by clicking the **Reference Variant** button in the **Control** tab of a costing variant, with Transaction OKYC, or via the IMG menu path **Controlling • Product Cost Controlling • Product Cost Planning • Material Cost Estimate with Quantity Structure • Costing Variant: Components • Define Reference Variants.** The screen shown in Figure 7.59 is displayed.

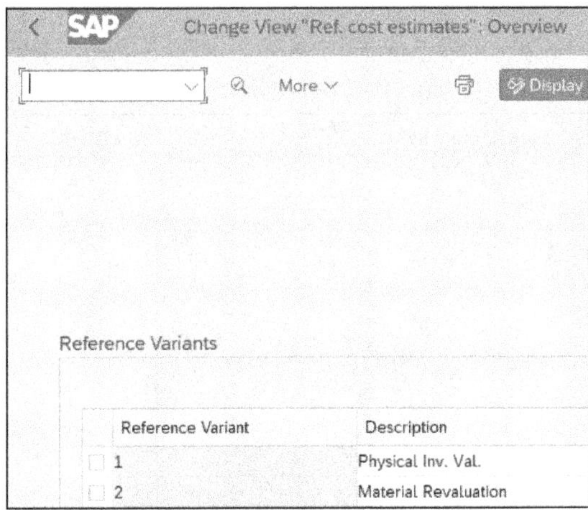

Figure 7.59 Reference Cost Estimate Overview

The overview displays a list of existing reference variants. You can use an existing variant or copy one and create your own. Double-click **Reference Variant 2** in this example to display the details shown in Figure 7.60.

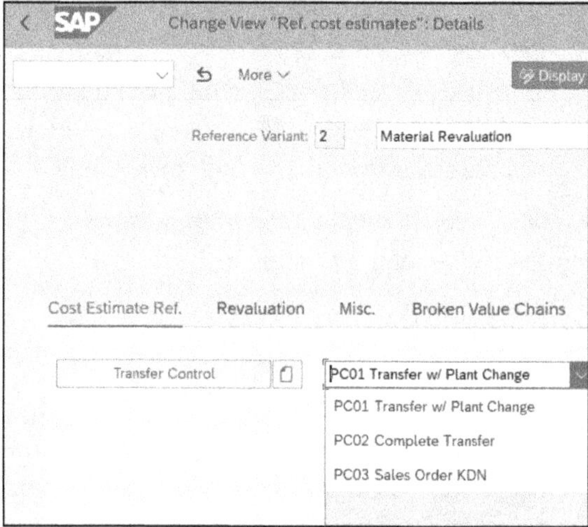

Figure 7.60 Reference Variant Cost Estimate

We'll discuss the details in the **Cost Estimate Ref.**, **Revaluation**, and **Misc.** tabs in the following sections.

7.7.1 Cost Estimate Reference

Transfer control allows you to choose which cost estimate to use as a reference. Click the **Transfer Control** description to display a list of possible entries as shown in Figure 7.60.

You can either click the **Transfer Control** button to maintain the existing transfer control or click the new page icon to create your own. In this example, click the **Transfer Control** button to display details of the existing transfer control **PC01**, as shown in Figure 7.61.

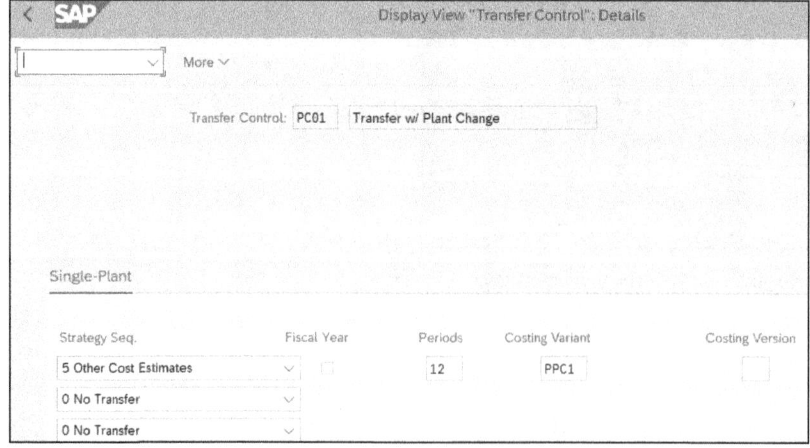

Figure 7.61 Reference Variant Transfer Control Details

You can only search for a reference cost estimate within the same plant because there is only the **Single-Plant** tab and no **Cross-Plant** tab as in the transfer control component described in Section 7.6.2.

In the **Strategy Seq.** (strategy sequence) fields, you determine the sequence in which the system searches for types of existing cost estimates. Click a strategy sequence field to display a list of possible entries of cost estimate types, as shown in Figure 7.60.

In the subsequent columns, you specify if the reference cost estimate must be any of the following:

- Valid within the current fiscal year
- Valid within a specified period age
- Based on a certain costing variant
- Based on a certain costing version

Now let's examine the next tab of the reference variant.

7.7.2 Revaluation

Select the **Revaluation** tab in Figure 7.60 to display the screen shown in Figure 7.62.

Figure 7.62 Reference Variant Revaluation

You select an option in the **Revaluation** tab to specify whether an item should be revaluated when referencing a cost estimate. In this example, only the **Material Components** of a reference cost estimate will be revalued.

Now let's look at the final tab of the reference variant.

7.7.3 Miscellaneous

Click the **Misc.** (miscellaneous) tab shown previously in Figure 7.60 to display the screen shown in Figure 7.63.

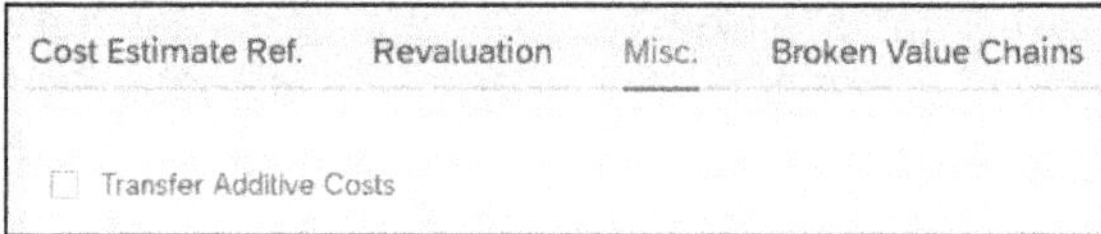

Figure 7.63 Reference Variant Transfer Additive Costs

You select **Transfer Additive Costs** to specify that additive costs should be transferred when referencing a cost estimate. Let's see an example of how you can use a reference variant.

Example: Reference Variant

You create a standard cost estimate and use it as a reference for an inventory cost estimate, changing only the value of the material components. You use the quantity structure of the standard cost estimate when it calculates the inventory cost estimate without having to redetermine the quantity structure.

7.8 SAP S/4HANA Cloud Configuration (Scope Item BEG)

The best-practice settings for product cost planning are delivered with scope item BEG (standard cost calculation). In SAP S/4HANA Cloud, this is a default scope item, so the settings will automatically be available for you. If you work with SAP S/4HANA Private Cloud or on-premise SAP S/4HANA, you can choose to use these settings to accelerate your initial implementation.

To make these changes in SAP S/4HANA Cloud, you'll need to be assigned to the role SAP Business Process Expert. To access the user interface for configuration, choose **Manage Your Solution • Configure Your Solution** and select **Finance • Product Costing** as shown in Chapter 5, Figure 5.17. In SAP S/4HANA Private Cloud or on-premise SAP S/4HANA, you can access the configuration using the menu paths described in the previous sections and make whatever changes you need.

To display the detailed settings for Product Cost Planning, select **Product Costing • Product Cost Planning** in Chapter 5, Figure 5.18 and choose **Define Costing Variants**, or enter the ID **106410** as shown in Figure 7.64. Notice that you cannot change the costing types, but you can define your own valuation variants (ID **102607**), assign valuation variants per plant (ID **102608**), and adjust the transfer control settings (ID **106207**).

In the following sections, we'll revisit the costing variant settings to explain the settings for costing variants, valuation variants and transfer control delivered with SAP S/4HANA Cloud.

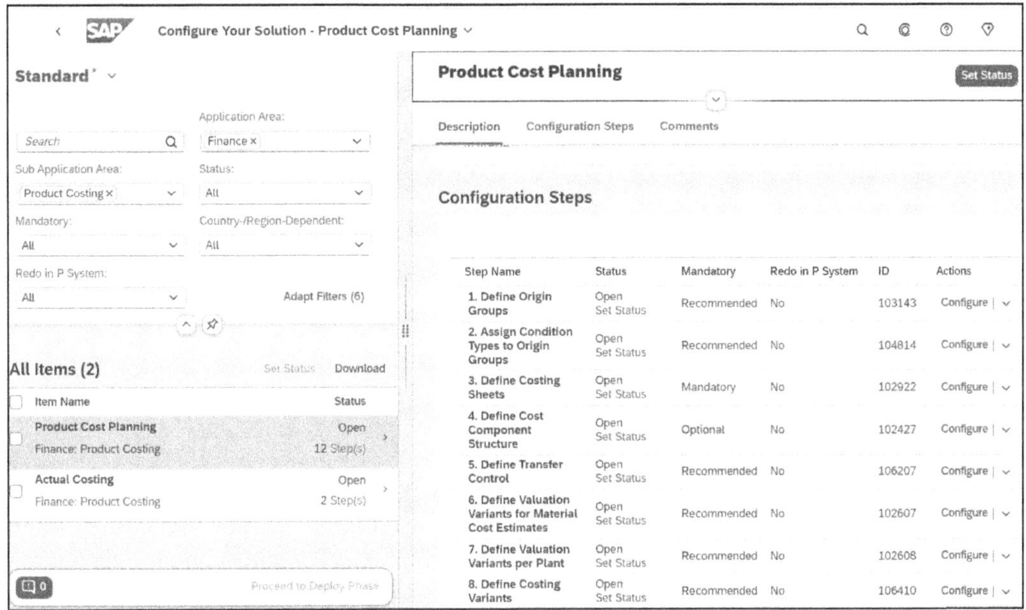

Figure 7.64 Configuration Settings in SAP S/4HANA Cloud

7.8.1 Define Costing Variants

In addition to the sample costing variants that we saw earlier in Figure 7.1, cloud systems and systems using best-practice content include the costing variants shown in Figure 7.65 with the naming convention **PY***. You can access these settings via **ID 106410** in Figure 7.64 in SAP S/4HANA Public Cloud, with Transaction OKKN, or via the IMG menu path **Controlling • Product Cost Controlling • Product Cost Planning • Material Cost Estimate with Quantity Structure • Define Costing Variants** in SAP S/4HANA Private Cloud or on-premise SAP S/4HANA .

Two of these costing variants are used to set standard costs, the idea being that costing variant **PYC1** is used to set the standard costs for inventory valuation and works in combination with valuation variant **0Y1** and transfer control **PYC1**, while costing variant **PYC2** is used to set the standard costs during planning and budgeting and works in combination with valuation variant **0Y2** and transfer control **PYC2**. In addition, **PYRM** is used to set the standard costs for a product cost collector and works in combination with valuation variant **0Y6**.

If you work with multiple ledgers, you will also find further costing variants to set the standard costs by ledger, a topic that we will discuss in more detail when we look at universal parallel accounting in Chapter 19.

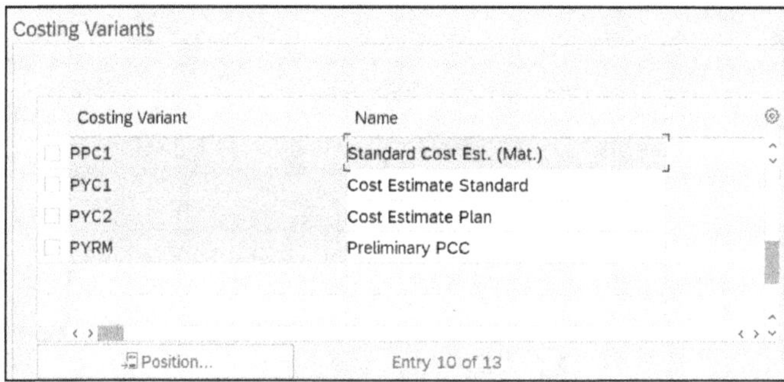

Figure 7.65 Sample Costing Variants

7.8.2 Define Valuation Variants

In addition to the sample valuation variants that we saw earlier in Figure 7.18, cloud systems and systems using best-practice content include the valuation variants shown in Figure 7.66 with the naming convention **0Y***, **0Y1** being used for inventory valuation and **0Y2** for planning and budgeting purposes. You can access these settings via **ID 102607** in Figure 7.64 in the public cloud or with Transaction OKK4 or via the IMG menu path **Controlling • Product Cost Controlling • Product Cost Planning • Material Cost Estimate with Quantity Structure • Costing Variant: Components • Define Valuation Variants** in SAP S/4HANA Private Cloud or on-premise SAP S/4HANA. It is also possible to set valuation variants per plant using the settings in ID **102608**.

If you work with multiple ledgers, you will also find valuation variants to select the material and activity prices by ledger. We will discuss this topic in more detail when we examine universal parallel accounting in Chapter 19.

Valuation Variant	Name
0Y1	Valuation: Standard
0Y2	Valuation: Plan
0Y6	Prod. Order: Planned
0Y7	Prod. Order: Actual

Figure 7.66 Sample Valuation Variants

The strategy sequence for **0Y1** is to use the information in the purchase info record, together with any additive costs for freight and transport, as the default and the material price according to the price control in the material master if this is not available.

The strategy sequence for **0Y2** is to use the value in **Planned Price 1** if available and the material price according to the price control in the material master if this is not available.

7.8.3 Define Transfer Control

In addition to the sample settings for transfer control that we saw earlier in Figure 7.54, cloud systems and systems using best-practice content include the transfer control settings shown in Figure 7.67 with the naming convention **PYC***, **PYC1** being used for inventory valuation and **PYC2** for budgeting purposes. You can access these settings via **ID 106207** in Figure 7.64 in SAP S/4HANA Public Cloud, with Transaction OKKM, or via the IMG menu path **Controlling • Product Cost Controlling • Product Cost Planning • Material Cost Estimate with Quantity Structure • Costing Variant: Components • Define Transfer Control** in SAP S/4HANA Private Cloud or on-premise SAP S/4HANA.

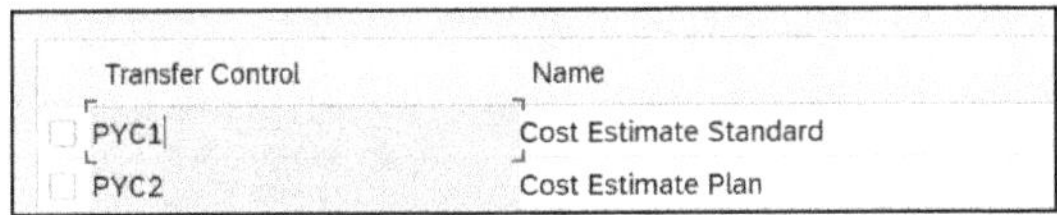

Figure 7.67 Sample Settings for Transfer Control

In both cases, a transfer control is used to reference an existing cost estimate if one has already been created for the relevant costing variant. As shown in Figure 7.68, cross-plant transfer control is not active by default but can be activated as required.

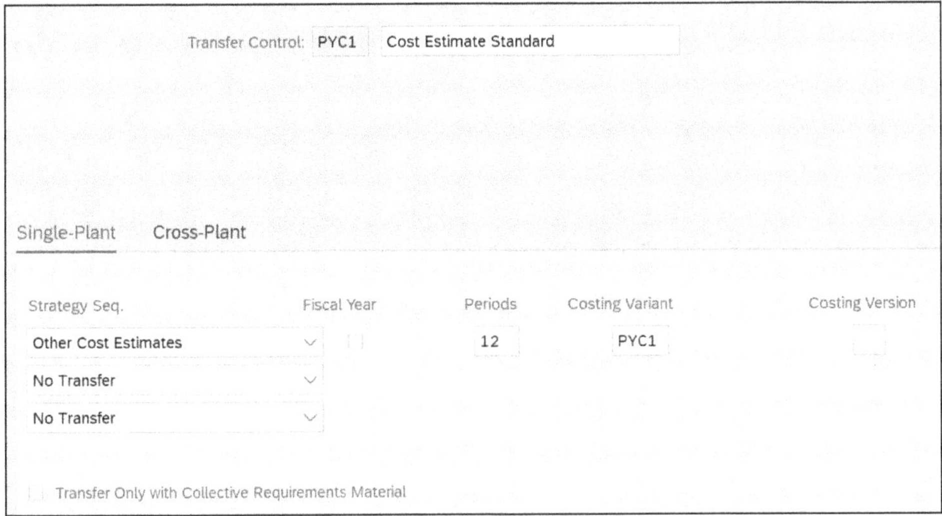

Figure 7.68 Detailed Settings for Transfer Control in a Single Plant

7.9 Summary

In this chapter, we discussed costing variant components and how they are used to create cost estimates. We looked at the definition of a costing variant and then examined each of the six components.

We first saw how the costing type determines whether a cost estimate can update the material master. We then looked at how the valuation variant selects prices for the cost estimate. We examined how to use date control to set default dates when creating a cost estimate. Next, we looked at how quantity structure control determines which BOM and routing are selected for a cost estimate. Finally, we looked at how transfer control and the reference variant allow you to use existing cost estimates to create new cost estimates.

Now that we've examined the six costing variant components in this chapter, we'll look at the costing variant tabs in Chapter 8.

Chapter 8
Costing Variant Tabs

A costing variant contains information on how a cost estimate calculates the standard price.

We examined costing variant components in the previous chapter. In addition to the six costing variant components, there are six costing variant tabs containing general configuration settings:

- **Control**
- **Quantity Structure**
- **Additive Costs**
- **Update**
- **Assignments**
- **Miscellaneous**

We'll examine each of the tabs in detail in this chapter. First, let's look at the costing variant **Control** tab.

8.1 Control

You access costing variant configuration settings with Transaction OKKN or via the IMG menu path **Controlling • Product Cost Controlling • Product Cost Planning • Material Cost Estimate with Quantity Structure • Define Costing Variants**. Double-click SAP standard costing variant **PPC1** to display the details shown in Figure 8.1.

Settings for the Costing Variant in Public Cloud and Best-practice Systems

If you work with the costing variants in a public cloud system, these settings are delivered as standard. You can change the settings under **Quantity Structure** and **Additive Costs**, but all other settings are hidden and cannot be changed. If you use best-practice content to accelerate your implementation in SAP S/4HANA Private Cloud or on-premise SAP S/4HANA, you can change the settings as described here.

The **Control** tab lists the six costing variant control parameters that we discussed in detail in Chapter 7.

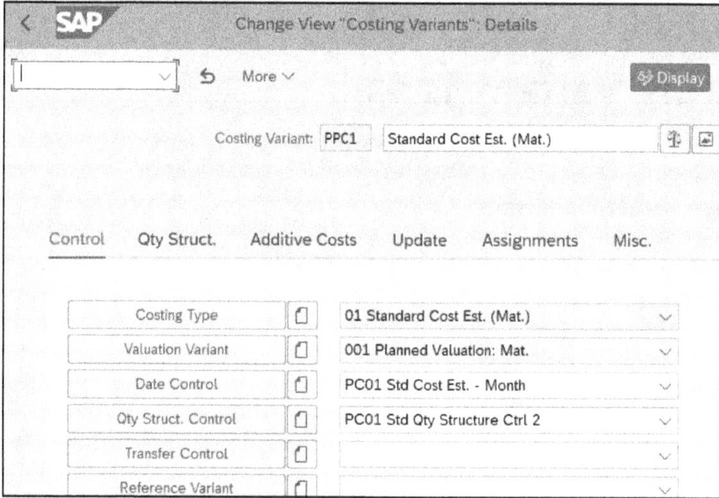

Figure 8.1 Costing Variant Control Tab

Tip: Overview Icon

You can see an overview of all costing variant component settings by clicking the **Overview** icon at the right.

Let's now examine the settings in the **Qty Struct.** tab.

8.2 Quantity Structure

Select the **Qty Struct.** (quantity structure) tab to display the details shown in Figure 8.2.

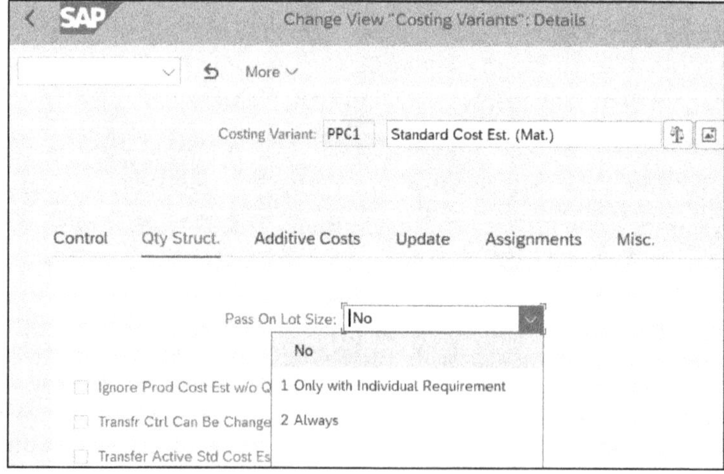

Figure 8.2 Costing Variant Qty Structure Tab

We'll walk through the details of this tab in this section. First, let's examine how **Pass On Lot Size** works.

8.2.1 Pass On Lot Size

The **Pass On Lot Size** field determines whether you base the costing lot size of all components on the costing lot size of the highest-level material. Click this field to display a list of possible entries, as shown in Figure 8.2. We'll review the three possible options for this setting:

- **No: Do Not Pass On Lot Size**
  This indicates that materials further down the bill of materials (BOM) are costed according to the lot size in the **Costing 1** view of each component. This is the default setting for the PPC1 costing variant.

Example: Do Not Pass on Lot Size

A finished product has a costing lot size of 10, whereas materials at the component (lower) level have a costing lot size of 100. A cost estimate for the finished product initially creates cost estimates for the lowest-level materials based on a costing lot size of 100.

You create the cost estimate for the finished product based on a costing lot size of 10 and access the component cost estimate prices based on a costing lot size of 100 to determine material costs.

- **1: Pass On Lot Size Only with Individual Requirements**
  In the **MRP 4** view, you can specify if a material is planned as an individual requirement, as discussed in Chapter 3. If you add this material to another material in a BOM, the cost estimate uses the lot size of the highest-level material.

Example: Pass On Lot Size with Individual Requirements

A component and finished product both have an individual requirements setting in the **MRP 4** view. A cost estimate for the component uses the costing lot size of the finished product and ignores the costing lot size of the component material. You base the cost estimate on a costing variant with the **Pass on Lot Size Only with Individual Requirements** setting.

- **2: Always Pass On Lot Size**
  All materials in a multilevel BOM are costed using the costing lot size of the finished product. You typically use this setting for sales order costing. When creating products for a specific customer requirement, you procure components just for

this requirement. You don't necessarily get the purchasing benefits of pooling component requirements, so ensure this is the best setting.

Note: System-supplied PPC4 is for Sales Order Cost Estimates

PPC4 is a system-supplied costing variant for sales order cost estimates. This costing variant contains **1: Pass on Lot Size Only with Individual Requirements** selected by default. You access this cost costing variant with Transaction OKY9 or via the IMG menu path **Controlling • Product Cost Controlling • Cost Object Controlling • Product Cost by Sales Order • Preliminary Costing and Order BOM Costing • Product Costing for Sales Order Items/Order BOMs • Costing Variants for Product Costing • Check Costing Variants for Product Costing.**

Now that we've examined **Pass On Lot Size**, let's examine the first indicator.

8.2.2 Ignore Cost Estimate without Quantity Structure

As shown in Figure 8.2, **Ignore Prod Cost Est w/o Qty Structure** (ignore product cost estimate without quantity structure) determines whether a cost estimate with a quantity structure can access data produced by a cost estimate without a quantity structure. Selecting this saves the system time searching for data produced by cost estimates without quantity structures. You may create many cost estimates without quantity structures during new product development that don't need to be considered when creating standard cost estimates for existing products.

Note: Cost Estimates Without Quantity Structure

Cost estimates without quantity structure aren't based on a BOM and routing. You manually enter the components, activities, and overhead to manufacture an assembly in a screen like a spreadsheet. You can also refer to this as *unit costing*. You can read more about unit costing in Chapter 9.

You typically use this functionality during new product development before you create a BOM and routing necessary for production. A unit cost estimate is useful for estimating the cost viability of a new product. You can easily compare the cost-effectiveness of designs by substituting components and activities before prototyping or production occurs.

You can copy existing BOMs and routings into unit cost estimates and determine a new unit cost estimate based on proposed changes to an existing product design.

Cost estimates without quantity structure are not supported in public cloud systems, so this option is unavailable in SAP S/4HANA Public Cloud.

If you accept the default setting for the costing variant PPC1 and leave this deselected, standard cost estimates will search for all cost estimates, with and without quantity structure. This costing variant option takes priority over the ignore the **With Quantity Structure** indicator in the **Costing 1** view.

Now, let's review the next option in the costing variant **Qty Struct.** tab.

8.2.3 Transfer Control Can Be Changed

Transfer Ctrl Can Be Changed determines whether you can change the transfer control parameter when creating a cost estimate. Transfer control requires a higher-level cost estimate to retrieve recently created standard cost estimates for lower-level materials. You can display the selection screen when creating an individual cost estimate with Transaction CK11N or via the menu path **Accounting • Controlling • Product Cost Controlling • Product Cost Planning • Material Costing • Cost Estimate with Quantity Structure • Create**. The details shown in Figure 8.3 are displayed.

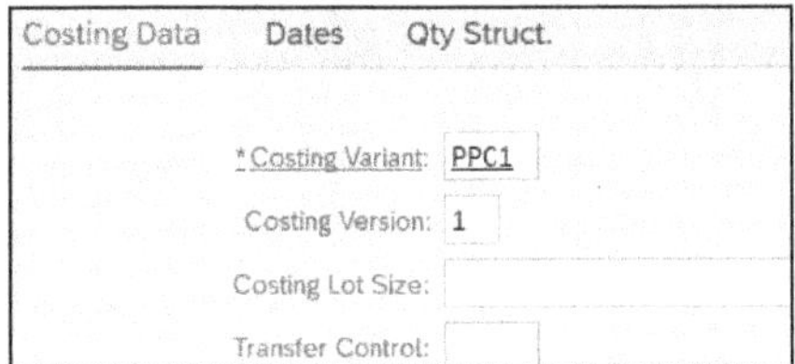

Figure 8.3 Transfer Control Changeable

You can change the **Transfer Control** because we selected **Transfer Ctrl Can Be Changed** in costing variant PPC1.

> **Note: Costing Run Transfer Control**
>
> **Transfer Control** appears in the header when creating cost estimates with costing run Transaction CK40N.

Let's see how **Transfer Control** appears when you use costing variant **PPC2**, as shown in Figure 8.4.

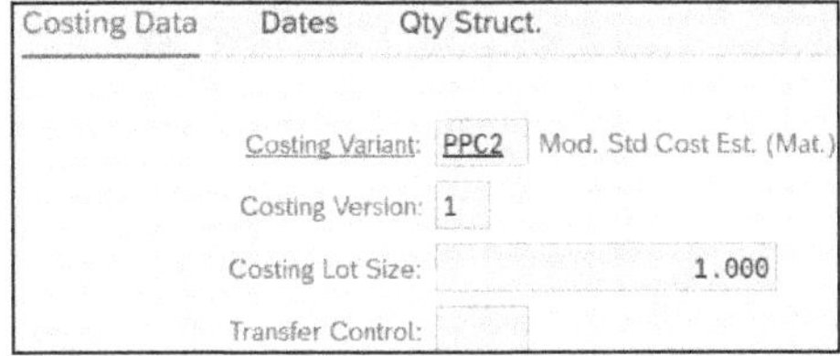

Figure 8.4 Transfer Control Not Changeable

Because **Transfer Ctrl Can Be Changed** is deselected for PPC2, you cannot change the default **Transfer Control** when creating a cost estimate. The default **Transfer Control** is set in the **Control** tab of the costing variant.

Case Study: Transfer Control Strategy

You conduct a main costing run annually to update standard prices and inventory valuation. Typically, new standard cost estimates are required for every material during a full costing run. To help meet this requirement, you can copy a new costing variant, ZPC1, from the standard PPC1, with these settings:

- In the **Qty Struct.** tab, **Transfer Ctrl Can Be Changed** shown previously in Figure 8.2 is deselected.
- In the **Control** tab, the optional **Transfer Control** component shown previously in Figure 8.1 is blank.

These settings ensure that, when you carry out a costing run with costing variant ZPC1, **Transfer Control** will be blank. You will create new standard cost estimates for every material selected in the costing run.

Standard cost estimates need to be created for new products, subassemblies, and components when you create new products. If new products include subassemblies or components that already exist, you don't want to overwrite the corresponding existing standard estimates. You can avoid overwriting these estimates by including transfer control in standard cost estimates created for new products during the year.

You can copy a new costing variant, ZPC5, from the standard PPC1 with the following settings:

- In the **Qty Struct.** tab, **Transfer Ctrl Can Be Changed** shown previously in Figure 8.2 is deselected.
- In the **Control** tab, you populate the optional **Transfer Control** component field shown previously in Figure 8.1 with a copy of standard transfer control ZC02—Complete Transfer.

When you create standard cost estimates for new products and components with costing variant ZPC5 during the year, the **Transfer Control** field populates with ZC02. This ensures that existing standard cost estimates created during the main costing run with ZPC1 will remain in place for twelve months and you won't overwrite them during the year.

You can control which costing variant you allow for release with price update Transaction CK24. Click the **Marking Allowance** button to display a screen in which you can enter an allowed costing variant per period and company code. You need to mark a standard cost estimate before you can release it to update the material standard price. Add costing variant ZPC1 for the first period and ZPC5 for the remaining periods for each fiscal year, and you have successfully controlled the creation of standard cost estimates in line with company policy. You can restrict the number of people with access to Transaction CK24 and to marking allowance with authorizations.

Now let's review the remaining indicator in the **Qty Struct.** tab shown earlier in Figure 8.2.

8.2.4 Transfer Active Standard Cost Estimate If Errors

If **Transfer Active Std Cost Est. if Mat. Costed w/Errors** is selected, the status "KF" (costed with errors) of a cost estimate for a component or subassembly is normally passed up to the highest-level material. You select this indicator to ensure that you use active standard cost estimates for any materials with errors, and costing continues up to the highest-level material without errors. If the system cannot find an active standard cost estimate, it searches for a price in the material master according to the valuation variant.

Case Study: Lower-Level Errors

You typically require cost estimates to access the most recent supplier-quoted prices in purchasing info records during a costing run. If cost estimate errors occur for components without current info record prices, for example, you need these errors to be passed up to the highest-level materials. This ensures that the highest-level standard cost estimates cannot be marked and released until you correct the lower-level cost estimates.

You typically provide the list of materials that need info records updated with the latest supplier quotation prices to purchasing. You then carry out another costing run and continue eliminating errors until you are ready to mark and release the standard cost estimates.

To ensure that you correct all lower-level standard cost estimate errors before you can release a higher-level standard cost estimate, deselect **Transfer Active Std Cost Est. if Mat. Costed w/Errors** in the **Qty Struct.** tab of costing variant ZPC1. See the previous case study for details regarding costing variant ZPC1.

For cost estimates that don't update standard prices, it may be acceptable to use active standard cost estimates when you encounter errors. If you are developing new products by copying existing BOMs and routings and making proposed improvements or adjustments, you can create a new cost variant with the indicator selected. This may shorten the product development process by not requiring cost estimates to be completely error-free to obtain an estimated price.

We've now reviewed all the indicators in the **Qty Struct.** tab shown in Figure 8.2. There's another indicator in this tab for sales order costing variants, so let's examine it before moving on to the next costing variant tab.

8.2.5 Transfer Cost Estimate of an Order Bill of Materials

You access sales order costing variants with Transaction OKY9 or via the IMG menu path **Controlling • Product Cost Controlling • Cost Object Controlling • Product Cost by Sales Order • Preliminary Costing and Order BOM Costing • Product Costing for Sales Order Items/Order BOMs • Costing Variants for Product Costing • Check Costing Variants for Product Costing**. Select the **Qty Struct.** tab to display the details shown in Figure 8.5.

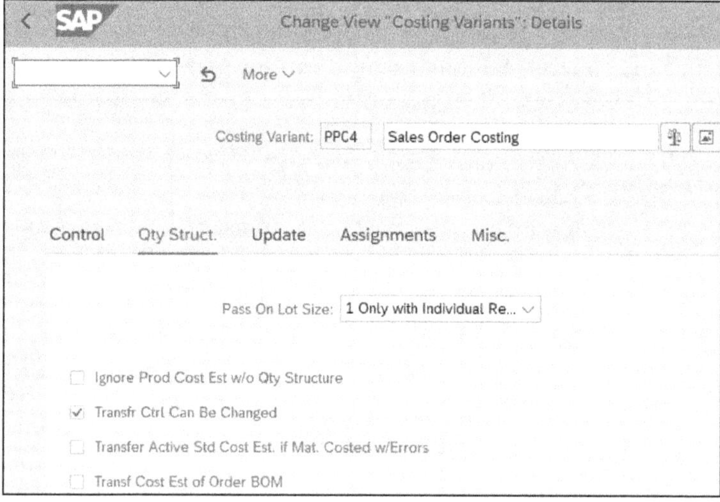

Figure 8.5 Sales Order Costing Variant Quantity Structure Tab

Transf Cost Est of Order BOM (transfer cost estimate of order BOM) controls whether you transfer existing order BOM cost estimates. An *order BOM* is a sales order BOM typically used in make-to-order and configurable product scenarios. These scenarios involve manufacturing products specifically for individual sales order line items. Customers specify certain required options for components and subassemblies for configurable products. You should transfer order BOM cost estimates for complex make-to-order production. Transferring order BOM cost estimates improves system performance.

Now that we've discussed the indicators in the **Qty Struct.** tab, let's move on to the **Additive Costs** tab.

8.3 Additive Costs

Select the **Additive Costs** tab shown previously in Figure 8.2 to display the screen shown in Figure 8.6.

First, let's examine how **Additive Cost Comps** (additive cost components) work.

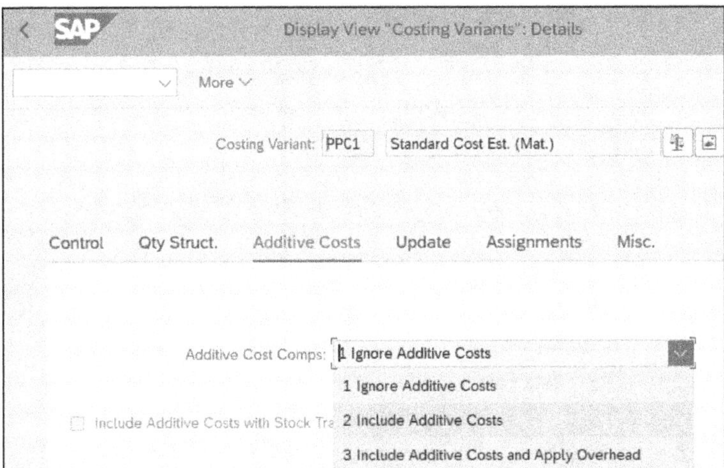

Figure 8.6 Costing Variant Additive Costs Tab

8.3.1 Additive Cost Components

You can add costs manually to an additive cost estimate, which is similar in functionality to a spreadsheet. The **Additive Cost Comps** flag determines whether costing items created using an additive cost estimate will be included in costing. The **Additive Cost Comps** also specifies if you include additive costs when applying overhead. You can see all three options displayed in the list of possible entries shown in Figure 8.6. We'll examine each in detail:

- **1: Ignore Additive Costs**
  You cannot manually add cost components with additive cost estimates for cost estimates created with this costing variant. The cost estimate will ignore any existing additive cost estimates. Choose this option if you need to ensure that you cannot adjust cost estimates with manual costs.
- **2: Include Additive Costs**
  You can include manual costs with additive cost estimates created with this costing variant. You can create and include additive costs for freight and handling charges, for example.

- **3: Include Additive Costs and Apply Overhead**
  You can include additive costs and calculate overhead using these costs as part of the calculation base. This decision depends on how significant the additive costs are and on the method of overhead calculation.

8.3.2 Include Additive Costs with Stock Transfers

The next item to consider in the **Additive Costs** tab is shown in Figure 8.7.

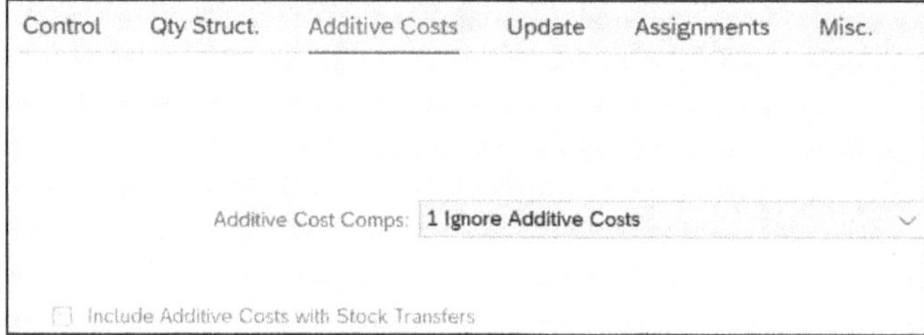

Figure 8.7 Costing Variant Additive Costs Tab

Selecting **Include Additive Costs with Stock Transfers** allows you to include additive cost estimates for stock transfers between plants. This may be useful if you include transportation costs for stock transfers.

You consider the following special procurement types with this option:

- Transfer to another plant
- Production in another plant

The special procurement type indicates to material requirements planning (MRP) and cost estimates how you produce and transfer materials across plants. The standard price in the receiving plant includes the valuation you calculated for the material in the issuing plant, together with transportation costs in the additive cost estimate.

Settings in SAP S/4HANA Public Cloud

Additive cost estimates are supported in SAP S/4HANA Public Cloud for the handling of freight and transportation costs. The default setting is that they are included in the cost estimate and for cross-plant movements.

Now, let's look at the next costing variant tab.

8.4 Update

Select the **Update** tab shown in Figure 8.7 to display the details shown in Figure 8.8.

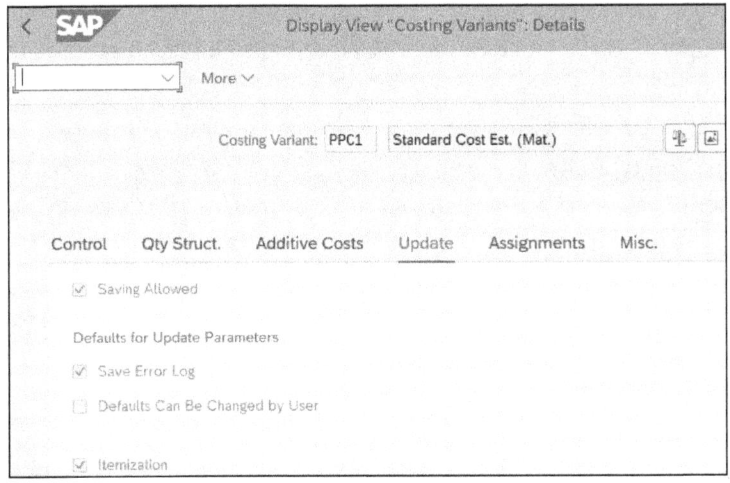

Figure 8.8 Costing Variant Update Tab

We'll examine **Saving Allowed** in the **Update** tab first.

8.4.1 Saving Allowed

Your choice for the **Saving Allowed** checkbox determines whether you can save cost estimates created with this costing variant. If you intend to update prices in the material master, such as the standard price or planned prices in the **Costing 2** view, you need to select this. You'll also need to select this if you intend to report on cost estimates or use the costing results in any of the following period-end activities:

- Variance calculation
- Work in process (WIP) calculation
- Results analysis

Let's consider the three options in the **Defaults for Update Parameters** section.

> **Note: Deselect Saving Allowed**
>
> If you deselect **Saving Allowed**, the remaining options in this tab automatically deselect. You need to save the cost estimate to save the error log and itemization. Itemization is a simple listing of components and resources.

8.4.2 Save Error Log

When you create a cost estimate, an error log with additional information is created. It's good practice to analyze all error log messages to ensure the integrity of the costing results. You can save a standard cost estimate with status "KF" (costed with errors), which gives you the opportunity to make corrections before rerunning and saving the standard cost estimate.

You need to reach a status "KA" (costed without errors) before you can mark and release a cost estimate. You typically save the error log to refer to messages and make corrections. There should be no storage capacity reasons preventing you from saving the error log. Selecting **Save Error Log** is the most common setting.

Now we'll consider the next checkbox in the **Update** tab.

8.4.3 Defaults Can Be Changed by User

When saving a cost estimate, you can specify whether the user can change the default settings for saving the message log and itemization. If you select the **Defaults Can Be Changed by User** checkbox, an **Update Parameters** dialog box will appear when you save the cost estimate, as shown in Figure 8.9.

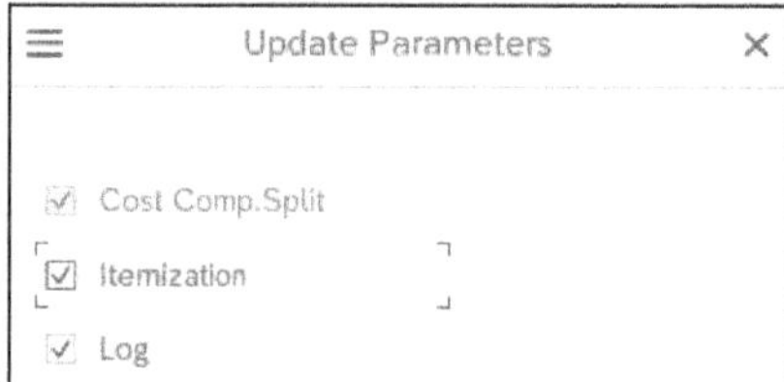

Figure 8.9 Update Parameters Dialog Box When Saving a Cost Estimate

Let's discuss each option further:

- **Cost Comp.Split (cost component split)**
  Cost Comp.Split is selected by default and cannot be changed. Cost components allow you to display costs in groups such as materials, labor, and overhead, as we discussed in Chapter 7.
- **Itemization**
  Itemization provides cost details about the resources required to manufacture a product at a line-item level. You typically retain **Itemization** for review when analyzing costs.
- **Log**
  The **Log** contains messages that you need to review to determine the integrity of cost estimates. It's good practice to retain messages for review.

Typically: Save Error Log and Itemization

You'll typically save itemization and the error log for analyzing costs and cost estimate integrity. You may decide that the user doesn't need to review the **Update Parameters** dialog box while saving every cost estimate. In that case, you'll deselect **Defaults Can Be Changed by User** and select **Save Error Log** and **Itemization** in the costing variant **Update** tab.

Now let's review the final indicator in the **Update** tab.

8.4.4 Itemization

A cost estimate creates an itemization list during cost calculation. You have the choice in the costing variant of whether to save itemization when you save the cost estimate. Itemization is useful when you analyze and review costs on a detailed level, whereas cost components provide summarized cost information.

Example: Itemization vs Cost Component View

A finished goods BOM contains 20 components and subassemblies. All 20 items are listed separately in itemization. Likewise, all activity items from the routing and overhead items from the costing sheet are listed separately. A costing sheet contains the rules for allocating overhead from cost centers to cost estimates, product cost collectors, and manufacturing orders.

The cost component view summarizes costs based on cost element groups or ranges. Typical cost components are materials, labor, and overhead.

You typically save the itemization for analysis purposes.

Now that we've reviewed the **Update** tab, let's move to the next tab.

Settings in Public Cloud Systems

These settings are hidden in the public cloud. All elements of the cost estimate, including the error log, are saved automatically.

8.5 Assignments

Select the **Assignments** tab shown in Figure 8.8 to display the details shown in Figure 8.10.

We'll discuss the options in the **Assignments** tab in the following sections.

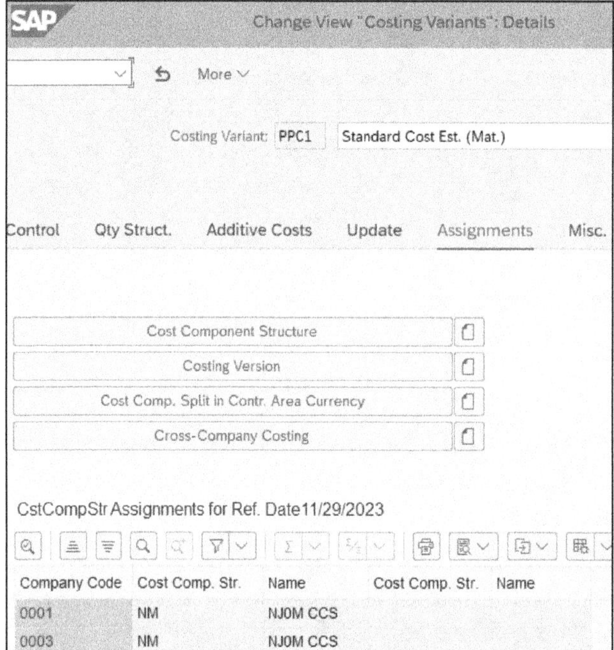

Figure 8.10 Costing Variant Assignments Tab

8.5.1 Cost Component Structure

You can check the assignment of cost component structures to organizational units for this costing variant by clicking the **Cost Component Structure** button. In the example shown in Figure 8.10, the main cost component structure assigned to **Company Code 0001** and **Company Code 0005** is **01**, and the auxiliary **Cost Comp. Str.** (cost component structure) column is blank.

For a listing of available cost component structures or to create your own, click the new page icon next to the **Cost Component Structure** button.

8.5.2 Costing Version

Click the **Costing Version** button shown in Figure 8.10 to display the screen shown in Figure 8.11.

Costing versions enable you to create multiple cost estimates for the same material. Because the costing variant contains all parameters for cost, creating new costing variants every time you want to make minor changes in the control parameters can be time-consuming. Costing versions enable you to simulate changes without creating new costing variants. Let's examine the columns in Figure 8.11:

- **Variant for TP (variant for transfer price determination)**
  You can define settings for the selection of transfer prices in customizing for enter-

prise controlling by following the IMG menu path **Enterprise Controlling • Profit Center Accounting • Transfer Prices.** When you use costing versions, you can specify how you select these transfer variants.

- **Exch. Rate Type (exchange rate type for currency translation)**
  You can use the exchange rate type to define a buying rate, selling rate, or average rate for translating foreign currency amounts.
- **QtyStruct. Type (quantity structure type for mixed costing)**
  Mixed-cost estimates are created with reference to a costing version. You can create more than one mixed-cost estimate for the same material with costing versions.

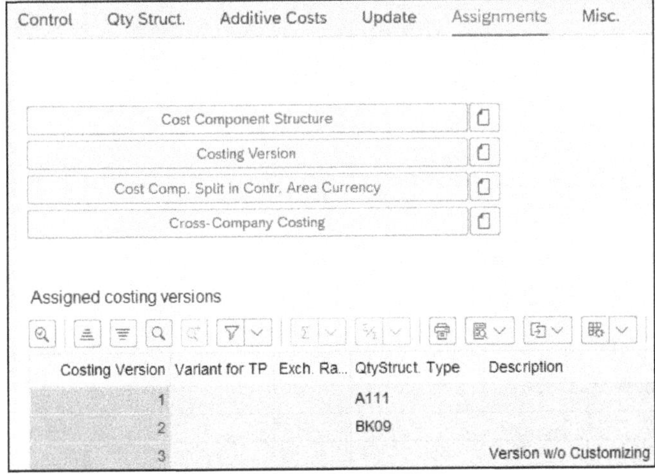

Figure 8.11 Costing Variant Costing Version Assignment

To list available costing versions, click the new page icon next to the button or use Transaction OKYD.

8.5.3 Cost Component Split in Controlling Area Currency

Click the **Cost Comp. Split in Contr. Area Currency** button shown in Figure 8.11 to display the screen shown in Figure 8.12.

You can specify that material cost estimate cost components are updated in the controlling area currency in the **Additional CCS** (cost component split) column, in addition to the company code currency in the main CCS. You must select the **All Currencies** checkbox for the controlling area with Transaction OKKP.

If the controlling area currency isn't the same as the company code currency, the system always updates the following:

- Cost component splits in the company code currency
- Itemizations in both currencies, provided that the costing variant allows itemizations to be saved

To list additional cost component splits in controlling area currency, click the new page icon next to the **Cost Comp. Split in Contr. Area Currency** button or use Transaction OKYW.

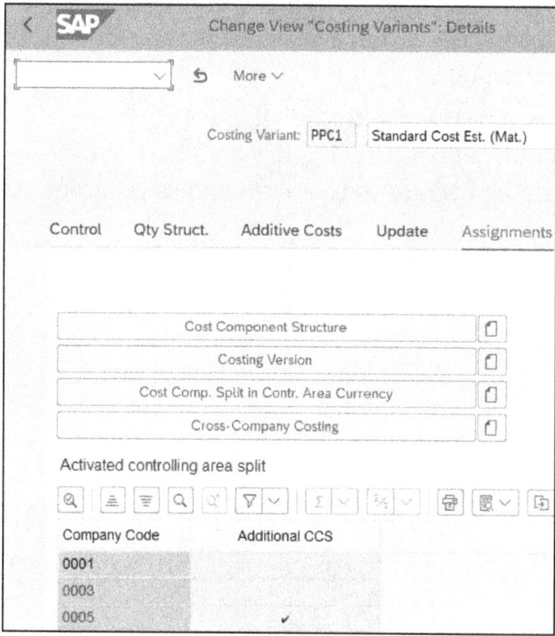

Figure 8.12 Costing Variant Controlling Area Currency Assignment

8.5.4 Cross-Company Costing

Click the **Cross-Company Costing** button shown in Figure 8.12 to display the screen shown in Figure 8.13.

You can allow cost estimates to access information in more than one company code. Either the system re-costs the materials in plants assigned to the company code of the plant in which you create the cost estimate, or it transfers an existing cost estimate following transfer control parameters.

The link for materials across plants is made with the special procurement key in the **MRP 2** view. We discussed the material master views in detail in Chapter 3.

For materials in plants that you assign to another company code, the **Cost Across Company Codes** checkbox in Transaction OKYV determines whether the materials are re-costed or whether the system uses an existing cost estimate in accordance with the valuation strategy. You can also access this transaction via the IMG menu path **Controlling • Product Cost Controlling • Product Cost Planning • Selected Functions in Material Costing • Activate Cross-Company Costing.**

For a listing of cross-company costing assignments, click the new page icon next to the button or use Transaction OKYV to create additional assignments.

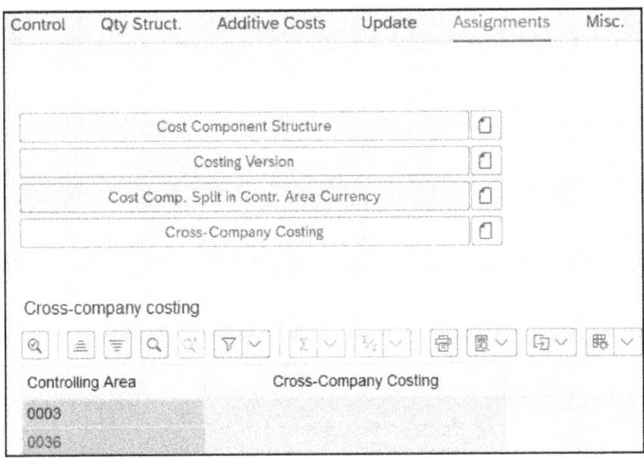

Figure 8.13 Costing Variant Cross-Company Costing

Settings in Public Cloud Systems

These settings are hidden in the public cloud. The cost component split is stored in both the Controlling area and company code currency. Since mixed costing is not yet supported, there are no costing versions for this purpose.

Now that we've discussed the checkboxes in the costing variant **Assignments** tab, let's move on to the last tab.

8.6 Miscellaneous

Select the **Misc.** (miscellaneous) tab shown in Figure 8.14 and click the **Error Management** button to display the screen shown in Figure 8.15.

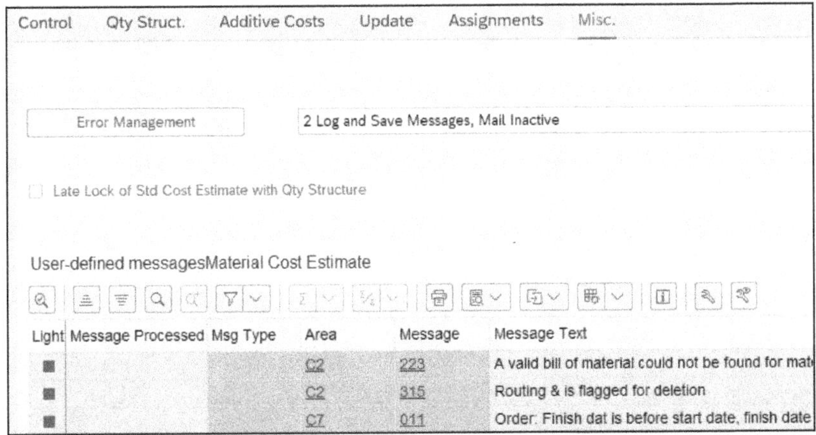

Figure 8.14 Costing Variant Miscellaneous Tab

This allows you to manage the processing of messages. Let's first look at how you can change message types.

8.6.1 Error Management

Click the **Error Management** button to display a list of messages. You can change a message type by clicking a message row in the **Msg Type** column to display the list of possible entries shown in Figure 8.15.

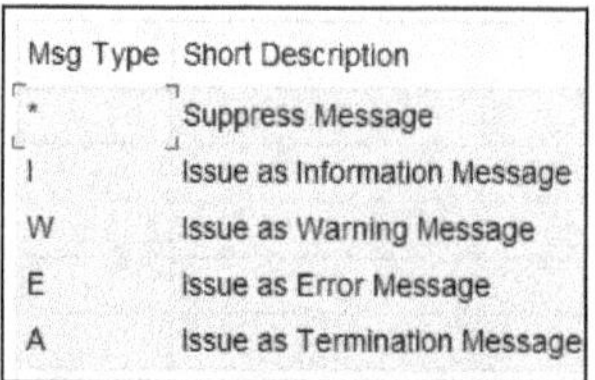

Figure 8.15 Allowed Message Types

The message type indicates the importance of a message. Double-click a **Msg Type** (message type) to change its message type.

> **Example: Message Type**
>
> The **Message Text** of the second message (**Message 315**) in Figure 8.14 indicates that a routing is flagged for deletion. Since this is a warning message, you can release a cost estimate. You can change the message type to an error to prevent the release of cost estimates with this message.

Now that we've examined message types, let's look at the next field.

8.6.2 Parameters for Error Management

Click the field next to the **Error Management** button shown in Figure 8.16 to display a list of possible entries.

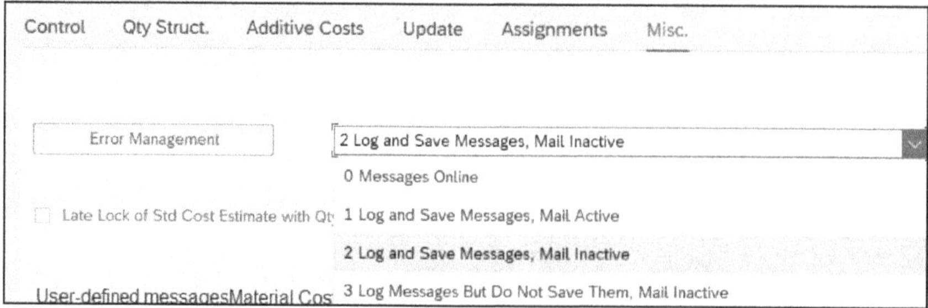

Figure 8.16 Parameters for Error Management

Let's examine each of the possible options:

- **0: Messages Online**
  The messages are issued individually from the status bar. The log function is inactive in a cost estimate.
- **1: Log and Save Messages, Mail Active**
  The messages are collected in a log, which can be saved. You can send the messages to the person responsible for correcting errors.
- **2: Log and Save Messages, Mail Inactive**
  The messages are collected in a log, which can be saved. You cannot send the messages. This setting is the most common and allows you to keep a record of all cost estimate messages that you can refer to when analyzing the messages at a later date as necessary.
- **3: Log Messages But Do Not Save Them, Mail Inactive**
  The messages are collected in a log, which can be processed online but not saved.

Settings in Public Cloud Systems

These settings are hidden in the public cloud. Messages are logged and saved, but mail is inactive.

8.7 Summary

In this chapter, we discussed the information contained in costing variant tabs. We first considered the **Control** tab, which contains the six costing variant components discussed in Chapter 7. We then considered the **Qty Struct.** tab, which allows you to make settings about passing on the costing lot size and which includes three checkboxes concerning the quantity structure and transfer control. Next, we considered the **Additive Costs** tab, which allows you to control whether additive costs can be included in the cost estimate. The **Update** tab allows you to determine if you can save message logs and if you can change the defaults. The **Assignments** tab allows you to check organizational assignments of the costing variant, and the **Misc.** tab contains error management and message log settings.

Now that we've completed preparation for cost estimates with master data and basic configuration settings in previous chapters, we'll look at how to create standard cost estimates in Chapter 9.

Chapter 9
Cost Estimates

A standard cost estimate calculates the planned cost to manufacture a product or purchase a component and is used to update inventory valuation.

We completed preparations for creating cost estimates in previous chapters by setting up master data and configuration and updating activity and purchase prices.

Cost estimates provide a plan cost to procure components and manufacture assemblies. Standard cost estimates are typically created several weeks or months before the start of the next fiscal year. System messages are analyzed, and corrections are made. For instance, you may need to enter missing info records or activity prices.

After you have made corrections, eliminated error messages, and reviewed the proposed inventory revaluation, you typically release standard cost estimates on the first day of the fiscal year. New material standard prices become the benchmarks for all production and purchasing activities for the next year.

Some companies with rapidly changing and developing products create and release standard cost estimates more frequently to keep pace with changes. Otherwise, variances would become so large toward the end of the year that they would provide no assistance during variance analysis.

Standard costing allows you to keep production and procurement prices stable over at least one or more periods to allow you to analyze differences between the planned costs by cost estimates and actual costs.

In this chapter, we'll first create, mark, and release a standard cost estimate and then, in the following section, process many cost estimates with a costing run.

9.1 Cost Estimate Types

Different types of cost estimates are available during the product life cycle. During the initial planning stages, with master data not yet available, you can begin planning by creating a base planning object and a manual unit cost estimate without requiring a production BOM. When the first material masters are created, you can then create a material cost estimate without quantity structure to manually plan the cost of goods manufactured.

When BOMs and routings become available, you create a material cost estimate with quantity structure, automatically calculating the cost with the master data and prices. The quality of the BOM and routing can evolve over time, so many organizations use separate engineering/design BOMs in the early stages and replace them with a production BOM as the BOM structure becomes more reliable.

In this chapter, we'll explore cost estimates with and without quantity structure. Let's start with a standard cost estimate and costing runs in the next section.

9.2 Standard Cost Estimates

Standard cost estimates are based on BOMs, routings, activity, and purchase prices. After you create them, you mark and release them to revalue inventory for the next 12 months. In this section, we'll explain how to create cost estimates for individual materials and how the BOM and routing are exploded to deliver the quantity structure.

9.2.1 Create

You create a standard cost estimate with Transaction CK11N or by following the menu path **Accounting • Controlling • Product Cost Controlling • Product Cost Planning • Material Costing • Cost Estimate with Quantity Structure • Create**. A selection screen is shown, as in Figure 9.1.

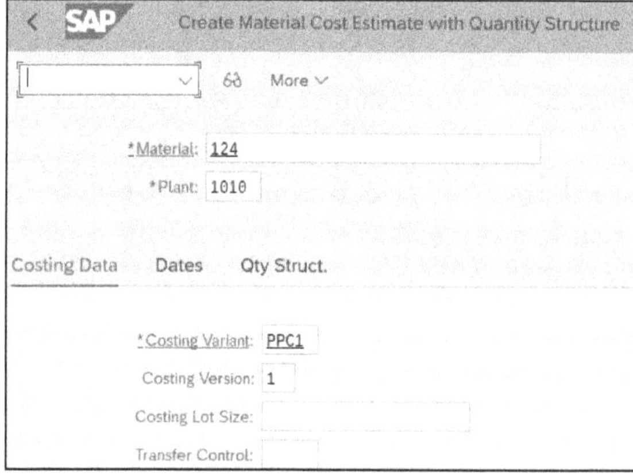

Figure 9.1 Create Standard Cost Estimate

You then enter the necessary data to create a cost estimate (**Material**, **Plant**, and **Costing Variant** are mandatory):

- **Costing Data tab**
  Let's review the fields in the **Costing Data** tab:

- **Costing Variant**
  The **Costing Variant** contains the configuration required for cost estimates to determine the standard price.
  For example, it determines if the info record price or the planned price 1 in the **Costing 2** view determines the price for purchased materials and which type of BOM will be used to determine the quantity structure.
  You typically create a standard cost estimate with **Costing Variant PPC1**. You can only use one costing variant to update the standard price in a **Plant**.
- **Costing Version**
  You can create many costing versions for scenario analysis. Typically, you only allow cost estimates created with **Costing Version 1** to update the standard price.
- **Costing Lot Size**
  Costing Lot Size determines the cost estimate quantity. It should be set close to actual purchase and production quantities to minimize lot size variance. We'll discuss lot size variance in detail in Chapter 16. If you leave the **Costing Lot Size** field blank, the cost estimate retrieves it from the **Costing 1** view.
- **Transfer control**
  Transfer control requires a higher-level cost estimate to use recently created lower-level estimates without re-creating them. If you leave the **Transfer Control** field blank, it defaults from the costing variant. It can be made not modifiable in costing variant configuration. We'll explain transfer control in more detail in Chapter 8.

- **Dates tab**
  After completing the **Costing Data** tab, press Enter to display the **Dates** tab shown in Figure 9.2.

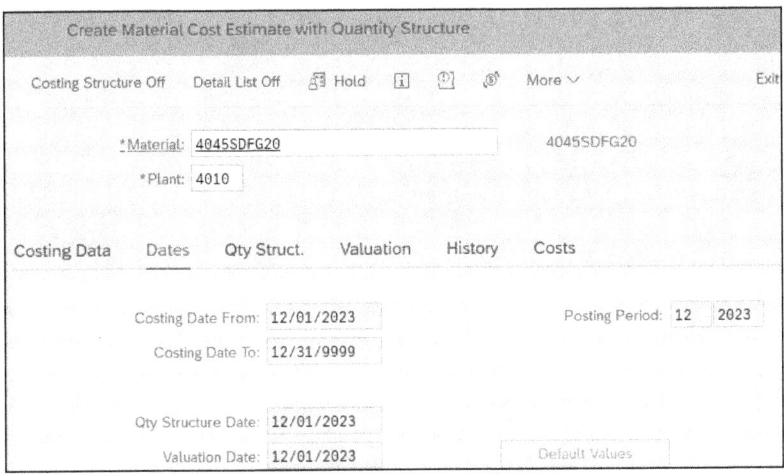

Figure 9.2 Standard Cost Estimate Dates Tab

The four default **Dates** are based on the costing variant date control component. You can either accept the default dates or change them if the date control allows (see Chapter 7 for more information on date control:

- **Costing Date From**
  Costing Date From determines the validity start date of the cost estimate. The cost estimate cannot be released until the start date.
- **Costing Date To**
  Costing Date To is the validity finish date of the cost estimate. Variance calculation requires a standard cost estimate valid for the entire fiscal year.
- **Quantity Structure Date**
  Qty Structure Date determines which bill of materials (BOM) and routing you select for the cost estimate.
- **Valuation Date**
  Valuation Date determines which material and activity prices you select for the cost estimate.

Press [Enter] to create the cost estimate immediately or check the details of the quantity structure on the next tab.

9.2.2 Quantity Structure

Select the **Qty Struct.** (quantity structure) tab in Figure 9.2 to display the details shown in Figure 9.3.

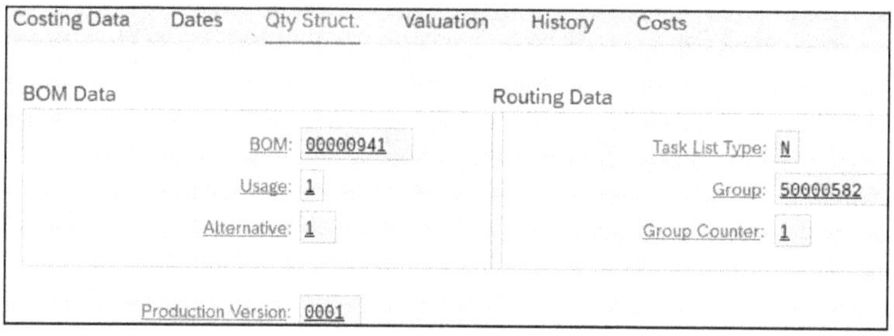

Costing Data | Dates | Qty Struct. | Valuation | History | Costs

BOM Data
BOM: 00000941
Usage: 1
Alternative: 1

Routing Data
Task List Type: N
Group: 50000582
Group Counter: 1

Production Version: 0001

Figure 9.3 Create Standard Cost Estimate Qty Struct.

If you don't change the entries in these fields, the standard cost estimate will be based on the **BOM**, **Routing**, or **Production Version** from the **Quantity structure data** section of the **Costing 1** view. If there are no entries in the **Quantity structure data** section of the **Costing 1** view, the quantity structure is determined by the quantity structure determination component of the costing variant, as discussed in Chapter 7.

Press [Enter] to create the cost estimate, as displayed in Figure 9.4.

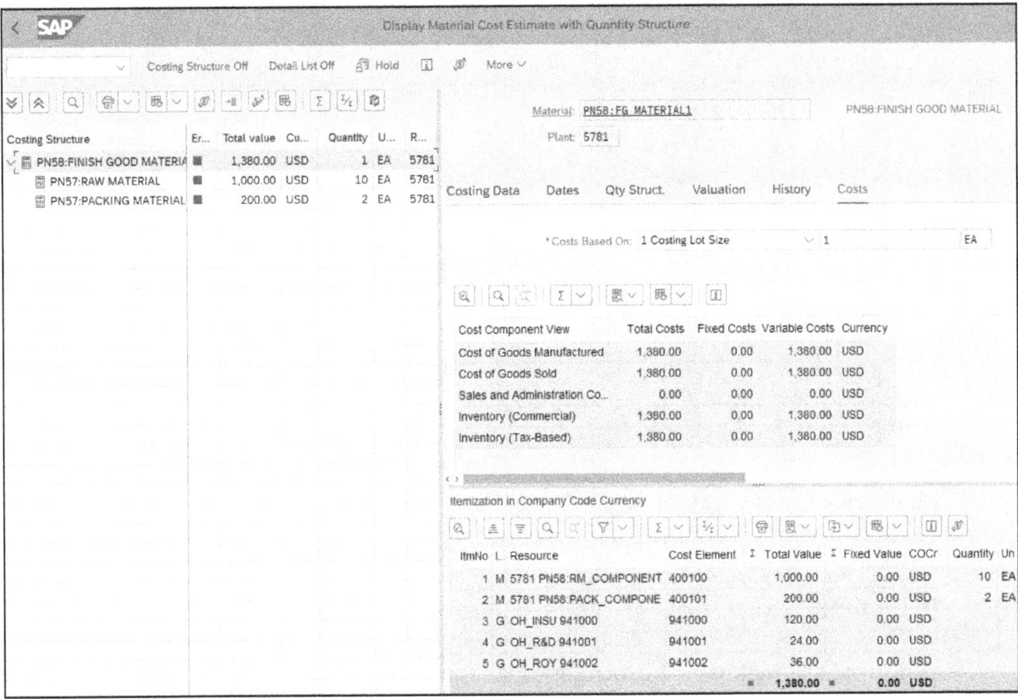

Figure 9.4 Material Cost Estimate with Quantity Structure

The **Costing Structure** on the left displays a costed multilevel BOM. Although we initially created a cost estimate for the top-level assembly, cost estimates were also created for all underlying components and subassemblies. Cost estimates are indicated by the **cost estimate** (calculator) icons in the costed BOM. Double-click any cost estimate in the costed BOM to display the cost estimate details on the right.

Costs Based On indicates the quantity of the displayed costs. You calculate the costs based on the **Costing Lot Size. Costs Based On** defaults to the **Costing Lot Size**. Purchasing or manufacturing in quantities different from the costing lot size causes variances because it's usually more efficient to purchase or manufacture items in larger quantities and less efficient to do so in smaller quantities.

To display costs based on a quantity of 1 (but still calculate them based on the costing lot size), click the **1 Costing Lot Size** text shown in Figure 9.4. You'll see the dropdown list shown in Figure 9.5.

Select **2 Price Unit** and inspect the quantity displayed in the field to the right of the text. If the price unit is **1 EA**, the cost estimate costs displayed are based on a quantity of 1. Otherwise, click **3 User Entry** and manually enter the quantity.

Now that we've created a standard cost estimate, mark and release are the next steps. During release, inventory is revalued if there is stock.

Material: PN58:FG MATERIAL1 PN58:FINISH GOOD MATERIAL

Plant: 5781

Costing Data | Dates | Qty Struct. | Valuation | History | Costs

*Costs Based On: 2 Price Unit 1 EA

1 Costing Lot Size
2 Price Unit
3 User Entry
4 Input Quantities from Multilevel BOM

Cost Component View			Costs	Currency
Cost of Goods Manufacture			0.00	USD
Cost of Goods Sold	1,380.00	0.00	1,380.00	USD
Sales and Administration Costs	0.00	0.00	0.00	USD
Inventory (Commercial)	1,380.00	0.00	1,380.00	USD
Inventory (Tax-Based)	1,380.00	0.00	1,380.00	USD

Figure 9.5 Change Cost Basis to Unit Entry

9.3 Mark and Release

You mark a standard cost estimate with Transaction CK24 or by following the menu path **Accounting • Controlling • Product Cost Controlling • Product Cost Planning • Material Costing • Price Update.** A selection screen displays, as shown in Figure 9.6.

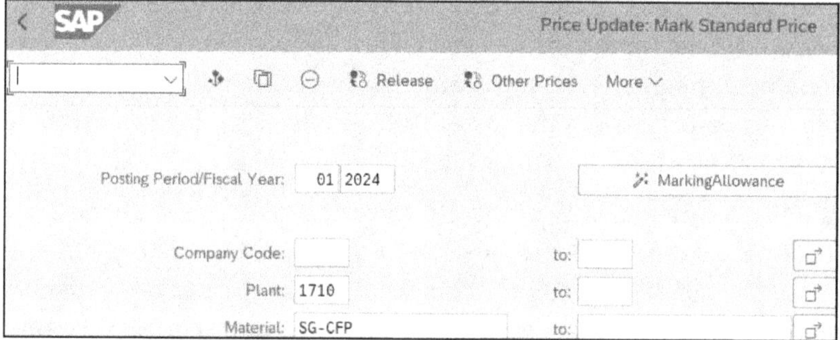

Figure 9.6 Mark Standard Cost Estimate Selection

We first create a marking allowance in addition to the mark and release steps here. Click the **MarkingAllowance** button to display the details in Figure 9.7.

A marking allowance specifies the **Company Code** and **Posting Period/Fiscal Year** in which you can mark a standard cost estimate with a valuation variant. You can't mark cost estimates with different valuation variants in this period. To create a marking allowance, click a company code with a round red traffic light in Figure 9.7. The dialog box in Figure 9.8 displays.

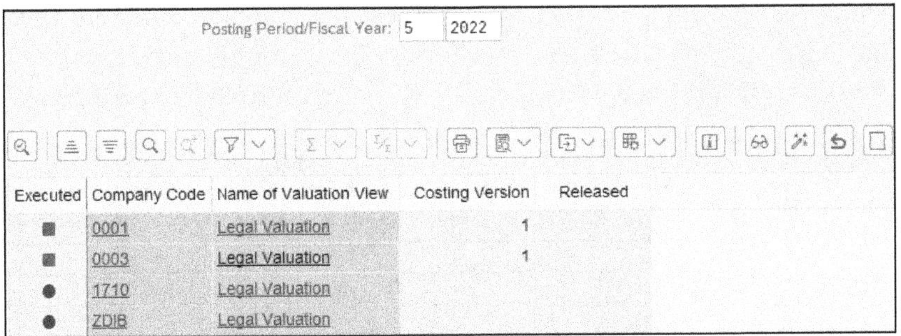

Executed	Company Code	Name of Valuation View	Costing Version	Released
■	0001	Legal Valuation	1	
■	0003	Legal Valuation	1	
●	1710	Legal Valuation		
●	ZDIB	Legal Valuation		

Figure 9.7 Marking Allowance Screen

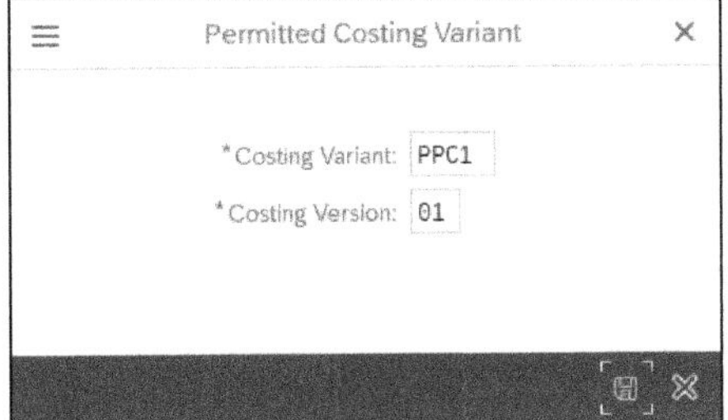

Figure 9.8 Create Marking Allowance

Enter a **Costing Variant** and **Costing Version,** then press [Enter] to create the marking allowance. A square green traffic light icon appears next to the company code, and the **Costing Version** column is populated, as shown earlier in Figure 9.7. Click a company code with a square green traffic light or for a marking allowance already created. You may see a dialog box with more permitted costing variants than entered in the dialog box in Figure 9.8. Because marking allowance is based on the valuation variant, a component of the permitted costing variant, there may be more than one costing variant that contains the permitted valuation variant.

Now that we've created a marking allowance, we mark a cost estimate by completing **Posting Period/Fiscal Year**, **Company Code**, **Plant**, and **Material** shown earlier in Figure 9.6, and then press [F8] to execute. The results are shown in Figure 9.9.

Price Update: Mark Standard Price

More

Exec...	Material	Plant	Valuation Type	Cos...	Fut. plnd price	Standard price	Price u...	Curr...	Valuation View	With Qty Struct.	Description	Ba...	Ledger	Legal Currency Type
■	SG-CFP	1710		VO	270.00	270.00	1	USD	Legal Valuation	✔	Centrifug...	EA	0L	10
■	SG-CFP	1710		VO	270.00	270.00	1	USD	Legal Valuation	✔	Centrifug...	EA	0L	30

Figure 9.9 Mark Standard Cost Estimate Results

A square green traffic light icon, together with **Cost** (costing status) **VO**, indicates that the cost estimate was marked. There are no inventory revaluations or financial postings during marking. The proposed standard price is copied to the **Standard Cost Estimate: Future** column in the **Costing 2** view, as shown in Figure 9.10.

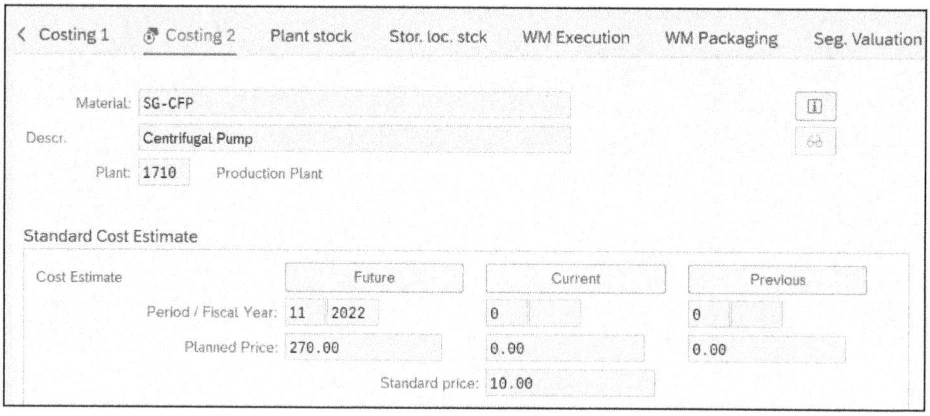

Figure 9.10 Marked Cost Estimate in the Future Column

You can create and mark standard cost estimates many times before release. Within the same fiscal period, new standard cost estimates overwrite existing ones you have not released. If you don't want to overwrite a standard cost estimate, create it with a different costing version, as shown previously in Figure 9.1. You may be unable to release cost estimates created with different costing versions. They are for information only.

After you have marked a cost estimate, check that the proposed standard price is correct. To release it, click the **Release** button shown earlier in Figure 9.10 and then execute. The results are shown Figure 9.11.

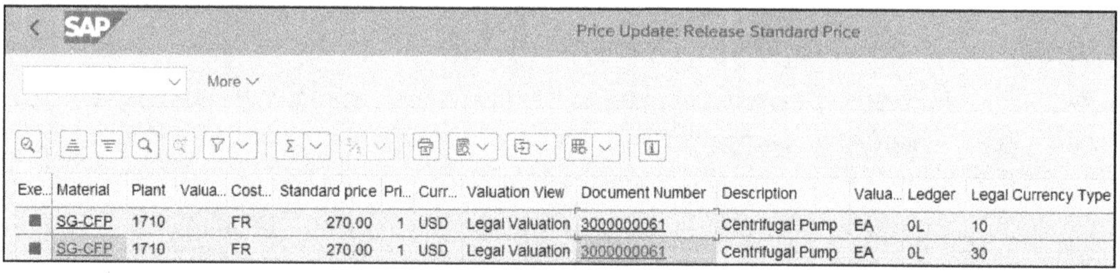

Figure 9.11 Release Standard Cost Estimate Results

A square green traffic light icon, together with **Cost** (costing status) **FR**, indicates that you have successfully released the standard cost estimate. You can see that with universal parallel accounting, you have **Valuation View** and **Ledger** columns. You can display the price change document by clicking the underlined **Document Number.** Any valuated stock will be revalued, and a financial posting will occur during release. The

Future Planned Price moves to the **Current Planned price** and the **Standard price** in the **Costing 2** view, as shown in Figure 9.12.

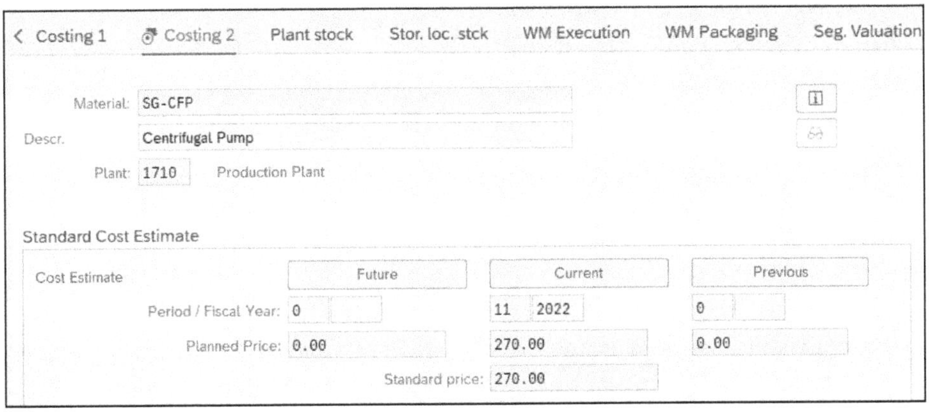

Figure 9.12 Released Cost Estimate in Current Column

When you release a standard cost estimate, the previous standard cost estimate moves to the **Previous** column to the right of the **Current** column.

Standard cost estimates can be released only once per period. As a rule, you should release cost estimates less frequently, for example, once every 12 months, which provides greater visibility to purchase price and production variances. Releasing cost estimates more frequently reduces variances, although inventory revaluations increase. Industries with rapidly moving purchase prices or short product development cycles may need to release cost estimates more frequently. In some cases, it can make sense to use actual costing in these circumstances to ensure that the value of inventory reflects extreme volatility either in the purchase prices of the raw materials or of the quantity structures.

9.4 Costing Run

A costing run will create, mark, and release many cost estimates—for example, all the materials in a plant or company code. You should start the costing run before the required release date because you might need to correct master data errors, such as missing info record prices. You also need time to analyze the differences between proposed and existing standard prices and obtain permission for release.

You maintain a costing run with the Manage Costing Runs - Estimated Costs app (SAP Fiori ID F1865), or Transaction CK40N, or via the menu path **Accounting • Controlling • Product Cost Controlling • Product Cost Planning • Material Costing • Costing Run • Edit Costing Run**. The selection screen is shown in Figure 9.13.

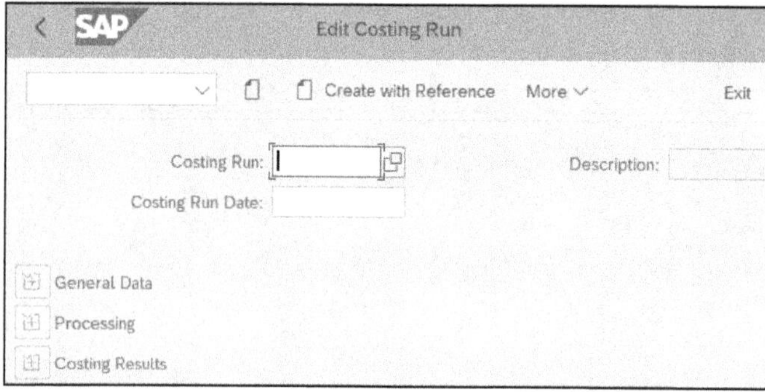

Figure 9.13 Edit Costing Run

Choose an existing **Costing Run** and press `Enter` to maintain the costing run. Before maintaining an existing costing run, we'll first follow an example of how to create a new costing run.

9.4.1 Costing Data Tab

To create a new costing run, click the **create** (new page) icon. The initial screen is shown in Figure 9.14.

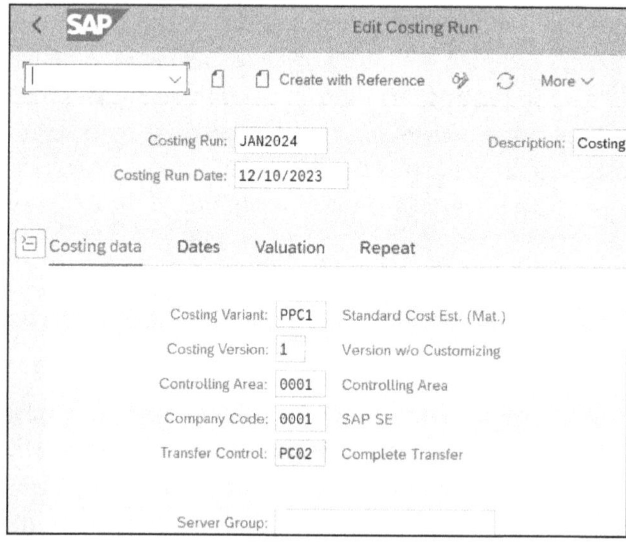

Figure 9.14 Create Costing Run

Complete the new **Costing Run** name, **Description, and Costing Run Date**. Complete the remaining fields in the **Costing data** tab. These are similar to the fields required to create an individual cost estimate initial fields, as described previously in Figure 9.1. The difference is that the **Create Costing Run** screen does not require the costing lot size field because the costing run retrieves it from the **Costing 1** view of each material master.

The **Server Group** in Figure 9.14 allows you to process a costing run with several servers in parallel. This reduces processing time for costing runs with many materials. Save your entries and click the **Dates** tab to edit the costing run in Figure 9.15.

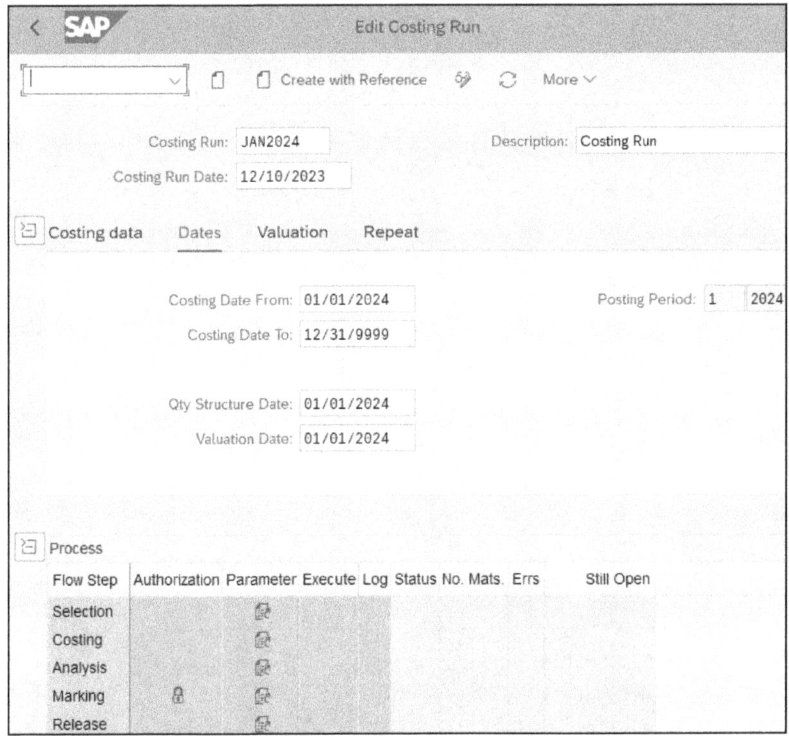

Figure 9.15 Costing Run Dates Tab

The **Dates** default from the costing variant. Depending on the costing variant configuration, you can change the default dates.

Select the **Valuation** tab in Figure 9.15 to display the valuation variant and costing sheet derived from the costing variant. This information is for display only and is not modifiable. You don't need to click this tab, so let's continue processing.

Now let's look at the **Process** section, which lists the five flow steps.

9.4.2 Costing Run Flow Steps

The costing run steps are listed in the Figure 9.15 **Flow Step** column. Let's examine each step.

Selection

To carry out the first step, click the icon in the **Parameter** column of the **Selection** row. The parameters are displayed in Figure 9.16.

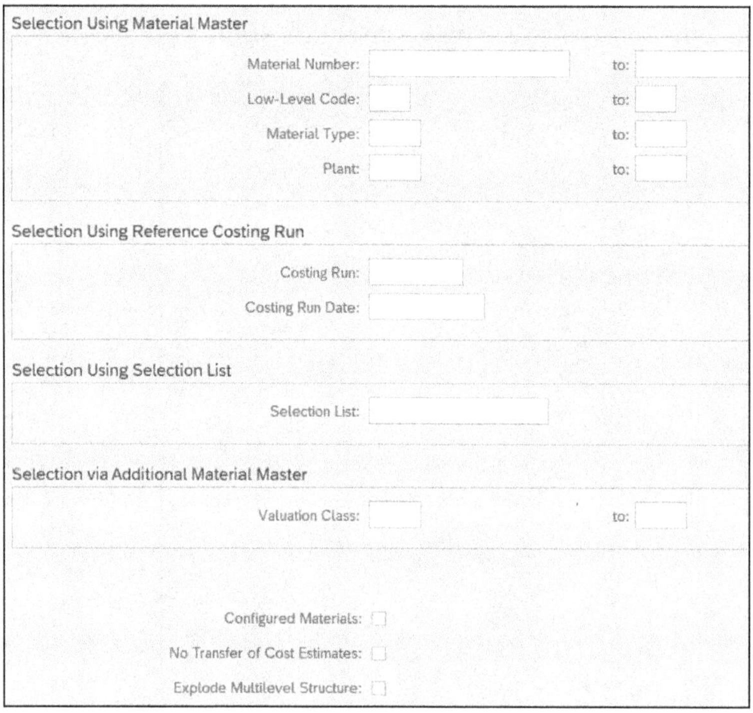

Selection Using Material Master

Material Number: to:
Low-Level Code: to:
Material Type: to:
Plant: to:

Selection Using Reference Costing Run

Costing Run:
Costing Run Date:

Selection Using Selection List

Selection List:

Selection via Additional Material Master

Valuation Class: to:

Configured Materials:
No Transfer of Cost Estimates:
Explode Multilevel Structure:

Figure 9.16 Costing Run Selection Parameters

Let's first consider the checkboxes in the lower part of the screen, described as follows:

- **Configured Materials** allows you to select only configured materials at the finished goods level in the costing run. Other materials aren't selected. We recommend selecting **Explode Multilevel Structure** when you select **Configured Materials** to include configured materials lower in the BOM. If you don't set this checkbox, no configured materials are selected.
- Select **No Transfer of Cost Estimates** to ignore transfer control in the costing run. If a BOM is determined for selected materials, transfer control becomes active. For materials determined via BOM explosion, the system looks for an existing cost estimate according to transfer control.
- The **Explode Multilevel Structure** checkbox was introduced with SAP S/4HANA 1809. This replaces the previous Structural Explosion costing run step, which followed the Selection step we are presently examining. When you select finished goods in the **Material Number** field, selecting **Explode Multilevel Structure** ensures all lower-level semi-finished products and raw materials are costed. Special procurement in the **MRP 2** view also selects all corresponding materials for stock transfer.

When you select all materials in a **Plant** or **Material Type** by leaving the **Material Number** or **Material Type** blank, you do not need to select the **Explode** checkbox, which increases the runtime unnecessarily because all materials are costed anyway.

Now that we've explored the three checkboxes, let's continue the selection step. You enter **Parameters** to select materials to include in the costing run. Complete the **Material Number** and **Plant** fields, save, and press F3. The costing run steps' updated icons are displayed in Figure 9.17.

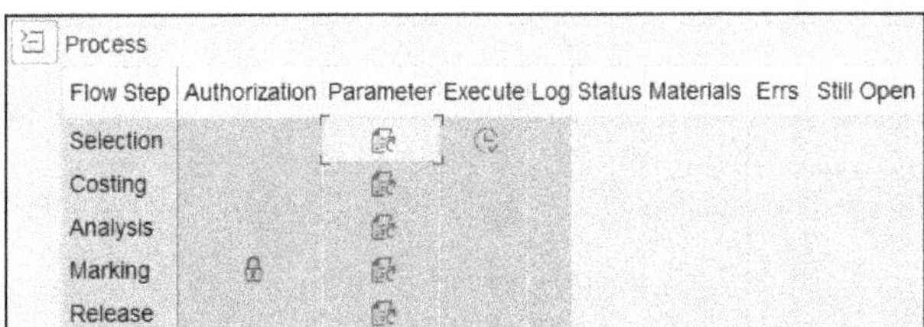

Figure 9.17 Selection Step Parameters Saved

The icon in the **Selection** row of the **Execute** column indicates that selection parameters are saved. Click the **Execute** icon to run the selection step. The screen in Figure 9.18 is then displayed.

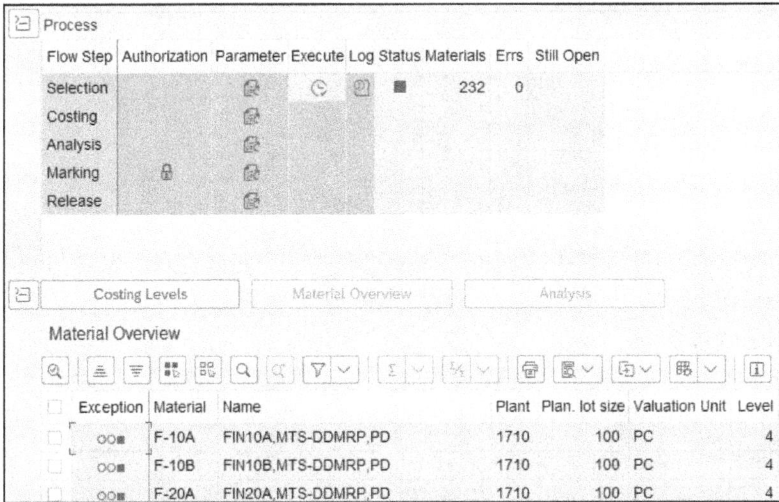

Figure 9.18 Selection Step Results

The square green traffic light in the **Selection** step row indicates that the selection step is complete. The materials selected are listed in **Material Overview** in the lower part of the screen. Now, let's examine the **Costing** step.

Execute Icons

After you enter and save parameters for each flow step, an icon displays in the **Execute** column.

Costing

In the **Costing** step, cost estimates are created for all selected materials, starting at the lowest BOM level. Click the icon in the **Costing** row of the **Parameter** column to display the parameters in Figure 9.19.

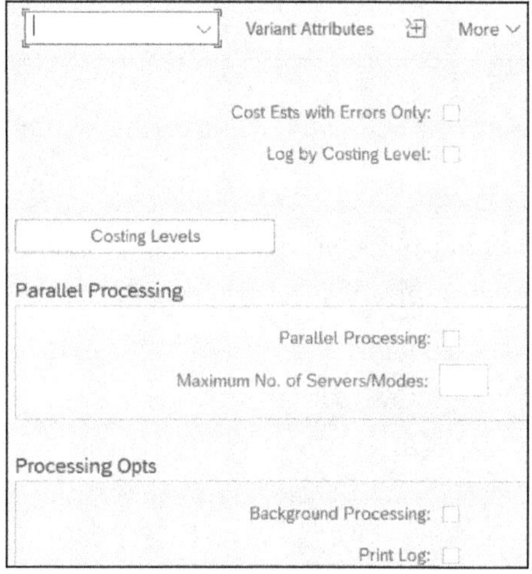

Figure 9.19 Costing Step Parameters

For large costing runs that generate many messages, these two options can simplify analysis:

- **Cost Ests with Errors Only**
  Since you cannot release cost estimates with errors, select this option. Correct the errors, deselect the checkbox, and execute the costing step again.
- **Log by Costing Level**
  A costing run creates cost estimates at the lowest BOM level first and then progressively upward through the BOM to the highest-level assemblies. Select this checkbox to correct the lowest-level messages first. This then reduces the number of messages and simplifies analysis at higher levels.

Click the **Costing Levels** button to display the cost estimates by costing level. Select the **Parallel Processing** checkbox for large costing runs to reduce processing time. Then, save the costing parameters, press F3, and click the **Execute** icon to display the results shown in Figure 9.20.

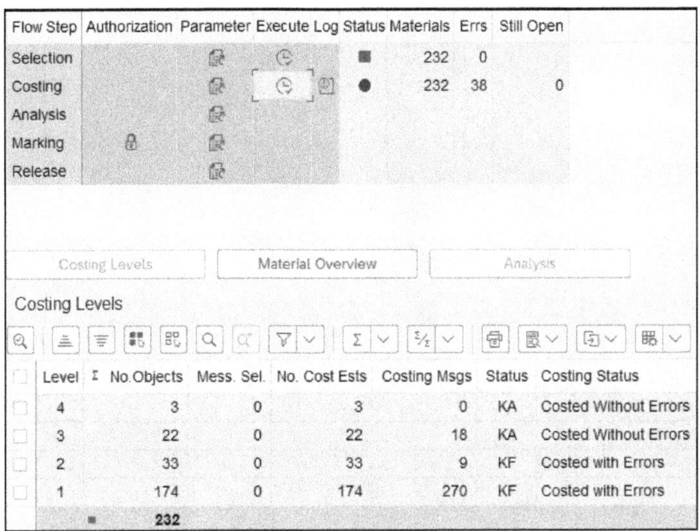

Figure 9.20 Costing Step Results

Click the icon in the **Log** column of the **Costing** step to display messages. Click the **Material Overview** button to display the cost estimates by **Costing Level**.

Now, let's analyze the results in the next step.

Analysis

Click the icon in the **Analysis** row of the **Parameter** column to display the parameters shown in Figure 9.21.

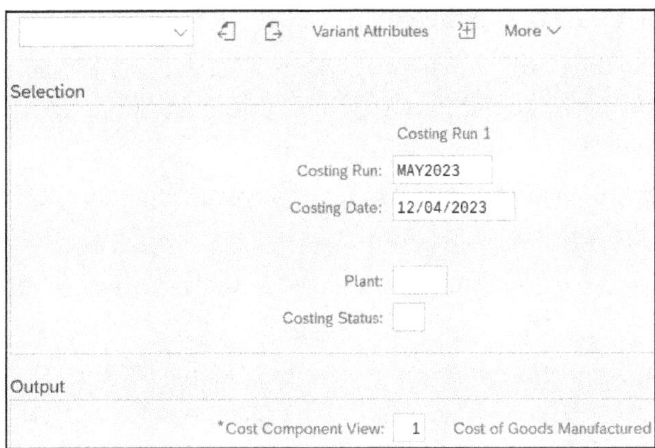

Figure 9.21 Analysis Parameters

The latest costing run you maintained defaults into **Costing Run** and **Costing Date**. To display more selection fields, click the plus sign icon. Extra fields display, as shown in Figure 9.22.

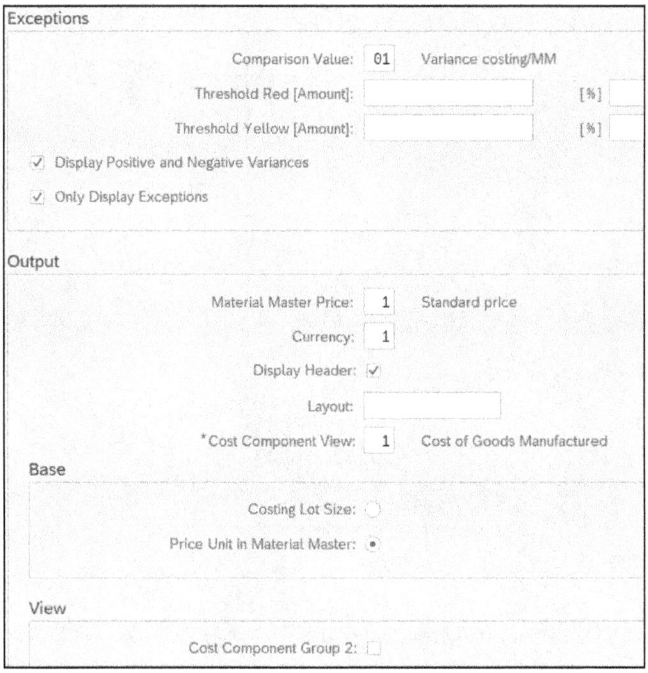

Exceptions

Comparison Value: 01 Variance costing/MM

Threshold Red [Amount]: [%]

Threshold Yellow [Amount]: [%]

✓ Display Positive and Negative Variances

✓ Only Display Exceptions

Output

Material Master Price: 1 Standard price

Currency: 1

Display Header: ✓

Layout:

*Cost Component View: 1 Cost of Goods Manufactured

Base

Costing Lot Size: ○

Price Unit in Material Master: ◉

View

Cost Component Group 2: ☐

Figure 9.22 Extra Analysis Fields

Exceptions allow you to define threshold colors and amounts to highlight certain variances.

In the **Output** section, you choose which **Material Master Price** to compare with the standard cost estimate. This comparison allows you to estimate the total change in inventory valuation before you release the costing run.

In the **Base** section, you choose the cost estimate display units. Change the default **Costing Lot Size** to the **Price Unit in Material Master** radio button. It's more useful because it's usually in multiples of 10.

In the **View** section, select **Cost Component Group 2** to display cost component summary groups. We discuss cost components in Chapter 6.

Save your entries, press F3, and then execute to display the **Analysis** step results shown in Figure 9.23.

Material Cost Estimates created during the costing step are listed. Double-click any line to display the cost estimate. Click the **Choose Layout Grid** icon at the right to display available standard layouts, as shown in Figure 9.24.

Choose **Layout 1SAP03** to display the report shown in Figure 9.25.

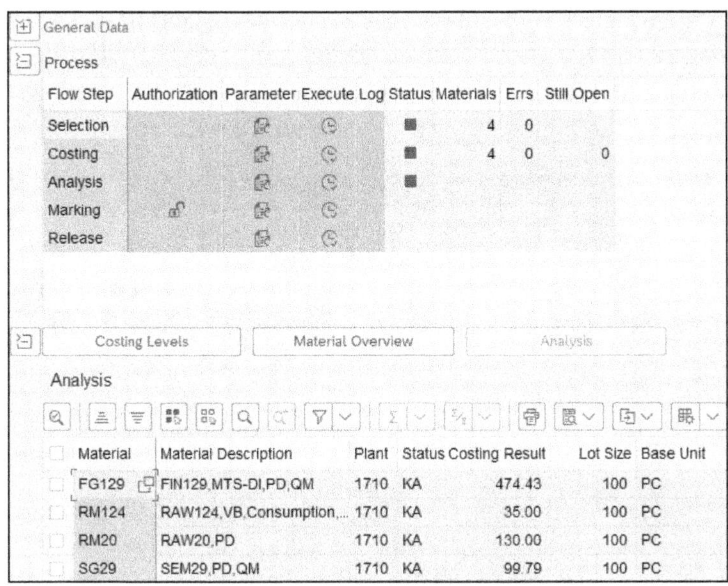

Figure 9.23 Analyze/Compare Material Cost Estimates

Choose Layout

Layout Setting: A All

Layout	Layout description	Default Setting
1SAP01	Results of costing run	
1SAP02	SAP basic list	
1SAP03	Standard price vs cost est/revaluation	
1SAP04	Variances between costing runs	
1SAP10	SAP standard: Mark/release	

Figure 9.24 Standard Layouts for Analysis

Costing Levels | Material Overview | Analysis

Analysis

Material	Material Description	Lot Size	per	BUn	%Var. costing/...	Anticip. reval.	Total Stock	Val. MatMs	Costing Result
FG129	FIN129,MTS-DI,PD,QM	100	1	PC	52.46-	14,148.00-	2,700	998.00	474.43
RM124	RAW124,VB,Consumption,...	100	1	PC			7,823	35.00	35.00
RM20	RAW20,PD	100	1	PC			12,047	130.00	130.00
SG29	SEM29,PD,QM	100	1	PC	58.93-	21,448.57-	14,999	243.00	99.79

Figure 9.25 Standard Layout with Anticipated Revaluation

This standard layout includes anticipated inventory revaluation per material in the **Anticip. reval.** column. Adjust prices or quantities as necessary and rerun the previous steps. After any adjustments, press F3 and then click the **Analysis** button to display the screen shown previously in Figure 9.23. Double-click any line in the **Analysis** section to display the cost estimate. After analysis and management approval, you'll mark and release the cost estimates for the whole costing run, which we'll discuss next.

Additional Analysis Options

You can also analyze costing run results with Transaction S_P99_41000111 or via the menu path **Accounting • Controlling • Product Cost Controlling • Product Cost Planning • Information System • Object List • For Material • Analyze/Compare Material Cost Estimates.**

Mark

Before marking cost estimates, you first allow marking in the period. Click the lock icon in the **Marking** row shown earlier in Figure 9.23 to display a list of company codes. Click on a **Company Code** to display the marking allowance dialog box shown in Figure 9.26.

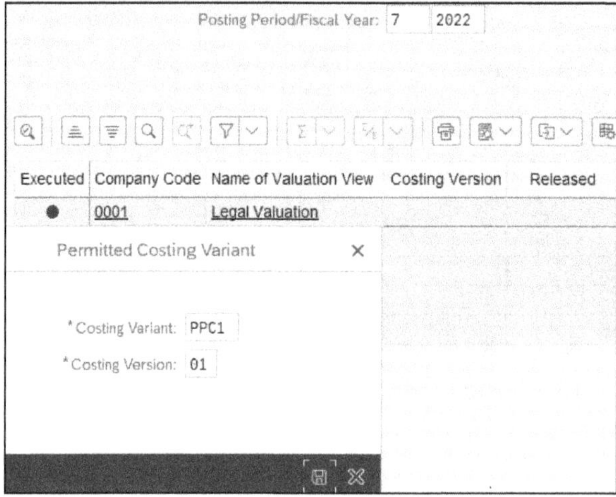

Figure 9.26 Create Marking Allowance

Complete the **Costing Variant** and **Costing Version** fields to create a marking allowance for the corresponding cost estimates. Click the highlighted Save icon, and then press F3 to return to the costing run steps.

Marking Allowance

Marking allowances are based on valuation variants, which are components of permitted costing variants. More than one costing variant containing the permitted valuation variant may exist.

Click the **Parameter** icon in the **Marking** row shown earlier in Figure 9.23, to display the parameters in Figure 9.27.

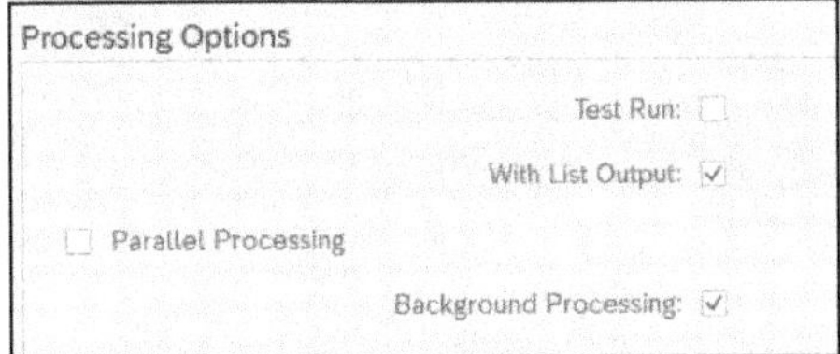

Figure 9.27 Costing Run Marking Parameters

Complete the marking parameters as follows:

- **Test Run**
  Select this option to analyze the costing results before marking and releasing.
- **Parallel Processing** and **Background Processing**
  Select these options for cost runs with many materials.
- **With List Output**
  Select this option to view more output details.

Save, and then press F3. Click the execute icon in the marking step. Press F3 to display the marking results shown in Figure 9.28.

Executed	Material	Plant	Valuation Type	Costing Status	Fut. plnd price	Standard price
■	AVC_RBT_BASE_FRAME	1710		VO	100.00	100.00
■	AVC_RBT_BASE_FRAME	1710		VO	100.00	100.00
■	AVC_RBT_BRD_PAINT	1710		VO	3,100.00	3,100.00
■	AVC_RBT_BRD_PAINT	1710		VO	3,100.00	3,100.00
■	AVC_RBT_BRD_SEAL	1710		VO	3,100.00	3,100.00

Figure 9.28 Costing Run Marking Step Complete

Costing Status VO indicates that the cost estimates were marked without errors.

After marking, the next step is release.

Release

Release parameters are similar to mark parameters, with the exception of the additional **Other Prices** button shown in Figure 9.29.

Figure 9.29 Release Other Prices Button

Click the **Other Prices** button to display the other price selections shown in Figure 9.30.

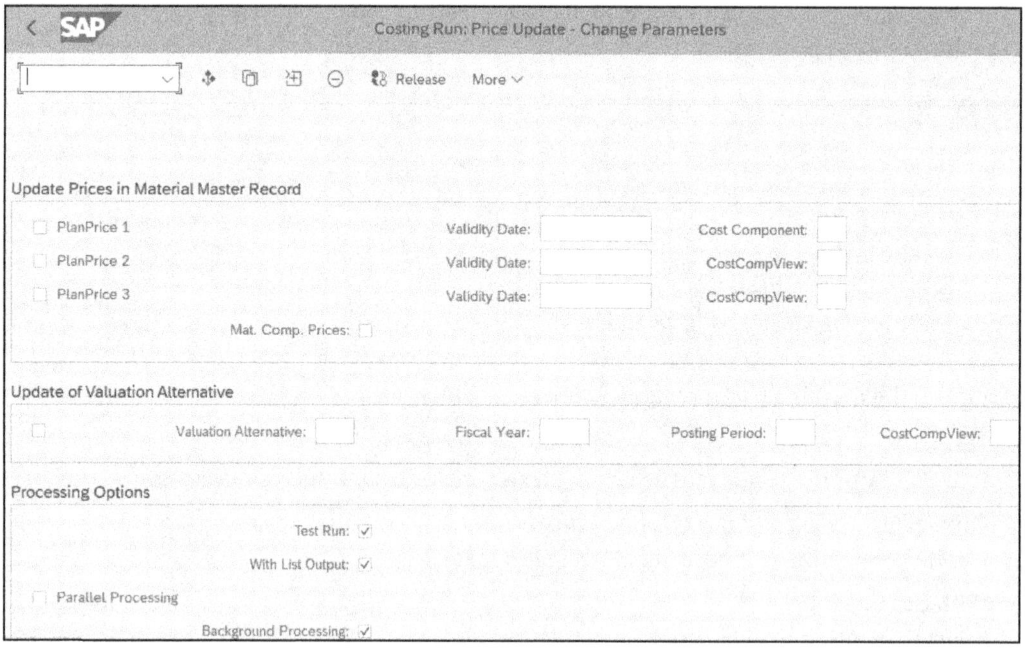

Figure 9.30 Release Other Prices Selection

These entries allow you to update the Plan Price and Validity Date in the **Costing 2** view and **CostCompView** (cost component view). Select the **Mat. Comp. Prices** checkbox to update the selected planned prices for material components (materials without a BOM).

Click the **Release** button to toggle to the selection screen in Figure 9.29, save your entries, press F3, and then execute to complete the release step. A list of updated materials is displayed.

Cost estimates with **FR** in the **Status** column were successfully released, which indicates that any existing inventory was revalued to reflect the values determined during costing.

Manage Costing Run—Estimated Costs App

You can also use the Manage Costing Run—Estimated Costs app (SAP Fiori ID F1865) to manage costing runs. You can cost, mark, and release materials in multiple plants or companies simultaneously. You can quickly review costing runs and create new runs directly from the list.

For more information on SAP Fiori apps, see *http://s-prs.co/v562901*.

Now that we've reviewed standard cost estimates and costing runs, let's look at other types of cost estimates, starting with the mixed-cost estimate.

9.5 Mixed-Cost Estimates

When you have more than one procurement alternative for the same material, such as two production lines or two suppliers, you can use *mixed-cost estimates* when you need inventory valuation to reflect the mixed procurement cost. You create a procurement alternative for each production version, supplier, or stock transfer characteristic and then define a mixing ratio to reflect the relative weight of the various alternatives. The mixed-cost estimate calculates a mixed price, which you create and release as the standard price, as we'll explore.

Mixed-price variance results from the difference between the *target credit* (actual quantity at the standard cost of the procurement alternative) determined in the variance calculation and the actual credit posted at the time of the goods receipt (actual quantity at the standard price). The standard price equals the mixed price.

9

Mixed-Price Cost Estimate Scenarios

Examples where you may use mixed-cost estimates:

- You procure the same item from different suppliers at different prices.
- You manufacture the same material on different production lines with different costs.
- You usually manufacture a material, but if your own facilities are unable to meet the demand for the material, you procure it externally.
- A mining company mines the same mineral from different sites with different extraction costs.

Mixed-cost estimates are important when inventory valuation should reflect the actual procurement value.

Let's examine the four steps required in preparation to create mixed-cost estimates in the following sections.

9.5.1 Define Quantity Structure Types

First, you access quantity structure types with Transaction OMXA or via the IMG menu path **Controlling • Product Cost Controlling • Product Cost Planning • Selected Functions in Material • Mixed Costing • Define Quantity Structure Types**. The quantity structure type configuration is shown in Figure 9.31.

QtStT (quantity structure type) controls how mixed costing is applied. You determine which procurement alternatives will be used with a mixing ratio for the materials involved in the costing.

The **Qty Struct.Cat.** (quantity structure category) indicates whether it is an actual (**I**), planned (**P**) quantity structure, or a quantity structure for mixed (**M**) costing. The **Time**

Dependency 0 indicates no time dependency; **1** is based on fiscal year, and **2** is based on period.

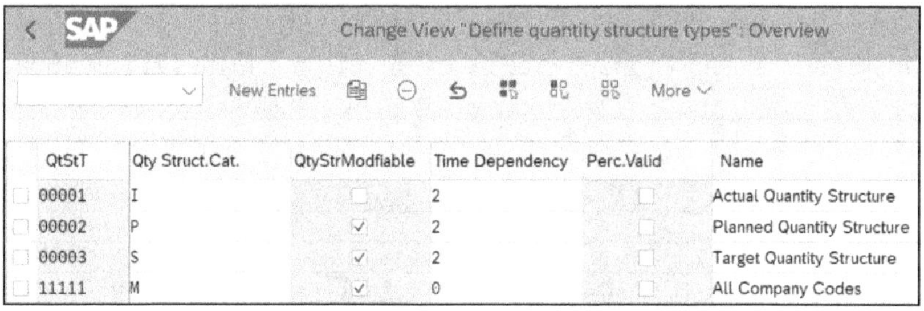

Change View "Define quantity structure types": Overview

New Entries More

	QtStT	Qty Struct.Cat.	QtyStrModfiable	Time Dependency	Perc.Valid	Name
☐	00001	I	☐	2	☐	Actual Quantity Structure
☐	00002	P	☑	2	☐	Planned Quantity Structure
☐	00003	S	☑	2	☐	Target Quantity Structure
☐	11111	M	☑	0	☐	All Company Codes

Figure 9.31 Define Quantity Structure Types

Perc.Valid (percent validation for mixed costing) controls whether checks are made so that the sum of the mixing ratios is 100% when a mixing ratio is created with this quantity structure category.

9.5.2 Define Costing Versions

In the second step, you assign a costing version to each quantity structure type with Transaction OKYD or via the IMG menu path **Controlling • Product Cost Controlling • Product Cost Planning • Selected Functions in Material • Mixed Costing • Define Costing Versions.** Costing versions are listed in Figure 9.32.

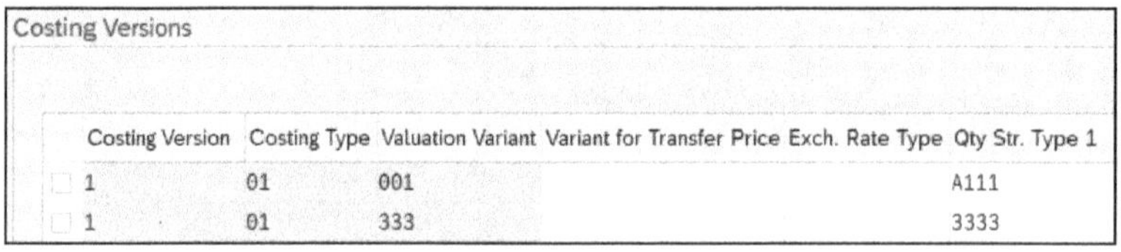

Costing Versions

	Costing Version	Costing Type	Valuation Variant	Variant for Transfer Price	Exch. Rate Type	Qty Str. Type 1
☐	1	01	001			A111
☐	1	01	333			3333

Figure 9.32 Costing Versions Configuration

You assign a **Qty Str. Type 1** (quantity structure type) to a **Costing Version** (first column). You create mixed-cost estimates with reference to a **Costing Version**. By creating costing versions, you can create more than one mixed-cost estimate for the same material.

9.5.3 Create Procurement Alternatives

In the third step, you create procurement alternatives with Transaction CK91N or via the menu path **Accounting • Controlling • Product Cost Controlling • Product Cost Planning • Material Costing • Master Data for Mixed Cost Estimate • Edit Procurement Alternatives.** Complete the **Material** and **Plant** fields and click the new page icon. Click on the **Process Cat.** field to display a dropdown list of possible entries, as shown in Figure 9.33. The same procurement alternatives are used to assign costs in actual costing. Be sure to define procurement alternatives that match the granularity required for product cost analysis in actual costing.

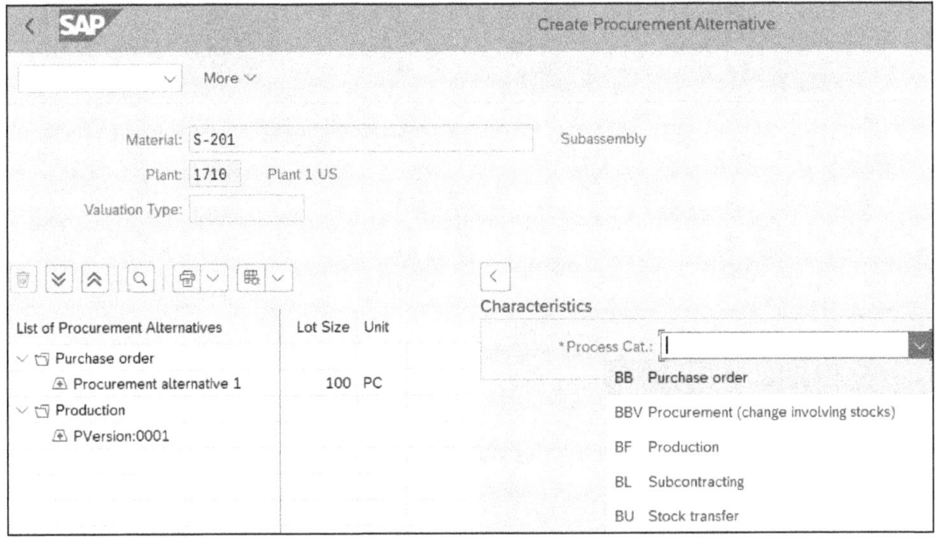

Figure 9.33 Procurement Alternative Process Category Possible Entries

The **Process Cat.** (process category) option selected determines the fields that appear instead of the dropdown list. In this example, we'll select **BB Purchase order**, which displays the purchasing fields shown Figure 9.34.

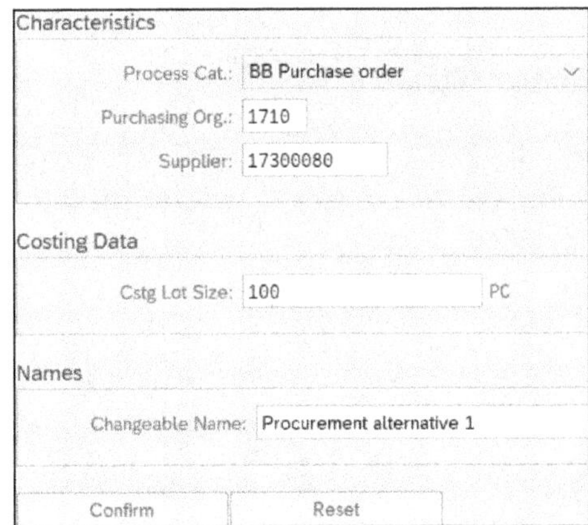

Figure 9.34 Characteristics for Procurement Alternative

Complete the screen as follows:

1. Complete the **Purchasing Org.**, **Supplier**, **Costing Lot Size**, and **Changeable Name** fields.
2. Click the **Confirm** button.

The new procurement alternative (**Procurement alternative 1**) appears in the **List of Procurement Alternatives** under the **Purchase order** folder in Figure 9.35.

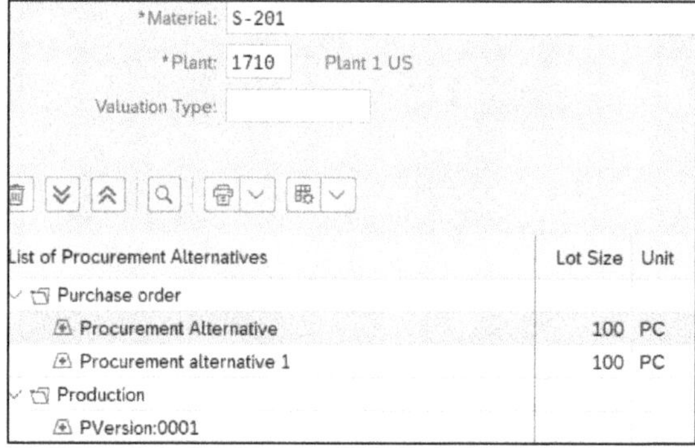

Figure 9.35 List of Procurement Alternatives

You can create many **Procurement Alternatives.** Cost estimates will only access those with mixing ratios.

9.5.4 Define Mixing Ratios

In the fourth step, you create mixing ratios with Transaction CK94 or via the menu path **Accounting • Controlling • Product Cost Controlling • Product Cost Planning • Material Costing • Master Data for Mixed Cost Estimate • Mixing Ratios • Create/ Change.** Mixing ratios are shown in Figure 9.36.

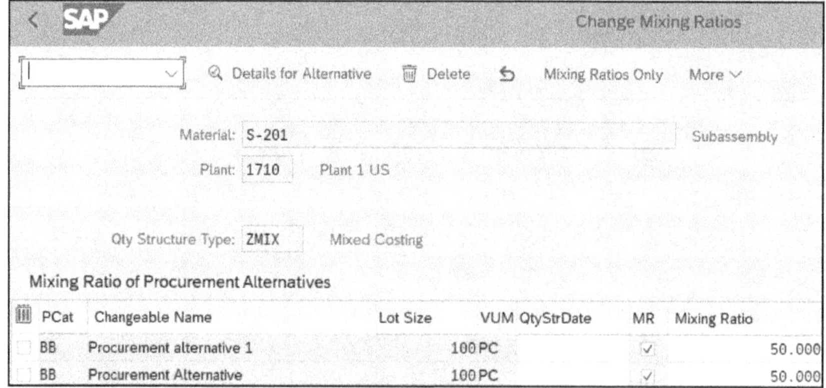

Figure 9.36 Change Mixing Ratios

Enter the **Mixing Ratio**, in this example, 50/50 for the procurement alternatives and save.

Now that the configuration and master data settings are in place, we can create a mixed-cost estimate.

9.5.5 Create Mixed-Cost Estimate

When you create a cost estimate for the material with the costing version configured in Figure 9.32, you create a mixed-cost estimate. You create a mixed standard cost estimate with Transaction CK11N or via the menu path **Accounting • Controlling • Product Cost Controlling • Product Cost Planning • Material Costing • Cost Estimate with Quantity Structure • Create**. A selection screen displays, as shown in Figure 9.37.

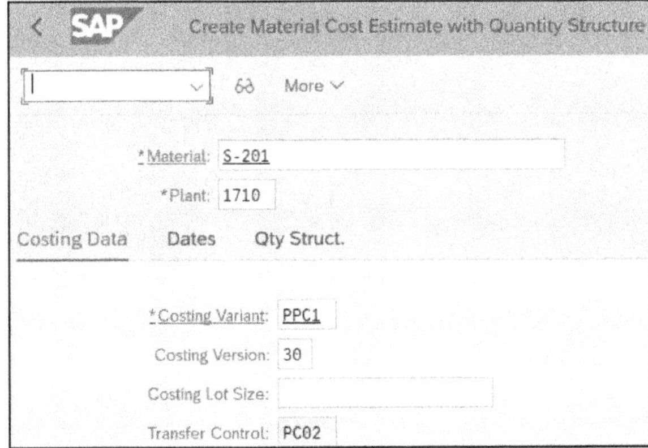

Figure 9.37 Create Mixed Cost Estimate

Complete the **Material**, **Plant**, and **Costing Variant** fields and change the **Costing Version** from the default entry of **1** to the costing version we defined in Figure 9.32 (in this example, **30**). Press Enter, check that the default date entries are correct, and press Enter again to create the mixed-cost estimate, as shown in Figure 9.38.

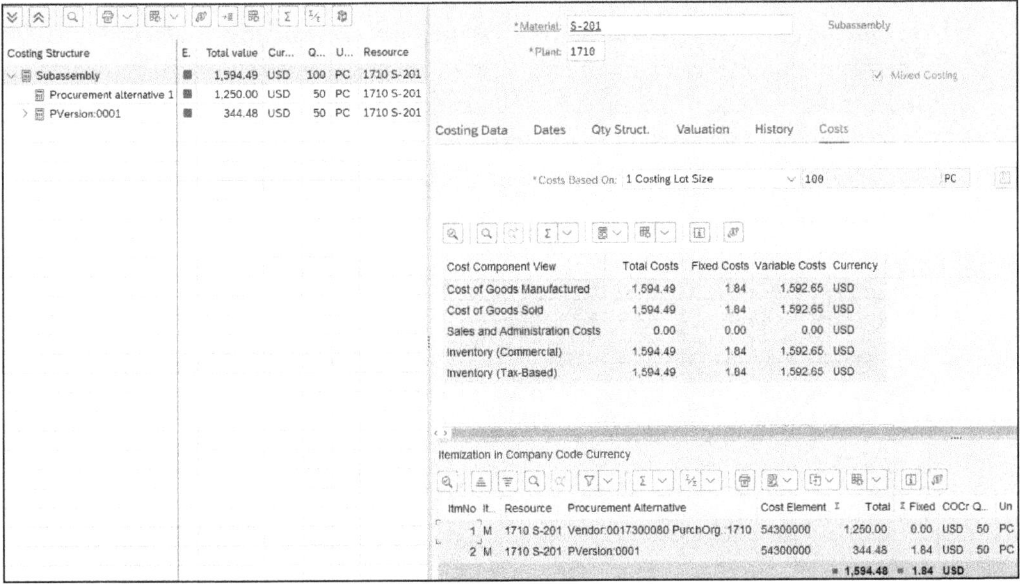

Figure 9.38 Mixed-cost Estimate

The **Mixed Costing** checkbox in the top right confirms that this is a mixed-cost estimate. You can compare the cost of each procurement alternative and the total value by inspecting the costed BOM in the **Costing Structure** section on the left. The **Cost Estimate** (calculator) icons in the costed BOM structure on the left represent the following, in order from top to bottom:

- **Subassembly**
  Cost 1,594.49 USD **Total value** to manufacture 100 PC.
- **Procurement alternative 01**
  Cost 1,250.00 USD to procure 50 PC.
- **PVersion:0001**
  Cost 344.48 USD to manufacture 50 PC.

We defined the portions of the two procurement alternatives in the **Mixing Ratio** in Figure 9.36.

Mixed-Cost Estimates in SAP S/4HANA Cloud

It is not yet possible to configure and use mixed cost estimates in SAP S/4HANA Cloud. This means that only one procurement alternative is supported for any given material.

Now that we've examined mixed-cost estimates, we'll next look at preliminary cost estimates.

9.6 Preliminary Cost Estimates

A *preliminary cost estimate* differs from a standard cost estimate in that it is created for a specific production version rather than at the material level. The product cost collector describes the actual process of manufacturing the material and may be slightly different from the more general assumptions behind the standard cost estimate. In this section, we'll show you how to create a preliminary cost estimate for a product cost collector and walk you through the main parameters that control the costing process.

9.6.1 Create

You create a preliminary cost estimate for a product cost collector with Transaction KKF6N or by following the menu path **Accounting • Controlling • Product Cost Controlling • Cost Object Controlling • Product Cost by Period • Master Data • Product Cost Collector • Edit**. A product cost collector is shown in Figure 9.39.

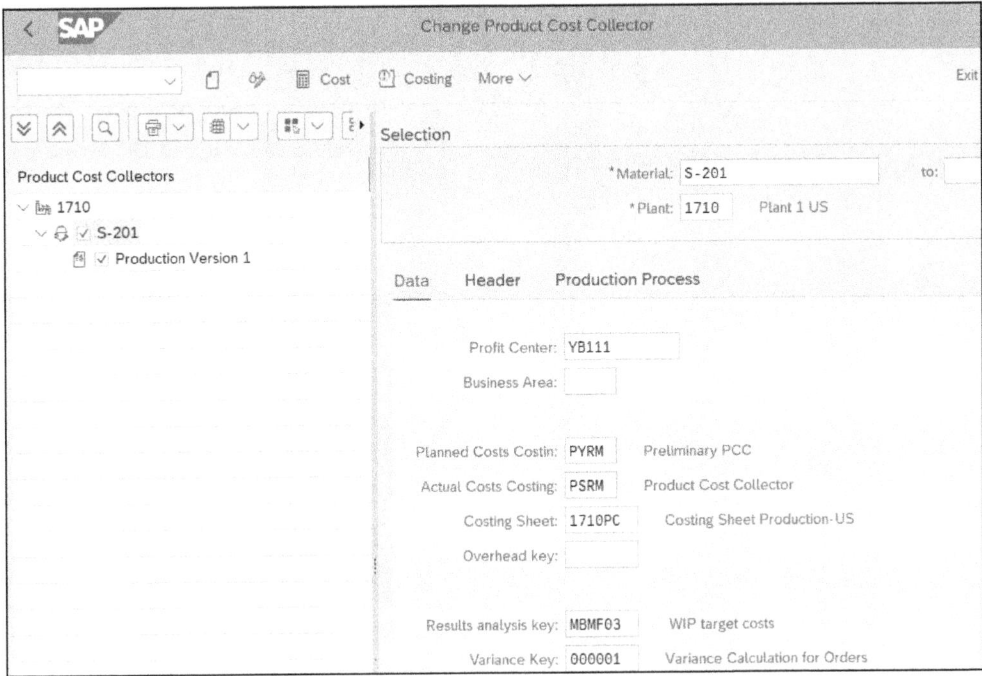

Figure 9.39 Product Cost Collector

To create a preliminary cost estimate for a **Product Cost Collector**, follow these steps:

1. Complete the **Material** and **Plant** fields and press `Enter`.
2. Select a **Production Version** checkbox (**Production Version 1**) to display the details on the right.
3. Click the pencil-and-glasses icon to display the **Cost** button to the right of the pencil and glasses icon.
4. Click the **Cost** button to create a preliminary cost estimate.
5. Select the **Header** tab and scroll down to display the **Cost Estimate** button shown in Figure 9.40.
6. Click the **Cost Estimate** button to display the preliminary cost estimate as shown in Figure 9.41.
7. Click the **Cost Estimate** button to display the most recent preliminary cost estimate, as shown in Figure 9.41.
8. Click the **Costs** button in Figure 9.40 to display an actual cost report by cost element.

Data | Header | Production Process

Description: Production Version 1
Company Code: 1710 Company Code 1710
Material: S-201
Subassembly
Order Type: YBMR Product Cost Collector
Order: 700000
External order no.:
Functional Area: YB20 Production

Created By: STUDENT025 M2B STUDENT025
Created on: 06/05/2022
Last Changed By:
Changed On:

System status REL PRC SETC VCAL

Orders/PV | Cost Estimate | Settlement Rule
Costs

Figure 9.40 Product Cost Collector Header Tab

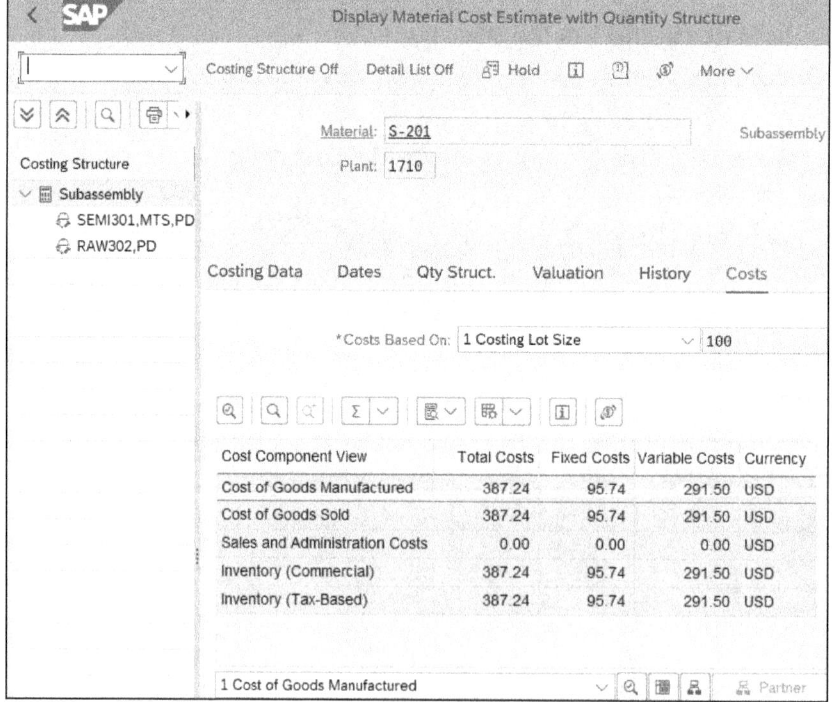

Figure 9.41 Preliminary Cost Estimate for Product Cost Collector

The preliminary cost estimate looks like the standard cost estimate we examined earlier in this chapter. Two important differences, however, are the number of preliminary cost estimates possible per material and transfer control. We'll discuss these differences in the following subsections.

9.6.2 Production Process

Only one released standard cost estimate per material is allowed. There can be many preliminary cost estimates because you base preliminary cost estimates for product cost collectors on production versions. A *production version* is a unique combination of BOM, routing, and production line. Because there can be many methods of manufacturing a material, there can be many production versions and many preliminary cost estimates.

You create product cost collectors with reference to a production process. A production process describes how you produce a material, that is, which quantity structure is used. You take the quantity structure from the production version. You can determine the production process by the following characteristics: material, production plant, and production version. You can create one production process for each production version.

Controlling Level

When you use a product cost collector with the Controlling level of production plant/ planning plant, you cannot create a preliminary cost estimate because the product cost collector doesn't have a quantity structure.

Now that we've discussed the production process, let's examine transfer control.

9.6.3 Transfer Control

Preliminary cost estimates for product cost collectors use transfer control, a component of the costing variant. Transfer control enables a top-level cost estimate to use previously created standard cost estimates for lower-level materials. The quickest way to create many standard cost estimates is with a costing run, as discussed earlier in this chapter.

Let's examine transfer control in more detail, because it's important to create preliminary cost estimates for product cost collectors without errors. In the preliminary cost estimate in Figure 9.41, select the **Costing Data** tab to display the screen shown in Figure 9.42.

Note that **Costing Variant PYRM** contains **Transfer Control PC02.** Double-click the underlined text **PYRM** to display the costing variant details shown in Figure 9.43.

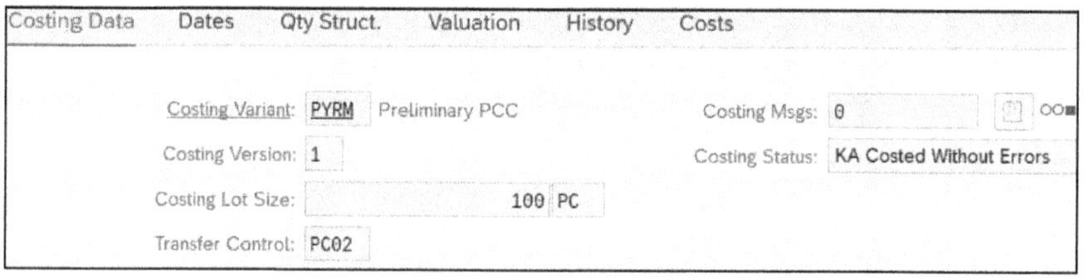

Costing Data | Dates | Qty Struct. | Valuation | History | Costs

Costing Variant: PYRM Preliminary PCC
Costing Version: 1
Costing Lot Size: 100 PC
Transfer Control: PC02
Costing Msgs: 0
Costing Status: KA Costed Without Errors

Figure 9.42 Preliminary Cost Estimate Costing Data Tab

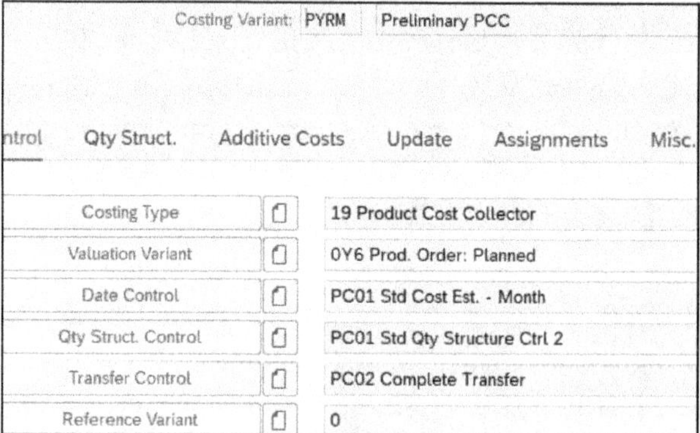

Figure 9.43 Costing Variant PYRM Details

The components of **Costing Variant PYRM** are listed. Click the **Transfer Control** button to display the transfer control details shown in Figure 9.44.

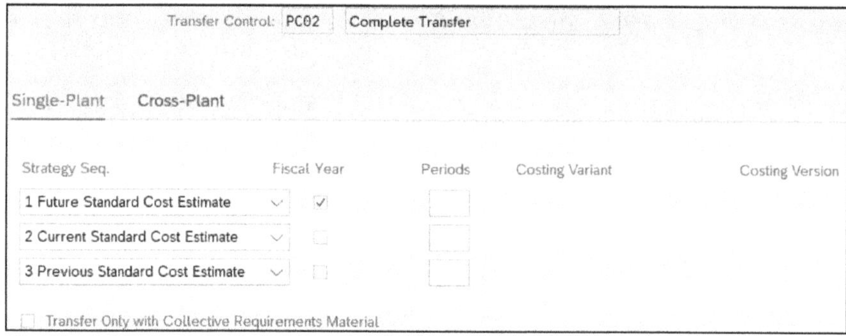

Figure 9.44 Transfer Control PC02 Details

Figure 9.44 shows the **Strategy Seq** (strategy sequence) for **Transfer Control PC02** for existing cost estimates within a single plant:

1. A **Future Standard Cost Estimate** with a start date within the current **Fiscal Year**
2. A **Current Standard Cost Estimate** with a start date within the current **Period**
3. A **Previous Standard Cost Estimate** with a start date within the current **Period**

Preliminary cost estimates created in the same period that standard cost estimates are released for lower-level materials use the standard cost estimates based on the second transfer control strategy sequence shown in Figure 9.44.

To demonstrate this, in the cost estimate shown previously in Figure 9.41, double-click on the second cost estimate on the left side of the screen, immediately below the highlighted **top-level cost estimate.** The right side of the cost estimate screen now refers to the second cost estimate. Click on the **Costing Data** tab to display the screen shown in Figure 9.45.

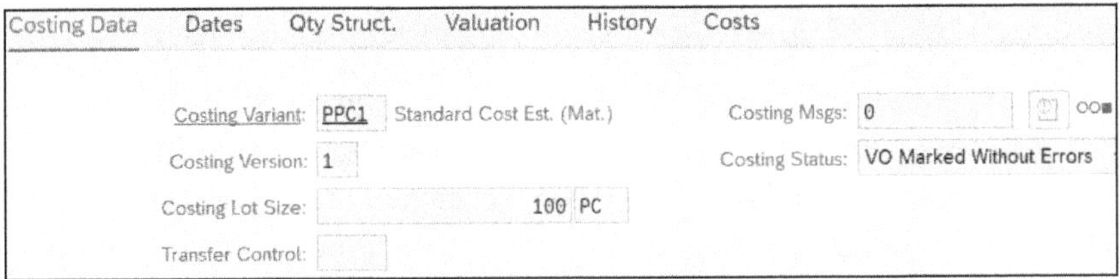

Figure 9.45 Lower-Level Cost Estimate - Costing Data Tab

Notice that the lower-level cost estimate was previously created with **Costing Variant PPC1**. To find out when the standard cost estimate was created, click on the **History** tab.

Select Fiscal Year

To increase the chance of a preliminary cost estimate successfully retrieving previously created standard cost estimates with transfer control, select **Fiscal Year** in the second step in the sequence (**Current Standard Cost Estimate**) shown in Figure 9.44.

9.6.4 Mass Processing

We've now created individual preliminary cost estimates and examined how they access existing standard cost estimates at lower levels in the BOM. Because you typically use preliminary cost estimates to value work in process (WIP) and scrap during period-end processing in product cost by period, you must mass-create preliminary cost estimates immediately following a costing run. This ensures the preliminary cost estimates are current during the first period-end processing following the costing run. In this section, we'll look at how to mass-process preliminary cost estimates.

You create multiple preliminary cost estimates for product cost collectors with Transaction MF30 or via the menu path **Accounting • Controlling • Product Cost Controlling • Cost Object Controlling • Product Cost by Period • Planning • Preliminary Costing for Product Cost Collectors**. The selection screen is shown in Figure 9.46.

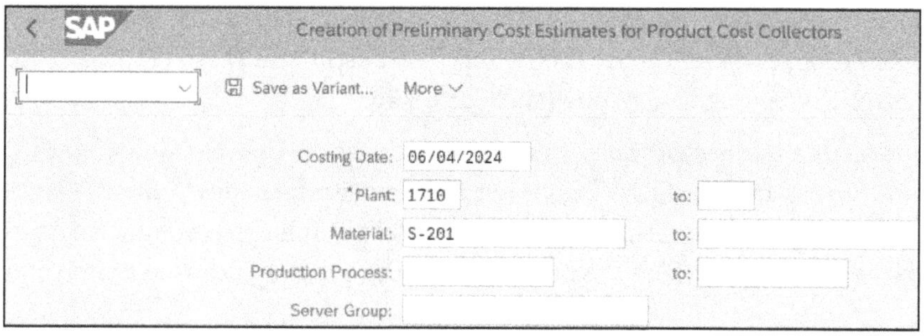

Figure 9.46 Preliminary Costing for Product Cost Collectors

The selection screen allows you to enter a range or list of materials or production processes. To mass-create preliminary cost estimates, complete the selection fields (**Costing Date**, **Plant**, and either a **Material** or **Production Process** range), and click the **Execute** icon or press F8 to run the transaction. The resulting screen displays a list of messages you should analyze and resolve errors where necessary.

9.7 Unit Cost Estimates

A unit cost estimate calculates the planned costs for materials without a BOM or a routing. When you first develop a new product or modify an existing product, there are several development stages:

1. If you have not yet developed master data, you can carry out initial cost planning by creating a base planning object.
2. After creating the first material master, you can create a material cost estimate without a quantity structure to manually plan costs for the new material. You can use the base planning cost estimate developed in step 1 as a reference.
3. After you have created BOMs and routings, you can create a material cost estimate with a quantity structure.

We'll examine each of the product lifecycle stages in the following sections, starting with the base planning object.

9.7.1 Base Planning Object

You maintain a base planning object for a new product with Transaction KKE2 or via the menu path **Accounting • Controlling • Product Cost Controlling • Product Cost Planning • Reference and Simulation Costing • Change Base Planning Object**. Enter a base planning object and press Enter to display the screen shown in Figure 9.47.

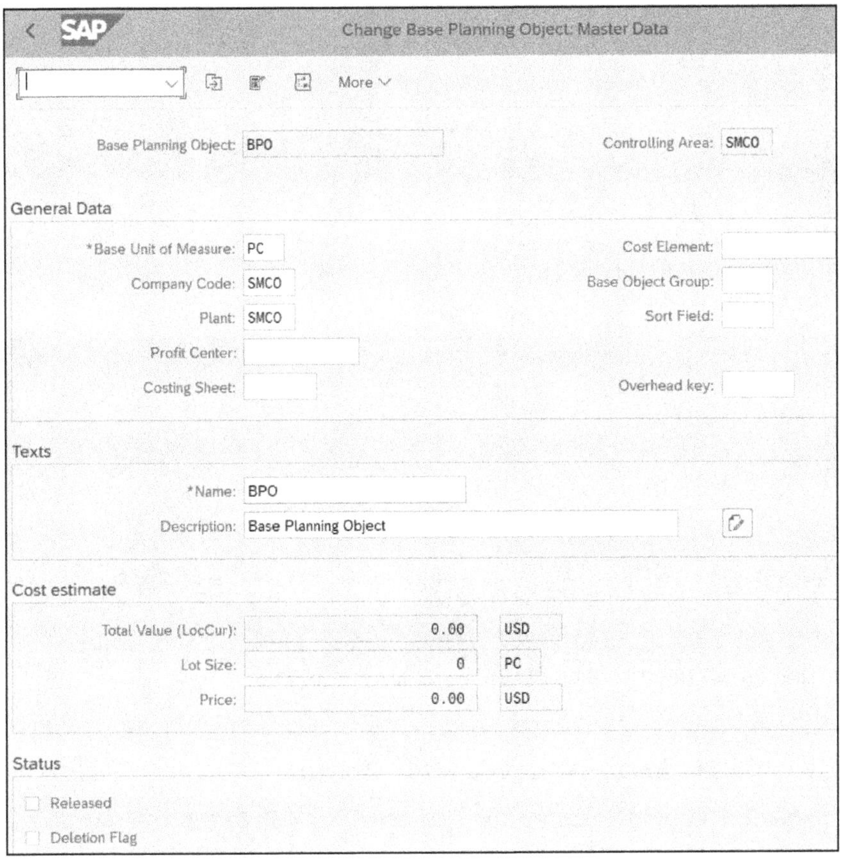

Change Base Planning Object: Master Data

More

Base Planning Object: BPO

Controlling Area: SMCO

General Data

*Base Unit of Measure: PC

Company Code: SMCO

Plant: SMCO

Profit Center:

Costing Sheet:

Cost Element:

Base Object Group:

Sort Field:

Overhead key:

Texts

*Name: BPO

Description: Base Planning Object

Cost estimate

Total Value (LocCur): 0.00 USD

Lot Size: 0 PC

Price: 0.00 USD

Status

Released

Deletion Flag

Figure 9.47 Base Planning Object

Base Planning Objects

Base planning objects are on the simplification list. They are no longer considered strategic, and the transactions have been removed from the menu. Visit the following URL for more information: *http://s-prs.co/v583201*.

Complete the mandatory fields: **Base Unit of Measure** and **Name**. and save your data. Click the calculator-and-pencil icon at the top to display the Create Cost Estimate dialog box, as shown in Figure 9.48.

Complete the **Costing Variant** and **Lot Size** fields, and press Enter to create the unit cost estimate shown in Figure 9.49. It's in spreadsheet format, so it's easy to modify and analyze changes in total value.

Create Cost Estimate

Costing Variant: PG

Lot Size: 1 PC

Copy from

Base Object:

Figure 9.48 Create Unit Cost Estimate for Base Planning Object

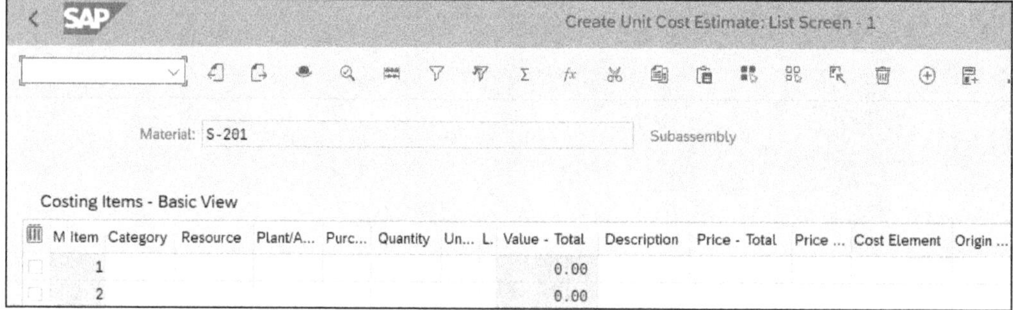

Figure 9.49 Unit Cost Estimate for Base Planning Object

9.7.2 Material Cost Estimate without Quantity Structure

After you've carried out initial research on new products with base planning objects, you may decide to create material masters and continue the development process. You may also copy existing materials and the quantity structures and manually adjust and conduct a cost analysis.

You create a material cost estimate without quantity structure for a new or modified product with Transaction KKPAN or via the menu path **Accounting • Controlling • Product Cost Controlling • Product Cost Planning • Material Costing • Cost Estimate Without Quantity Structure • Create**. Enter a material, plant, and costing variant, and then press Enter twice to display the cost estimate shown in Figure 9.50.

Create Unit Cost Estimate: List Screen - 1

Material: 1048 Avocado

Costing Items - Basic View

Category	Resource	Plant/Activity	Purc...	Quantity	Un...	Lot-Size Indep.	Value - Total	Description	Price - Total	Price unit	Cost Element
							0.00				
							0.00				

Figure 9.50 Cost Estimate without Quantity Structure

The cost estimates in Figure 9.49 and Figure 9.50 have a similar spreadsheet layout because they are both unit cost estimates. Let's examine relevant columns in Figure 9.50:

- **Item**
  You can insert a new **Item** by clicking the plus sign icon on the right.
- **Category**
  Click the **Category** column and press F4 to display the possible entries shown in Figure 9.51.

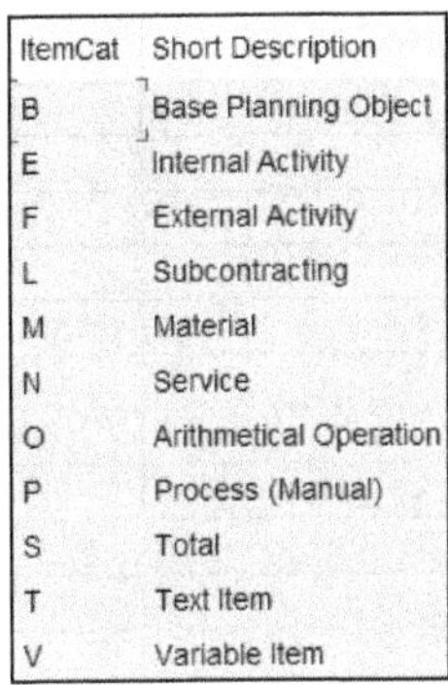

ItemCat	Short Description
B	Base Planning Object
E	Internal Activity
F	External Activity
L	Subcontracting
M	Material
N	Service
O	Arithmetical Operation
P	Process (Manual)
S	Total
T	Text Item
V	Variable Item

Figure 9.51 Item Category Possible Entries

  The item category you choose influences the entries you can make in the following fields for each item. Let's look at some example item categories to see how they work:

  - **B: Base Planning Object**
    If you select this category, you enter a previously created base planning object in the **Resource** column.
  - **M: Material**
    This category requires you to enter an existing material in the **Resource** column.
  - **S: Total**
    For this category, you leave the **Resource** column entry blank and instead enter a formula to, for example, total the value of previous items.
  - **V: Variable Item**
    Use this category if you want to enter an item without any restriction, for example, to add some more value as a contingency.
- **Resource**
  The resource you enter in the **Resource** column corresponds to the item **Category** entered in the **Category** column. For example, if you enter item category **M**, you'll be restricted to entering an existing material in the corresponding field in the **Resource** column. Some item categories, such as V, don't require an entry in the **Resource** column.

- **Plant/Activity**
  An entry you make in the **Plant/Activity** column corresponds to the entries you made in the Item **Category** and **Resource** columns.
- **Quantity**
  This column refers to the quantity of resources required for this estimate. You can easily change quantities and see how this affects the total value of the base planning object.
- **Value – Total**
  Value – Total (total value) for each item is calculated by multiplying the **Quantity** by the **Price – Total**. You should be aware of the value in the **Price Unit** column when considering base planning object prices. Divide the **Price – Total** entry by the **Price Unit** entry to calculate the unit cost. You can increase the price unit to improve the accuracy of the price.
- **Cost Element**
  The **Cost Element** identifies the type of cost. You can see standard base planning object detailed reports by following the menu path **Accounting • Controlling • Product Cost Controlling • Product Cost Planning • Information System • Detailed Reports • For Base Planning Object**.

9.8 Cost Estimate User Exits

Five user exits are available for cost estimates. A *user exit* is a point in the standard program where you call your own program. User exits use *includes*, which are criteria you use to group fields and insert them in a table or structure.

Because you integrate product costing with other modules, you should use standard functionality whenever possible. The following cost estimate user exits are available:

- COPCP001
  Cross-company costing
- COPCP002
  Material valuation of valuated sales order stock
- COPCP003
  Costing production resources/tools
- COPCP004
  Costing bulk materials
- COPCP005
  Valuation strategy U
- SAPLXCKA
  Costing reports

In addition to user exits, you can influence standard SAP valuation functionality with these BAdIs:

- MAT_SELECTION_CK for material selection in costing
- SUBCONTRACTING_VAL_CK for valuation of subcontracting
- SUBCONTRACTING_VAL_CK for valuation of external processing
- DELIVERY_COSTS_ASSIGNMENT_CK for raw material costs

We'll first examine how to configure a user exit and then examine each of the above user exits in detail. You access cost estimate user exits with Transaction CMOD or by following the IMG menu path **Controlling • Product Cost Controlling • Product Cost Planning • Selected Functions in Material Costing • Develop Enhancements for Material Costing**. Type in the name of your user exit, click the **Create** button, type in the short text, click the **Enhancement assignments** button, and click the **Local Object** button to display the screen shown in Figure 9.52.

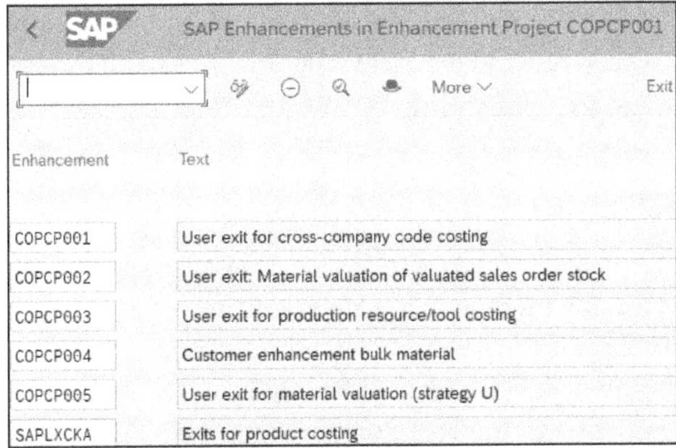

Figure 9.52 Cost Estimate User Exits

9.8.1 Cross-Company Code Costing

Enhancement COPCP001 allows you to define prices for materials transferred between company codes where you have not activated cross-company code costing. Let's compare this with cross-company code costing activated with Transaction OKYV or via the IMG menu path **Controlling • Product Cost Controlling • Product Cost Planning • Selected Functions in Material Costing • Activate Cross-Company Costing**. The screen shown in Figure 9.53 is displayed.

Cost Across Company Codes allows you to cost and release a material in more than one company code within a controlling area. If you enter a special procurement type with transfer from another plant, proceed as follows:

- To re-cost materials in another company code or transfer an existing cost estimate using transfer control, select the checkboxes as shown in Figure 9.53.

- To transfer the current valuation price from the material master with the valuation strategy, don't select the checkbox. The system will treat the material as externally procured.

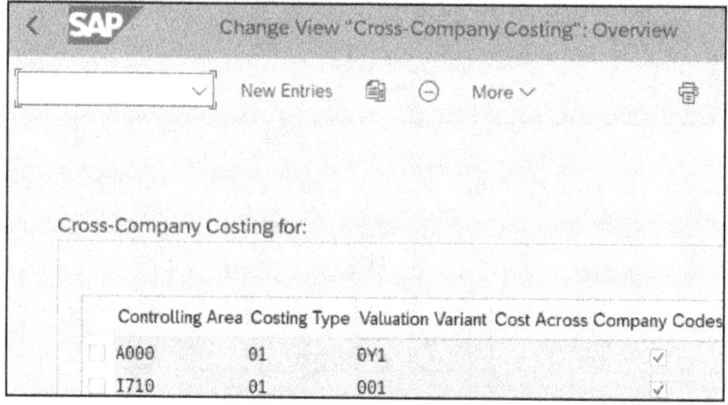

Figure 9.53 Activate Cross-Company-Code Costing

You can develop enhancement COPCP001 to define a price if costing across company codes isn't active. In include LXCKAF, you can find example program code, but it cannot be directly copied and used.

9.8.2 Production Resource/Tool Costing

A *production resource* or *tool* is a movable operating resource used in production or plant maintenance. You call the function module if you have activated enhancement COPCP003. You calculate the costs for production resources and tools as a flat rate within production overhead. This enhancement enables you to plan the costs for production resources as separate cost components. It contains component EXIT_SAPLCK01_001, which contains example program code that you can modify or use as-is.

9.8.3 Bulk Material Costing

Bulk materials, such as washers and grease, are normally charged directly to a cost center and included in overhead cost components. If you attempt to make an entry in the **Relevant to Costing** field in a BOM item for a bulk material, you will receive an error message. Enhancement COPCP004 allows you to include bulk materials in the material cost component and itemization.

9.8.4 Material Valuation

Enhancement COPC005 allows you to specify prices for valuation of materials with valuation strategy U in a valuation variant with Transaction OKK4 or by following the IMG menu path **Controlling • Product Cost Controlling • Product Cost Planning • Material**

Cost Estimate with Quantity Structure • Costing Variant: Components • Define Valuation Variants. Double-click a costing variant to display the screen shown in Figure 9.54.

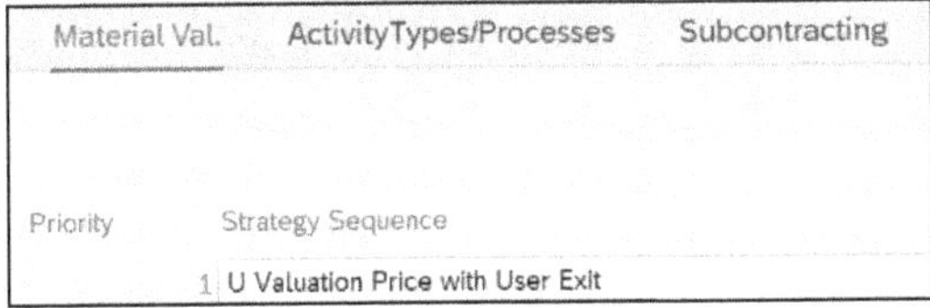

Figure 9.54 Valuation Variant with Valuation Strategy U

You use this user exit to determine a valuation price, such as a fixed price, that isn't available with standard valuation strategies.

9

9.8.5 Costing Reports

You can display itemization and cost component reports directly when you view a cost estimate. You can add columns to these reports using standard layouts. If you need to display a nonstandard or calculated column, you can create your own custom report and add it to the costing report user exit with enhancement SAPLXCKA. You can run the reports directly from the menu bar.

Display a cost estimate with Transaction CK13N or via the menu path **Accounting • Controlling • Product Cost Controlling • Product Cost Planning • Material Costing • Cost Estimate with Quantity Structure • Display**. You can see the menu path in Figure 9.55.

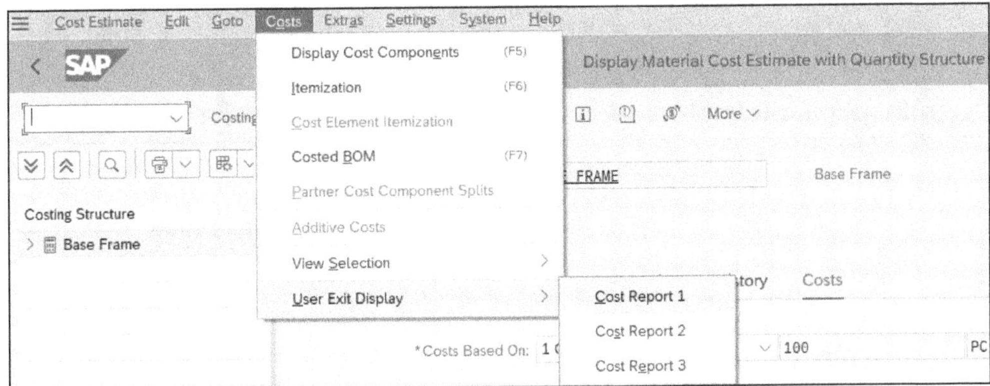

Figure 9.55 Cost Estimate User Exit Report Menu Path

Select **Costs • User Exit Display • Cost Report 1** from the menu bar to display the example report shown in Figure 9.56.

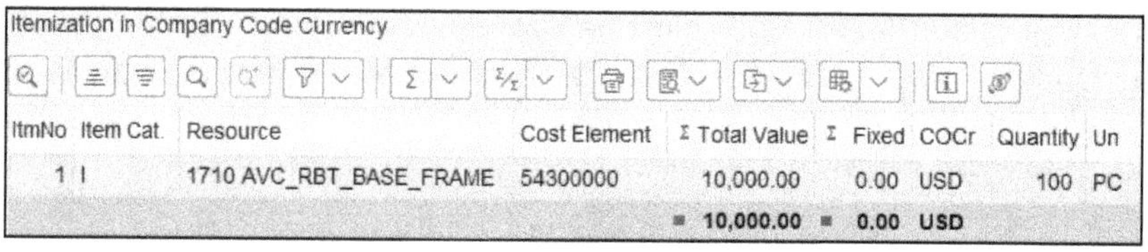

Itemization in Company Code Currency

ItmNo	Item Cat.	Resource	Cost Element	Σ Total Value	Σ Fixed	COCr	Quantity	Un
1	I	1710 AVC_RBT_BASE_FRAME	54300000	10,000.00	0.00	USD	100	PC
				■ **10,000.00**	■ **0.00**	**USD**		

Figure 9.56 Cost Estimate Report with User Exit

You can design this itemization report to display columns unavailable in standard reports. Cost estimate user exit SAPLXCKA contains the following components:

- **EXIT_SAPLCKAZ_001 (display/print itemization)**
  This enhancement contains example program code that you can modify or activate directly.
- **EXIT_SAPLCKAZ_002 (display/print cost components)**
  This enhancement contains example program code that you can modify or activate directly.
- **EXIT_SAPLCKAZ_003 (display/print cost components and itemization)**
  This enhancement doesn't contain example program code.

9.9 Summary

In this chapter, we discussed the different cost estimate types available during the product lifecycle, including how to create, mark, and release standard cost estimates. We examined how to create costing runs, and we also discussed mixed, preliminary, and unit cost estimates and cost estimate user exits.

In the next chapter, we'll look at how to create a preliminary cost estimate for a production order.

Chapter 10
Preliminary Costing on Production Orders

Preliminary costing refers to assigning planned debits and credits to manufacturing orders.

In previous chapters, we accomplished the following:

- Carried out initial planning to create cost estimates
- Set up Controlling, material, and logistics master data
- Configured costing sheets, cost components, and costing variants
- Created standard cost estimates to plan the manufacturing cost of products

In this chapter, we'll explore preliminary costing for manufacturing orders. Preliminary costing determines the planned costs of a manufacturing order. We discussed preliminary costing and cost estimates for product cost collectors in Chapter 9.

To recap, in preliminary costing, you valuate bill of materials (BOM) components and routing activities included in the production order using the valuation variant specified in the costing variant. You determine the costing variant by the order type, which is a required field when you create an order. The system calculates overhead expenses using the costing sheet if specified in the valuation variant.

We'll now look at reporting on the operation and work center and then production order reporting.

10.1 Work Center and Operation

SAP S/4HANA includes the work center and operation in the Universal Journal tables ACDOCA and ACDOCP for greater transparency in production reporting with two SAP Fiori apps: Product Cost Analysis and Analyze Costs by Work Center/Operation. We'll discuss each of these in the following sections.

10.1.1 Production Cost Analysis (App ID F1780)

With this app, you display actual production costs by cost component and cost component group for manufacturing orders. You analyze plan, target, and actual costs, as well as production variances. This app replaces SAP GUI order selection report S_ALR_87013127. A list of orders is displayed, as shown in Figure 10.1.

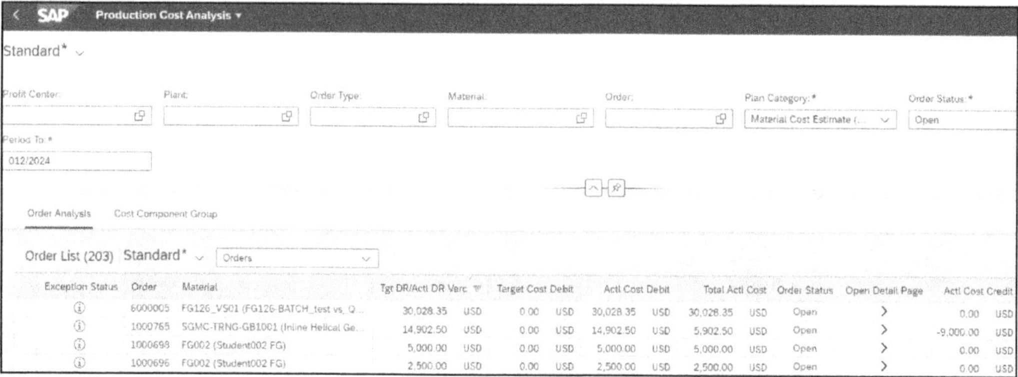

Figure 10.1 Production Cost Analysis Summary View

Select the primary cost filters at the top of the screen to display a list of orders with costs. Click in the **Plan Category** field to display a list of possible entries, as shown in Figure 10.2.

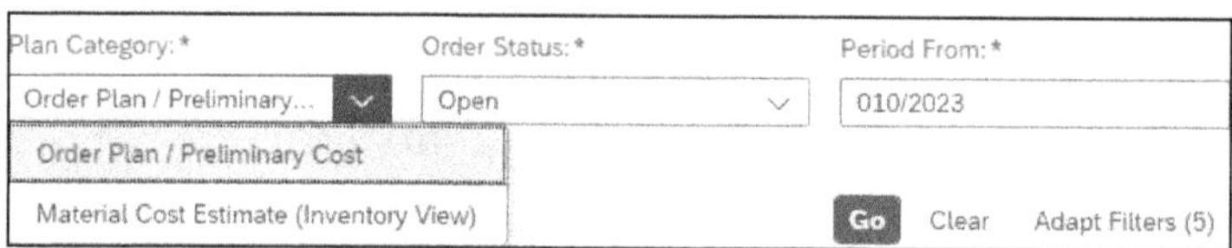

Figure 10.2 Plan Category Possible Entries

Order target costs calculated based on **Order Plan / Preliminary Cost** represent expected costs with the BOMs and routing assigned to the order. Variances using these targets represent manufacturing performance independent of the production line. The differences correspond to target cost version 1. We'll discuss target cost versions in Chapter 14.

Costs based on the **Material Cost Estimate (Inventory View)** represent total variances like the target cost version 0 in SAP GUI reports. The **Order Status** determines the differences that appear in the report:

- **Open**
  These orders are released (REL) but not fully delivered (DLV) or technically complete (TECO). Differences represent work in process (WIP).
- **Closed**
  These orders are either DLV or TECO, and differences represent variances.

Order Status selection narrows the list to either open or closed orders.

Double-click an item in Figure 10.1 to display the target and actual costs by order. This view is like the SAP GUI order detail report KKBC_ORD. Click on **>** in the **Open Detail Page** column to view an individual order. An example is shown in Figure 10.3, where you can see **Order Cost Detail** by **Business Transaction**.

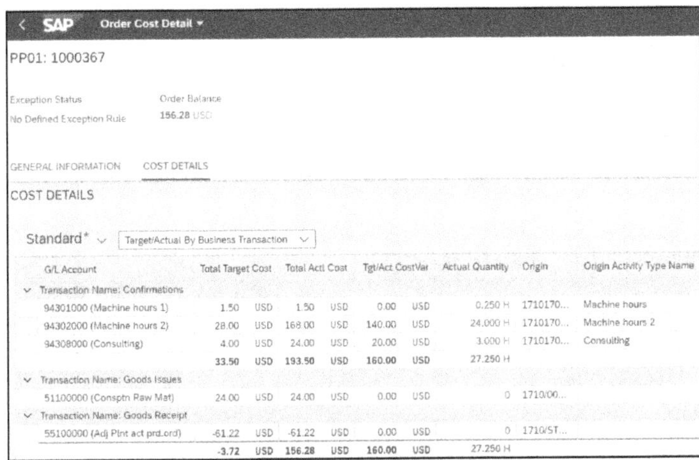

Figure 10.3 Product Cost Analysis Order Detail

10.1.2 Analyze Costs by Work Center/Operation (App ID F3331)

This app accesses the Universal Journal work center and operation fields to provide variance management at the work center level. This detailed reporting isn't available with SAP GUI reports. The report contains a bar chart and tabular report sections, displayed when the app is first run, as shown in Figure 10.4.

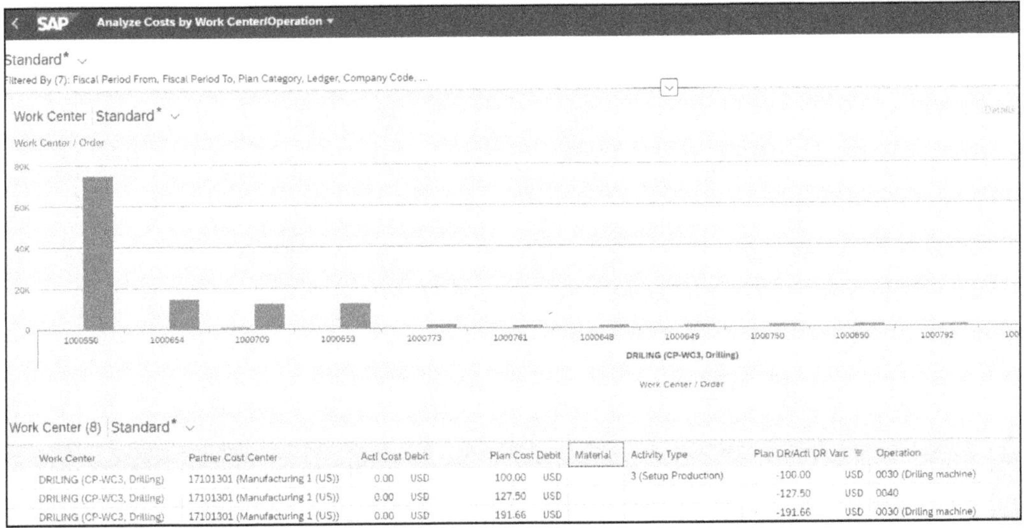

Figure 10.4 Analyze Costs by Work Center/Operation Initial Screen

You can expand each section to display more information. The bar chart at the top is useful for comparing the performance of work centers and identifying areas for further review. This example uses a bar chart, but you can also select pie charts, stacked bar charts, line charts, and scatter charts.

10.2 Production Order Control

We'll now look at how you access a preliminary cost estimate for a production order. First, we'll select a production order from a list and then display the preliminary costs.

10.2.1 Select and Display Production Orders

You can display a list of production orders with Transaction COOIS or via the menu path **Logistics • Production • Shop Floor Control • Information System • Order Information System**. The screen shown in Figure 10.5 is displayed.

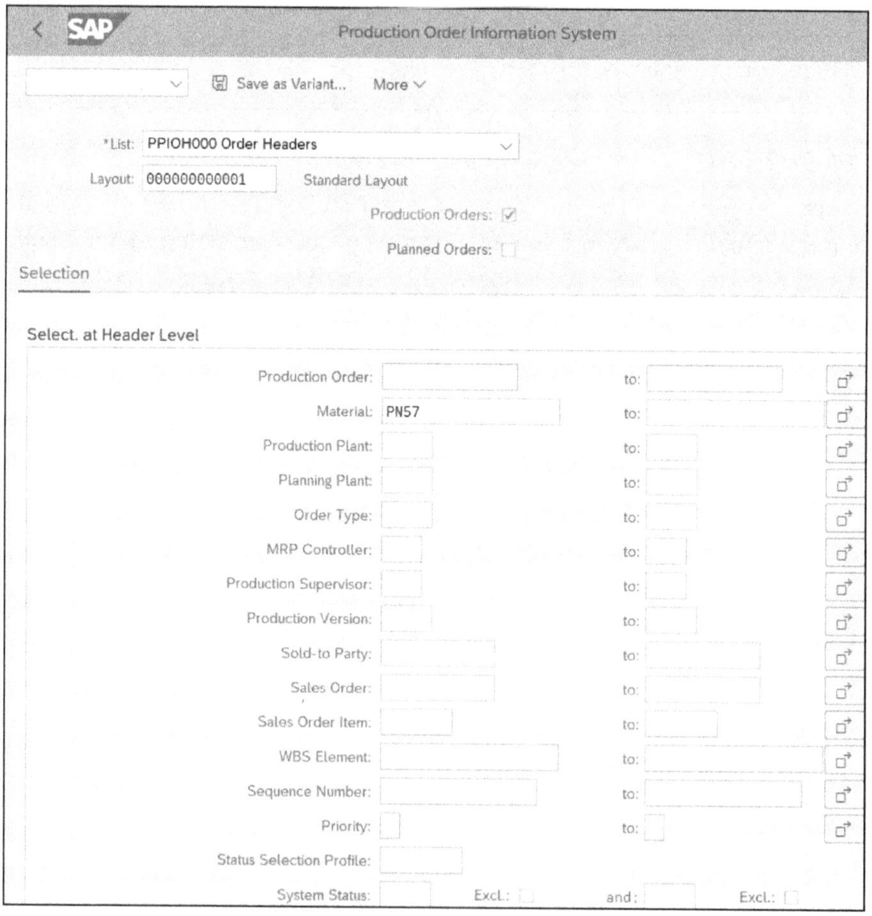

Figure 10.5 Production Order Information System

> **Logistics Information System**
>
> The shop floor information system is part of the logistics information system (LIS), which is part of the SAP S/4HANA compatibility scope. You can continue using LIS, and all LIS tables are available in SAP S/4HANA. However, you should not make major investments in this area because it will eventually be phased out. See *http://s-prs.co/v583202* for more details.

> **Process Order List**
>
> You display a list of process orders with Transaction COOISPI or via the menu path **Logistics • Production – Process • Process Order • Reporting • Order Information System • Process Order Information System.**

Many **Selection** fields are available to filter the production order list. Let's follow an example by entering "PN57" in the **Material** field. Click the **Execute** button or press F8 to display the screen in Figure 10.6.

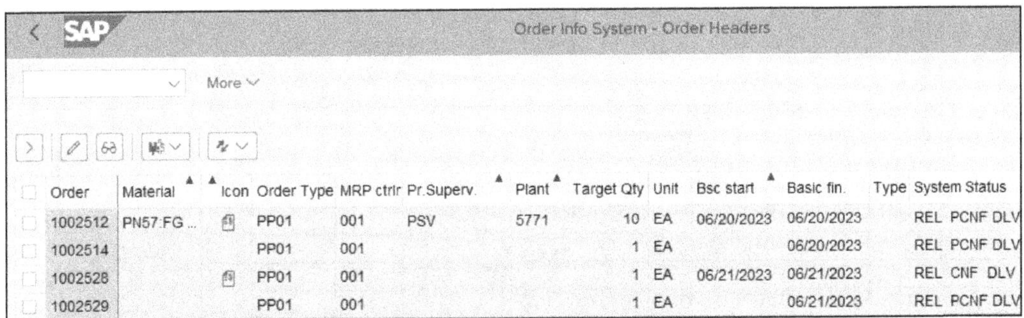

Order	Material	Icon	Order Type	MRP ctrlr	Pr.Superv.	Plant	Target Qty	Unit	Bsc start	Basic fin.	Type	System Status
1002512	PN57 FG ...		PP01	001	PSV	5771	10	EA	06/20/2023	06/20/2023		REL PCNF DLV
1002514			PP01	001			1	EA		06/20/2023		REL PCNF DLV
1002528			PP01	001			1	EA	06/21/2023	06/21/2023		REL CNF DLV
1002529			PP01	001			1	EA		06/21/2023		REL PCNF DLV

Figure 10.6 Production Order List

Figure 10.6 shows a list of production orders created for **Material PN57**. Select the first line, click the pencil icon to edit the production order, and select the **Control** tab to show the screen displayed in Figure 10.7.

The **Control** tab displays production order fields relevant to costing. Let's discuss the fields in the **Costing** section.

10.2.2 Costing Variant for Planned Costs

The **PlnCstgVar:** (plan costing variant) field specifies which costing variant you use to determine plan costs. The order type determines the default costing variant, which in turn determines the valuation variant. You maintain the default values for order types with Transaction OKZ3 or via the IMG menu path **Controlling • Product Cost Controlling • Cost Object Controlling • Product Cost by Order • Manufacturing Orders • Define Cost-Accounting-Relevant Default Values for Order Types and Plants.** Double-click a **Plant** and **Order type** combination to display the screen shown in Figure 10.7.

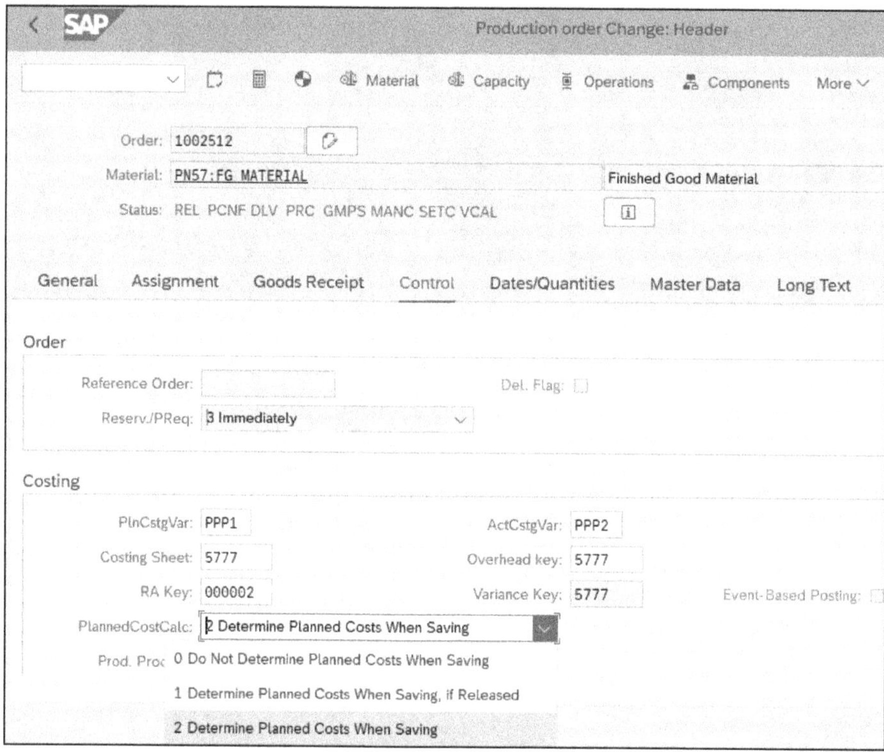

Figure 10.7 Production Order Control Tab

This screen allows you to maintain default values for a combination of **Plant** and **Order Type**. In this example, the default value for the costing variant is **PPP1**, which determines the valuation variant **006**.

10.2.3 Costing Variant for Actual Costs

The **ActCstgVariant** (actual costing variant) in Figure 10.7 specifies which costing variant you use to determine production order actual costs. The order type determines the default costing variant **PPP2**, which in turn determines the **Val. Var.** (valuation variant) **007** as shown in Figure 10.8.

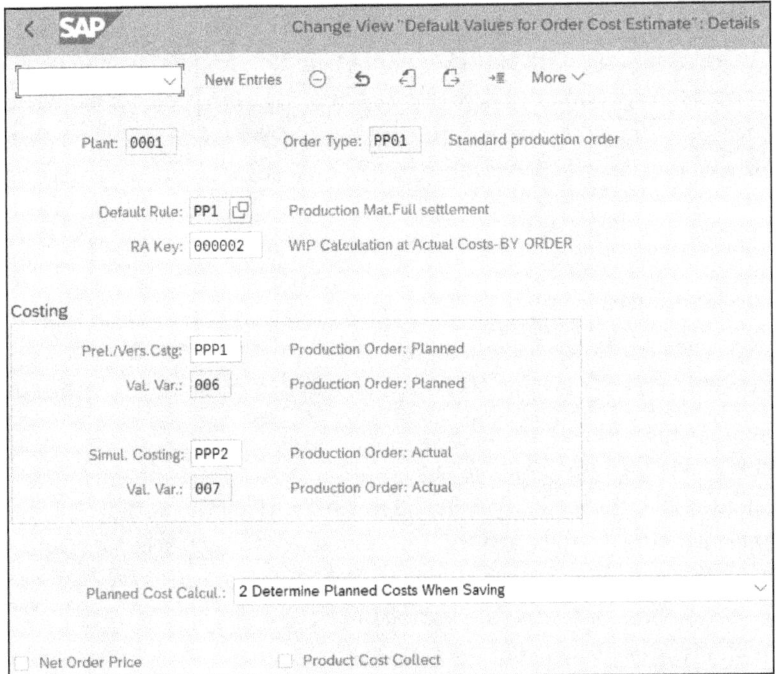

Figure 10.8 Default Values for Order Type and Plant

10.2.4 Costing Sheet

A *costing sheet* determines the allocation of overhead costs from cost centers to manufacturing orders. The **Costing Sheet** entry shown earlier in Figure 10.7 defaults from the valuation variant, which, in turn, you determine by the costing variant. You store the costing variant as a default value per order type and plant, as shown earlier in Figure 10.7. The same costing sheet is used for the planned and actual costs.

10.2.5 Overhead Key

An *overhead key* allows you to assign specific overhead rates to individual or groups of orders. This functionality gives you flexibility in assigning overhead rates to orders. But more overhead keys might mean more time for frequently calculating and maintaining overhead rates. Refer to Chapter 5 for more information on overhead key maintenance.

10.2.6 Results Analysis Key

The results analysis key (**RA Key** field) determines period-end work in process (WIP) calculation. The system only calculates WIP on orders with an entry in this field.

10.2.7 Variance Key

The system calculates variances at the period end on manufacturing orders and product cost collectors containing a variance key. This key defaults from the **Costing 1** view when you create manufacturing orders and product cost collectors. The variance key also determines if you subtract the scrap value from actual costs before determining variances.

10.2.8 Determine Plan Costs

The **Determine Plan Costs** field determines when the system saves manufacturing order plan costs. Click the field to display the three possible entries shown earlier in Figure 10.7:

- **0 Do Not Determine Planned Costs When Saving**
  If you specify that no planned costs are calculated when saving the manufacturing order, you can calculate the planned costs of the order at any time by selecting **Goto • Costs • Analysis** from the production order menu bar. This option can reduce the time required when saving many manufacturing orders.
- **1 Determine Planned Costs When Saving, if Released**
  A manufacturing order status of "Released" indicates that you can post actual costs to the order. If you create manufacturing orders well before releasing them, you can consider selecting this option to save time when initially saving many manufacturing orders.
- **2 Determine Planned Costs When Saving**
  This is the most common setting and indicates that planned costs are calculated when the manufacturing order is initially created and saved. This allows you to correct any costing errors earlier than the other two options.

The option of when to determine plan costs defaults from the order type and plant with Transaction OKZ3, as shown in Figure 10.8. With the **Planned Cost Calcul.** field entry, you maintain the default value for saving manufacturing order plan costs.

The **Event-Based Posting** checkbox in Figure 10.7 if selected, indicates that event-based WIP and variance posting is activated. The event-based posting checkbox setting is derived from the event-based results analysis key. We'll look at this topic in more detail in Chapter 18.

Now that we've discussed how and when planned production order costs are calculated, let's look at how to display the planned costs.

10.3 Display Planned Costs

Plan costs are posted to the Universal Journal planning table ACDOCP, which includes the work center and operation, and are only available in the Universal Journal. You can see this data via CDS views and the following apps, which we also discussed at the start of this chapter:

- Analyze Costs by Work Center/Operation app (SAP Fiori ID F3331)
- Production Cost Analysis app (SAP Fiori ID F1780)

There is a parallel update to legacy tables COSP and COSS, which you access via report KKB_ORD and directly in the production order. The data is aggregated to cost center/activity type in these reports.

You display production order planned costs by selecting **Goto • Costs • Analysis** from the production order menu bar. The screen shown in Figure 10.9 is displayed.

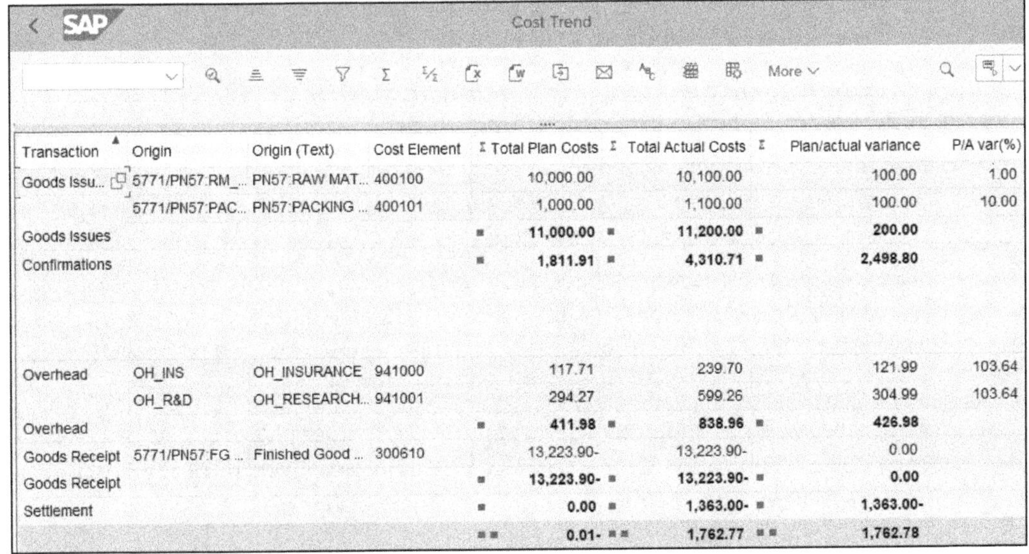

Transaction	Origin	Origin (Text)	Cost Element	Total Plan Costs	Total Actual Costs	Plan/actual variance	P/A var(%)
Goods Issu...	5771/PN57:RM_...	PN57:RAW MAT...	400100	10,000.00	10,100.00	100.00	1.00
	5771/PN57:PAC...	PN57:PACKING ...	400101	1,000.00	1,100.00	100.00	10.00
Goods Issues				11,000.00	11,200.00	200.00	
Confirmations				1,811.91	4,310.71	2,498.80	
Overhead	OH_INS	OH_INSURANCE	941000	117.71	239.70	121.99	103.64
	OH_R&D	OH_RESEARCH...	941001	294.27	599.26	304.99	103.64
Overhead				411.98	838.96	426.98	
Goods Receipt	5771/PN57:FG ...	Finished Good ...	300610	13,223.90-	13,223.90-	0.00	
Goods Receipt				13,223.90-	13,223.90-	0.00	
Settlement				0.00	1,363.00-	1,363.00-	
				0.01-	1,762.77	1,762.78	

Figure 10.9 Production Order Planned Costs

The production order planned costs are displayed in the **Total plan costs** column. Let's examine each section in the **Transaction** column of this report:

- **Goods Issues**
  As you create the production order, the BOM is copied to the production order. Additional materials can be added manually if necessary. The planned goods issues to the production order are determined from the BOM and displayed as a plant and material combination in the **Origin** column. If you use the work center/operation view, you should ensure that the materials in the BOM are assigned to the correct operation.

- **Confirmations**
  As the production order is created, the routing is copied to the production order. Additional operations can be added manually if necessary. The planned activity confirmations are determined from the operations in the routing and displayed as a cost center and activity combination in the **Origin** column.
- **Overhead**
  Overhead is determined by the costing sheet in the production order **Control** tab, as displayed earlier in Figure 10.7. A costing sheet allocates overhead costs from cost centers to manufacturing orders. The cost center that receives the credit is displayed in the **Origin** column. You can read more about costing sheets in Chapter 5.
- **Goods Receipt**
  You are typically required to enter a material when you create a production order. The production order contains all the information necessary to manufacture the material. When you place the manufactured material into inventory, the production order receives a credit based on the standard price of the plant and material shown in the **Origin** column of Figure 10.9.

A Production Order Cost Estimate Is Single-Level

A production order preliminary cost estimate is single-level since each level is delivered into stock and reissued to the next level. A product cost collector preliminary cost estimate (as discussed in Chapter 9) is multilevel.

To see a multilevel cost estimate for the material you are producing, go to the material master **Costing 2** view and click the **Current Cost Estimate** button. The system displays the costed multilevel BOM in the **Costing Structure** section at the left of the screen.

You can display three additional views of production order planned costs by selecting **Goto • Costs • Itemization**, **Goto • Costs • Cost Component Split**, or **Goto • Costs • Balance** from the production order menu bar, as shown in Figure 10.10.

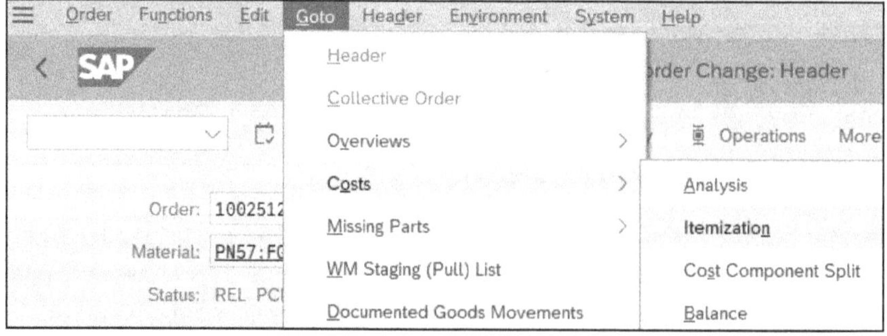

Figure 10.10 Production Order Cost Views Menu Path

The **Itemization** view provides a simple listing of components and activities. The **Cost Component Split** view allows you to display costs grouped by cost components, and the **Balance** view provides a simple order balance report in legal, group, and profit center valuations.

10.4 Summary

In this chapter, we examined preliminary costing for manufacturing orders. We examined relevant fields in the production order **Costing** tab and how to display a preliminary cost estimate. We also reviewed SAP Fiori apps that access the work center and operation in table ACDOCP. In Chapter 11, we'll examine simultaneous costing and actual postings during a period. In Chapter 14, we'll examine period-end postings, beginning with overhead calculation.

Chapter 11
Simultaneous Costing

You assign actual debits and credits to cost objects such as production and process orders. All postings are in the general ledger, including secondary postings.

In Chapter 10, we discussed calculating plan costs during preliminary costing for production orders. In this chapter, we'll discuss posting actual costs to cost objects, such as manufacturing orders and product cost collectors. In the following chapters, we'll discuss period-end and event-based processing, which compares target and actual costs.

Actual costs debit or credit manufacturing orders during goods movements and activity confirmations. We'll review these costs in this chapter. We'll also discuss how to report on the costs.

11.1 Goods Movements

Goods movements to and from production orders result in debits or credits. We'll examine each in turn.

11.1.1 Debits

When goods are issued from inventory, a general ledger balance sheet account is credited, and a profit and loss (P&L) consumption expense account is debited. When a primary cost element account is detected during a posting to a general ledger account, the P&L is extended to include controlling information (the production order).

The production order triggers a reservation for the required components, which is the basis for the material document and subsequent accounting document. The production order is debited automatically because the goods are issued to the production order. The debit value is calculated by multiplying the component standard price by the quantity issued from inventory. Figure 11.1 shows an example debit when components with a value of 100 are issued from raw material (RM) inventory to a production order.

Component allocations in routings determine how goods are moved to specific work centers and operations. These goods issues are displayed along with the activity postings at the work center level.

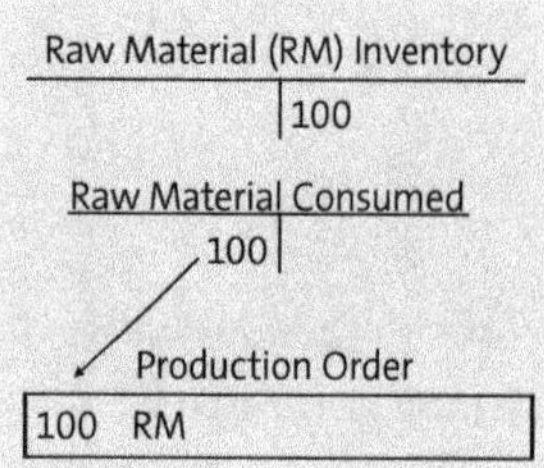

Figure 11.1 Goods Issue to Production Order

Universal Journal Posting

The account posting includes RM inventory (balance sheet) and RM consumed (expense) accounts. Cost object assignment in the Universal Journal allows you to analyze all expense postings with standard SAP S/4HANA and compatibility reports, as well as SAP Fiori reports.

We discuss reporting in detail in Chapter 20.

You don't need to manually enter general ledger accounts during goods issues to a production order because the accounts are configured in a table during initial implementation. These settings are delivered as business content in SAP S/4HANA Cloud and optionally in SAP S/4HANA Private Cloud and on-premise SAP S/4HANA. You display the table's contents with Transaction SE16N or SE16H or via the menu path **Tools • ABAP Workbench • Overview • Data Browser**. Enter **Table T030** and press [Enter] to display the selection screen shown in Figure 11.2.

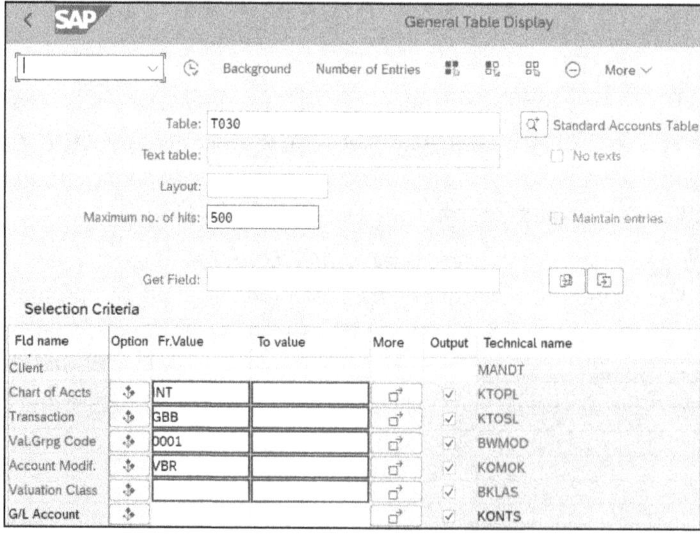

Figure 11.2 Table T030 Selection Screen

SE16 Access

Access to the SE16 transactions is restricted in many organizations.

Transaction SE16 User Parameters

If you use Transaction SE16 instead of Transactions SE16N or SE16H, you may need to select **Settings • User Parameters** from the menu bar and select **Field Label** instead of **Field name** to ensure that you display the **Fld name** (field name) text instead of the **Technical name**.

You typically filter the table entries displayed on the following results screen by making entries in the **Selection Criteria** section of Figure 11.2. You display accounts posted during production order goods movements by making the following entries:

- **Chart of Accts**
  You typically restrict your selection by entering a chart of accounts.
- **Transaction**
  You display entries relevant to inventory movements with Transaction GBB.
- **Val. Grpg Code**
  You assign company codes to different valuation grouping codes if automatic account determination is to run differently for different company codes. You make this assignment with Transaction OMWD.
- **Acct Modif.**
  This indicates the type of inventory movement. In Figure 11.2, **VBR** relates to goods issues to production orders.
- **Valuation Class**
  You assign a valuation class to the **Costing 2** view of each material, and you typically assign different valuation classes per material type, such as raw materials and finished goods. Assigning different valuation classes allows you to analyze inventory financial journal entries in different accounts. Refer to Chapter 3, Section 3.5.3 , for information on valuation classes.

After you've made the **Selection Criteria** entries, click the **Execute** icon to display the screen shown in Figure 11.3.

This screen displays the **G/L Acct** (general ledger account) debited or credited during goods issues for production orders. You can find out more information on table T030 entries by following the IMG menu path **Materials Management • Valuation and Account Assignment • Account Determination • Account Determination Without Wizard • Configure Automatic Postings**. The menu path is shown in Figure 11.4.

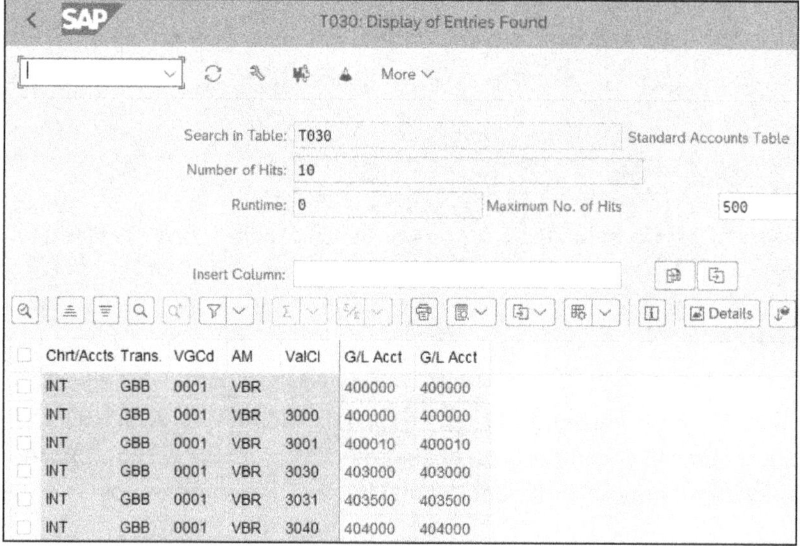

Figure 11.3 Table T030 Entries Found

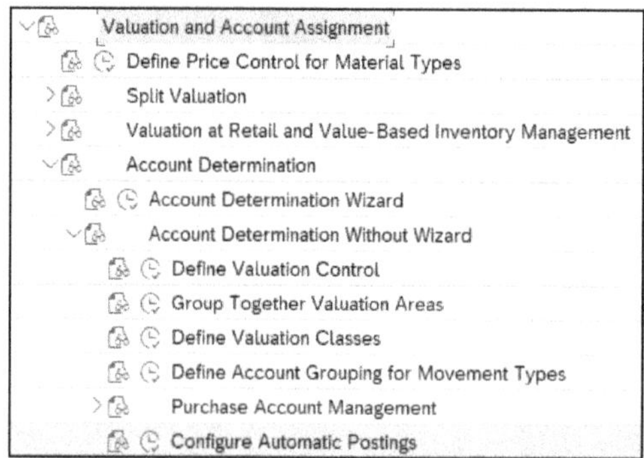

Figure 11.4 Configure Automatic Postings Menu Path

Click the paper-and-glasses icon to the left of **Configure Automatic Postings** to display detailed documentation on setting up automatic general ledger account determination. Click the **Execute** icon to display the initial screen shown in Figure 11.5.

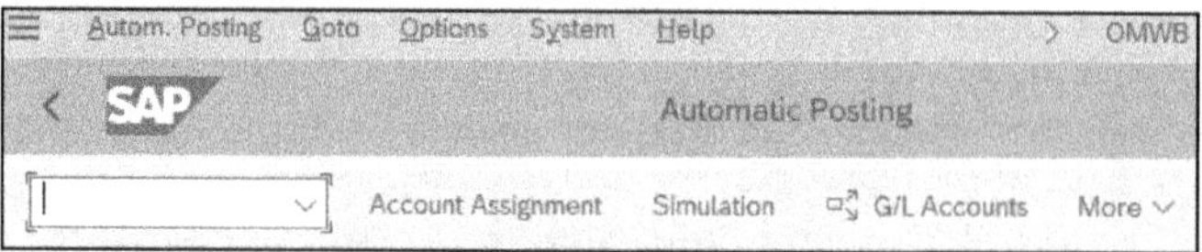

Figure 11.5 Automatic Account Assignment Initial Screen

The **Simulation** and **G/L Accounts** buttons allow you to analyze general ledger accounts for automatic postings. To configure automatic account determination, click the **Account**

Assignment button, double-click Transaction GBB, and enter the chart of accounts to display the screen shown in Figure 11.6:

- **AUA** is for offsetting entries for order settlement.
- **AUF** is for offsetting entries for order settlement if AUA is not maintained.

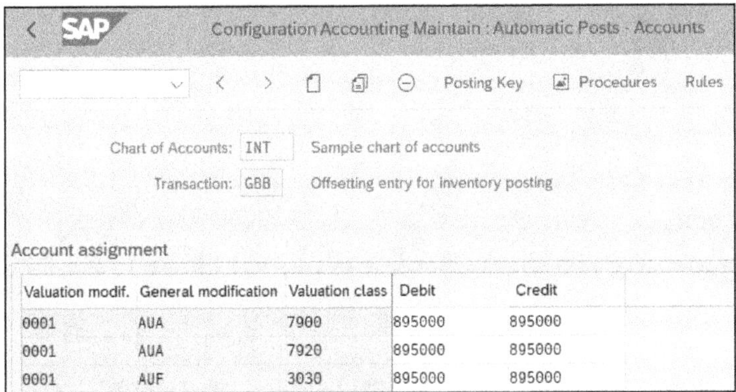

Valuation modif.	General modification	Valuation class	Debit	Credit
0001	AUA	7900	895000	895000
0001	AUA	7920	895000	895000
0001	AUF	3030	895000	895000

Figure 11.6 Configure Automatic Account Assignment

You can also search for the entries in table T030 with Transaction SE16N, as shown earlier in Figure 11.3. You can change the general ledger accounts at any time, but for consistent reporting, you may decide to restrict any changes to the beginning of a fiscal year, for example.

Now that we've examined debit postings during goods movements, let's look at credits.

11.1.2 Credits

As you deliver finished goods into inventory from a production order, you debit an inventory balance sheet account and credit a profit and loss production output account. Because a primary cost element corresponds to the production output account, you also credit a controlling cost object—the production order, in this case. The credit value is determined by multiplying the standard price by the finished goods delivered quantity. Figure 11.7 shows an example of a production order cost report with a primary credit of 250 for delivery to inventory.

As discussed in the previous section, the general ledger account credited during the delivery of the finished goods is determined automatically.

Variance is the production order balance or the difference between actual debits and credits at standard. To assist you in determining corrective actions, variance calculation divides the variance into categories based on the source of the variance. We discuss variance analysis in detail in Chapter 14.

The production order receives a secondary credit equal to the variance during settlement, resulting in a zero balance.

	Production Order	
Debits	100	Raw Materials
	100	Labor
	100	Overhead
Credits	(250)	Finished Goods
Balance	50	Variance

Figure 11.7 Production Order Delivery Credits

Production Order Credits

A production order can also receive credits for other reasons, such as the return of components to inventory or co- or byproducts in joint production.

Production variances settled to margin analysis are included at the gross profit margin level. Cost center under/overabsorption costs assessed to margin analysis are included at the operating profit level. The following box discusses the different levels of margin analysis.

Margin Analysis

You carry out margin analysis at different levels to understand the impact of the various contribution margins within the company. The most common levels are gross profit, operating profit, pretax profit, and net profit, as described here:

- **Gross profit**
  A company's cost of sales (COS) represents the expense related to labor, raw materials, and manufacturing overhead. You deduct this expense from the company's net sales/revenue, which results in a company's first level of profit, or *gross profit*. The *gross profit margin* indicates how efficiently a company uses raw materials, labor, and manufacturing-related fixed assets to generate profits. Generally, management cannot exercise complete control over these costs.
- **Operating profit**
  You subtract selling, general, and administrative (SG&A) expenses, often referred to as *operating expenses*, from gross profit to calculate the operating profit margin. Management has much more control over operating expenses than over COS. Positive and negative trends in operating profit are, for the most part, directly attributable to management decisions.
- **Pretax profit**
  A company has access to various tax-management techniques, which allow it to manipulate the timing and magnitude of its taxable income.

- **Net profit**
  This is often referred to as a company's profit margin or bottom line. The *net profit margin* is determined after considering all expenses and taxes.

In addition to primary journal entries to production orders corresponding to goods movements, secondary postings occur during activity confirmations, which we'll now discuss.

11.2 Confirmations

Activity confirmations involve accounts with secondary cost elements. In the following sections, we'll analyze the journal entries and then follow an example of an activity confirmation.

11.2.1 Secondary Costs

When you confirm production order activities, a production order is debited, and a cost center is credited. You identify the postings with accounts with secondary cost elements.

A production cost center receives debits due to primary costs such as payroll and electricity. You can manufacture many products at a work center, with labor and facilities paid for by the cost center. Confirmation of labor and overhead activities allocates these primary costs across many products. Figure 11.8 shows an example of labor allocation postings from a production cost center to a production order:

❶ Payroll paid: Expense to cost center

❷ Goods issue: Raw material expense to production order

❸ Activity confirmation: Allocate labor to production orders

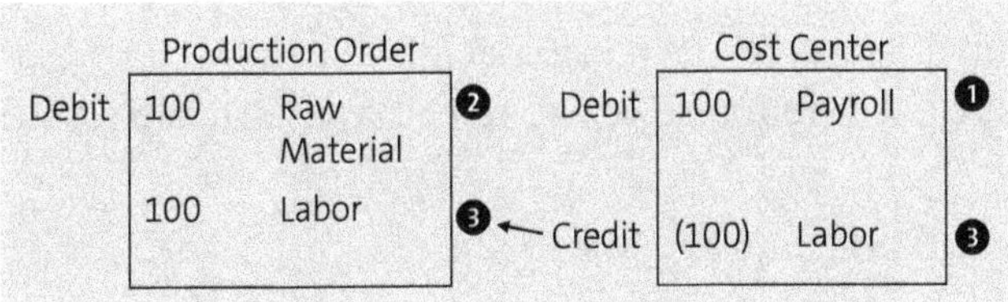

Figure 11.8 Production Cost Center Allocation during Activity Confirmation

In ❶, factory payroll is expensed directly to the production cost center via primary postings and then progressively allocated to each production order, with secondary postings during labor activity confirmation in ❸. All postings and assignments happen in real time in the Universal Journal.

You allocate a labor value of 100 from a production cost center to a production order in ❸ in Figure 11.8. The labor allocation occurs during activity confirmation.

Overhead Allocation

Overhead calculation allocates overhead costs across products from cost centers similarly to activity confirmation. Overhead costs are distributed with overhead calculation at period end, as we'll discuss in Chapter 12, and during events, as we'll discuss in Chapter 19.

Now that we've discussed how secondary postings occur during activity confirmations, let's follow an example of the postings by creating a production order and then confirming an activity.

11.2.2 Create Production Order

You create a production order with Transaction CO01 or by following the menu path **Logistics • Production • Shop Floor Control • Order • Create • With Material**. A selection screen is displayed, as shown in Figure 11.9.

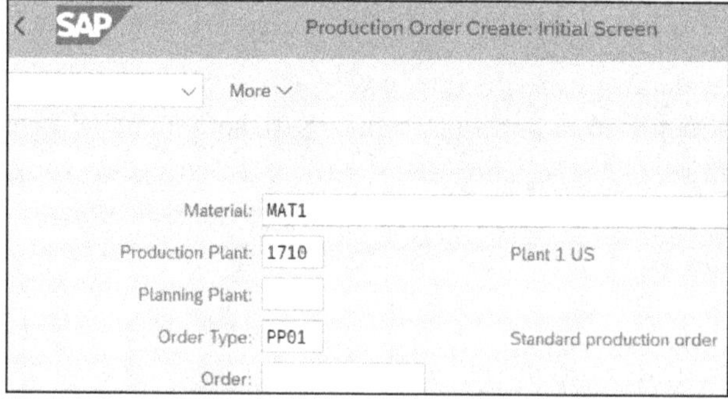

Figure 11.9 Create Production Order

You typically create production orders based on a material linked to BOMs and routings. All relevant master data is copied to the production order as you create it with the following steps:

1. Complete the **Material**, **Production Plant**, and **Order Type** fields.
2. Press [Enter] to display the screen shown in Figure 11.10.
3. Complete the **Total Qty** and **Basic End Dates** fields.
4. Click the release (green flag) icon to release the production order.
5. Save the production order.

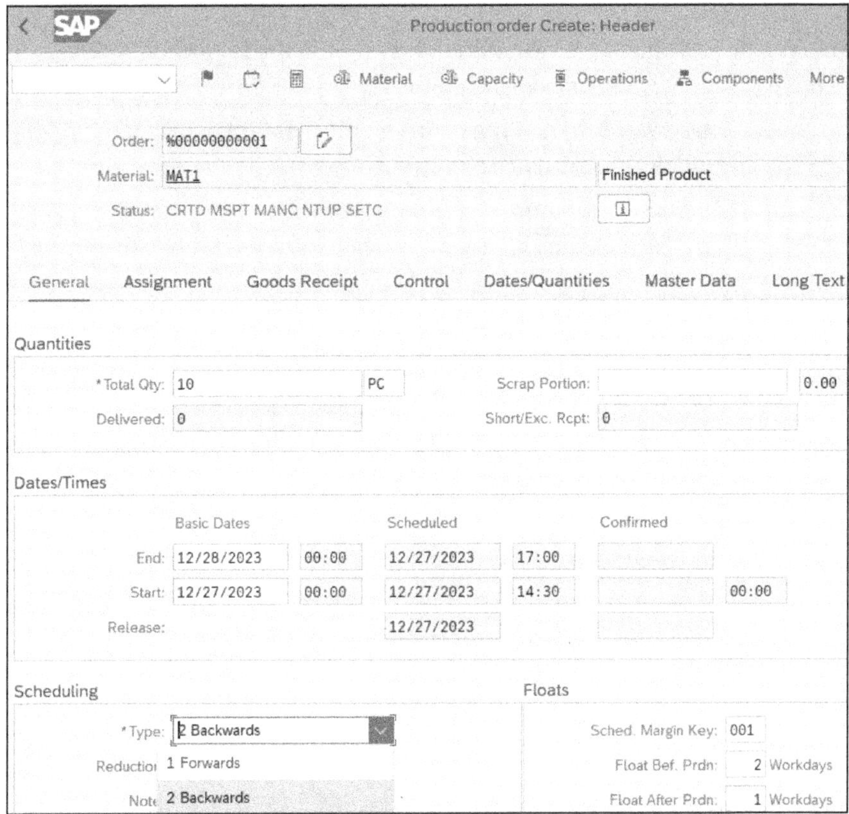

Figure 11.10 Create Production Order Header

Let's discuss the fields in this screen:

- **Total Qty (total quantity)**
  The **Total Qty** is the total quantity, including assembly scrap, that the order will produce.

- **Base unit of measure**
  To the right of the **Total Qty** field is the base unit of measure (**PC**, in this example), where you manage inventory, which defaults from the material master. You can change the unit of measure if you have entered the conversion factor in the **Additional Data** section of the material master. You can find the **Additional Data** button at the top section of every material master view. Then click the **Units of measure** tab to maintain **Conversion to Base Units of Measure.**

- **Scrap Portion**
  The **Scrap Portion** refers to the plan assembly scrap that defaults from the **MRP 1** view. The assembly scrap percentage automatically increases the order quantity. This field is optional.

- **Basic Dates (End and Start)**
  The **Dates** you enter depend on the choices you make in the following **Type** field in the **Scheduling** section. Here are two examples:
  - If you choose **2 Backwards** in the **Type** field, you must enter the **End** date, and the system will automatically calculate the **Start** date.
  - If you choose **1 Forwards**, you must enter the **Start** date, and the system will automatically calculate the **End** date.

  You can manually change the basic dates that are automatically calculated.
- **Scheduling Type**
  The **Type** field in the **Scheduling** section specifies the scheduling type for detailed scheduling, such as forward or backward. Click the **Type** field to display the possible entries, as shown in Figure 11.10. With production-rate and rough-cut scheduling, you always use backward scheduling.
- **Release**
  After you have completed the previous fields, you'll need to carry out the release step before you can collect costs on the production order. You release the production order by clicking the green flag icon at the top of the screen.
- **Operation Overv. (operation overview)**
  To review the operations copied from the routing to the production order, click the **Operations** button shown in Figure 11.10 to display the screen shown in Figure 11.11.

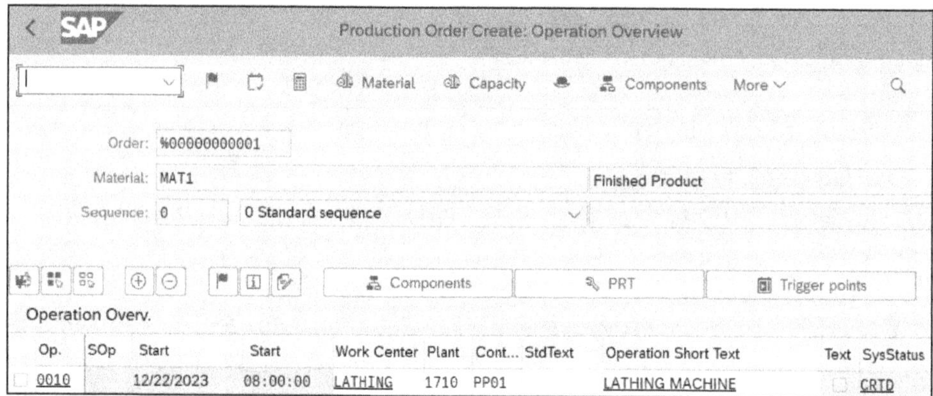

Figure 11.11 Production Order Operation Overview

  You can review and change the operations. Any changes will result in a variance because the standard price is based on the operations copied from the routing.
- **Component Overview**
  To review the components copied from the BOM to the production order, click the **Components** button shown at the top of Figure 11.11. The component overview in Figure 11.12 is displayed.

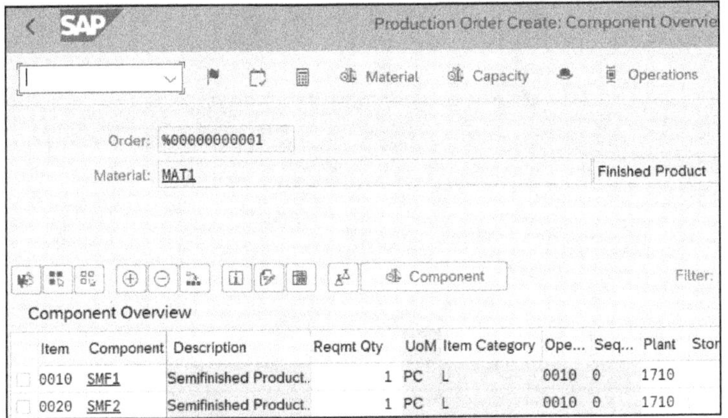

Figure 11.12 Production Order Component Overview

Review and change the components as needed. Any changes you make might result in a variance because the standard price is based on the components copied from the BOM.

Now that we've discussed how to create and release a production order, let's look at how to post actual costs.

11.2.3 Confirm Activities

We can now confirm activities for our released production order. The actual activity quantity is multiplied by the planned activity price to calculate the activity value. The production order is debited, and the production cost center is credited.

Backflushing and Automatic Goods Receipt

Goods issue during activity confirmation, known as *backflushing*, is a common practice. You can also post an automatic goods receipt during confirmation of the final activity. Companies routinely use backflushing and automatic goods receipts to reduce manual inventory transactions. Because you list all the components in the BOM, transferring this information to the production order and issuing the components automatically during activity confirmation are relatively simple. But a disadvantage of backflushing is that you won't see any component input quantity variance or resource usage variance because the actual component quantity is always the same as the plan quantity.

You post an automatic goods receipt via the final operation's control key. When you confirm the final operation, you specify the finished material for the goods receipt.

You confirm activity quantities per operation with Transaction CO11N (Time Ticket) or Transaction CO19 (Time Event) or via the menu path **Logistics • Production • Shop Floor**

Control • Confirmation • Enter • For Operation. A selection screen for a **Time Ticket** confirmation is shown in Figure 11.13.

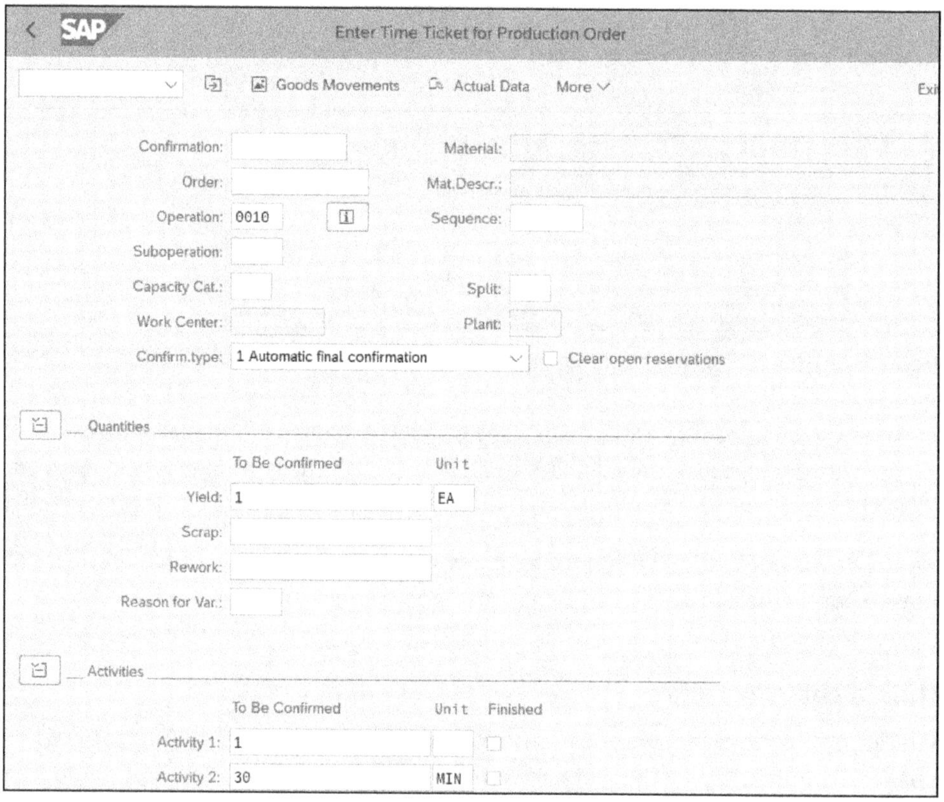

Figure 11.13 Time Ticket Activity Confirmation

You carry out a **Time Ticket** confirmation per operation. Complete the production **Order** and **Operation** fields, and then press [Enter] to default the expected **Yield** and **Scrap Quantities** and **Activities To Be Confirmed**. If you manually change the default quantities, you'll introduce a variance since the standard price is based on the quantities defaulted from the BOM and routing.

If no more confirmations are expected for the operation, select the **Finished** checkbox. If this is the final operation and you activated automatic goods receipt, this will occur during the confirmation based on the **Yield** quantity.

Click the **Goods Movements** button before saving to display plan component goods issues quantities. The component quantities are copied to the confirmation from the production order. Press [Ctrl]+[S] to save the confirmation.

Locate Production Orders with Transaction COOIS

When confirming activities, you can quickly locate a production order with Transaction COOIS or via the menu path **Logistics • Production • Shop Floor Control • Information System • Order Information System**. Complete the **Material** and **Production Plant** fields and click the **Execute** icon to display a list of production orders. Double-click the production order number to display the production order details. You display a list of process orders with Transaction COOISPI.

We'll explore production order reporting in more detail in Chapter 20.

11.2.4 Default Activities

You can specify a default yield and an activity quantity to confirm for each confirmation. For example, you can default **1 EA** of **Yield** and **30 MIN** of **Activity 2 To Be Confirmed,** as shown earlier in Figure 11.13. The order quantity and the quantity of previously confirmed operations are considered.

You configure the proposed yield and activity quantities with Transaction OPK4N or via the IMG menu path **Production • Shop Floor Control • Operations • Confirmation • Define Confirmation Parameters.** Double-click a plant and production order type combination and click the **Indiv. Entry of Operation w. Init. Screen** tab to display the screen shown in Figure 11.14.

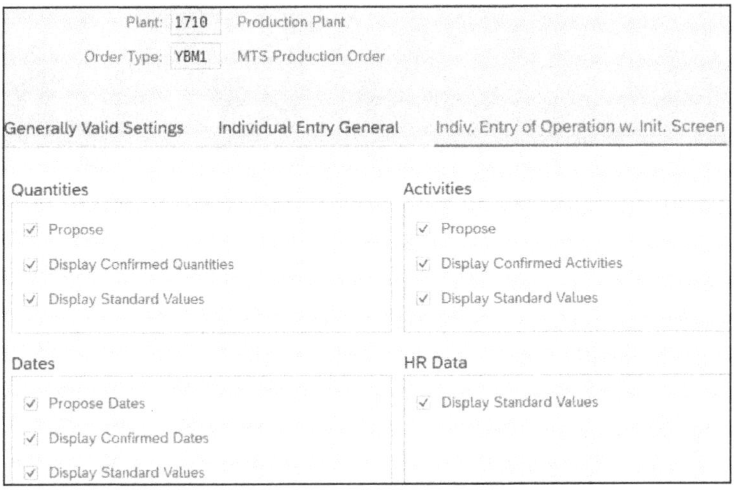

Plant: 1710 Production Plant
Order Type: YBM1 MTS Production Order

Generally Valid Settings | Individual Entry General | Indiv. Entry of Operation w. Init. Screen

Quantities
- [x] Propose
- [x] Display Confirmed Quantities
- [x] Display Standard Values

Activities
- [x] Propose
- [x] Display Confirmed Activities
- [x] Display Standard Values

Dates
- [x] Propose Dates
- [x] Display Confirmed Dates
- [x] Display Standard Values

HR Data
- [x] Display Standard Values

Figure 11.14 Propose Yield and Activity Quantities for Order Confirmation

Select the **Propose** checkbox in the **Quantities** section to specify the yield yet to be confirmed. For example, if the total yield you need to confirm is 100 pieces and the previously confirmed yield is 70 pieces, the proposed yield will be 30 pieces.

Select the **Propose** checkbox in the **Activities** section to specify the activity quantity yet to be confirmed. For example, if the total activity you need to confirm is 100 hours and the activity quantity previously confirmed is 70 hours, the proposed activity quantity to confirm will be 30 hours.

Default Activities and Activity Input Quantity Variance

If you accept the default activity quantity without change during a confirmation, you won't post an activity input quantity variance, and any resulting variance will be contained in cost center under- or overabsorption.

You may consider the option of not changing the default activities when the activity input quantity variance is small and reporting only on component input quantity variance and cost center under- or overabsorption.

11.2.5 Operation Sequence

Some companies require that operations be carried out in the sequence specified in the routing to reduce variances due to manually inserted operations. Specifying the operation sequence also ensures that quality inspection operations are carried out in the correct order. To specify the operation sequence, click the **Generally Valid Settings** tab in Figure 11.14 to display the screen shown in Figure 11.15.

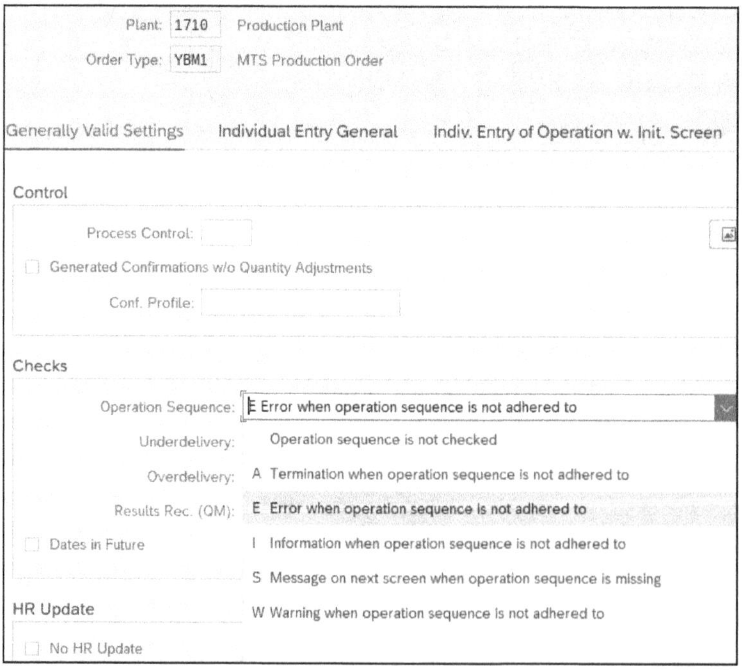

Plant: 1710 Production Plant
Order Type: YBM1 MTS Production Order
Generally Valid Settings | Individual Entry General | Indiv. Entry of Operation w. Init. Screen
Control
Process Control:
Generated Confirmations w/o Quantity Adjustments
Conf. Profile:
Checks
Operation Sequence: E Error when operation sequence is not adhered to
Underdelivery: Operation sequence is not checked
Overdelivery: A Termination when operation sequence is not adhered to
Results Rec. (QM): E Error when operation sequence is not adhered to
Dates in Future I Information when operation sequence is not adhered to
S Message on next screen when operation sequence is missing
HR Update W Warning when operation sequence is not adhered to
No HR Update

Figure 11.15 Check Operation Sequence in Production Order

Click in the **Operation Sequence** field to display a list of possible entries as shown. Choose **E Error when operation sequence is not adhered to** from the dropdown list to confirm the production order operations in the correct sequence.

Operation Sequence—Only Confirmations Per Operation

Operation sequence configuration only applies when you enter confirmations per operation, such as Transaction CO11N (Time Ticket) and Transaction CO19 (Time Event). The configuration doesn't apply when entering confirmations per production order with Transaction CO15.

11.3 Report Costs

11

Now that we've examined production order actual costs, let's report on the costs in this section. We'll discuss reporting in more detail in Chapter 20.

11.3.1 SAP GUI Transactions

You report on actual costs by displaying a detailed analysis of a production order with Transaction KKBC_ORD or by following the menu path **Accounting • Controlling • Product Cost Controlling • Cost Object Controlling • Product Cost by Order • Information System • Reports for Product Cost by Order • Detailed Reports • for Orders**. A selection screen is displayed, as shown in Figure 11.16. Use Transaction KKBC_PKO for product cost collectors.

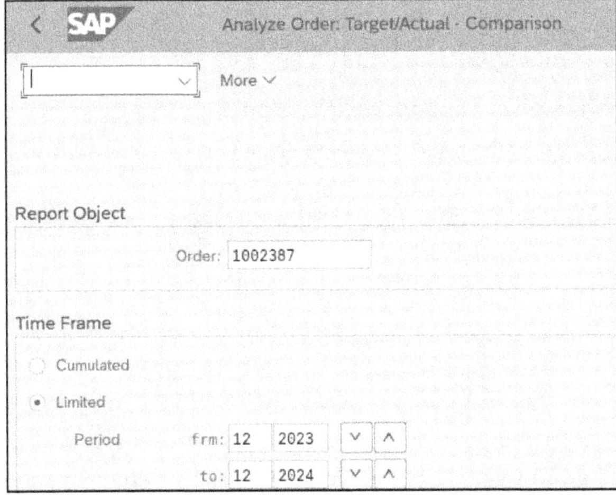

Figure 11.16 Analyze Production Order Selection

Follow these steps to run the production order analysis report:

1. Complete the **Order** field.
2. Select the **Limited** radio button in the **Time Frame** section.
3. Complete the **Period** fields.
4. Click the **Execute** button or press F8 to display the analyze order results in Figure 11.17.

Order	1002387 SMCO-Finished Material
Order Type	PI01 Process order (internal number assgnmnt)
Plant	SMCO SMCO-Manufacturing Plant
Material	SMCO-FG SMCO-Finished Material
Planned Quantity	10 EA each
Actual Quantity	10 EA each

Transaction	Cost Element	Cost Element (Text)	Origin	Σ Total Plan Costs	Σ Total Actual Costs	Σ Total Target Costs
Goods Issues	400200	RM Consumption	SMCO/SMCO-RM	1,880.00	2,068.00	1,880.00
Goods Issues				**1,880.00**	**2,068.00**	**1,880.00**
Confirmations	943004	Repairs & Maintenance	SMCO_1/REP	20.00	250.00	200.00
	943003	Power/Electricity	SMCO_1/POWER	200.00	3,142.86	200.00
	943002	Machinery	SMCO_1/MACH	3,240.00	6,480.00	3,240.00
	943005	Quality	SMCO_1/QA	36.00	500.00	360.00
	943001	Labor	SMCO_1/LAB	500.00	4,000.00	500.00
Confirmations				**3,996.00**	**14,372.86**	**4,500.00**
Overhead	941001	Frieght/Landed Costs	SMCO_2	188.00	206.80	188.00
Overhead				**188.00**	**206.80**	**188.00**
Goods Receipt	300500	Cost of Goods Mfg	SMCO/SMCO-FG	6,568.00-	6,568.00-	6,568.00-
Goods Receipt				**6,568.00-**	**6,568.00-**	**6,568.00-**
Settlement	300500	Cost of Goods Mfg		0.00	10,079.66-	0.00
Settlement				**0.00**	**10,079.66-**	**0.00**
				504.00-	**0.00**	**0.00**

Figure 11.17 Analyze Production Order Results

Double-click on any line to display actual line-item details. For example:

- Double-click on **RM Consumption** to display material documents.
- Double-click on **Labor** to display a list of activity confirmations.

Continue drilling down (double-clicking) on any line in Figure 11.17 to the original source document.

11.3.2 SAP Fiori Apps

SAP Fiori is a web-based interface that you can use in place of SAP GUI. You access SAP Fiori with a URL from your system administrator or manager, which takes you to the SAP Fiori launchpad. Alternatively, you can try Transaction /N/UI2/FLP, depending on your release. The launchpad displays a homepage with tiles. Each tile represents a business app that users can launch. The launchpad is role-based, where applications are grouped based on user roles. Only applications that the user has access to are displayed on the launchpad.

The example launchpad in Figure 11.18 demonstrates how SAP Fiori tiles are grouped based on roles—Production Accountant, in this example.

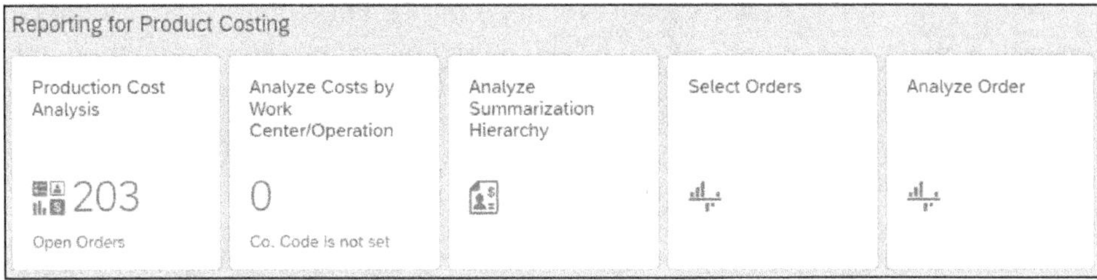

Figure 11.18 SAP Fiori Launchpad

11

You need to be assigned the Production Accountant role to access the **Production Cost Analysis** app and the **Analyze Costs by Work Center/Operation** app, which are the first two tiles on the **Reporting for Product Costing** SAP Fiori launchpad in Figure 11.18. We'll discuss these two apps in the following sections.

These two SAP Fiori apps access data directly from the Universal Journal tables ACDOCA (actual) and ACDOCP (plan). The Universal Journal contains the fields for the work center and operation, which the apps access to provide transparency for production reporting. The two SAP GUI reports, S_ALR_87013127 and KKBC_ORD, access summary tables COSS and COSP, which don't include work center and operation.

Production Cost Analysis App

With this app, you display overall and detailed production costs for production orders by cost component and cost component group. You can also compare actual, plan, and target costs, as well as analyze production variances. This app replaces SAP GUI order selection report S_ALR_87013127. A list of orders is displayed with target and actual costs, as well as variances displayed in columns, as shown in Figure 11.19.

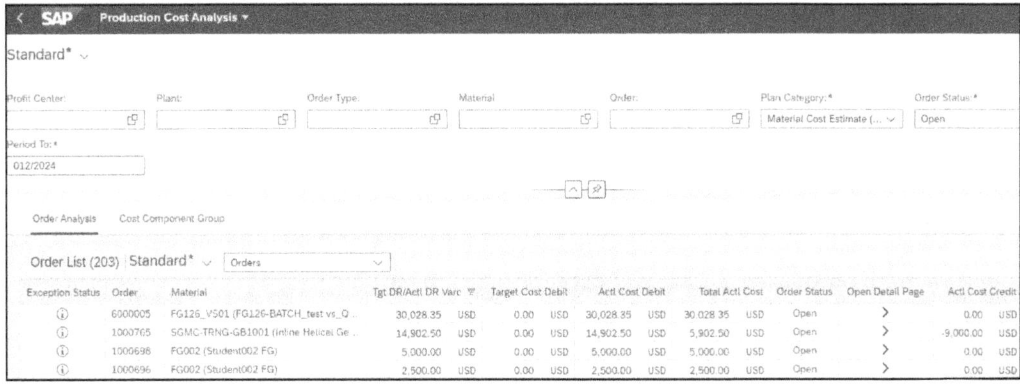

Figure 11.19 Production Cost Analysis - Summary View

The primary report filters at the top of the screen allow you to display a list of orders with costs as columns. Click in the **Plan Category** field to display the possible entries shown in Figure 11.20.

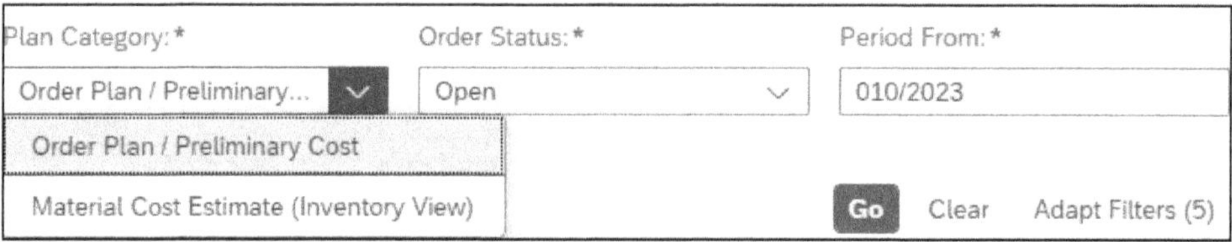

Figure 11.20 Choose How Target Costs Are Calculated

Order target costs calculated based on **Order Plan / Preliminary Cost** represent expected costs with the BOM and routing assigned to the order. These variances represent manufacturing efficiency independent of the production line chosen, which corresponds to target cost version 1. We'll discuss target cost versions in Chapter 14. After you choose a **Plan Category** in Figure 11.20, click the **Go** button to display the results as shown previously in Figure 11.19.

Target costs based on **Material Cost Estimate (Inventory View)** represent total variances or target cost version 0 in SAP GUI reports. **Order Status** determines the differences that appear in the report:

- **Open**

  This status refers to orders that are released (REL) but not fully delivered (DLV) or technically complete (TECO). Differences in these orders represent work in process (WIP).

- **Closed**

  This status refers to orders that are either DLV or TECO, and the differences represent variances.

The **Order Status** selection narrows the list to either open or closed orders. Double-click an item to display order detailed costs. This view is like the order detail report KKBC_ORD. Click on an **>** icon in the **Order Detail Page** column to view an individual **Order**. An example is shown in Figure 11.21, where you can see individual order costs by **Business Transaction**.

The variances are calculated using the following formula: **Total Target Cost** (total expected cost for the quantity produced) – **Total Actl Cost** (total actual cost for the quantity produced) = **Tgt/Act CostVar** (total cost variance).

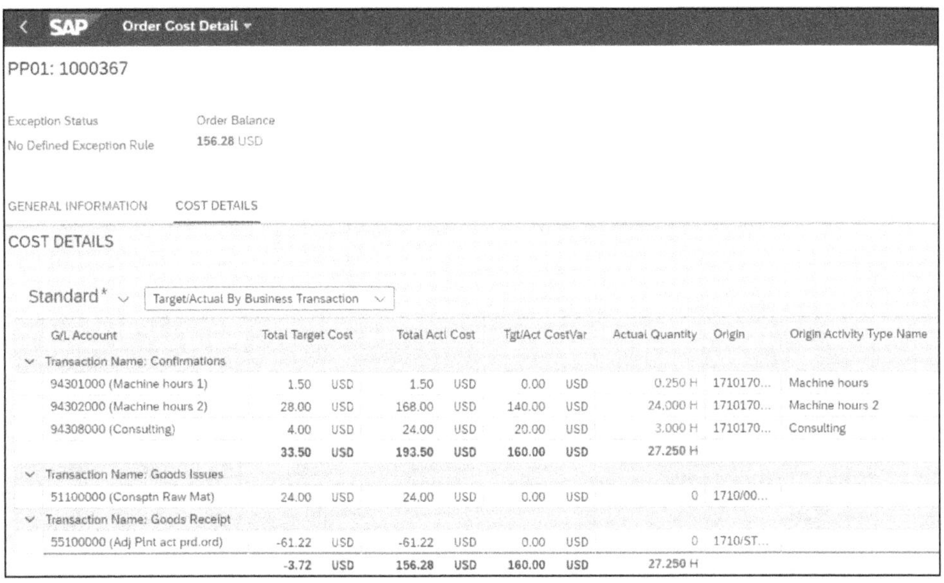

Figure 11.21 Production Cost Analysis: Order Details

Analyze Costs by Work Center/Operation App

The report contains chart and tabular (table) report sections, which are both displayed when the app is first run, as shown in Figure 11.22. You can expand each section for more information. The chart section at the top is useful for comparing the performance of work centers. Multiple measures can be assigned to the chart for analysis. This example shows a bar chart. You also have the option of selecting a pie chart, stacked bar chart, line chart, or scatter chart.

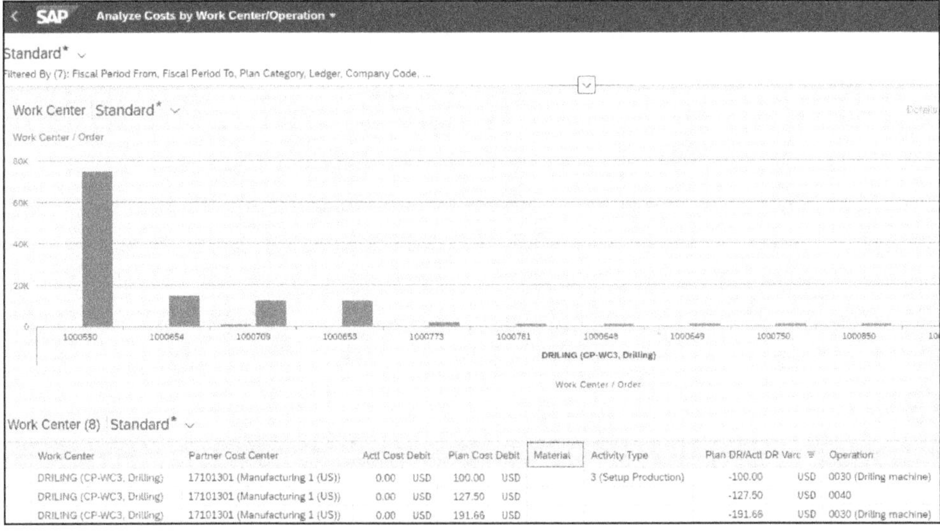

Figure 11.22 Analyze Costs by Work Center/Operation Initial Report Window

A bar chart view comparing **Act Cost Debit** with **Plan Cost Debit** by work center is shown in Figure 11.23.

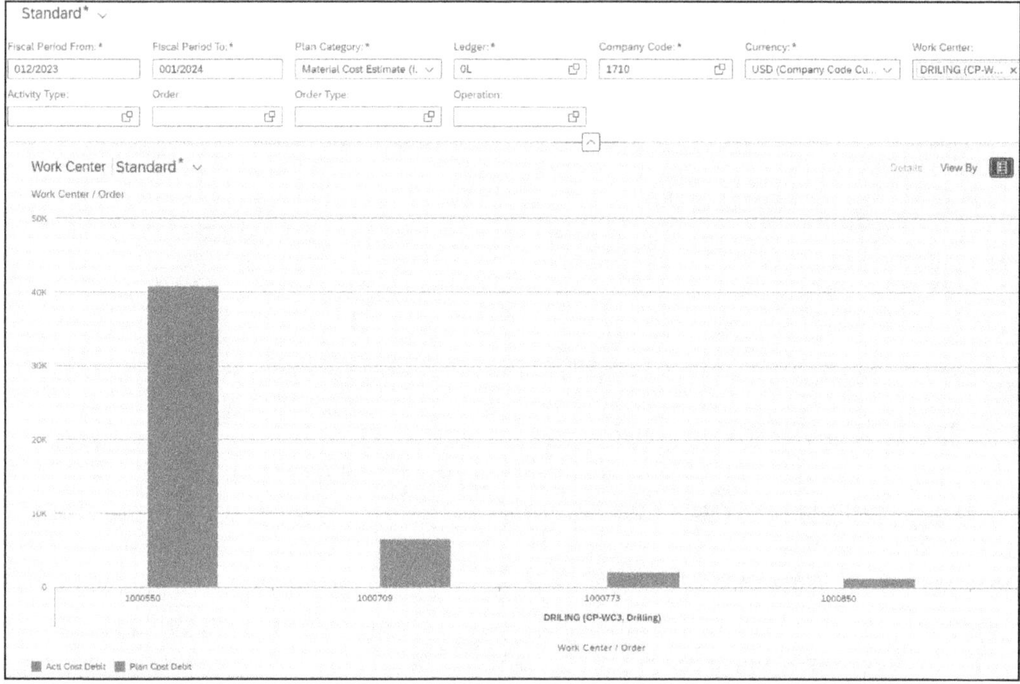

Figure 11.23 Analyze Costs by Work Center/Operation Chart View

You select the **Plan Category** when you execute the report, which allows you to use order-based targets rather than financial targets for analysis. The data is cumulative for a fiscal period, and the results of corrective actions may not be apparent until the next period. Aggregating this data by work center is useful for a production manager to control the production process.

In the **View By** section shown previously in Figure 11.22, you may also choose a tabular report, as shown in Figure 11.24.

Costs at each work center based on both activity and material goods issues are shown. Component allocations in routings and recipes determine how goods movements are assigned to work centers. These goods issues are displayed along with the activity postings at the work center level. Production managers view performance at the work center level and can use this information to find areas that need attention.

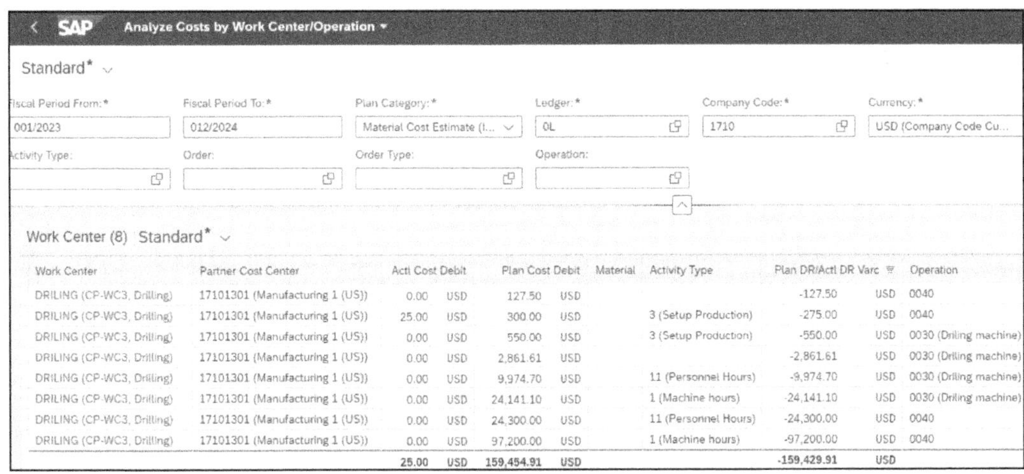

Work Center	Partner Cost Center	Actl Cost Debit		Plan Cost Debit		Material	Activity Type	Plan DR/Actl DR Varc		Operation
DRILING (CP-WC3, Drilling)	17101301 (Manufacturing 1 (US))	0.00	USD	127.50	USD			-127.50	USD	0040
DRILING (CP-WC3, Drilling)	17101301 (Manufacturing 1 (US))	25.00	USD	300.00	USD		3 (Setup Production)	-275.00	USD	0040
DRILING (CP-WC3, Drilling)	17101301 (Manufacturing 1 (US))	0.00	USD	550.00	USD		3 (Setup Production)	-550.00	USD	0030 (Driling machine)
DRILING (CP-WC3, Drilling)	17101301 (Manufacturing 1 (US))	0.00	USD	2,861.61	USD			-2,861.61	USD	0030 (Driling machine)
DRILING (CP-WC3, Drilling)	17101301 (Manufacturing 1 (US))	0.00	USD	9,974.70	USD		11 (Personnel Hours)	-9,974.70	USD	0030 (Driling machine)
DRILING (CP-WC3, Drilling)	17101301 (Manufacturing 1 (US))	0.00	USD	24,141.10	USD		1 (Machine hours)	-24,141.10	USD	0030 (Driling machine)
DRILING (CP-WC3, Drilling)	17101301 (Manufacturing 1 (US))	0.00	USD	24,300.00	USD		11 (Personnel Hours)	-24,300.00	USD	0040
DRILING (CP-WC3, Drilling)	17101301 (Manufacturing 1 (US))	0.00	USD	97,200.00	USD		1 (Machine hours)	-97,200.00	USD	0040
		25.00	USD	159,454.91	USD			-159,429.91	USD	

Figure 11.24 Analyze Costs by Work Center/Operation Tabular View

11.4 Summary

In this chapter, we discussed how controlling is completely integrated into financial accounting with the Universal Journal. All postings, including secondary postings, are in the general ledger.

We discussed how activity confirmations and assessments result in secondary postings. We carried out a production order activity confirmation, ran a report to display the actual postings, and examined configuration settings to determine default yield and activity quantities. In Chapter 12, we'll look at the overhead calculation.

Chapter 12
Overhead

Overhead calculation refers to allocating overhead from cost centers to manufacturing orders or product cost collectors with costing sheets, activity types, or templates.

In previous chapters, we created master data, configuration, and standard cost estimates and carried out preliminary and simultaneous costing. We are now ready to carry out period-end processing, which includes the following processing steps:

- Overhead
- Work in process (WIP)
- Variance calculation
- Settlement

Although other more advanced and specialized transactions are possible during period-end processing, the four explained in this and the following chapters are the most common. Using other period-end processes, you can apply the principles you learn in this section to those processes. We'll look first at the overhead configuration and the period-end process.

12.1 Configuration

During a fiscal period, you debit primary (external) costs, such as payroll and electricity, to cost centers. Some of these costs may be included as part of the planned activity rate and allocated to products from production cost centers during activity confirmations, as discussed in Chapter 11 on simultaneous costing.

Another method of allocating overhead costs to products is using period-end overhead calculations. Overhead calculations offer flexible allocation across products through costing sheet configuration, as discussed in Chapter 5. Allocating overhead with costing sheets requires an additional period-end activity but is straightforward.

12.2 Templates

Templates allow you to allocate overhead via drivers and formulas, which provides more flexibility than activity types and costing sheets. However, more flexibility means more configuration setup is required, which we'll now discuss.

Templates use functions to access data to determine the allocation proportions dynamically. They also use formulas to perform calculations on the data accessed by functions. For example, you can use the number of employees in a department to calculate personnel planned overhead costs.

You create a template with Transaction CPT1 or via IMG menu path **Controlling • Product Cost Controlling • Product Cost Planning • Basic Settings for Material Costing • Templates • Maintain Templates.** Double-click **Display Template**, choose a template, and press Enter to display the template, as shown in Figure 12.1.

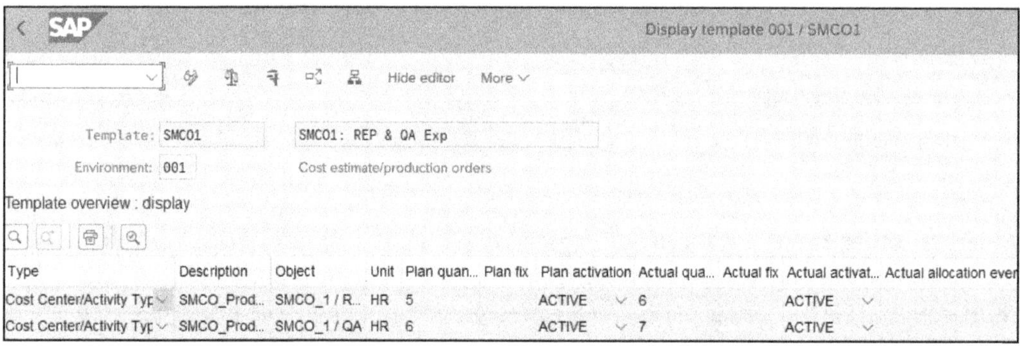

Figure 12.1 Display Template

An **Environment** determines the columns available in a template. **Environment 001 Cost estimate/production orders** is for setting up product cost planning for production orders, as you can see from the first column named **Type**, which combines **Cost Center** and **Activity Type**. To get an idea of the different environments available, click in the **Environment** field and press F4 to display the list of possible entries shown in Figure 12.2.

This displays the first 12 of 61 in this test system. You can define your own environment with Transaction CTU6 or by following the IMG menu path **Controlling • Product Cost Controlling • Product Cost Planning • Basic Settings for Material Costing • Templates • Define Environments and Function Trees.**

While we're still discussing template allocation, at period close, you run template allocation for production orders with Transaction CPTA or menu path **Accounting • Controlling • Product Cost Controlling • Cost Object Controlling • Product Cost by Order • Period-End Closing • Single Functions • Template Allocation.** Template allocation is not supported in SAP S/4HANA Cloud.

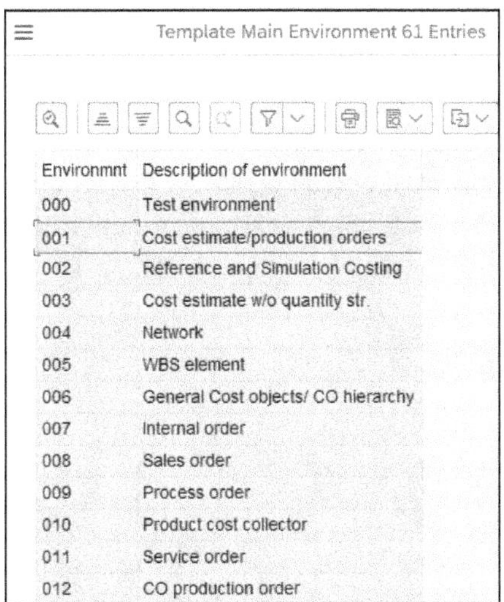

Template Main Environment 61 Entries

Environmnt	Description of environment
000	Test environment
001	Cost estimate/production orders
002	Reference and Simulation Costing
003	Cost estimate w/o quantity str.
004	Network
005	WBS element
006	General Cost objects/ CO hierarchy
007	Internal order
008	Sales order
009	Process order
010	Product cost collector
011	Service order
012	CO production order

Figure 12.2 Environment Possible Entries

12.3 Costing Sheets

Costing sheets offer flexibility in allocating overhead across products or product groups on a percentage rate basis. Configuration is required, as explained in Chapter 5. Costing sheets are the most common method for allocating overhead, and you should explore them as your first choice. If you need more flexibility, templates are also available, as we discussed in the previous section.

12.4 Activities

You also have the option of allocating overhead with activities. You can either increase the planned activity price to include overhead or create separate overhead activity types.

Advantages of using activities for overhead allocation include:

- Real-time overhead posting during activity confirmation
- Minimal configuration requirement

Disadvantages include:

- Increased activity type data setup in work centers and routings
- Increased maintenance during activity confirmations

While you should first investigate costing sheets to allocate overhead, activities might also be an option in your scenario.

Now, let's discuss period-end activities with costing sheets.

12.5 Overhead Period-End Processing

You run period-end overhead calculation for production orders with costing sheets with Transaction CO42 (individual) and Transaction CO43 (collective) or by following the menu path **Accounting • Controlling • Product Cost Controlling • Cost Object Controlling • Product Cost by Period • Period-End Closing • Single Functions: Product Cost Collector • Overhead**. A selection screen is displayed, as shown in Figure 12.3.

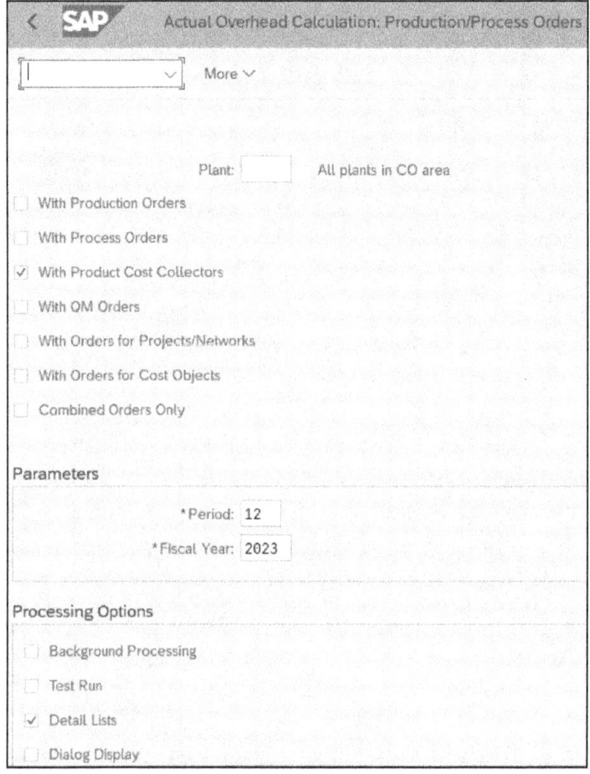

Figure 12.3 Actual Overhead Calculation Selection Screen

> **Note**
> You calculate overhead for production and process orders with Transaction KGI2 (individual) and Transaction CO43 (collective).

In this example, we'll calculate overhead for product cost collectors by selecting the **With Product Cost Collectors** checkbox. You can also select other objects depending on period-end processing time and the number of messages that require analysis following overhead calculation.

Select the **Dialog display** checkbox if you require detailed information when you analyze the overhead calculation and messages. You normally only need this level of detail when initially calculating overhead following system implementation. This detailed information is useful when analyzing overhead calculation for a single production order or product cost collector.

Complete the selection screen and click the **Execute** button or press F8 to display the screen shown in Figure 12.4.

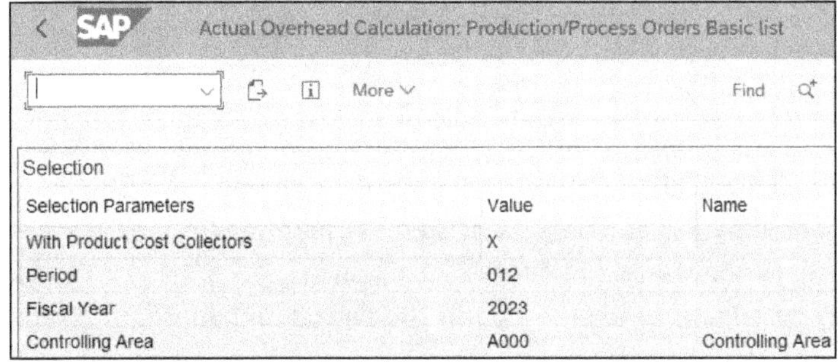

Figure 12.4 Actual Overhead Calculation Basic List

This screen provides basic information on the parameters entered in the previous selection screen, such as **Period**, **Fiscal Year**, and **Controlling Area**. Click the right-pointing arrow icon to proceed to a detailed overhead list (if you selected the **Detail Lists** checkbox in Figure 12.3), as shown in Figure 12.5.

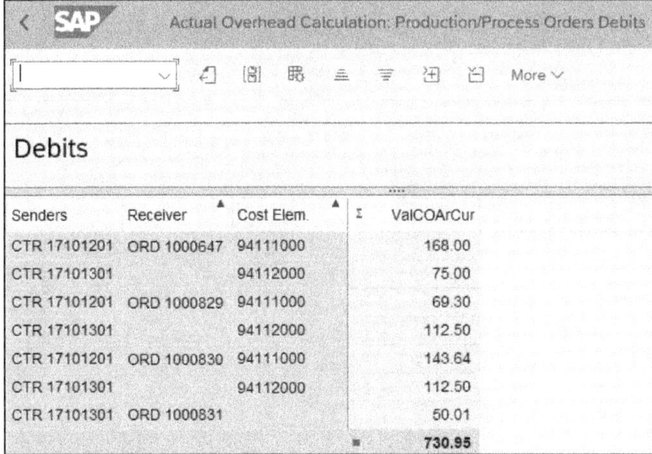

Figure 12.5 Actual Overhead Calculation Details

Let's review each column:

- The **Senders** column shows the cost center credited.
- The **Receiver** column shows the product cost collector or manufacturing order debited.

- The **Cost Elem.** column shows the secondary cost element identifying the type of overhead cost. The cost elements appear on cost center and product cost collector reports.
- The **ValCOArCur** column shows the overhead value allocated in controlling area currency. By selecting **Settings • Layout • Current** from the menu bar, you can display additional columns, such as overhead values in object (company code) currency.

12.6 Summary

In this chapter, we calculated and allocated actual overhead. In the next chapter, Chapter 13, we'll examine the next period-end processing step, calculating WIP. We'll discuss event-based overhead allocation later on in Chapter 19. If you use SAP S/4HANA Cloud, this is the primary way of allocating overhead, and the period-end approach (scope item BEI) is gradually being deprecated.

Chapter 13
Work in Process

Work in process (WIP) represents the production costs of incomplete assemblies moved temporarily to WIP general ledger accounts at period end.

In Chapter 12, we examined the configuration and period-end processing of overhead calculation. In this chapter, we'll analyze the configuration settings and period-end processing of WIP costs.

The system temporarily tracks production costs associated with manufacturing orders on profit and loss (P&L) financial statements. When you remove components from inventory to a manufacturing order, you expense them and remove them from the financial balance sheet. Production costs return to the balance sheet when you deliver assemblies and finished goods to inventory from the manufacturing order.

WIP represents the production costs of incomplete assemblies at period end. For balance sheet accounts to reflect company assets at period end, WIP costs are temporarily moved to the WIP balance sheet and P&L general ledger accounts. WIP postings are canceled during period-end processing following the delivery of associated assemblies or finished products to inventory.

There are two types of WIP valuations:

- **Target**
  WIP at target valuates based on a cost estimate.
- **Actual**
  WIP at actual is valued based on actual debits and credits to a manufacturing order or product cost collector.

We'll now discuss WIP configuration and then WIP period-end processing.

13.1 Work in Process Configuration

The WIP configuration for product cost by period and product cost by order are similar. We'll look at the configuration for product cost by period and highlight the differences along the way. In SAP S/4HANA Cloud and best practices, the configuration for WIP is delivered as part of scope item BEI (Period-End Closing—Plant).

WIP configuration for reporting and posting is a mapping exercise as follows:

1. Create results analysis keys and versions.
2. Create line IDs and map to source cost elements.
3. Map line IDs to WIP or results analysis cost elements (type 31) for reporting.
4. Create posting rules for settling WIP to general ledger accounts.

We analyze each configuration step in the following sections.

13.1.1 Define Results Analysis Keys

You define results analysis keys with Transaction OKG1 or by following the IMG menu path **Controlling • Product Cost Controlling • Cost Object Controlling • Product Cost by Period • Period-End Closing • Work in Process • Define Results Analysis Keys**. The screen shown in Figure 13.1 is displayed.

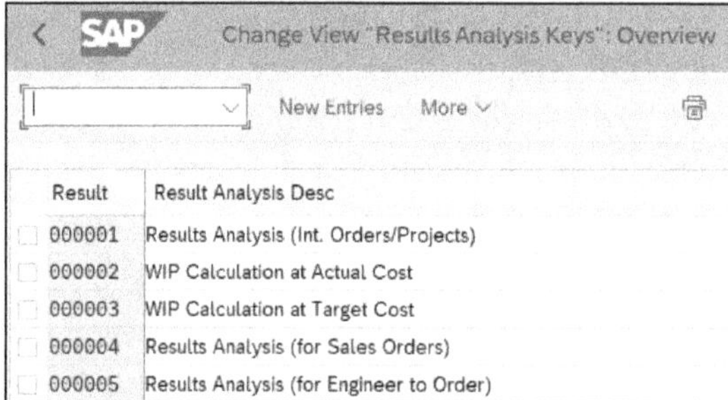

Figure 13.1 Define Results Analysis Keys

This **Overview** screen displays a list of available **Results Analysis Keys**. You can use an existing key or copy one and create your own. Each manufacturing order included in WIP processing must contain a results analysis key in the **Control** tab. Product cost collectors must contain the results analysis key in the **Data** tab for inclusion during WIP processing.

You specify the results analysis key as a default value for each order type and plant combination with Transaction OKZ3. It is then added by default when you create an order. Work in process can only be calculated for production orders that contain a results analysis key.

WIP Is a Simplified Version of Results Analysis

WIP is a simplified results analysis version because a manufacturing order does not have revenue. Results analysis calculates the cost of sales (COS) and revenue postings on a sales order item, work breakdown structure (WBS), or internal order.

13.1.2 Define Results Analysis Versions

You define results analysis versions with Transaction OKG9 or via the IMG menu path **Controlling • Product Cost Controlling • Cost Object Controlling • Product Cost by Period • Period-End Closing • Work in Process • Define Results Analysis Versions**. The screen shown in Figure 13.2 is displayed.

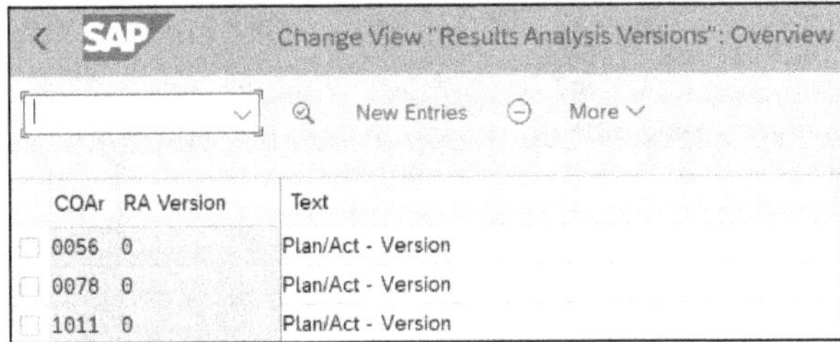

Figure 13.2 Define Results Analysis Versions

Results Analysis Versions allow you to define different methods of WIP calculation. For a simple WIP calculation, you generally capitalize all WIP costs. In advanced results analysis, you can define different amounts of WIP capitalization. Double-click a results analysis version in the **RA Version** column to display the **Details** screen shown in Figure 13.3. If you report according to multiple accounting principles, you might need to create multiple results analysis versions, as the rules for capitalization are different for each principle.

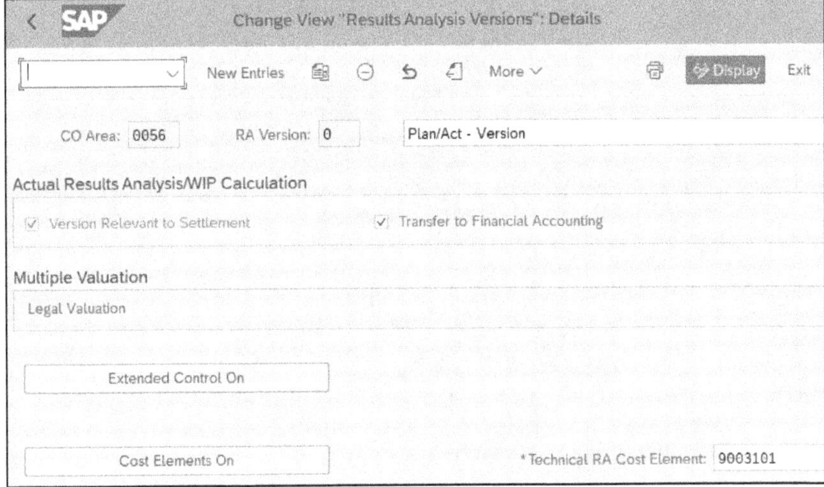

Figure 13.3 Results Analysis Version Details Screen

This screen allows you to define simplified and expert settings for an **RA Version**. We'll analyze each indicator on this screen:

- **Version Relevant to Settlement**
  The **Version Relevant to Settlement** checkbox is selected by default for results analysis version **0**. You consider a results analysis version relevant to settlement if the results analysis data of that results analysis version is settled according to the settlement rule specified in the object to be settled.
- **Transfer to Financial Accounting**
  Select the **Transfer to Financial Accounting** checkbox if the results analysis data saved under this results analysis version settles to general ledger accounting. WIP and results analysis can still be calculated without this selected, but you won't be able to reflect the result in the balance sheet.
- **Technical RA Cost Element**
  The **Technical RA Cost Element** field for results analysis data is only for internal purposes, so you don't need to create a separate cost element for each section of advanced results analysis data. You only need one technical results analysis cost element. Enter a technical results analysis cost element of category **31** (order/project results analysis).
- **Extended Control On and Cost Elements On**
  For more advanced results analysis version and WIP calculations, click the **Extended Control On** or **Cost Elements On** buttons. You'll see more fields as displayed in Figure 13.4.

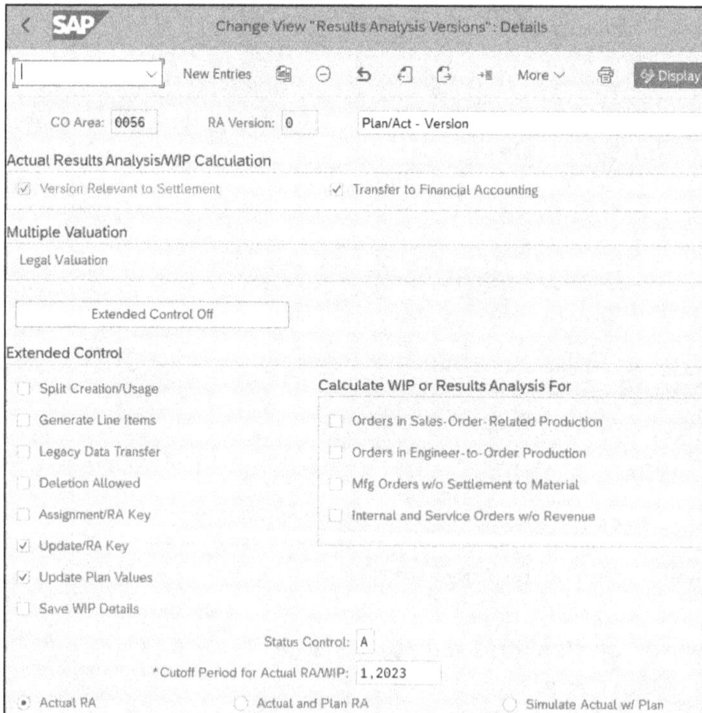

Figure 13.4 Advanced Results Analysis Version Settings

You can find out more information on these additional fields by following the IMG menu path **Controlling • Product Cost Controlling • Cost Object Controlling • Product Cost by Period • Period-End Closing • Work in Process • Define Results Analysis Versions**. The menu path is displayed in Figure 13.5.

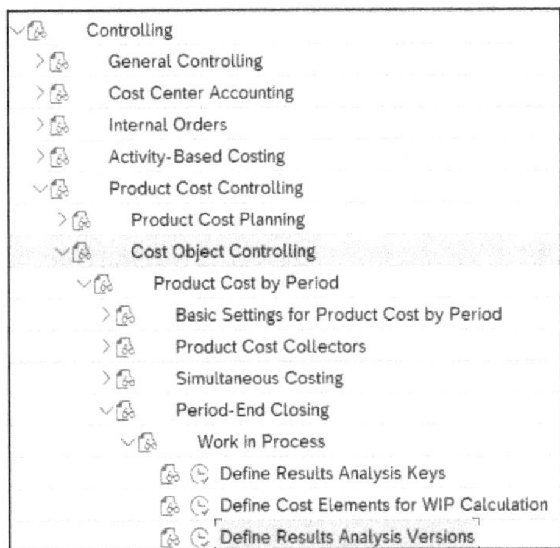

Figure 13.5 Define Results Analysis Versions Menu Path

Click the paper-and-glasses icon to the left of **Define Results Analysis Versions** to display standard documentation about setting up and leveraging advanced results analysis functionality. We'll discuss some of these fields in detail in the following sections.

Configuration Documentation Available in the IMG

You can find documentation on any configuration transaction using the same technique.

13.1.3 Define Valuation Method (Target Costs)

You define valuation methods with Transaction OKGD or by following the IMG menu path **Controlling • Product Cost Controlling • Cost Object Controlling • Product Cost by Period • Period-End Closing • Work in Process • Define Valuation Method (Target Costs)**. The screen shown in Figure 13.6 is displayed.

The first three lines in this screen are typical for WIP based on order status for actual costs:

- **Status REL** (released) results in WIP calculation based on actual costs at period end.
- **Status DLV** (delivered) or **TECO** (technically complete) results in cancellation of WIP and results analysis data at period end.

Change View "Valuation Method for Work in Process": Overview

New Entries | More | Display

CO Area	RA Version	RA Key	Status	Status Number	RA Type
0001	0	000002	REL	2	WIP Calculation on Basis of Actual Costs
0001	0	000002	DLV	3	Cancel Data of WIP Calculation and Results Ana
0001	0	000002	TECO	4	Cancel Data of WIP Calculation and Results Ana
0001	0	000003	REL	2	WIP Calculation on Basis of Target Costs

Figure 13.6 Valuation Method for WIP

This screen is relatively easy to configure. Click the **New Entries** button in Figure 13.6 to display the dialog box shown in Figure 13.7.

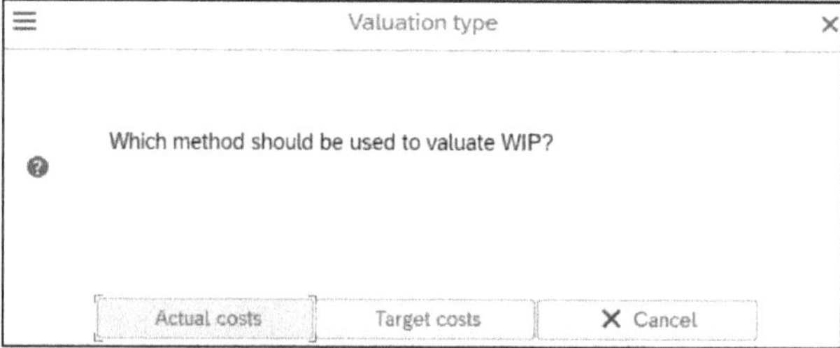

Figure 13.7 WIP Valuation Method Dialog Box

Click the **Actual costs** button to valuate WIP at actual costs or click the **Target costs** button to valuate WIP at target costs. The dialog box shown in Figure 13.8 is then displayed.

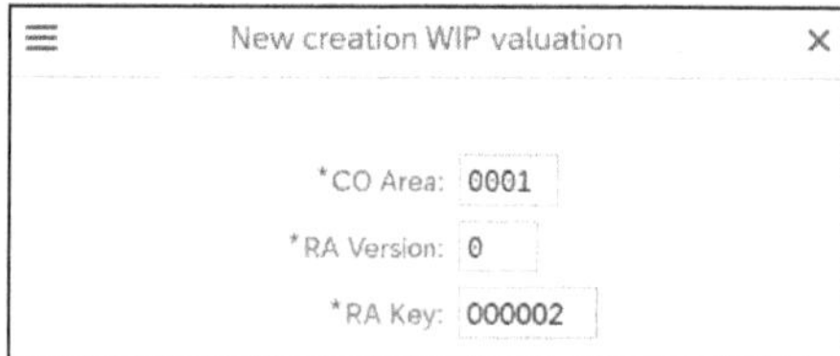

Figure 13.8 Create New WIP Valuation Entries

Complete the three fields and press [Enter], and the required entries appear automatically in the method for **WIP valuation** screen, as shown earlier in Figure 13.6. Now we'll consider the next WIP configuration transaction.

13.1.4 Define Valuation Variant for Work in Process and Scrap

The valuation variant allows you to choose the cost estimates to value scrap and WIP. If you change the routing structure after a costing run, you can't value WIP, resulting in an error message. All product cost collector costs are then posted as variances.

You can eliminate the error message by creating and releasing another standard cost estimate. Because releasing a standard cost estimate can change inventory valuation, many companies prefer to release standard cost estimates only during main costing runs in a controlled and managed environment. (We explained costing runs in detail in Chapter 9.) WIP at target eliminates the need to create and release new standard cost estimates because you base valuation on the preliminary cost estimate, which doesn't affect inventory valuation.

WIP at target allows the product cost collector preliminary cost estimate to value WIP, even if the routing structure is changed. If you don't define the valuation variant for scrap and WIP, you value scrap and WIP based on the current standard cost estimate.

You define the valuation variant for WIP and scrap (target costs) by following the IMG menu path **Controlling • Product Cost Controlling • Cost Object Controlling • Product Cost by Period • Period-End Closing • Work in Process • Define Valuation Variant for WIP and Scrap (Target Costs)**. The screen in Figure 13.9 is displayed.

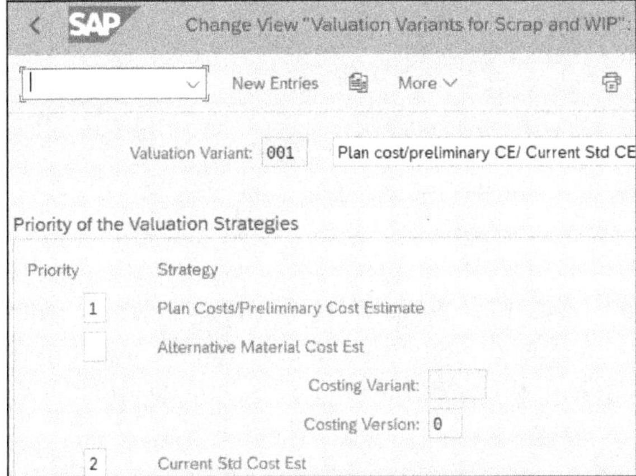

Figure 13.9 Define Valuation Variant for Scrap and WIP

The valuation strategies indicate that WIP calculation will first search for a **Preliminary Cost Estimate** and, if unsuccessful, the system will search for a **Current Std Cost Est** (current standard cost estimate). This strategy is useful because if for any reason you don't create a preliminary cost estimate, the system can still use the current standard cost estimate to value WIP, avoiding an error message that would otherwise be generated.

Product cost collectors in repetitive manufacturing use WIP at target. Repetitive manufacturing eliminates the need for production or process orders in environments with production lines and long production runs. It also reduces the work involved in production control and simplifies confirmations and goods receipt postings.

Assignment to a product cost collector occurs automatically when you create a production order in this scenario. At period end, the system moves target costs (planned costs adjusted for confirmed yield not yet delivered to inventory) from product cost collectors to WIP general ledger accounts. Variance is calculated and posted at the same time.

You can indicate that the system should link production orders to a product cost collector with configuration Transaction OPL8, where you set up order type-dependent parameters.

13.1.5 Assignment of Valuation Variant for Work in Process

You assign valuation methods by following the IMG menu path **Controlling • Product Cost Controlling • Cost Object Controlling • Product Cost by Period • Period-End Closing • Work in Process • Assignment of Valuation Variant for WIP**. The screen in Figure 13.10 is displayed.

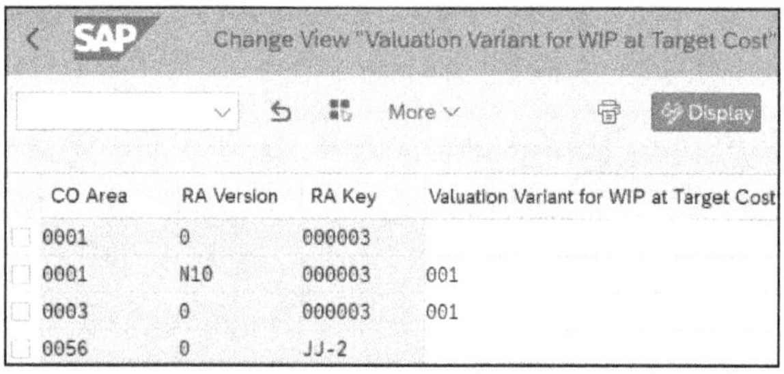

Change View "Valuation Variant for WIP at Target Cost"

CO Area	RA Version	RA Key	Valuation Variant for WIP at Target Cost
0001	0	000003	
0001	N10	000003	001
0003	0	000003	001
0056	0	JJ-2	

Figure 13.10 Assignment of Valuation Variant for WIP

In this screen, you assign a **Valuation Variant for WIP at Target Cost** to a combination of **CO Area** (controlling area), **RA Version** (results analysis version), and **RA Key** (results analysis key). The only results analysis keys that are relevant here are those used to calculate WIP at target costs.

13.1.6 Define Line IDs

You create line IDs by following the IMG menu path **Controlling • Product Cost Controlling • Cost Object Controlling • Product Cost by Period • Period-End Closing • Work in Process • Define Line IDs**. The screen shown in Figure 13.11 is displayed.

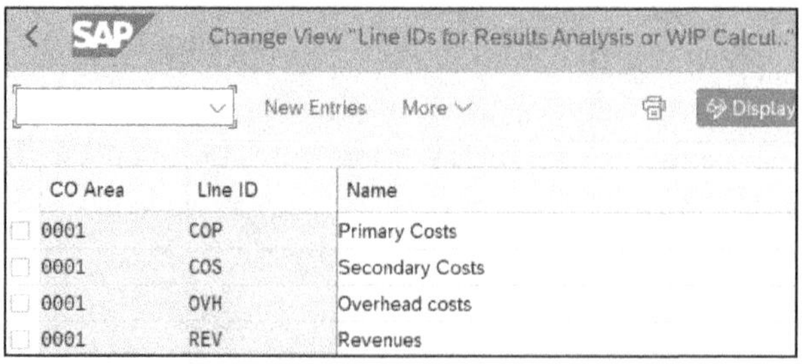

Change View "Line IDs for Results Analysis or WIP Calcul..."

CO Area	Line ID	Name
0001	COP	Primary Costs
0001	COS	Secondary Costs
0001	OVH	Overhead costs
0001	REV	Revenues

Figure 13.11 Define Line IDs for Results Analysis and WIP Calculation

Line IDs allow you to group together similar types of costs for results analysis or WIP calculation and postings. WIP calculation updates the WIP and reserves for unrealized costs on the order, grouped by line ID. We'll define posting rules that associate the data with general ledger accounts as described in Section 13.1.9.

13.1.7 Define Assignment

You assign source cost elements to line IDs with Transaction OKG5 or by following the IMG menu path **Controlling • Product Cost Controlling • Cost Object Controlling • Product Cost by Period • Period-End Closing • Work in Process • Define Assignment**. The screen shown in Figure 13.12 is displayed.

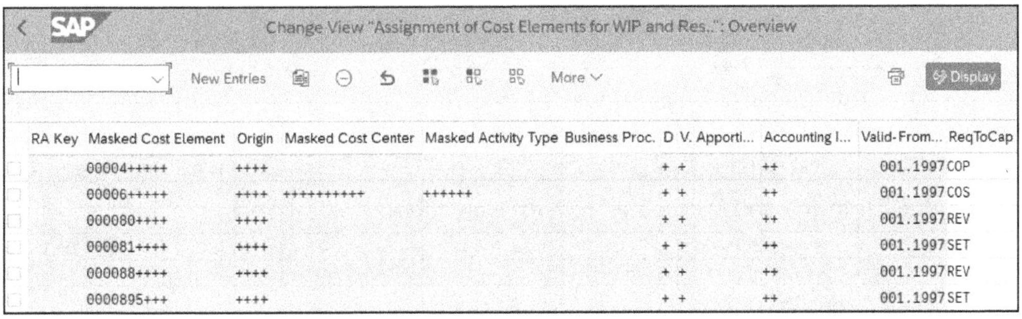

Change View "Assignment of Cost Elements for WIP and Res..": Overview

RA Key	Masked Cost Element	Origin	Masked Cost Center	Masked Activity Type	Business Proc.	D	V. Apporti...	Accounting I...	Valid-From...	ReqToCap
	00004+++++	++++				+	+	++	001.1997	COP
	00006+++++	++++	++++++++++	++++++		+	+	++	001.1997	COS
	000080++++	++++				+	+	++	001.1997	REV
	000081++++	++++				+	+	++	001.1997	SET
	000088++++	++++				+	+	++	001.1997	REV
	0000895+++	++++				+	+	++	001.1997	SET

Figure 13.12 Assignment of Cost Elements for WIP and Results Analysis

Completing this screen is typically an exercise in mapping source cost elements in the **Masked Cost Element** column to line IDs in the **ReqToCap** (requirement to capitalize) column:

- **Masked Cost Element**
  Source cost elements typically identify costs that debit objects, such as manufacturing orders and product cost collectors. You can enter individual source cost elements or masked cost elements, which is a method of grouping cost elements. For example, in the first line in Figure 13.12, **Masked Cost Element 00004+++++** represents all cost elements that start with the numbers 00004.
- **Masked Cost Center**
  When more detail is required in order detail reports such as KKBC_ORD, KKBC_PKO, or the Production Cost Analysis app (SAP Fiori ID F1780), you can also enter individual cost centers in the **Masked Cost Center** column. This level of detail isn't normally required, but you can enter individual entries in any of the masked columns in Figure 13.12.
- **ReqToCap (requirement to capitalize)**
  Usually, you must capitalize the total WIP and enter the line IDs corresponding to the cost elements in the **ReqToCap** column. In complex projects, you may not be

legally able to capitalize certain costs, and you must make entries in columns to the right in Figure 13.12, as shown in Figure 13.13.

In more complicated cases, you can place these line IDs in the **OptToCap** (option to capitalize) and **CannotBeCap** (cannot be capitalized) columns together with the relevant percentages in the last two columns in the screen shown in Figure 13.13. You always update WIPs and reserves for unrealized costs with reference to a line ID, as this provides the link to the cost element used for the update.

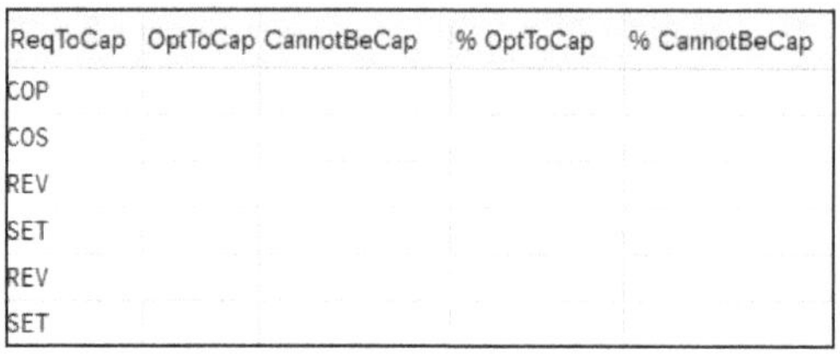

ReqToCap	OptToCap	CannotBeCap	% OptToCap	% CannotBeCap
COP				
COS				
REV				
SET				
REV				
SET				

Figure 13.13 Option to Capitalize WIP Columns

Now, we'll consider the next WIP configuration transaction, which involves assigning line IDs to WIP and results analysis cost elements.

13.1.8 Define Update

You assign line IDs to WIP and results analysis cost elements with Transaction OKGA or by following the IMG menu path **Controlling • Product Cost Controlling • Cost Object Controlling • Product Cost by Period • Period-End Closing • Work in Process • Define Update**. The screen in Figure 13.14 is displayed.

You map line IDs to WIP cost elements for reporting purposes. Let's discuss the fields in this screen:

- **LID (line ID)**
  You assign each line ID in the **LID** column to a category shown in the next column to the right. Click a category and press F4 to display the list of categories shown in Figure 13.14. Certain categories, such as **N Costs Not to Be Included,** don't require assigning WIP cost elements, and fields are not available to enter cost elements.
- **Creation**
  You enter **WIP** cost elements in the **Creation** column for categories such as **K** (**Costs**).
- **Usage**
  If you selected **Split Creation/Usage** in the **Extended Control** section of the advanced results analysis version settings shown previously in Figure 13.4, you must enter a usage WIP cost element (type 31) in Figure 13.14, which provides detailed WIP reporting.

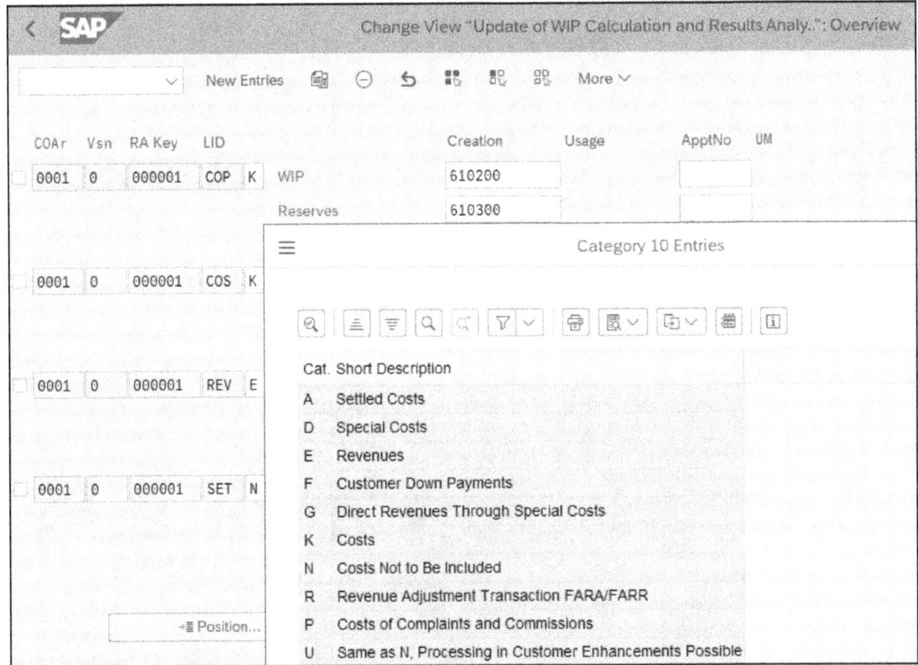

Figure 13.14 Update of WIP Calculation and Results Analysis

Assign WIP Cost Elements as Cost Components

You can assign WIP cost elements to cost components described in Chapter 6 and report WIP as cost components. If you don't assign the WIP cost elements to cost components, the costs appear in an unassigned category in standard reports.

Assigning line IDs to WIP and results analysis cost elements is for reporting purposes only. The next configuration step defines which general ledger accounts you post to during settlement.

13.1.9 Define Posting Rules for Settling Work in Process

You define posting rules for settling WIP to general ledger accounts with Transaction OKG8 or by following the IMG menu path **Controlling • Product Cost Controlling • Cost Object Controlling • Product Cost by Period • Period-End Closing • Work in Process • Define Posting Rules for Settling Work in Process**. The screen shown in Figure 13.15 is displayed.

In this screen, you define the WIP general ledger accounts for settlement in the **P&L Acct** and **BalSheetAcct** columns. You can enter six main results analysis categories in the **RA category** column:

- **WIPR**
  WIP with a requirement to capitalize costs
- **WIPO**
  WIP with an option to capitalize costs
- **WIPP**
  WIP with a prohibition to capitalize costs
- **RUCR**
  Reserves for unrealized costs (group must be capitalized)
- **RUCO**
  Reserves for unrealized costs (group can be capitalized)
- **RUCP**
  Reserves for unrealized costs (group cannot be capitalized)

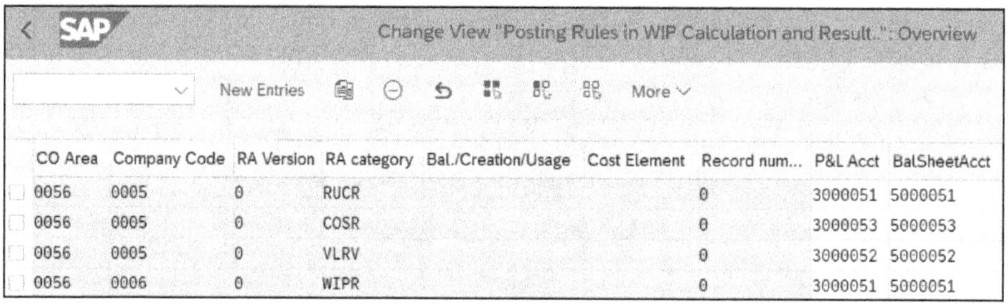

Change View "Posting Rules in WIP Calculation and Result..": Overview

New Entries More

CO Area	Company Code	RA Version	RA category	Bal./Creation/Usage	Cost Element	Record num...	P&L Acct	BalSheetAcct
0056	0005	0	RUCR			0	3000051	5000051
0056	0005	0	COSR			0	3000053	5000053
0056	0005	0	VLRV			0	3000052	5000052
0056	0006	0	WIPR			0	3000051	5000051

Figure 13.15 Posting Rules for WIP and Results Analysis

You can define separate reserve accounts if production credit exceeds costs (more costs are expected).

You Can Calculate WIP as Often as Required

You can calculate WIP as often as required during a period. Posting to WIP general ledger accounts only occurs during settlement at period end for period-end processing.

Now that we've discussed the WIP configuration for reporting and posting, we'll look at the results analysis configuration.

13.2 Results Analysis Configuration

You calculate WIP on make-to-stock manufacturing orders and product cost collectors. An additional step is required to settle sales order costs and nonvaluated customer stock. Results analysis calculates COS and revenue on sales order line items.

Accounting principles require you to match the recognition of COS and revenue. Results analysis provides various methods for calculating COS and matching revenue.

You define valuation methods for results analysis with Transaction OKG3 or by following the IMG menu path **Controlling • Product Cost Controlling • Cost Object Controlling • Product Cost by Sales Order • Period-End Closing • Results Analysis • Define Valuation Methods for Results Analysis**. The screen shown in Figure 13.16 is displayed.

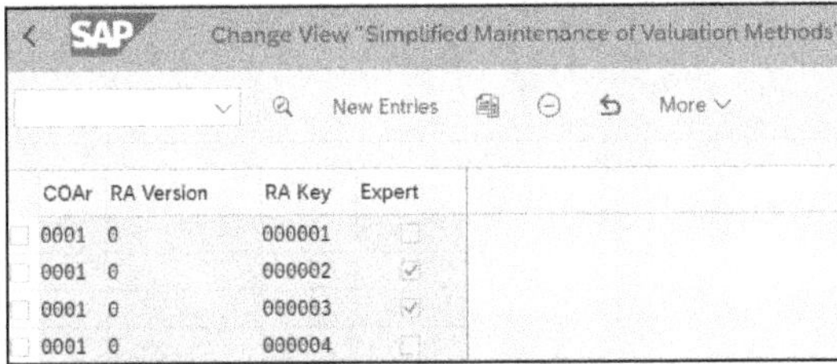

Figure 13.16 Simplified Maintenance List of Valuation Methods

You can configure results analysis in either simplified or expert mode as indicated by the checkbox in the **Expert** column:

- In simplified mode, you can choose from a list of standard valuation methods.
- In expert mode, you can modify any of the standard valuation methods to suit any special requirements you may have.

Double-click the first line in Figure 13.16 to display the screen shown in Figure 13.17.

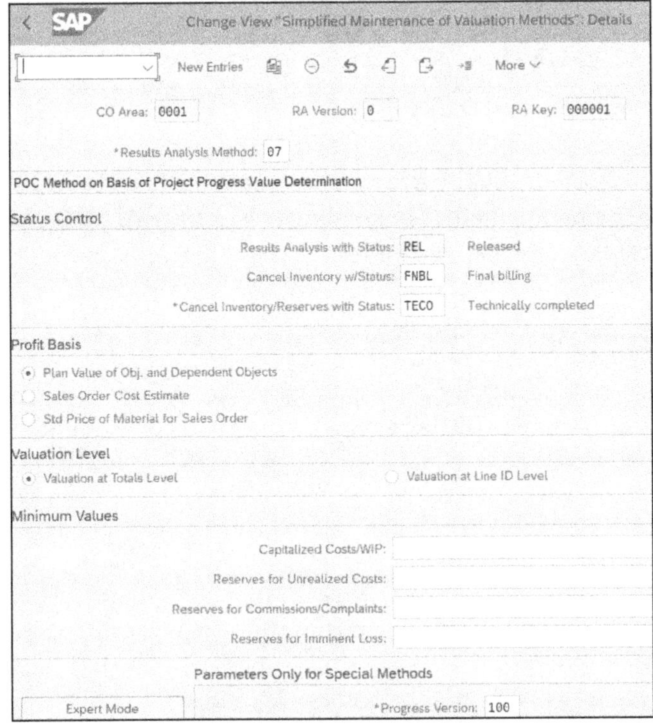

Figure 13.17 Simplified Maintenance of Valuation Methods

In this screen, you choose a **Results Analysis Method**, and you can change the sales order line item **Status Control**, which determines when results analysis is calculated and when inventory and reserves are canceled. You can display a list of standard results analysis methods by clicking the field and pressing F4. The screen in Figure 13.18 is displayed.

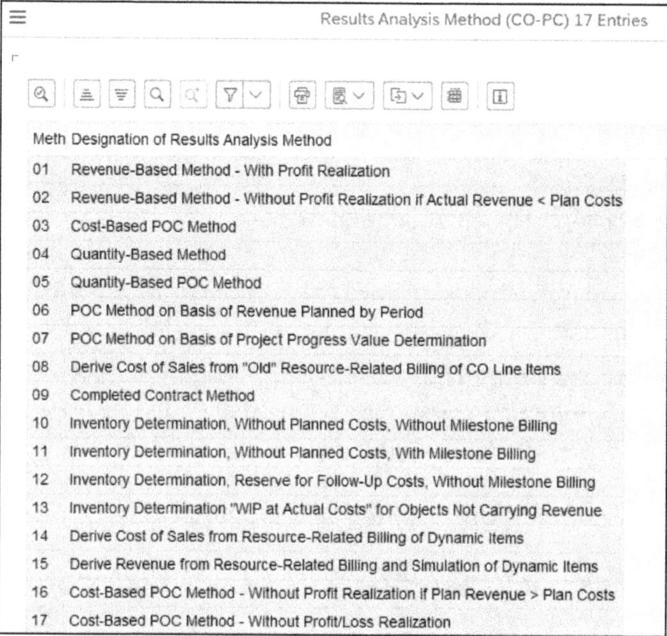

Figure 13.18 Results Analysis Method Possible Entries

Select the **Results Analysis Method** you require from the descriptions shown in Figure 13.18. If none of the standard valuation methods meets your requirements, click the **Expert Mode** button in Figure 13.17 to display an information dialog box as shown in Figure 13.19.

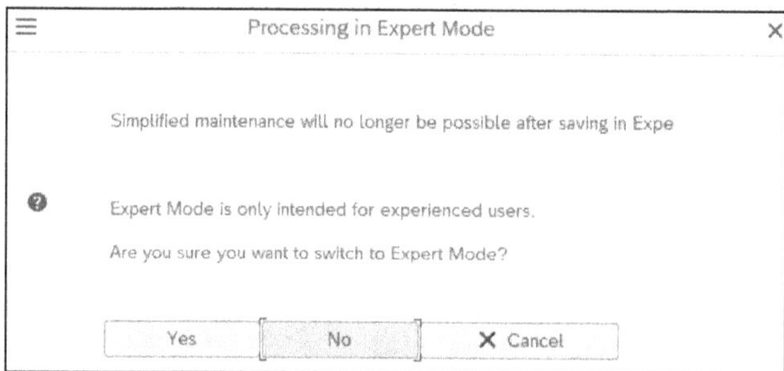

Figure 13.19 Expert Mode Warning Message

This message informs you that expert mode is complex and only for experienced users. You also cannot revert to simplified maintenance after saving in expert mode. If you agree to proceed, click the **Yes** button to display the screen shown in Figure 13.20.

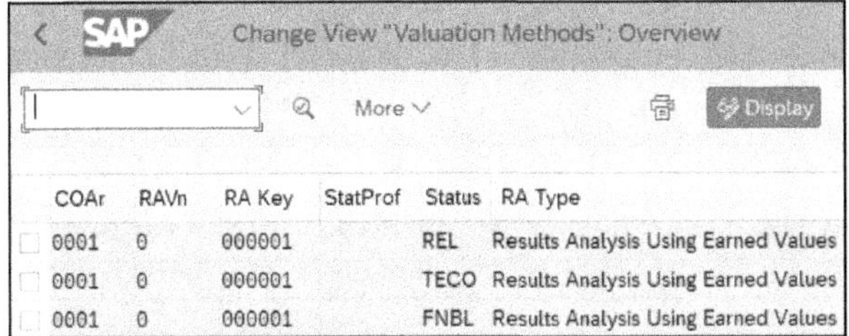

COAr	RAVn	RA Key	StatProf	Status	RA Type
0001	0	000001		REL	Results Analysis Using Earned Values
0001	0	000001		TECO	Results Analysis Using Earned Values
0001	0	000001		FNBL	Results Analysis Using Earned Values

Figure 13.20 Valuation Methods Overview

The three lines correspond to the three **Status Control** entries shown earlier in Figure 13.17 for results analysis method **07**. The configuration behind these entries determines when results analysis is calculated or when inventory and reserves are canceled depending on sales order line item **Status**. Double-click the first line in this example to display the expert mode screen shown in Figure 13.21.

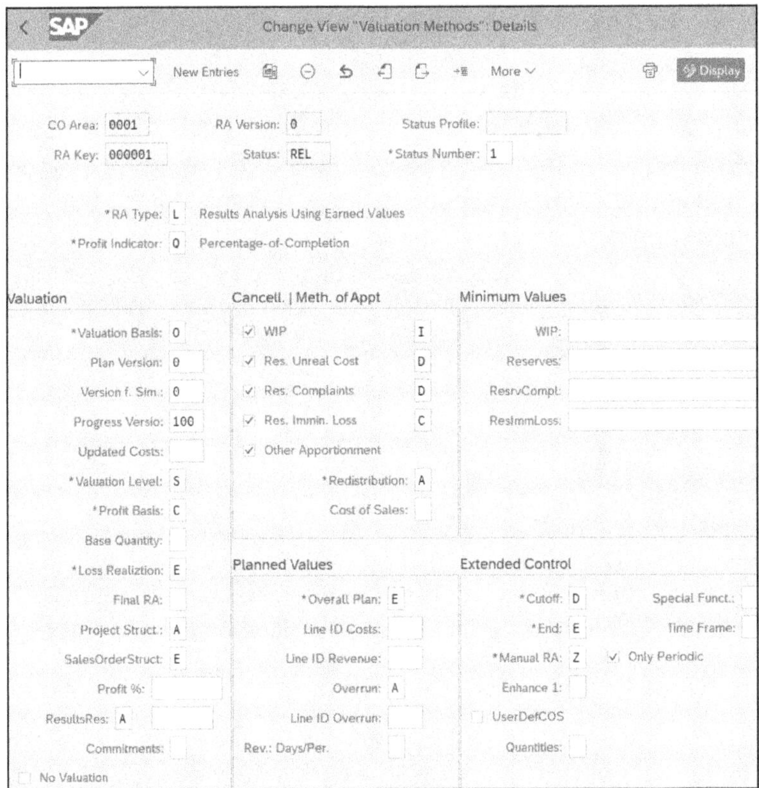

Figure 13.21 Valuation Methods Expert Mode

In this screen, you configure how you calculate results analysis for a sales order line item with a **Status** of **REL** (released). The standard settings on this screen represent how you calculate results analysis by the simplified settings shown earlier in Figure 13.17.

Keep in mind that as you change the settings, you move away from the standard settings. This expert mode screen has many specialized fields and checkboxes. If you aren't experienced with configuring results analysis, you may need to test different settings until they meet your requirements.

Now that we've configured WIP and results analysis, we'll examine the WIP period-end process.

13.3 SAP S/4HANA Cloud Configuration (Scope Item BEI)

In SAP S/4HANA Cloud, scope item BEI is gradually being deprecated, as it is not compatible with universal parallel accounting, where WIP is calculated with every goods movement and confirmation (see Chapter 19). You will, however, still find the configuration if you use best practices in SAP S/4HANA Private Cloud or on-premise SAP S/4HANA without universal parallel accounting. The transactions and menu paths to reach these settings are the same as we described earlier.

13.3.1 Delivered Results Analysis Keys

The best-practice systems include two results analysis keys, as shown in Figure 13.22, where **MBMF01** is intended for use with production orders, WIP is calculated based on actual costs and canceled when the order is complete, and **MBMF03** is intended for use with product cost collectors where WIP is calculated at target costs using the cost estimate.

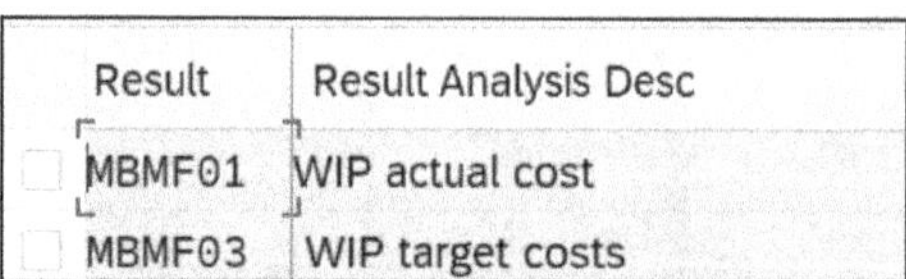

Result	Result Analysis Desc
MBMF01	WIP actual cost
MBMF03	WIP target costs

Figure 13.22 Results Analysis Keys for Best Practices

The results analysis keys are linked with the relevant statuses, as shown in Figure 13.23. The status with the largest status number determines the valuation method (creation or cancellation of WIP).

CO Area	RA Version	RA Key	Status	Status Number	RA Type
A000		MBMF01	REL	2	WIP Calculation on Basis of Actual Costs
A000	0	MBMF01	DLV	3	Cancel Data of WIP Calculation and Results Ana
A000	0	MBMF01	PREL	1	WIP Calculation on Basis of Actual Costs
A000	0	MBMF01	TECO	4	Cancel Data of WIP Calculation and Results Ana
A000	0	MBMF03	REL	2	WIP Calculation on Basis of Target Costs
A000	0	MBMF03	DLV	3	Cancel Data of WIP Calculation and Results Ana
A000	0	MBMF03	PREL	1	WIP Calculation on Basis of Target Costs
A000	0	MBMF03	TECO	4	Cancel Data of WIP Calculation and Results Ana

Figure 13.23 Valuation Methods for the Creation and Cancellation of Work in Process (WIP)

13.3.2 Define Assignment

The work in process is grouped according to the categories **MAT** (material costs), **LBR** (labor costs), and **OVH** (overhead costs), which are represented by the line IDs shown in Figure 13.24. Work in process is reduced for any costs assigned to the category **STL** (settled costs). This includes the value of the delivery to the finished inventory and settlement on completion.

13

CO Area	Line ID	Name
A000	LBR	Labor Costs
A000	MAT	Material Costs
A000	OVH	Overhead Costs
A000	REV	Revenues
A000	STL	Settled Costs

Figure 13.24 Line IDs to Classify Work in Process

The cost element assignment shown in Figure 13.25 determines which accounts are considered part of each category. Here we see an example of the assignment of costs to line ID **MAT**. To ensure that the WIP is correctly reduced when the finished good is delivered to stock and the order settled, these cost elements are assigned to line ID **STL**. This configuration is based on the standard chart of accounts delivered as part of the best-practice settings.

WIP is assigned initially to the accounts shown in Figure 13.26, where we see the WIP accounts for labor (**LBR**), material (**MAT**), and overhead (**OVR**). These cost elements will be shown in reporting once you have calculated WIP. If some of the goods movements or confirmations have not been captured when WIP is calculated, the system will calculate reserves for unrealized costs for the difference between the value of the goods delivered to stock and the values accumulated on the production order until that point.

CO Area	RA Ve...	RA Key	Masked Cost E...	Origin	Masked Cost ...	Masked...	Bu...	D. V. Account...	Valid-From ...	ReqToCap
A000	0	MBMF01	0051100000	++++	++++++++++	++++++		+ + ++	001.2007	MAT
A000	0	MBMF01	0051500000	++++	++++++++++	++++++		+ + ++	001.2007	MAT
A000	0	MBMF01	0051600000	++++	++++++++++	++++++		+ + ++	001.2007	MAT
A000	0	MBMF01	0051700000	++++	++++++++++	++++++		+ + ++	001.2007	MAT
A000	0	MBMF01	0051950000	++++	++++++++++	++++++		+ + ++	001.2007	MAT
A000	0	MBMF01	0052001000	++++				+ + ++	001.2007	STL
A000	0	MBMF01	0052021000	++++				+ + ++	001.2007	STL
A000	0	MBMF01	0052501000	++++				+ + ++	001.2007	STL
A000	0	MBMF01	0052511000	++++				+ + ++	001.2007	STL
A000	0	MBMF01	0054051000	++++				+ + ++	001.2007	STL
A000	0	MBMF01	0054053000	++++				+ + ++	001.2007	STL

Figure 13.25 Assignment of Cost Elements to Line IDs MAT and STL

COAr	Vsn	RA Key	LID			Creation	Usage
A000		MBMF01	LBR	K	WIP	93113000	
					Reserves	93114000	
A000	0	MBMF01	MAT	K	WIP	93111100	
					Reserves	93112000	
A000	0	MBMF01	OVH	K	WIP	93115000	
					Reserves	93116000	

Figure 13.26 Update of Results Analysis Cost Elements for WIP and Reserves

When you run settlement, the system will create journal entries for the work in process (or reserves for unrealized costs) under the accounts shown in Figure 13.27. You can run settlement once per period.

CO Area	Compan...	RA Version	RA category	Bal./Crea...	Cost Element	Record n...	P&L Acct	BalSheetAcct
A000	1010	0	WIPR			0	54200000	13200000
A000	1010	0	WIPO			0	54200000	13200000
A000	1010	0	RUCR			0	54040000	24[illegible]98600

Figure 13.27 Postings Rules for Work in Process and Reserves for Unrealized Costs

13.4 Work in Process Period-End

We'll discuss WIP calculation for product cost collectors and manufacturing orders in detail in this section.

13.4.1 Product Cost by Period

You value WIP at target cost for product cost by period and product cost collectors. You valuate operation quantities confirmed for manufacturing orders at the target cost of the operation, minus scrap and goods receipt quantities. You don't base WIP at target value on actual costs. Instead, you base WIP at target on what value WIP should be according to a cost estimate. You can specify which cost estimate to calculate target costs in the valuation variant for scrap and WIP, as discussed in Section 13.1.4. For product cost collectors, SAP recommends calculating target costs based on the product cost collector preliminary cost estimate, which we discussed in Chapter 9.

One of the main advantages of WIP at target is that variance and WIP post simultaneously. If production orders remain open for multiple periods, variance reconciliation is easier with WIP at target.

Another advantage of WIP at target is that variance analysis is based on a material or product, usually a key reporting requirement. Comparing variances of different products allows you to analyze which product you make more efficiently, improving product profitability. This analysis is usually more beneficial to management than analyzing variance per production order.

You run period-end WIP calculation with Transaction KKAS (individual) and Transaction KKAO (collective) or by following the menu path **Accounting • Controlling • Product Cost Controlling • Cost Object Controlling • Product Cost by Period • Period-End Closing • Single Functions: Product Cost Collector • Work in Process**. A selection screen is displayed, as shown in Figure 13.28.

WIP Calculation for Production Orders

You calculate WIP for production and process orders with Transaction KKAX (individual) and Transaction KKAO (collective). You can also run results analysis calculations with Transaction KKA3 (individual) and Transaction KKAK (collective).

To carry out collective WIP calculation, complete the following steps:

1. Complete the **Plant** field or leave the **Plant** field blank to calculate WIP for **All Plants**.
2. Choose which objects to include in the calculation with the three checkboxes below the **Plant** field.
3. Complete the **WIP to Period** and **Fiscal Year** fields.
4. Complete the **RA Version** field with RA version 0 (typically).
5. When you have many production orders, you can select the **Background Processing** checkbox.
6. You can select the **Test Run** checkbox to review messages before calculating the WIP.
7. Select the **Log Information Messages** checkbox to review the information messages log.

8. Select the **Output Object List** checkbox to output a list with WIP per order.
9. Select the **Display Orders with Errors** checkbox to display orders with errors.
10. You can **Hide Orders for Which WIP = 0** by selecting this checkbox.
11. Press the **Execute** button or press F8 to display an output list as shown in Figure 13.29.

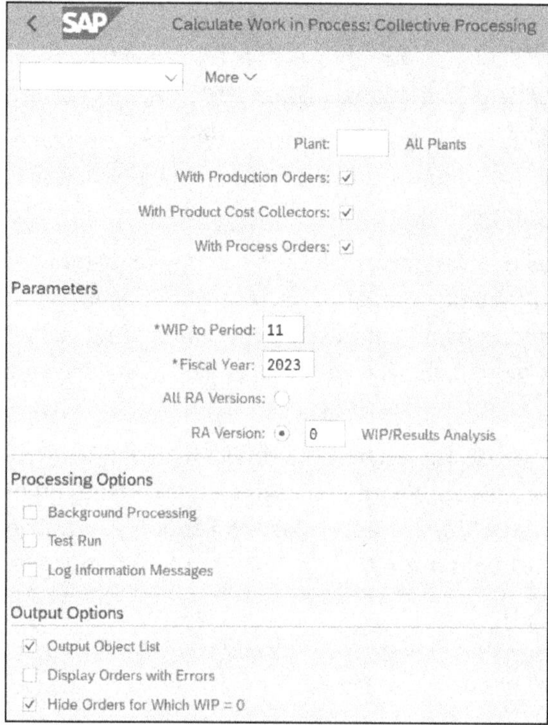

Figure 13.28 WIP Selection Screen

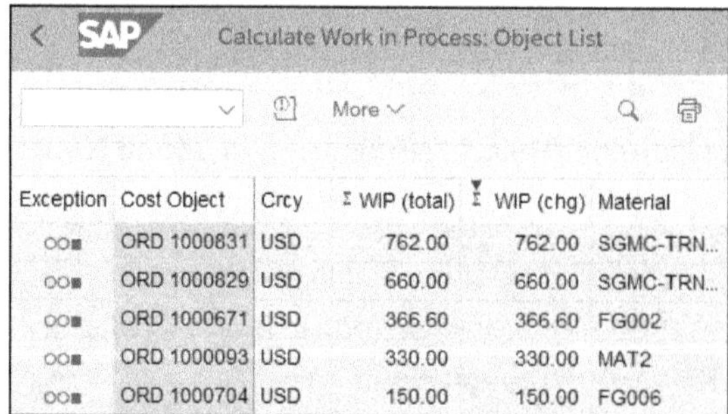

Exception	Cost Object	Crcy	WIP (total)	WIP (chg)	Material
OO■	ORD 1000831	USD	762.00	762.00	SGMC-TRN...
OO■	ORD 1000829	USD	660.00	660.00	SGMC-TRN...
OO■	ORD 1000671	USD	366.60	366.60	FG002
OO■	ORD 1000093	USD	330.00	330.00	MAT2
OO■	ORD 1000704	USD	150.00	150.00	FG006

Figure 13.29 Calculate WIP Object List

You display messages by clicking the exclamation point icon to the left of **More**. You should analyze all messages and take corrective action where necessary. Identify specific materials with several messages. It can be easier to calculate WIP for the individual orders or materials separately with Transaction KKAS and then analyze messages for a specific order or material.

In Figure 13.29, the **WIP (chg)** column is sorted in descending order. The sort order is indicated by the small red inverted triangle just to the left of the **WIP (chg)** column header text. Descending order provides visibility of product cost collectors or production orders with the largest WIP accumulated during the period of WIP calculation. Analyze the product cost collectors with the six largest positive and negative values of change in WIP during each period. You'll find that the number and severity of messages progressively reduces at each period end. You can use the same technique on the **WIP (total)** column if you find this helpful.

How to Clear Product Cost Collector WIP

You cannot cancel WIP by changing the status of underlying manufacturing orders to technically complete (**TECO**). This status prevents further processing or cost posting to the product cost collector through the manufacturing order.

However, the status of **TECO** doesn't remove existing WIP, which remains associated with the product cost collector. Setting the manufacturing order deletion flag, corresponding with the status **DLFL**, will cancel the existing WIP. You can set the deletion flag while viewing a production order in change mode by selecting **Functions • Deletion Flag • Activate** from the menu bar. The deletion flag can be revoked by selecting **Functions • Deletion Flag • Revoke** from the menu bar.

No financial postings occur during WIP calculation. You can run the WIP transaction as often as you like, and you can carry out analyses and fixes progressively during a period. Financial postings only occur during settlement, which you normally carry out as the last step at the period end.

13.4.2 Product Cost by Order

The product cost-by-order scenario involves working with manufacturing orders. We'll now examine the differences between the WIP calculation for manufacturing orders and product cost collectors.

You can value WIP at actual cost in product cost by order. All order debits are WIP until valuated goods receipt into inventory occurs. At period end, the actual balance of incomplete manufacturing orders not fully delivered to inventory is determined during WIP calculation. During settlement, you post calculated WIP to a WIP balance sheet account and an offsetting P&L account.

WIP calculations are based on the following manufacturing order status:

- **REL (released)**
  Calculate WIP
- **DLV (delivered)**
  Cancel WIP
- **TECO (technically complete)**
  Cancel WIP

You calculate WIP at the end of each period until the order status is set to **DLV** or **TECO**, then WIP is canceled, and variance is calculated.

13.5 Summary

In this chapter, we configured and calculated WIP and results analysis. This process calculates the cost of incomplete assemblies on the production floor. We'll look at the next period-end process, variance calculations, in Chapter 14.

Chapter 14
Variance Calculations

Variance calculations assign the difference between order debits and credits into categories for detailed analysis of the cause of the variance.

In Chapter 12 and Chapter 13, we studied the period-end processing steps of overhead and work in process (WIP), respectively. In this chapter, we'll discuss variance calculation in detail.

Variance calculation provides information to assist you in analyzing production variance and can help you determine the reason for the difference between order debits and credits. It does this by analyzing the causes of the variance and assigning categories. The following sections present an overview of the types of variance calculation, configuration, and period-end processing.

SAP S/4HANA Cloud Configuration (Scope Item BEI)

In SAP S/4HANA Cloud, scope item BEI is gradually being deprecated because it is not compatible with universal parallel accounting, where production variances are calculated with the final goods receipt (see Chapter 19). You will, however, still find the configuration if you use best practices in SAP S/4HANA Private Cloud or on-premise SAP S/4HANA without universal parallel accounting. The transactions and menu paths to reach these settings are the same as described earlier. Variance key 000001 is delivered as a default and supports all variance categories with the exception of scrap variances.

Note

Further detailed information on variance analysis, case studies, examples, scrap variance analysis, and advanced functionality can be found in *Production Variance Analysis in SAP S/4HANA* (SAP PRESS, 2023) at *https://www.sap-press.com/5629*.

14.1 Types of Variance Calculation

There are three main types of variance categories: total, production, and planning variance. We will discuss them in the following sections.

14.1.1 Total Variance

The *total variance* is the difference between the actual cost debited to the order and credits from deliveries to inventory at standard. You calculate total variance with target cost version 0, determining the basis for calculating target costs. *Target costs* are the expected costs when you deliver a quantity to inventory. The total variance is the only variance relevant to settlement. You settle the difference between debits and credits in financial accounting, profit center accounting, costing-based profitability analysis, or margin analysis.

Example: Total Variance

A production order has a balance of $100 at period end. During variance analysis, with target cost version 0, $40 is assigned to input price variance and $60 to lot size variance.

During settlement, $100 is assigned to Financial accounting and profit center accounting, $40 settles to an input price variance value field, and $60 is settled to a lot size variance value field in costing-based profitability analysis. The variances settle to margin analysis via the variance split, which we discuss in Chapter 15. You can report on the calculations and postings with standard reports, as described in Chapter 20.

14.1.2 Production Variance

Production variance is the difference between net actual costs debited to the order and target costs based on the preliminary cost estimate and quantity delivered to inventory. You calculate production variances with target cost version 1. Production variances are for information only and aren't relevant for settlement.

Example: Production Variance

A production order has a balance of $100 at period end. During variance analysis, with target cost version 1, $30 is assigned to input price variance and $50 to lot size variance. Target cost version 1 calculations aren't relevant for settlement. You can report on the calculations with standard reports, as described in Chapter 20.

14.1.3 Planning Variance

Planning variance is the difference between costs on the preliminary cost estimate for the order and target costs based on the standard cost estimate and planned order quantity. You calculate planning variances with target cost version 2. Planning variances are for information only and aren't relevant for settlement.

Now that we've examined the three main types of variance calculations, we'll examine the configuration required for variance analysis to help you understand the calculations during variance analysis.

Example: Planning Variance

A production order has a balance of $100 at period end, and during variance analysis, with target cost version 2, $10 is assigned to input price variance and $10 to lot size variance. During settlement, target cost version 2 calculations aren't relevant. You can report on the calculations with standard reports, as described in Chapter 20.

14.2 Variance Configuration

The variance configuration for product cost by period and product cost by order are similar. We'll examine the variance configuration for product cost by period and consider the differences along the way.

14.2.1 Define Variance Keys

You define variance keys with Transaction OKV1 or by following the IMG menu path **Controlling • Product Cost Controlling • Cost Object Controlling • Product Cost by Period • Period-End Closing • Variance Calculation • Variance Calculation for Product Cost Collectors • Define Variance Keys**. The screen in Figure 14.1 is displayed.

14

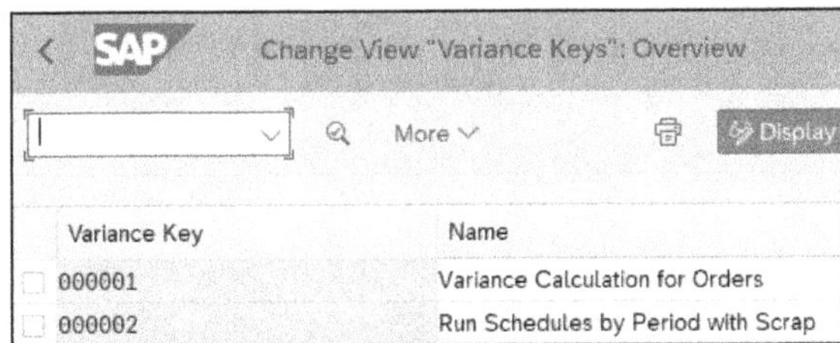

Figure 14.1 Define Variance Keys

This **Overview** screen displays a list of available variance keys. Double-click a variance key in the **Variance Key** column to display the details screen shown in Figure 14.2.

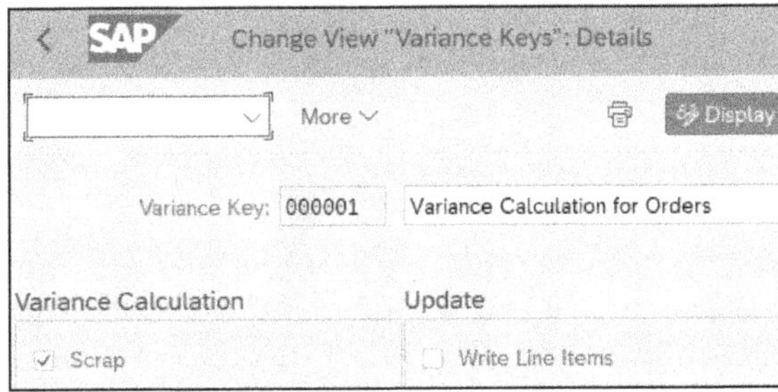

Figure 14.2 Edit Variance Keys

This screen lets you change the variance key description and maintain two checkboxes:

- **Scrap**
  Select the **Scrap** checkbox to ensure that the scrap value is calculated and subtracted from total variances during variance calculation. Any difference between planned and actual scrap appears as scrap variance. If planned scrap equals actual scrap, no scrap variance exists. Target costs for scrap valuation are calculated according to the valuation variant for WIP and scrap, as discussed in Chapter 13. If you do not define a valuation variant for WIP and scrap, target costs are valued based on the standard cost estimate.

Error Messages During Variance Calculation

You may encounter many error messages during period-end variance calculation that indicate routings have changed since the last costing run. In this case, you have three options to reduce the number of error messages:

- Implement product cost collectors by defining order types with Transaction OPJH and Transaction OPL8. Product cost collectors allow you to valuate WIP at target based on the preliminary cost estimate.
- Implement costing routings with Transaction CA01. Costing routings remain unchanged when the actual routings are changed. A disadvantage of costing routings is the extra maintenance required.
- Deselect the **Scrap** checkbox in Figure 14.2. This eliminates the error messages, but the scrap columns in variance reports will not be populated.

- **Write Line Items**
  Select the **Write Line Items** checkbox shown in Figure 14.2 to ensure a document is created when variances or target costs are calculated. The line-item document records when the target costs or variances are calculated and who created them. The document also displays which target costs or variances are changed. The **Write Line Items** checkbox isn't selected by default because you generally don't need this level of detail, except when calculating variances initially, and it increases system load requirements.

Now that we've defined variance keys, let's examine the next variance configuration step.

14.2.2 Define Default Variance Keys for Plants

You can define default variance keys per plant with Transaction OKVW or by following the IMG menu path **Controlling • Product Cost Controlling • Cost Object Controlling • Product Cost by Period • Period-End Closing • Variance Calculation • Variance Calculation for Product Cost Collectors • Define Default Variance Keys for Plants**. The screen shown in Figure 14.3 is displayed.

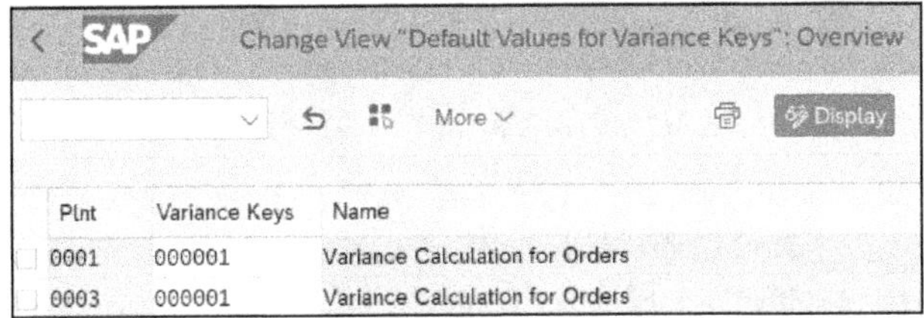

Figure 14.3 Default Variance Keys per Plant

When you create a material master, the system proposes a default variance key for the **Costing 1** view field, based on the **Variance Keys** field entry in Figure 14.3. When you create a manufacturing order or product cost collector, a variance key is proposed based on the default variance key entered in the **Costing 1** view.

14.2.3 Define Variance Variants

You define variance variants with Transaction OKVG or by following the IMG menu path **Controlling • Product Cost Controlling • Cost Object Controlling • Product Cost by Period • Period-End Closing • Variance Calculation • Variance Calculation for Product Cost Collectors • Define Variance Variants**. The screen shown in Figure 14.4 is displayed.

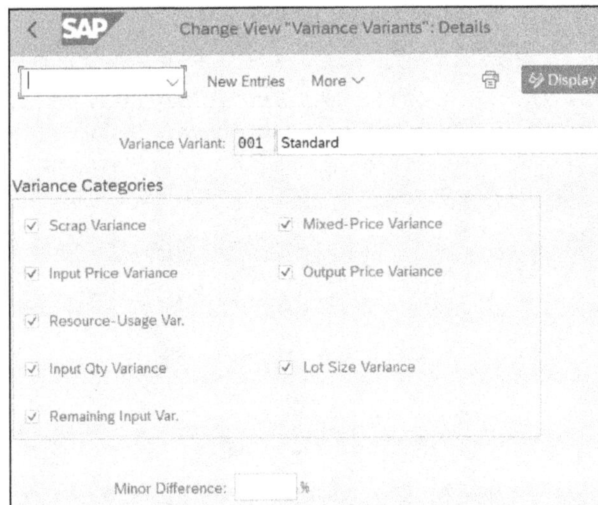

Figure 14.4 Define Variance Variants

Variance Variants determine which **Variance Categories** are calculated. The system calculates variances for all categories you select in this view. If you don't select a variance category, variances of that category are assigned to the remaining variances. Scrap variances are the only exception to this rule. If you don't select **Scrap Variance**, these variances enter all other variances on the input side.

You can specify whether you *calculate* scrap variances by selecting the **Scrap** checkbox in the **Variance Key** in Figure 14.2. You control whether you *display* scrap variances by selecting the **Scrap Variance** checkbox in the **Variance Variant** in Figure 14.4.

You can assign different variance variants to each target cost version. If you have specified that scrap variances will be calculated in the variance key, you could use a variance variant with the **Scrap Variance** checkbox selected for target cost version 0 and deselected for target cost version 3. Variance variants allow you to have one view of variances in target cost version 0 with scrap variances displayed separately and another in target cost version 3 without scrap variances displayed separately.

The **Minor Difference** field determines what percentage of the difference between the target costs and the control costs is interpreted as a minor difference and updated as a remaining variance.

Variance Categories

Even if you aren't aware that postings occur to a particular variance category in your system, it's a good idea to select all the checkboxes in Figure 14.4, as we've done, at least initially. For example, suppose you aren't aware that component substitution occurs, and you don't select the **Resource-Usage Variance** checkbox if component substitution does occur. In that case, the resulting variances will be posted to **Remaining Input Variance**.

If you are concerned with variance calculation runtime, ensuring that the **Write Line Items** checkbox in Figure 14.2 is deselected may improve system performance more than deselecting a checkbox in Figure 14.4. You can also improve system period-end performance by activating the deletion flag on old production orders. You set the deletion flag while viewing a production order in change mode by selecting **Functions • Deletion Flag • Activate** from the menu bar. The deletion flag can be revoked, if necessary, by selecting **Functions • Deletion Flag • Revoke** from the menu bar.

We'll discuss individual variance categories in more detail in Section 14.3.

14.2.4 Define Valuation Variant for Work in Process and Scrap

You define the valuation variant for WIP and unplanned scrap (target costs) by following the IMG menu path **Controlling • Product Cost Controlling • Cost Object Controlling • Product Cost by Period • Period-End Closing • Variance Calculation • Variance Calculation for Product Cost Collectors • Define Valuation Variant for WIP and Scrap (Target Costs)**. The screen shown in Figure 14.5 is displayed.

The **Valuation Strategy** column indicates that unplanned scrap calculation will first search for a **Plan Costs/Preliminary Cost Estimate** and, if unsuccessful, will then search for a **Current Std Cost Est.** This strategy is useful because, if you don't create a preliminary cost estimate, the current standard cost estimate can still be used to valuate unplanned scrap, avoiding an error message that would otherwise be generated.

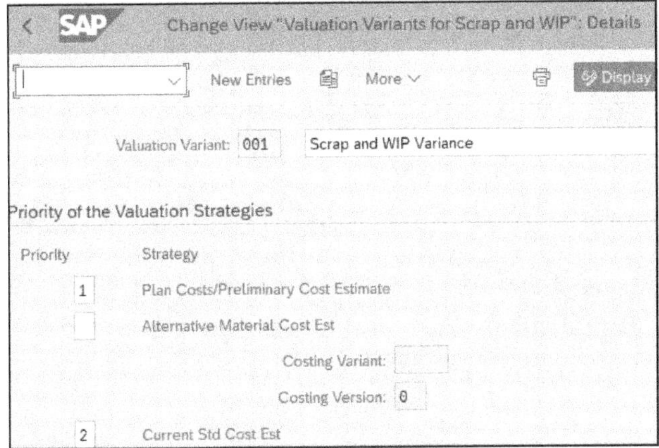

Figure 14.5 Define Valuation Variant for Scrap and WIP

14.2.5 Define Target Cost Versions

You can define target cost versions with Transaction OKV6 or by following the IMG menu path **Controlling • Product Cost Controlling • Cost Object Controlling • Product Cost by Period • Period-End Closing • Variance Calculation • Variance Calculation for Product Cost Collectors • Define Target Cost Versions.** The overview screen shown in Figure 14.6 displays.

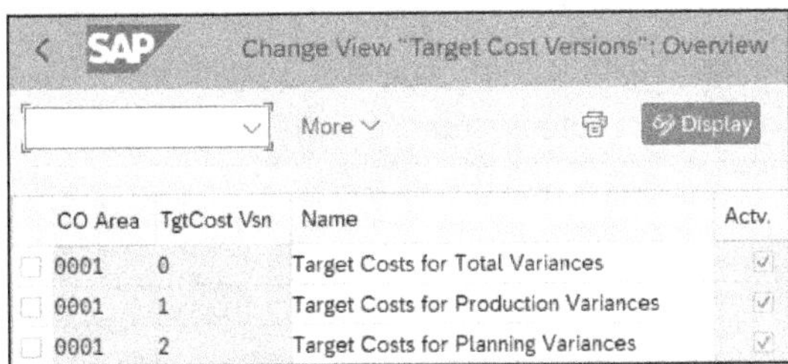

Figure 14.6 Define Target Cost Versions

This overview screen presents a list of available **Target Cost Versions**. We'll examine each of these target cost versions in detail in the following sections.

Target Cost Version 0

Double-click target cost version **0** in the **TgtCostVsn** column to display the details screen shown in Figure 14.7.

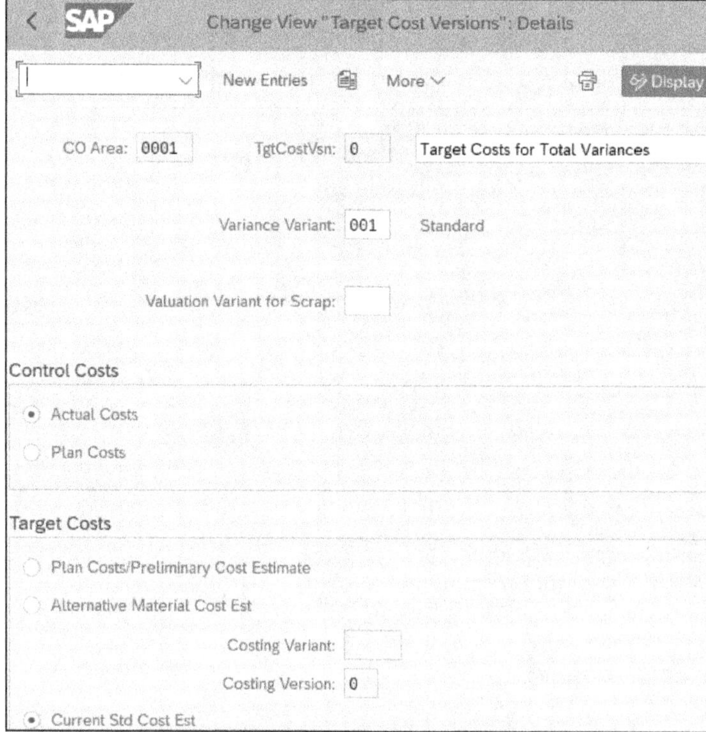

Figure 14.7 Target Cost Version 0 Details Screen

You base **Control Costs** on **Actual Costs** or, in other words, actual debits. **Target Costs** are based on the current standard cost estimate (**Current Std Cost Est** radio button in the **Target Costs** section) or, in other words, actual credits.

TgtCostVsn 0 (target cost version 0) calculates total variance, and you use it to explain the difference between actual debits and credits on an order. It's the only target-cost version that you can settle.

You can specify the **Valuation Variant for Scrap** (and WIP) option with target cost version 0, allowing you to control which cost estimate you use to value scrap, as discussed in Section 14.2.4. The valuation variant for scrap isn't changeable in other target cost versions.

In addition to maintaining the settings for variance calculation, you must select the checkbox **Variances to Costing-Based PA** in the settlement profile to ensure that the calculated variances are assigned to costing-based profitability analysis during settlement. If you don't set this flag, variances will be updated in total. We discuss this in detail in Chapter 15 and Chapter 20.

You also must assign variance categories to accounts in the variance split for variance reporting in margin analysis. We'll discuss this configuration in Chapter 15.

Now that we've examined how to configure target cost version 0, let's look at other target cost versions.

Target Cost Version 1

Double-click target cost version **1** under the **TgtCostVsn** column in the screen shown in Figure 14.6 to display the details screen shown in Figure 14.8.

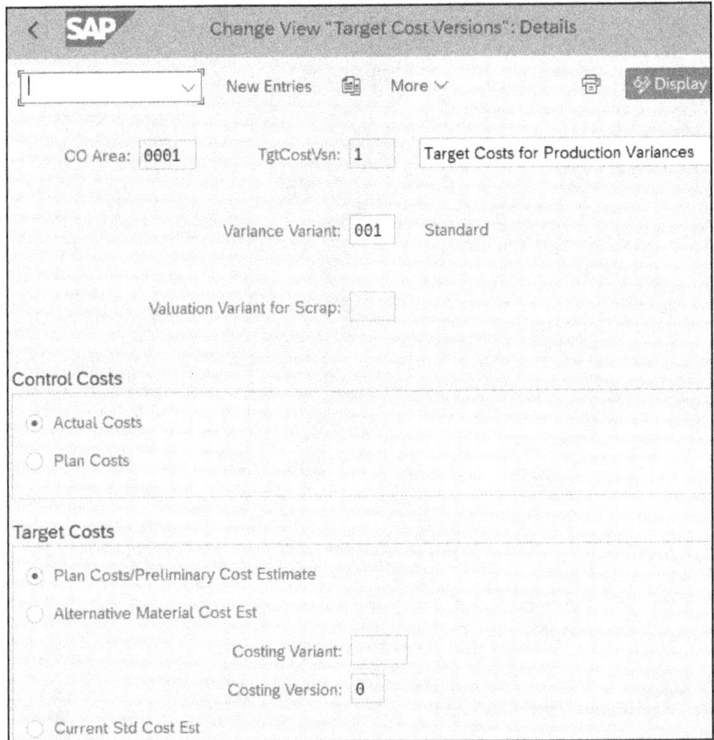

Figure 14.8 Target Cost Version 1 Details Screen

Target cost version 1 calculates *production variance*, which is the difference between net **Actual Costs** and target costs based on the **Preliminary Cost Estimate**. This allows you to exclude variances that occurred because you used a different quantity structure (e.g., production line) during production compared to the standard cost estimate.

Although you cannot settle with this target cost version, you can use it for analyzing production performance and efficiency.

Target Cost Version 2

Double-click target cost version **2** under the **TgtCostVsn** column in the screen shown in Figure 14.6 to display the details shown in Figure 14.9.

Target cost version **2** calculates *planning variance*, which is the difference between plan costs based on the **Preliminary Cost Estimate** of a manufacturing order and target costs based on the current standard cost estimate (**Current Std Cost Est** radio button).

You can use target cost version **2** to decide whether to manufacture orders with a particular quantity structure. Although this target cost version is for information only, it is useful for analyzing production planning performance and efficiency.

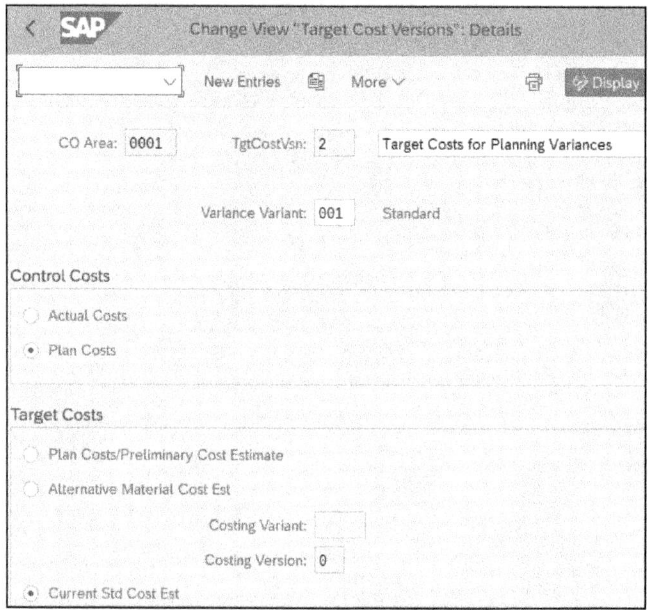

Figure 14.9 Target Cost Version 2 Details Screen

The system doesn't allow the calculation of a planning variance between a current standard cost estimate and a preliminary cost estimate for product cost collectors. Therefore, you cannot calculate variances with target cost version **2** for product cost collectors.

So far, in this chapter, we've analyzed types of variance calculation and configuration. Next, we'll discuss variance categories and follow a typical period-end processing scenario.

14.3 Variance Categories

During variance calculation, the order balance is divided into categories on the input and output sides. Variance categories provide reasons for the cause of the variance, which you can use when deciding the corrective action to take. No financial postings occur during variance calculation, and you can run the transaction as often as necessary to analyze and control production processes. The frequency can be daily if variances are high and many corrective actions are necessary. Continual improvements in master data and user knowledge and skills through frequent variance analysis usually result in reduced variances over time. First, we'll discuss the four input variance categories and the four output variance categories.

14.3.1 Input Variances

Variances on the input side are based on goods issues, internal activity allocations, overhead allocation, and general ledger account postings. Input variances are divided into the following four possible categories during variance calculation, according to their source:

- **Input price variance**
  Input price variance occurs due to component price changes after the higher-level assembly cost estimate is released, which may occur in the following scenarios:
  - If you base the component valuation on standard price control, a standard cost estimate for the component could change after you release the cost estimate for the assembly.
  - If you base the component valuation on moving average price control, a goods receipt of the component could change the component price after you release the cost estimate for the assembly.

  Transfer control can reduce input price variance by using existing component cost estimates instead of creating new ones. You can read more on the details of transfer control in Chapter 7.
- **Resource-usage variance**
  A *resource-usage variance* occurs due to substituting components, which could occur if a component isn't available and you use another component with a different material number instead. The costs for both components are reported as resource-usage variances.
- **Input quantity variance**
  Input quantity variance occurs due to a difference between planned and actual quantities of materials and activities consumed.
- **Remaining input variance**
  Remaining input variance occurs when you cannot assign input variances to any other variance category.

14.3.2 Output Variances

Variances on the output side result from too little or too much planned order quantity delivered or because the delivered quantity was valued differently. Output variances are broken down into the following four possible categories during variance calculation:

- **Output price variance**

  Output price variance can occur in three situations:

  - If you change the standard price after delivery to inventory and before you calculate variance.
  - If you value the material at the moving average price and the material is not delivered to inventory at the standard price during target value calculation. You control how the target value is calculated for delivery to stock when the price control checkbox is set to **V** (moving average price) in customizing with Transaction OPK9 or by following the IMG menu path **Controlling • Product Cost Controlling • Cost Object Controlling • Product Cost by Period • Simultaneous Costing • Define Goods Received Valuation for Order Delivery**.
  - If you don't select the **Mixed-Price Variance** checkbox in the variance variant, as discussed in Section 1.2, an output variance will occur.

- **Mixed-price variance**

  A *mixed-price variance* occurs when you valuate inventory using a mixed-cost estimate for the material. If you want to perform mixed costing, you create a procurement alternative for each production version and then define a mixing ratio to weigh the alternatives. Refer to Chapter 9 for more information on the procurement alternative.

  The mixed-cost estimate calculates a mixed price, which you can write to the material master as the standard price. You base the target credit on the confirmed quantity multiplied by the standard cost of the procurement alternative. You base the actual cost on the confirmed quantity multiplied by the standard price, where the standard price corresponds to the mixed price.

  The mixed-price variance results from a difference between target and actual costs. If you don't select the **Mixed-Price Variance** checkbox in the variance variant, as discussed in Section 14.2, mixed-price variances display as output price variances.

- **Lot size variance**

  Lot size variance occurs if a manufacturing order lot size is different from the standard cost estimate costing lot size. Setup time doesn't usually change with lot size, so a different lot size will either increase or decrease the unit cost. Whenever a portion of manufacturing cost doesn't change with output quantity, such as setup or teardown time, lot size variance can occur.

- **Remaining variance**

  Remaining variance occurs if variances cannot be assigned to any other variance category. Rounding differences or overhead applied to costs that don't vary with lot size report as remaining variances.

Remaining variance also occurs when you cannot calculate target costs, such as when a standard cost estimate doesn't exist, or if a goods receipt for the order has not taken place. If you didn't select any variance categories in the variance variant, the variance is reported as remaining variance. Scrap variances are an exception to this rule. If you did select the **Scrap Variance** checkbox in the variance variant, you can report scrap variances against any other relevant variance on the input side.

Now that we've examined types of variance calculations, configurations, and variance categories, we'll follow a typical period-end processing scenario.

14.4 Variance Period End

There are differences in variance calculation for product cost collectors and manufacturing orders. You calculate variance every period end for product cost collectors, whereas the timing is dependent on the order status for manufacturing orders. WIP and scrap variances are subtracted from actual costs to determine control costs for product cost collectors. The system only subtracts scrap variances from actual costs to determine control costs for manufacturing orders.

We'll now discuss variance calculation for product cost collectors in detail. Then we'll discuss only the differences for variance calculation for manufacturing orders because there are many similarities between the two processes.

14.4.1 Product Cost by Period

You run period-end variance calculation with Transaction KKS6 (individual) and Transaction KKS5 (collective) or by following the menu path **Accounting • Controlling • Product Cost Controlling • Cost Object Controlling • Product Cost by Period • Period-End Closing • Single Functions: Product Cost Collector • Variances**. A selection screen is displayed, as shown in Figure 14.10.

Variance Analysis for Production and Process Orders

You can also carry out variance analysis for production and process orders with Transaction KKS2 (individual) and Transaction KKS1 (collective).

In this example, we'll calculate variance only for product cost collectors by selecting the **W/Product Cost Collectors** checkbox. You can include other objects in the calculation, without running another transaction, by selecting the relevant indicator in Figure 14.10.

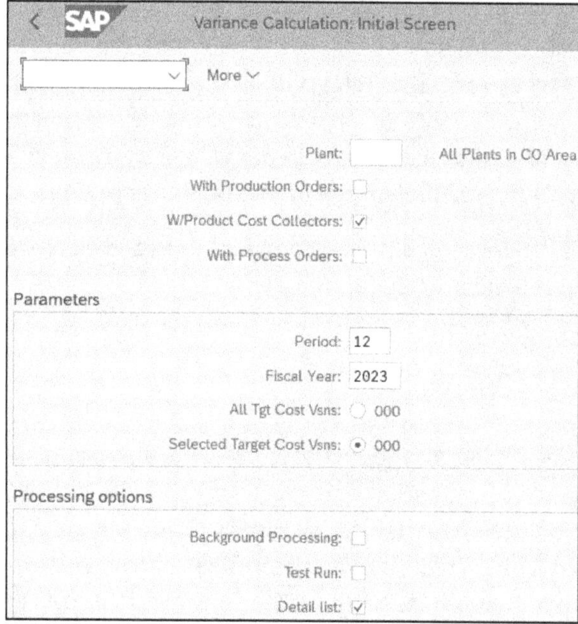

Figure 14.10 Variance Calculation Selection Screen

Complete the selection screen as follows:

1. Since **Plant** is not mandatory, you can enter a plant or leave it blank to run **Variance Calculation** for **All Plants in CO Area**, as shown in Figure 14.10.
2. You can select one, two, or all three checkboxes below the **Plant** field. You can use trial-and-error to choose the checkboxes that produce the results that meet your needs. For example, if you only use **Production Orders**, you may only need to select that checkbox.
3. Complete the **Period** and **Fiscal Year** fields.
4. You can choose **All Tgt Cost Vsns** (all target cost versions) or **Selected Target Cost Vsns** (selected target cost versions). The target cost versions selected in this screen determine whether only total variances (target cost version **000**) or all selected variances are calculated. You can adjust the selected target cost versions by selecting **Extras • Set Target Cost Versions** from the menu bar.
5. Select the **Background Processing** checkbox for variance calculation with many objects. Try it without this checkbox selected at first. If the calculation times out, select the checkbox and try again.
6. Select the Test Run checkbox for the first calculation to check for errors and make corrections.
7. You can reduce the collective processing time by deselecting the **Detail list** indicator. Variances are calculated, though you won't see any variance calculation details in

the following screens. You can then use summarized reporting to analyze the aggregated data. You can identify the orders that caused high variances by sorting by variance columns or using exception rules. To find detailed information for an order that caused high variances, recalculate the variance for the individual order with the **Test Run** and **Detail list** checkboxes selected.

8. Click the **Execute** button or press F8 to display a results screen as shown in Figure 14.11.

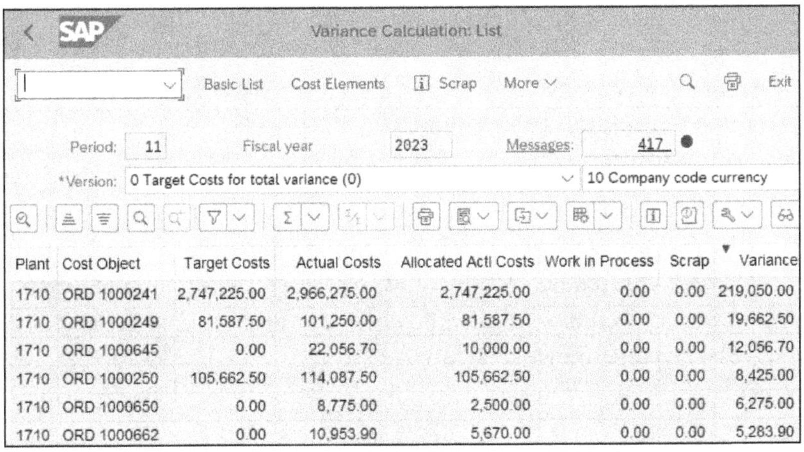

Variance Calculation: List

Basic List | Cost Elements | Scrap | More | Exit

Period: 11 Fiscal year 2023 Messages: 417

*Version: 0 Target Costs for total variance (0) | 10 Company code currency

Plant	Cost Object	Target Costs	Actual Costs	Allocated Actl Costs	Work in Process	Scrap	Variance
1710	ORD 1000241	2,747,225.00	2,966,275.00	2,747,225.00	0.00	0.00	219,050.00
1710	ORD 1000249	81,587.50	101,250.00	81,587.50	0.00	0.00	19,662.50
1710	ORD 1000645	0.00	22,056.70	10,000.00	0.00	0.00	12,056.70
1710	ORD 1000250	105,662.50	114,087.50	105,662.50	0.00	0.00	8,425.00
1710	ORD 1000650	0.00	8,775.00	2,500.00	0.00	0.00	6,275.00
1710	ORD 1000662	0.00	10,953.90	5,670.00	0.00	0.00	5,283.90

Figure 14.11 Variance Calculation Results Screen

You should analyze all messages and take corrective action where necessary. You display messages by clicking the underlined number to the right of **Messages**. If specific materials have several messages, calculating variances for the individual materials separately with Transaction KKS6 and then analyzing the messages for the specific materials might be easier.

The formula for calculating variance in Figure 14.11 is:

Variance = Actual Costs – Actual Costs Allocated (credits) – WIP – Scrap

Sort the **Variance** column in descending order, as indicated by the small red inverted triangle just to the left of the **Variance** column header. Sorting provides visibility of product cost collectors or orders with large variances during the period. Let's follow an example demonstrating how to analyze the cause of the largest variances. We'll analyze the **ORD** (production order) corresponding to the first line in Figure 14.11 because it has the largest unfavorable variance.

To display more details of the product cost collector or order variance calculation, click the first row shown and then click the **Cost Elements** button. The screen shown in Figure 14.12 is displayed. If you don't see the variance category columns as shown in Figure 14.12, click the **Select Layout** (grid) icon and select the **Variance Categories** layout.

Variances | Target Costs

Cost El...	Cost Element (Text)	Origin	Variance	Price Var.	ResUsgVar	Qty V...	Remin...	Mxd...	OutP...	OtptQt...	LotSize...	Rem. Var.
511000...	Consumption - Raw Mat...	1710/KE...	0.00	0.00	0.00	0.00	0.00	0.00	0.00	0.00	0.00	0.00
511000...	Consumption - Raw Mat...	1710/RA...	0.00	0.00	0.00	0.00	0.00	0.00	0.00	0.00	0.00	0.00
511000...	Consumption - Raw Mat...	1710/RA...	0.00	0.00	0.00	0.00	0.00	0.00	0.00	0.00	0.00	0.00
516000...	Consumption - Trading ...		0.00	0.00	0.00	0.00	0.00	0.00	0.00	0.00	0.00	0.00
941110...	Overhead Material		138,775.00-	0.00	138,775.00-	0.00	0.00	0.00	0.00	0.00	0.00	0.00
Debit			**138,775.00-**	**0.00**	**138,775.00-**	**0.00**	**0.00**	**0.00**	**0.00**	**0.00**	**0.00**	**0.00**
551000...	Adjustment Plant Activity...	1710/BA...	0.00	0.00	0.00	0.00	0.00	0.00	0.00	0.00	0.00	0.00
Delivery			**0.00**	**0.00**	**0.00**	**0.00**	**0.00**	**0.00**	**0.00**	**0.00**	**0.00**	**0.00**

Figure 14.12 Cost Elements Breakdown of Variance Calculation

You can display a more detailed view of variances and target costs by clicking the **Variances** and **Target Costs** buttons shown in Figure 14.12.

You can further analyze input quantity variance by displaying a detailed analysis of the product cost collector or order (see note below Figure 14.13) for the period with Transaction KKBC_PKO or by following the menu path **Accounting • Controlling • Product Cost Controlling • Cost Object Controlling • Product Cost by Period • Information System • Reports for Product Cost by Period • Detailed Reports • For Product Cost Collectors.** A selection screen is displayed, as shown in Figure 14.13.

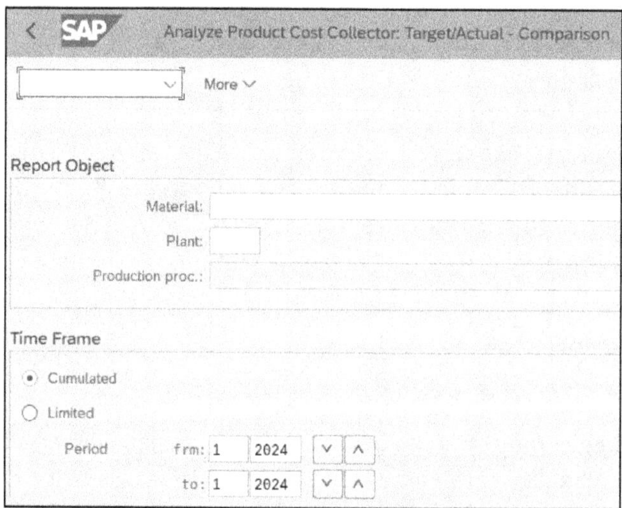

Figure 14.13 Analyze Product Cost Collector Selection Screen

> **Production and Process Orders**
>
> Use Transaction KKBC_ORD for production and process orders.

Complete the **Material** and **Plant** fields, select **Cumulated,** and click the **Execute** icon or press [F8] to display the screen shown in Figure 14.14.

Cost Element	Cost Element (Text)	Origin	Σ Total Actual Costs	Total Actual Qty	Σ Total Plan Costs	Σ Plan Qty	Σ Plan/actual variance	Σ Target/actual var.	T/I var(%)	Currency
50000001	Other Income		1,500.00		0.00		1,500.00	1,500.00		USD
50000001	Other Income		1,500.00-		0.00		1,500.00-	1,500.00-		USD
		■	**0.00**	■	**0.00**	■	**0.00** ■	**0.00**		**USD**
		■■	**0.00**	■■	**0.00**	■■	**0.00** ■■	**0.00**		**USD**

Figure 14.14 Analyze Product Cost Collector Results Screen

Double-click the first row to display the line-item details as shown in Figure 14.15.

Cost Elem.	Cost element name		Σ Val.in rep.cur.	Total quantity	PUM	O	Offsetting Acct	Name of Offsetting Account
50000001	Other Income		1,500.00			K	46	General Approver
Order 300008 Test BB_2		■	**1,500.00**					
		■■	**1,500.00**					

Figure 14.15 Line Items

We've discussed variance calculation for product cost collectors in detail. We'll now discuss the differences in variance calculations for manufacturing orders because the two processes have many similarities.

14

14.4.2 Product Cost by Order

In this section, we'll examine variance calculation for product cost by order, which means working with manufacturing orders. We'll analyze the differences between variance calculation for manufacturing orders and product cost collectors.

Variance calculation for manufacturing orders is also known as *cumulative variance*, which means that the variances can accumulate over several accounting periods before being settled once the order is complete. Cumulative variance compares target costs and cumulative control costs. As previously shown in the configuration for target cost version 0 in Section 14.2.5, you base target costs for total variance on the standard cost estimate or, in other words, the valuated goods receipt. Control costs are equal to actual costs, less scrap variances. The manufacturing order must meet the following two conditions to calculate variance:

- Settlement type **FUL** (full settlement) in the settlement rule
- Status **DLV** (delivered) or **TECO** (technically complete)

Settlement type **FUL** allows all unsettled WIP and variance from the current and previous periods to settle in the current settlement period. WIP and variances are calculated based on order status, not period. You typically calculate WIP and variances and settle manufacturing orders every period.

Status **DLV** is determined automatically when posting valuated goods receipts during manufacturing order confirmation. Status **TECO** is determined manually and indicates that processing is complete, even though the order isn't status **DLV**. When the system detects either status during period-end processing, WIP cancels, and variance is calculated.

During variance calculation, the system compares target and control costs and assigns variance categories in the following sequence:

1. Input price variance
2. Resource-usage variance
3. Input quantity variance
4. Remaining input variance
5. Mixed-price variance
6. Output price variance
7. Lot size variance
8. Remaining variance

We discussed variance categories in detail in Section 14.3.

SAP Note

SAP Note 2027639 describes the current functionality and limitations of variance calculation for production orders, process orders, and product cost collectors.

14.5 Summary

In this chapter, we studied the types of variance calculation, configuration, and period-end processing. We'll examine the final period-end settlement process next in Chapter 15 and event-based processing later on in Chapter 19.

Chapter 15
Settlement

Settlement transfers work in process (WIP) and variances to general ledger accounting, costing-based profitability analysis, and margin analysis as part of period-end-processing.

In this chapter, we'll discuss settlement configuration, including the settlement profile, allocation structure, source structure, profitability analysis transfer structure, margin analysis variance split, and cost component split. We'll then discuss settlement period-end processing, including parameters and processing options.

In Chapter 19, we'll discuss universal parallel accounting and event-based processing, which creates WIP, overhead, and variance journal entries in real time and removes the need for period-end settlement when activated.

15.1 Settlement Configuration

Settlement configurations for product cost by period and product cost by order are similar. We'll examine settlement configurations for product cost by period and highlight the differences.

15.1.1 Variance Split

Margin analysis configuration in SAP S/4HANA allows you to assign each variance category to a general ledger account. You first create a general ledger account with cost element category 1 (primary costs/cost-reducing revenues) for each variance category. We discuss creating general ledger accounts in Chapter 3. You then create a splitting profile for the variance split. Follow the IMG menu path **Financial Accounting • General Ledger Accounting • Periodic Processing • Integration • Materials Management • Define Accounts for Splitting Price Differences**. The screen shown in Figure 15.1 is displayed.

Select a splitting profile (**Price Diff. Profile**), **Z100** in this example, and double-click **Detailed Price Difference Accounts** at the left. Click the **New Entries** button (not shown) to assign a general ledger account to each variance category, as shown in Figure 15.2.

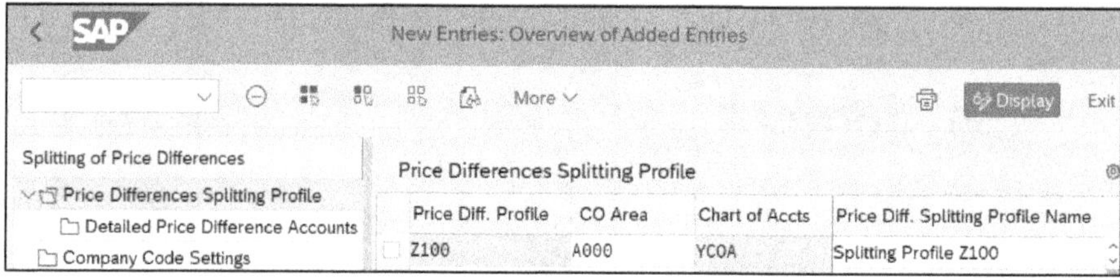

Figure 15.1 Define Splitting Profile

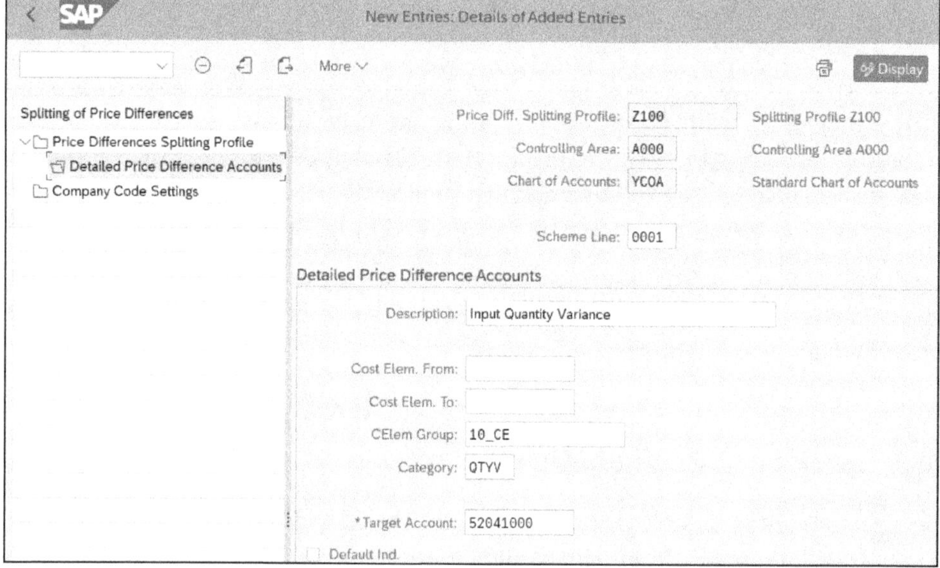

Figure 15.2 Maintain Price Difference Accounts

In this screen, you assign a variance **Category**, **Input Quantity Variance** (**QTYV**) in this example, to a **Target Account** or the general ledger account to which the variance account will be posted when the production order is settled.

In the **Cost Elem. From** field, you enter a single cost element as the source cost element for the variance calculation. In the **Cost Elem. To** field, you can enter a cost element to maintain a cost element interval. In the **CElem Group** field, you can enter a cost element group as the source cost elements.

You must select the **Default Ind.** for one variance category to post rounding differences.

15.1.2 Settlement Profile

A *settlement profile* contains settlement control parameters on how orders settle. You enter the settlement profile in the order type definition with Transaction KOT2 or by following the IMG menu path **Controlling • Product Cost Controlling • Cost Object Controlling • Product Cost by Order • Manufacturing Orders • Check Order Types**. When you create an order, the settlement profile in the order type defaults to the order settlement rule. To display the settlement profile in a production order, follow these steps:

1. Display a production order with Transaction CO02.
2. Select **Header • Settlement Rule** from the menu bar.
3. You'll see a list of distribution rules that define where the costs are to settle. Select **Goto • Settlement Parameters** from the menu bar to display the settlement parameters shown in Figure 15.3.

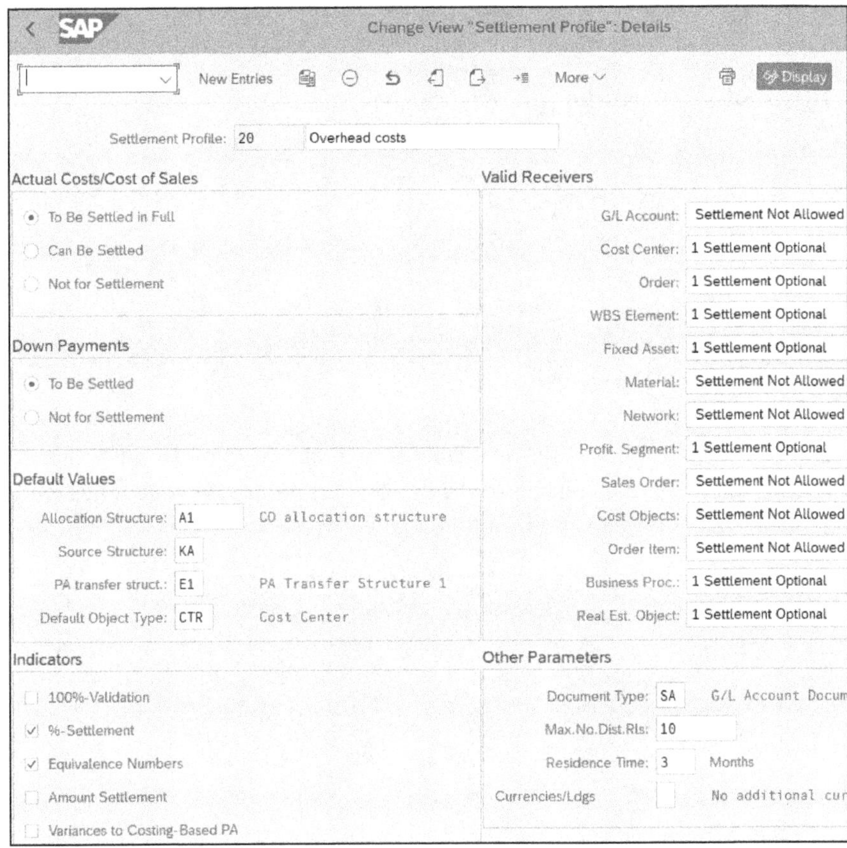

Figure 15.3 Settlement Rule Parameters

The **Settlement Profile** in the **Parameters** section defaults from the order type you configure with Transaction KOT2. The **Allocation Structure**, **PA transfer struct.**, and **Source Structure** default from the settlement profile, which we'll examine in detail in this chapter.

Manually Change Default Values

The settlement rule parameters default from the order type and settlement profile. You can manually change the default values.

You create and maintain settlement profiles with Transaction OKO7 or by following the IMG menu path **Controlling • Product Cost Controlling • Cost Object Controlling • Product Cost by Period • Period-End Closing • Settlement • Create Settlement Profile**. The screen shown in Figure 15.4 is displayed.

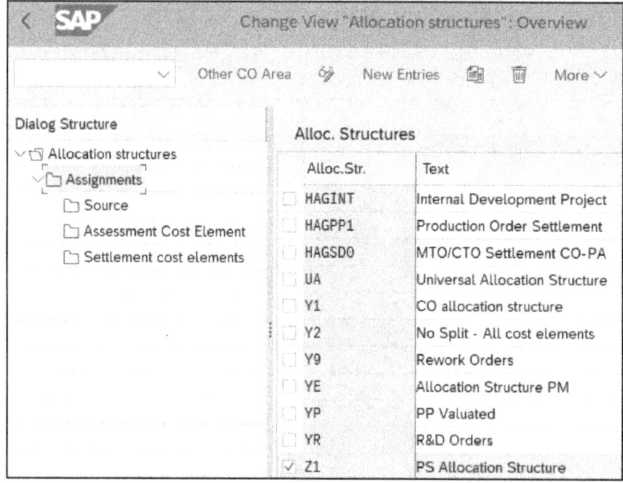

Figure 15.4 Settlement Profile Overview

The **Overview** screen displays a list of available settlement profiles. Double-click an entry in the **Settlement Profile** column to display the details screen shown in Figure 15.5.

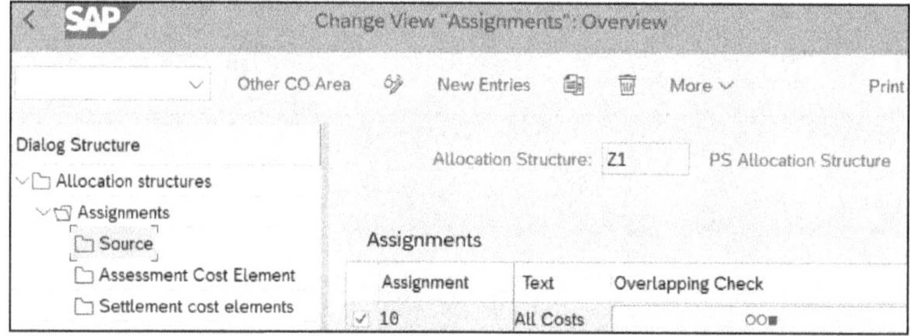

Figure 15.5 Settlement Profile Details

You maintain **Settlement Profile Details** in this screen. Let's examine the fields and checkboxes of each section:

- **Actual Costs/Cost of Sales**

  The **Actual Costs/Cost of Sales** section controls how actual costs or COS will be settled.

  - **To Be Settled in Full**

    This is the default setting. If you try to close an order or set the deletion flag, an error message displays if the balance of the object isn't zero.

- **Can Be Settled**
  You can settle actual costs and COS, but you don't have to. When you try to close the object or set the deletion flag, you'll receive a warning message if the balance in the object isn't zero. You can set the deletion flag on an order with a balance.
- **Not for Settlement**
  When you close the object or set the deletion flag, the system doesn't display a message if the balance in the object isn't zero.

Orders with Remaining Balance

If you need to set the deletion flag on an order with a remaining balance, you'll receive an error message until you do one of the following:

- Settle the balance. This can be difficult if the order has problems or has not been settled for a long time.
- Change the **To Be Settled in Full** setting to **Can Be Settled** in the settlement profile, set the deletion flag on the order, and then reselect **To Be Settled in Full**.

- **Down Payments**
  To Be Settled is the default setting. Down payments are settled to an asset under construction. Before selecting **Not for Settlement**, ensure the relevant down payment clearings or reversals of open down payments are settled for all down payments that have already been settled.
- **Default Values**
  The **Default Values** section contains default values for control structures that we'll discuss in more detail in the following sections.
- **Indicators**
  The **Indicators** section shown in Figure 15.5 contains checkboxes that control the amount you'll settle:
  - **100%-validation**
    This checkbox controls whether percentages in the settlement rule are checked. A settlement rule determines which portions of a sender's costs are allocated to each receiver. A settlement rule is contained in a manufacturing order or product cost collector header data.

    Suppose you have defined percentage distribution rules for a particular settlement rule. In that case, the total percentage is checked either when you save the settlement rule or if you use the percentage check function:
    - If the checkbox is selected, a warning is issued if the total is less or more than 100%.
    - If the checkbox isn't selected, a warning is issued if the total is more than 100%.

15

In both cases, you can ignore the warning and save the settlement rule. However, to run the settlement, you must first correct the settlement rule. The settlement run prevents you from settling more than 100%.

- **%-Settlement**
  If you select this checkbox, you use the settlement rule to determine the distribution rules governing the percentage costs to be settled.
- **Equivalence numbers**
  If you select this checkbox, you can define distribution rules in the settlement rule according to the costs you settle proportionally.
- **Amount settlement**
  If you select this checkbox, you can define distribution rules in the settlement rule, which allows you to settle costs by amount. For example, you can settle a specific amount, such as $5,000 of the costs incurred on an order to cost center 4711.
- **Variances to Costing-Based PA**
  If you select this checkbox, variances are sent to costing-based profitability analysis during order settlement.

Setting the Deletion Flag for Orders with Unsettled Variances

If you need to set the deletion flag on an order with unsettled variances, you'll receive an error message until you do one of the following:

- If possible, settle the variances to profitability analysis. This can be difficult if the order has problems or has not been settled for a long time.
- Deselect the **Variances to Costing-Based PA** checkbox in the settlement profile, set the **Deletion** checkbox on the order, and then reset the **Variances to Costing-Based PA** checkbox in the settlement profile.

- **Valid Receivers**
  In this section of the settlement profile, you choose whether settlement to certain types of receivers is optional, required, or not allowed.

Example: Valid Receivers

In Figure 15.5, we've allowed production orders to settle to a cost center, an order, a WBS element, a fixed asset, a profitability segment, a business process, or a real estate object. We don't want the settlement to occur to other object types, so they are set to **Settlement Not Allowed**.

- **Other Parameters**
  In this section, you maintain the following fields:
  - **Document type**
    You can assign individual document types for each settlement rule to assist in identifying settlement documents separately. You create and maintain document types by customizing Transaction OBA7.
  - **Max.No.Dist.Rls (maximum number of distribution rules)**
    This field determines the maximum number of distribution rules possible for each settlement rule.
  - **Residence time**
    This field determines the minimum time between setting the order deletion flag (revocable) and deletion indicator (not revocable—cannot be changed).
  - **Currencies/Ldgs (settlement of freely defined currencies and parallel ledgers)**
    Defines how the settlement works concerning the currencies and ledgers of the Universal Journal. Refer to Chapter 19 for more details on this setting. Also, refer to SAP Note 2894297 to explain how settlement in multiple currencies works before universal parallel accounting.

Now that we've examined settlement profiles, let's examine the next settlement configuration transaction: the allocation structure.

15.1.3 Allocation Structure

An *allocation structure* allocates the costs incurred on a sender by cost element or cost element group. You can use the allocation structure for settlement and assessment. Each allocation structure contains several assignments of cost elements (or cost element groups) to settlement or assessment cost elements. All accounts/cost elements posted to the order must be included in the allocation structure before settlement.

Each allocation structure must fulfill the following criteria:

- **Completeness**
  You assign an allocation structure to each order you settle. You must represent all cost elements that incur costs (source cost elements) in the allocation structure.
- **Uniqueness**
  Each cost element that incurs costs can only appear once in an allocation structure. You can assign only one settlement cost element to a source within a particular allocation structure.

In this section, we explain how to create an allocation structure and build the mapping from the original cost elements to the settlement cost elements.

Creating and Adjusting Allocation Structures

You create an allocation structure with Transaction OKO6 or by following the IMG menu path **Controlling • Product Cost Controlling • Cost Object Controlling • Product Cost by Period • Period-End Closing • Settlement • Create Allocation Structure**. The screen shown in Figure 15.6 is displayed.

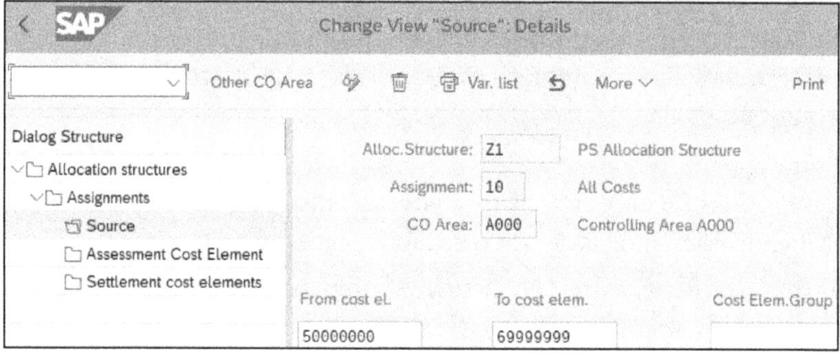

Figure 15.6 Allocation Structures Overview

Available **Allocation Structures** are listed at the right of the **Overview** screen. You can use any allocation structure or copy one and create your own. Let's choose allocation structure **Z1**, as shown selected in Figure 15.6, and examine its components.

Double-click **Assignments** on the left to display the screen shown in Figure 15.7.

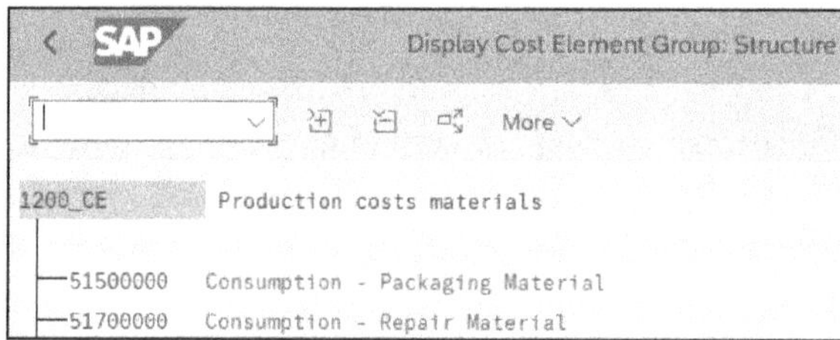

Figure 15.7 Allocation Structure Assignments Overview

Available **Allocation Structure Assignments** are listed on the right side of the **Overview** screen. Select an assignment and double-click **Source** to show the screen displayed in Figure 15.8.

Three methods are available to assign source cost elements:

- A single cost element
- A range of cost elements
- A cost element group

In Figure 15.8, we've entered the cost element range **From cost el. 50000000 To cost elem. 69999999**.

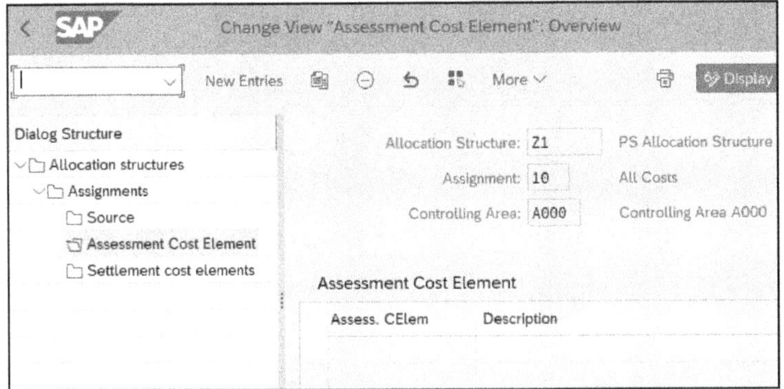

Figure 15.8 Source Cost Element Details

Cost Element Group Master Data

If you introduce a new source cost element, for example, to identify a new external processing cost, you must ensure that the source cost element is included in the allocation structure, or an error message will occur when you run the settlement transaction.

The easiest way to include new source cost elements in an allocation structure is to assign a cost element group, even if the group includes only one cost element. A cost element group is master data that you can maintain with Transaction KAH2 or by following the menu path **Accounting • Controlling • Cost Center Accounting • Master Data • Cost Element Group • Change**. A cost element group is shown in Figure 15.9.

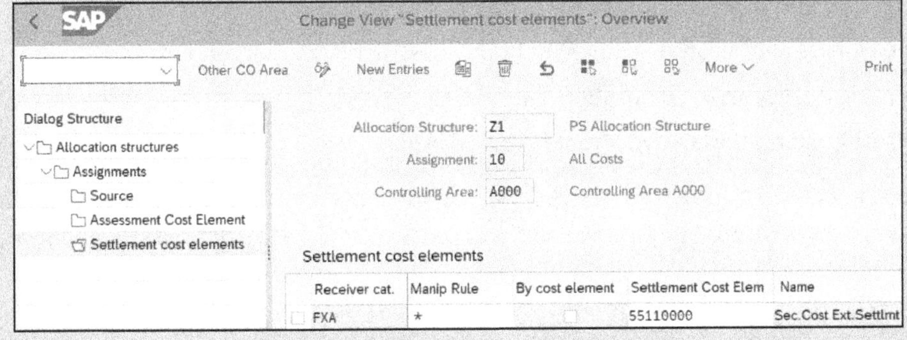

Figure 15.9 Cost Element Group

Allocation Structure Cost Element Group

Suppose you don't assign a cost element group. In that case, you may need to add the new cost element as a configuration change in the allocation structure, which generally requires more documentation, procedures, and time than a simple master data change to a cost element group.

Double-click **Assessment Cost Element** in Figure 15.8 to display the screen shown in Figure 15.10.

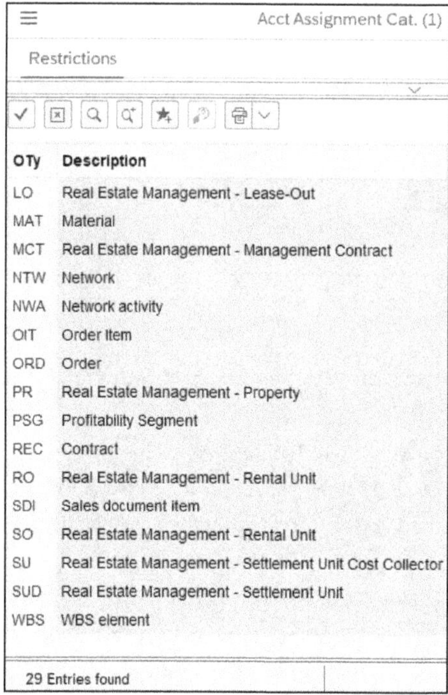

Figure 15.10 Assessment Cost Element Overview

You can use allocation structures for both assessments and settlements:

- Assessments allocate costs between cost centers.
- Settlements allocate costs from orders to other objects.

You generally set up allocation structures for either assessments or settlements. In this case, because we are configuring a settlement, there is no need to enter assessment cost elements.

Double-click **Settlement cost elements** in Figure 15.10 to display the screen shown in Figure 15.11.

We'll examine the **Settlement cost elements** section of this settlement profile in the following sections, beginning with the first column.

Account Assignment Category

When assigning settlement cost elements, the first field you maintain is **Receiver cat.** (account assignment category), which specifies the object types allowable for the settlement receiver. Click a category and press F4 to display the screen shown in Figure 15.12.

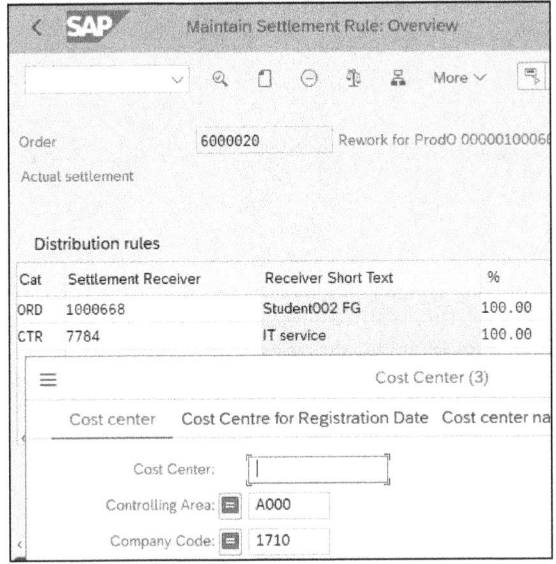

Figure 15.11 Settlement Cost Elements Overview

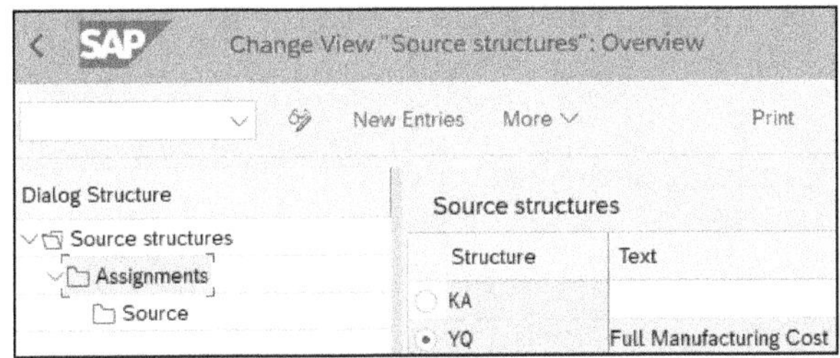

Figure 15.12 Account Assignment Categories Possible Entries

This screen lists the possible entries for the account assignment category. By entering the account assignment category in the settlement rule, you specify how you interpret the entry in the **Settlement Receiver** field shown in Figure 15.13.

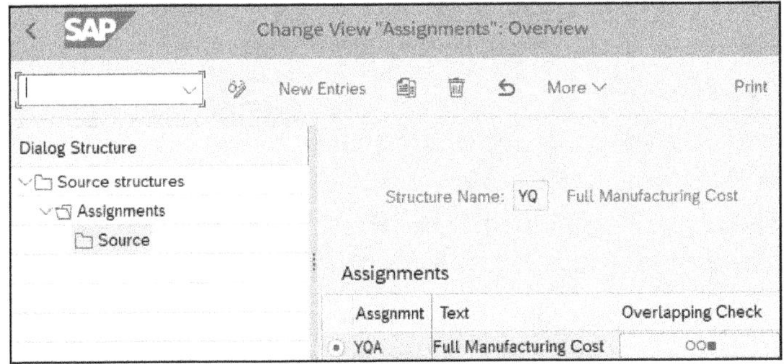

Figure 15.13 Settlement Rule Receiver Possible Entries by Category

Example: Account Assignment Category

If you enter "CTR" as the account assignment category and "1000" as the receiver, the system will settle to cost center 1000. If, however, you enter "ORD" and "1000," the order with the number 1000 is the settlement receiver.

An example of when you may not want to settle a production order to a material is when you manufacture free samples that will not be received into inventory.

The account assignment category influences the F4 help in the general receiver field. If, for example, you enter "CTR" as the account assignment category, the system will offer you F4 help to select cost centers.

Example: Production Order Settlement Receiver

Display a production order with Transaction CO02 and select **Header • Settlement Rule** from the menu bar. Enter "CTR" in the first available account assignment rule field. Then click the **Settlement Receiver** field and press the F4 key to display the screen shown in Figure 15.13.

The list of possible entries is restricted to **Cost Center** because the category **Cat** (category) entered is "CTR" (cost center).

Let's now examine the next columns in the **Settlement cost elements** section of the **Allocation structure** shown in Figure 15.11.

Manipulation Rule

A **Manip Rule** (manipulation rule) is used for recovery indicator manipulations during controlling allocations in combination with joint venture accounting.

By Cost Element

If you select the **By cost element** checkbox in Figure 15.11, the settlement cost element will be the same as the source or debit cost element. You only select this checkbox if you need the original debit cost element to appear on the settlement object, such as a cost center, which isn't normally necessary.

You typically group source cost elements to settle with one settlement cost element. If you require more details of the source costs when viewing the object settled to, you simply double-click the settlement cost element to take you directly to the sender details.

When you select the **By cost element** checkbox, you cannot enter a settlement cost element in the next field because you have already defined it as the source cost element.

Settlement Cost Element

The settlement cost element credits the sender and debits the receiver during settlement. Settlement cost elements are secondary cost elements of category **22**. If you press the F4 key in the **Settlement cost elem** field, you'll see a list of secondary cost elements of category **22**.

Name

The text in the **Name** field is the name of the settlement cost element that is automatically transferred from the cost element.

Now that we've examined allocation structures, let's look at the next settlement configuration transaction: the source structure.

15.1.4 Source Structure

You define source structures when settling and costing joint products. Joint production involves the simultaneous production of many materials in a single production process. A *source structure* contains several source assignments, each containing the individual cost elements or cost element intervals you settle using the same distribution rules. We discussed the joint structure in the **MRP 2** view in Chapter 3.

In the settlement rule for the sender, you can define one distribution rule, in which you specify the distribution and the receivers for the costs for each source assignment.

Source Structures

Check whether you need to use source structures in your settlement procedures. If settling for cost elements, you don't need a source structure. Otherwise, create one using the following procedure.

Create a source structure with Transaction OKEU or by following the IMG menu path **Controlling • Product Cost Controlling • Cost Object Controlling • Product Cost by Period • Period-End Closing • Settlement • Create Source Structure**. The screen shown in Figure 15.14 is displayed.

You'll see available **Source structures** listed on the right side of this overview screen. You can use any source structure or copy one and create your own. We'll choose the source structure **YQ**, shown in Figure 15.14, and examine its components. Select **Source structure YQ** and double-click **Assignments** to display the screen shown in Figure 15.15.

You'll see available source structure **Assignments** listed on the right side of the **Overview** screen. Select **Assgnmnt** (assignment) **YQA** and double-click the **Source** folder to show the screen displayed in Figure 15.16.

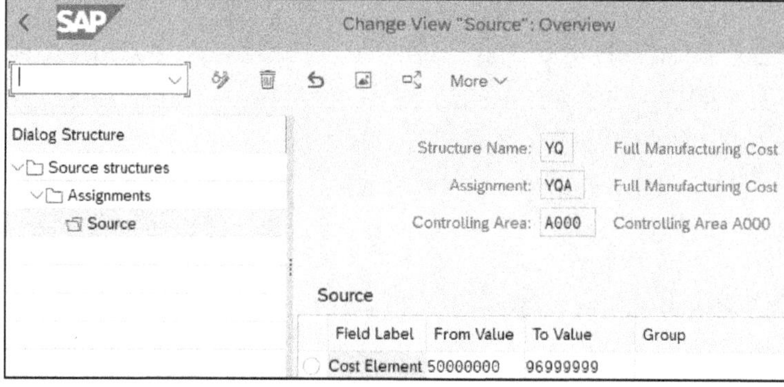

Figure 15.14 Source Structures Overview

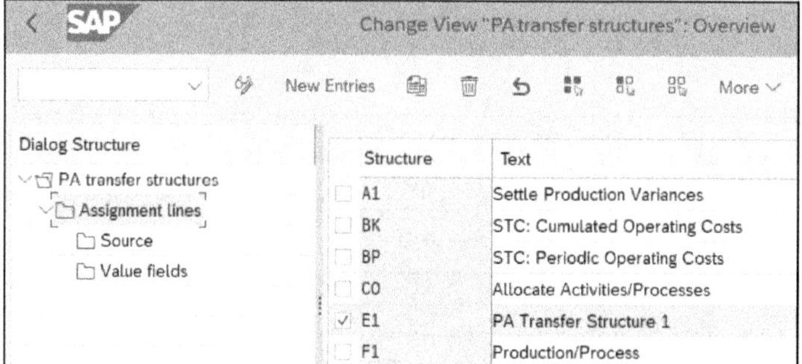

Figure 15.15 Source Structure Assignments

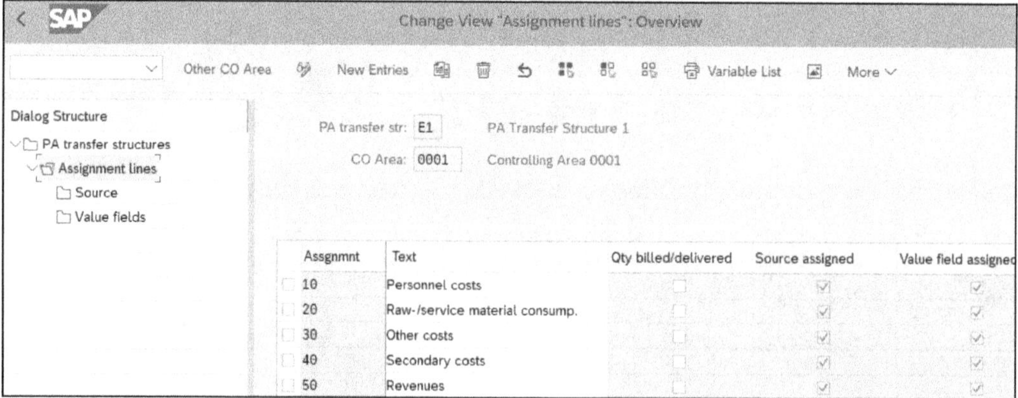

Figure 15.16 Source Structure Source

You enter the source structure source in this screen. Let's examine the rows in detail:

- **Acctg Indicator (accounting indicator)**
  The **Acctg Indicator** row (not shown) serves as another criterion besides the cost element, discussed in the next section. You can identify incurred costs and earned profits by sales volume, warranty, or goodwill, for example.

The accounting indicator was introduced with service management to distinguish billable and non-billable costs. Thus, you can define goodwill and warranty as non-billable costs compared with normal spare parts that would be billable.

- **Cost Element**
  You can enter an individual cost element, range, or **Group** in the **Cost Element** row. You maintain cost element groups with Transaction KAH2.

Source Structure Assignments Example

An order has incurred both direct and overhead costs. The direct costs are to be divided 50% each between a fixed asset and a cost center, whereas you'll settle the overhead in full to an administration cost center. To do this, you create a source structure with two source assignments:

- Direct cost elements
- Overhead cost elements

Now that we've examined settlement profiles, allocation, and source structures, let's look at the profitability analysis transfer structure.

15

15.1.5 Profitability Analysis Transfer Structure

Settlement lets you transfer costs, revenues, sales deductions, and production variances to costing-based profitability analysis. The profitability analysis transfer structure defines which quantities or values of a sender you transfer to which value fields in costing-based profitability analysis. You use value fields to report on costing-based profitability analysis values.

With SAP S/4HANA margin analysis, you also need to set up the configuration to map production variance categories to accounts in the variance split via the IMG menu path **Financial Accounting • General Ledger Accounting • Periodic Processing • Integration • Materials Management • Define Accounts for Splitting Price Differences**. We discussed the variance split previously in Section 15.1.1 and will again in the margin analysis section in Chapter 20.

Example: Marketing Order

For a marketing order, you can assign the cost element group **Personnel Costs** to the value field **VTRGK** (sales overhead).

Costing-based profitability analysis enables you to evaluate market segments, which you can classify according to products, customers, orders, or any combination of these, or strategic business units, such as sales organizations or business areas, with respect to your company's profit or contribution margin.

You maintain a profitability analysis transfer structure with Transaction KEI1 or by following the IMG menu path **Controlling • Product Cost Controlling • Cost Object Controlling • Product Cost by Period • Period-End Closing • Settlement • Create PA Transfer Structure**. The screen shown in Figure 15.17 is displayed.

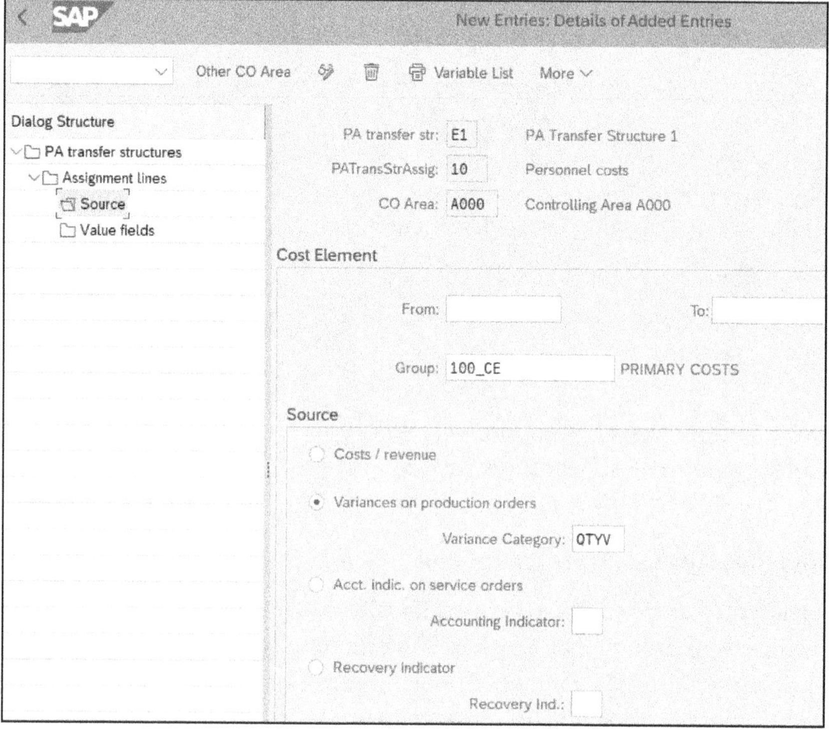

Figure 15.17 Profitability Analysis Transfer Structures Overview

Available **PA transfer structures** are listed on the right side of the **Overview** screen. You can use any profitability analysis transfer structure or copy one and create your own. Let's choose profitability analysis transfer structure **E1**, shown selected in Figure 15.17, and examine its components.

Select profitability analysis transfer structure **E1** and double-click **Assignment lines** to display the screen shown in Figure 15.18.

A profitability analysis transfer structure allows you to define the level of detail of production variances you transfer to costing-based profitability analysis. You can transfer all production variance categories to separate value fields or just one value field, depending on the detail you require in variance reporting in costing-based profitability analysis. Select an assignment and double-click **Source** to show the screen displayed in Figure 15.19.

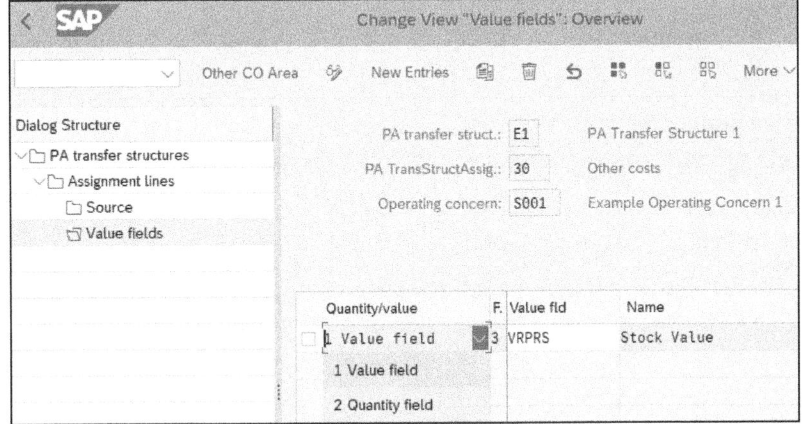

Figure 15.18 Transfer Structure Assignment Lines Overview

Field Name	Description
ERLOS	Revenue
MRABA	Qty discount
PRABA	Price reduction
RABAT	Other Discounts
AUSFR	OutgoingFreight
VSVP	DispatchPackag.
UMSLZ	Licensing Fees
PROVV	Ext. Commission
VWGK	Admin. Overhead
VTRGK	Sales Overhead
VRPRS	Stock Value

Figure 15.19 Assignment Lines Source Details

Let's examine the fields of this source details screen:

- **Cost Element**
  In the **Cost Element** fields, you can enter an individual source cost element, range, or **Group**. Double-click the group **100_CE** to display the cost elements in the cost element group.

Maintain Master Data Rather Than Configuration

Maintaining a cost element group is easier than maintaining the individual cost elements or ranges entered in this screen because a cost element group is master data, which is relatively easy to maintain. Maintaining values directly is a configuration change, which requires more documentation and testing than master data changes.

- **Source**
  In this section, you define the source information transferred to costing-based profitability analysis depending on your selection of the following:

- **Costs/revenue**
  Costs posted to the cost elements in the **Cost Element** section will transfer to a value field.
- **Variances on production orders**
  You transfer production variances by variance category to individual value fields.

Material and Labor Input Quantity Variance

You differentiate between material and labor input quantity variance by assigning material and labor cost elements to different value fields. This is one of the only ways you can do so in standard reporting.

- **Acct. indic. on service orders**
  Costs posted by the accounting indicator defined in service orders post to a value field.

To maintain the value fields to which you post the source costs, double-click the **Value fields** folder in Figure 15.19 to display the screen shown in Figure 15.20. This screen lets you maintain details of the costing-based profitability analysis fields to which you transfer the source information.

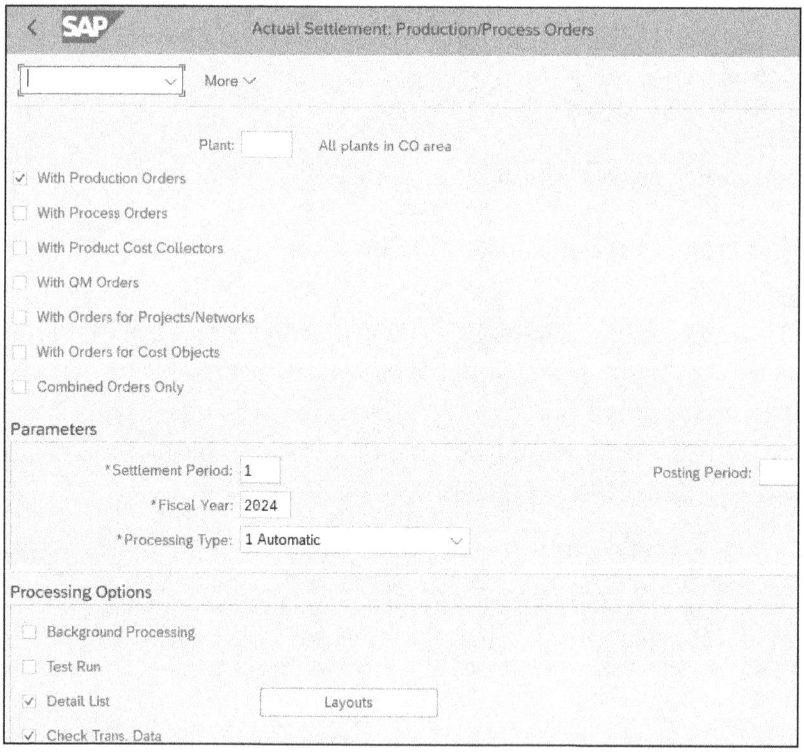

Figure 15.20 Assignment Lines Value Fields

- **Quantity/Value**
  Click this field to see the two possible entries as shown in Figure 15.20. You have the following choices, as shown in Figure 15.20:
  - **Value field**
    You transfer monetary amounts to this field.
  - **Quantity field**
    You transfer quantity amounts to this field.
- **Fixed/Variable**
  This column (**F.**) allows you to enter the following three options depending on the level of detail you require to transfer to the individual value field in costing-based profitability analysis:
  - 1
    Fixed amounts
  - 2
    Variable amounts
  - 3
    The sum of fixed and variable amounts

  This fixed/variable split in the split in margin analysis is often more reliable if this is an option for you.
- **Value fld (value field)**
  In **Value fld**, you enter the value field to receive the amount or quantity from the source. You can display a list of available value fields, as shown in Figure 15.21, by clicking the field and pressing the F4 key.

Field Name	Description
ERLOS	Revenue
MRABA	Qty discount
PRABA	Price reduction
RABAT	Other Discounts
AUSFR	OutgoingFreight
VSVP	DispatchPackag.
UMSLZ	Licensing Fees
PROVV	Ext. Commission
VWGK	Admin. Overhead
VTRGK	Sales Overhead
VRPRS	Stock Value

Figure 15.21 Value Field Possible Entries

Standard value fields are available, or you can freely define your own. Value fields you manually create typically have a VVxxx format. You can maintain value fields with Transaction KEA6 or by following the IMG menu path **Controlling • Profitability Analysis • Structures • Define Operating Concern • Maintain Value Fields.**

15

Now that we've finished examining settlement configuration settings, let's look at cloud configuration and the period-end process. We discuss event-based processing in Chapter 19.

15.2 SAP S/4HANA Cloud Configuration

In SAP S/4HANA Cloud, the use of scope item BEI for the close activities for production orders, process orders, and product cost collectors is gradually being phased out (deprecated), as it is not compatible with the use of universal parallel accounting where there is no separate settlement step but rather work-in-process and production variances are recognized at the same time as the underlying journal entries. We'll explain the use of scope item 3FO for event-based product costing in Chapter 19. You will, however, still find the settings for scope item BEI if you use best practices in SAP S/4HANA Private Cloud or on-premise SAP S/4HANA. The transactions and menu paths to reach these settings are the same as we described earlier in Section 15.1.

The settlement profile groups together all parameters needed for settlement, and several different settlement profiles are delivered as best practices, as shown in Figure 15.22.

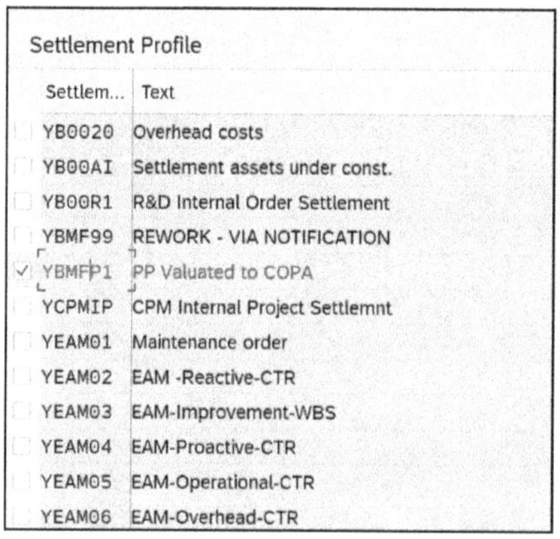

Settlement Profile

Settlem...	Text
YB0020	Overhead costs
YB00AI	Settlement assets under const.
YB00R1	R&D Internal Order Settlement
YBMF99	REWORK - VIA NOTIFICATION
YBMFP1	PP Valuated to COPA
YCPMIP	CPM Internal Project Settlemnt
YEAM01	Maintenance order
YEAM02	EAM -Reactive-CTR
YEAM03	EAM-Improvement-WBS
YEAM04	EAM-Proactive-CTR
YEAM05	EAM-Operational-CTR
YEAM06	EAM-Overhead-CTR

Figure 15.22 Sample Settlement Profiles

The first settlement profiles in the list shown are not intended for use in production but cover various types of internal orders that settle overhead costs, research and development costs, and capital expenses. The two settlement profiles used in production are **YBMFP1** and **YBMF99**. We'll look at the detailed settings for **YBMFP1.** We then see further settlement profiles for projects, maintenance orders, and enterprise asset management. We discuss rework orders in Chapter 11, where we create different production order types, including rework orders without reference to material.

Figure 15.23 shows the detailed settings for settlement profile **YBMFP1**.

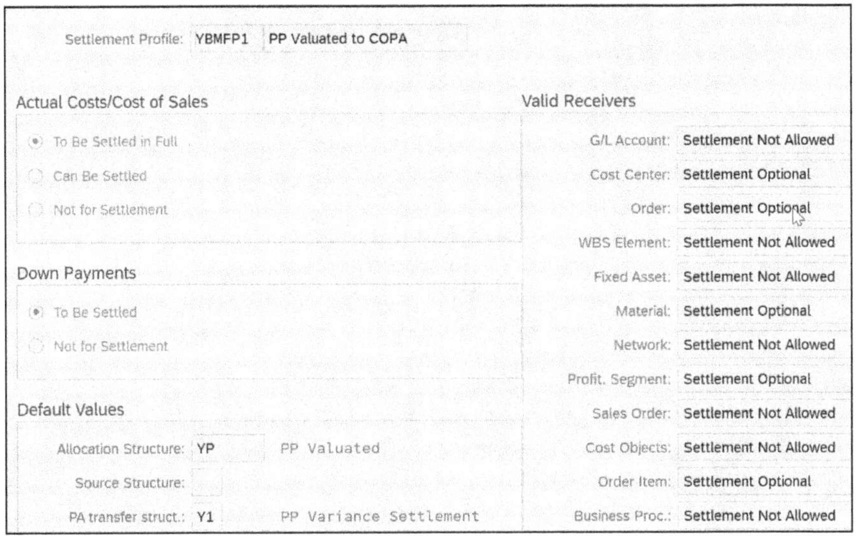

Figure 15.23 Details of the Settlement Profile YBMFP1

Notice the link to the **Allocation Structure** (**YP**) and the list of **Valid Receivers**. The default receivers for a production order are the **Material** (the inventory account for the material produced) and the **Profit. Segment** (the profitability segment for the material produced and any derived characteristics, such as the product group), but it is also possible to settle to an **Order Item** (in the case of joint production) and to add additional distribution rules to settle to a **Cost Center** or **Order** if you do not want to assign the production costs to inventory but rather to reflect the costs, for example, of building a sample product or performing test activities on a product.

The allocation structure ensures that all accounts used to record costs to the production order are correctly collected for the purposes of settlement, and you will receive an error message if you try to settle the costs on an account that has not been included in this structure. Again, the allocation structures are also used outside the production environment, so Figure 15.24 shows allocation structures for many different purposes. As we saw in Figure 15.23, the **Alloc. Str.** (allocation structure) **YP** is used for production orders, and if you work with best practices, you only need to ensure that all relevant general ledger accounts are assigned to a source since all costs should flow to inventory (material) during settlement.

In Figure 15.24, we selected **YP** and double-clicked the **Source** folder. The details are displayed in Figure 15.25, and you can see that **Alloc. Structure** (allocation structure) **YP** includes only one **Assignment** (**100**), and all accounts in the **Cost Elem. Group YB_ALL** are included in this assignment. If you work with the standard chart of accounts, you will not normally need to change this assignment.

Display View "Create Allocation Structure": Overview

Other CO Area | Var. list | More

Dialog Structure
- Allocation structures
  - Assignments
    - Source
    - Assessment Cost Element
    - Settlement cost elements

Alloc. Structures

Alloc.Str.	Text
UA	Universal Allocation Structure
Y1	CO allocation structure
Y2	No Split - All cost elements
Y9	Rework Orders
YE	Allocation Structure PM
YI	PS Project Settlement
YO	PS Overhead Project Settl
YP	PP Valuated
YR	R&D Orders
YSERV	Settlement Service

Figure 15.24 List of Allocation Structures

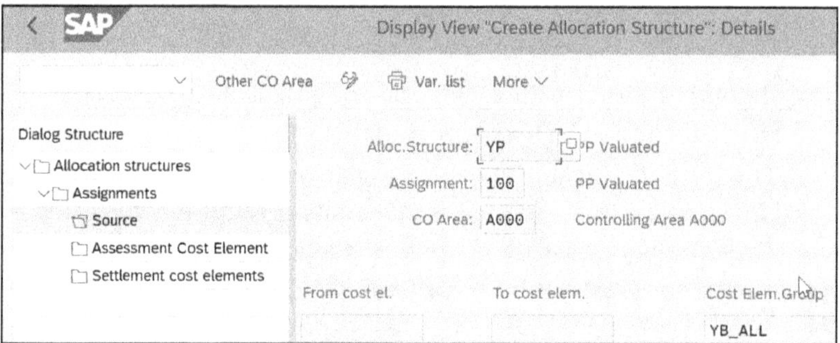

Figure 15.25 Sources in Allocation Structure

With the relevant general ledger accounts collected under a source, we must now link them with a settlement cost element that can be different depending on the receiver of the settlement. Figure 15.26 shows the **Receiver cat.** (receiver categories): **PSG** (profitability segment), **CTR** (cost center), and **ORD** (order). These are the **Valid Receivers** shown earlier Figure 15.23. However, the receivers **MAT** (material) and **OIt** (order item) are missing from the list. When settling on these receivers, you use the materials management account assignment settings to find the correct receiver account.

Although a profitability analysis transfer structure is shown in Figure 15.23, costing-based profitability analysis is not used in the best-practice settings, so there is no need to map the general ledger accounts additionally to value fields. You instead map accounts to cost components in the *cost component split*.

SAP S/4HANA enables you to split the cost of goods sold (COGS) into individual cost components according to the cost component structure of the material cost estimates. COGS are posted in financial accounting and margin analysis during the goods issue to a delivery.

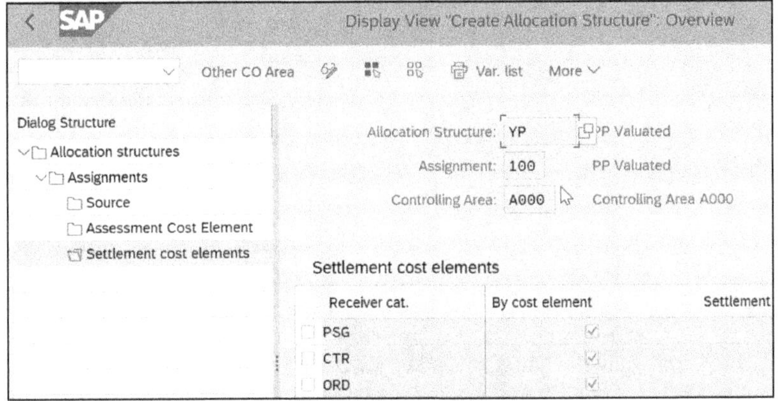

Figure 15.26 Allocation Structure Settlement Cost Elements

The first step in creating the split is to create a general ledger account with cost element category 1 for each cost component of the cost component structure. We discussed how to create general ledger accounts in Chapter 3.

To configure the COGS split in margin analysis, follow the IMG menu path **Financial Accounting • General Ledger Accounting • Periodic Processing • Integration • Materials Management • Define Accounts for Splitting the Cost of Goods Sold**. The screen shown in Figure 15.27 is displayed.

15

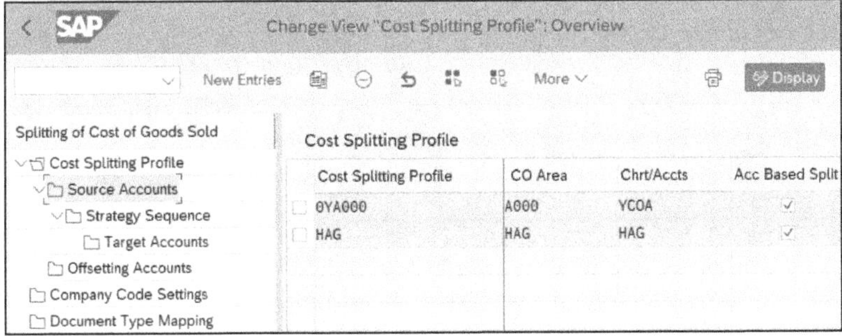

Figure 15.27 Cost Splitting Profile

Select the checkbox in the **Acc Based Split** (account-based split) column to ensure the system always splits costs into cost components when posting to the **Source Accounts**. We'll examine this in detail in the next section. Don't select the checkbox if you only want the system to split the COGS based on the movements of the sales order, which is a common setting.

Choose a **Cost Splitting Profile**, **HAG** in this example, and double-click the **Source Accounts** folder to display the screen shown in Figure 15.28.

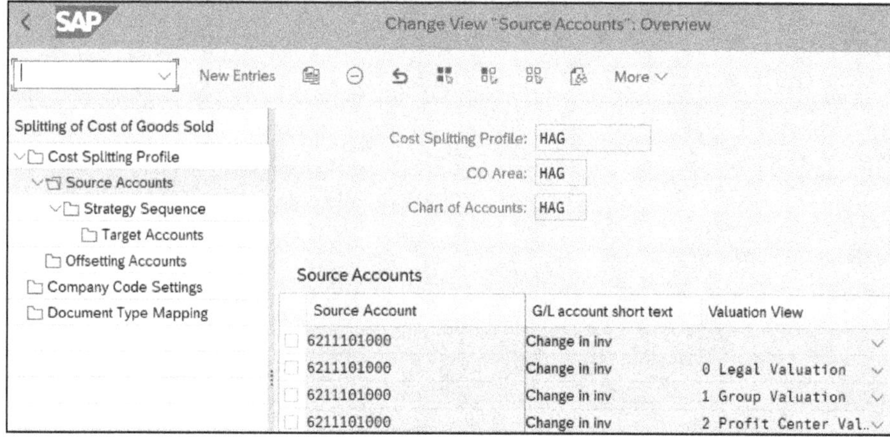

Figure 15.28 Maintain Source Accounts

The **Source Accounts** are the general ledger accounts posted during the goods issue for delivery. You maintain the accounts for the goods issue in account determination with Transaction GBB/VAV. We discussed automatic account determination in Chapter 11. In this example, we maintain general ledger account **6211101000** for Change in Inventory COGS as the **Source Account**.

In the **Valuation View** column, you maintain a specific valuation, such as **Legal**, **Group**, or **Profit Center Valuation**. If you don't maintain a valuation, the split will happen for all existing valuations. Choose a source account by selecting the checkbox at the beginning of the line and double-clicking the **Strategy Sequence** folder to display the screen shown in Figure 15.29.

Change View "Strategy Sequence": Overview

New Entries More

Splitting of Cost of Goods Sold
- Cost Splitting Profile
  - Source Accounts
    - Strategy Sequence
      - Target Accounts
  - Offsetting Accounts
- Company Code Settings
- Document Type Mapping

Cost Splitting Profile: 5777
Source Account: 400610
Valuation View:
CO Area: 5777
Chart of Accounts: 5777

Strategy Sequence

Sequence Number	Strategy	Strategy Type
10	10	3 Current Standard Cost Estimate
20	20	5 Cost Component Split from Actual Costin...
30	30	1 Released Cost Estimates
40	40	2 Upcoming Released Cost Estimates
50	50	4 Future Standard Cost Estimate

Figure 15.29 Strategy Sequence

In the **Strategy Sequence**, you define a **Strategy** for every **Source Account**, after which the system tries to split the cost. If the system doesn't find a material cost estimate in the first sequence, it moves on to the next sequence. In Figure 15.29, according to **Strategy 10** with **Sequence Number 10**, the system will split the cost with the **Current Standard Cost Estimate**. This is the cost estimate active in the **Costing 2** view at the time of the goods issue.

Next, you maintain a target account for every cost component of the cost component structure. The target account must be a general ledger account with cost element type 1 (primary costs/cost-reducing revenues). Double-click the **Target Accounts** folder on the left to display the assignment screen in Figure 15.30.

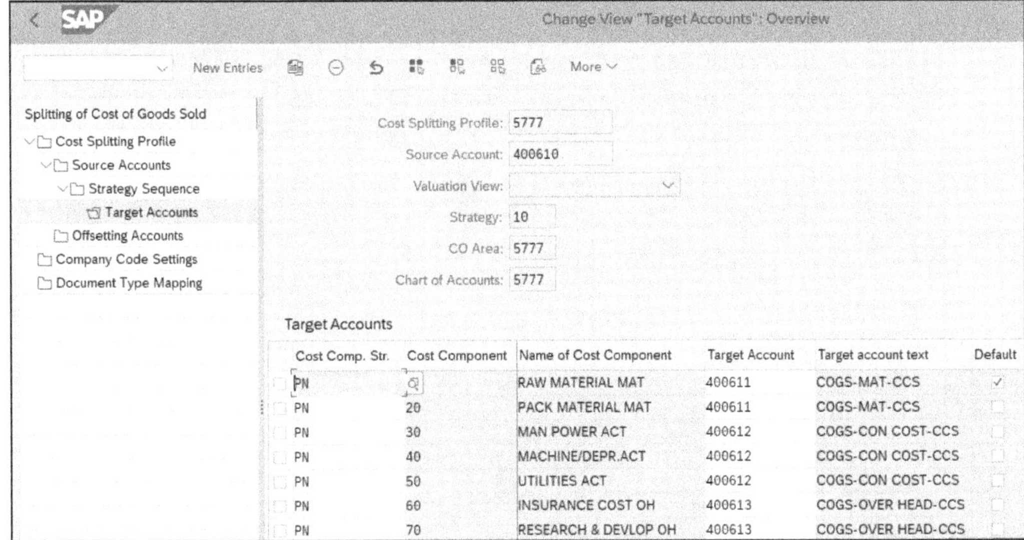

Figure 15.30 Maintain Target Accounts

In this screen, you assign a **Target Account** to each **Cost Component** of the **Cost Comp. Str.** (cost component structure). We discussed cost components in Chapter 6.

Assign one cost component as the **Default** so that if the system can't determine a target account for a cost component according to the cost-splitting profile, this cost component is assigned to the default cost component.

Now, let's examine offsetting accounts. Usually, the COGS account is used as an offsetting account for the COGS component split. You can compare the originally posted costs with the actual costs according to the Material Ledger. Double-click the **Offsetting Accounts** folder on the left to display the screen in Figure 15.31.

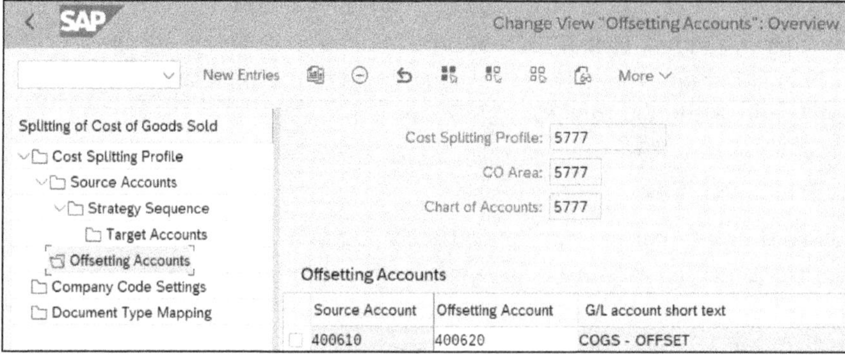

Figure 15.31 Maintain Offsetting Accounts

In this screen, you maintain an **Offsetting Account** for each **Source Account**. Before saving the cost-splitting profile, you must activate it for a company code. Double-click the **Company Code Settings** folder on the left to display the screen in Figure 15.32.

Change View "Company Code Settings": Overview

New Entries More

Splitting of Cost of Goods Sold
Cost Splitting Profile
Source Accounts
Strategy Sequence
Target Accounts
Offsetting Accounts
Company Code Settings
Document Type Mapping

Company Code Settings

Company Code	Valid From	Cost Splitting Profile
5777	01.01.2023	5777

Figure 15.32 Company Code Settings

You assign your **Cost Splitting Profile** to a **Company Code** in Figure 15.32. You can assign a different document type for each posting of the COGS splitting. Select a **Company Code** and double-click the **Document Type Mapping** folder on the left to display the screen in Figure 15.33.

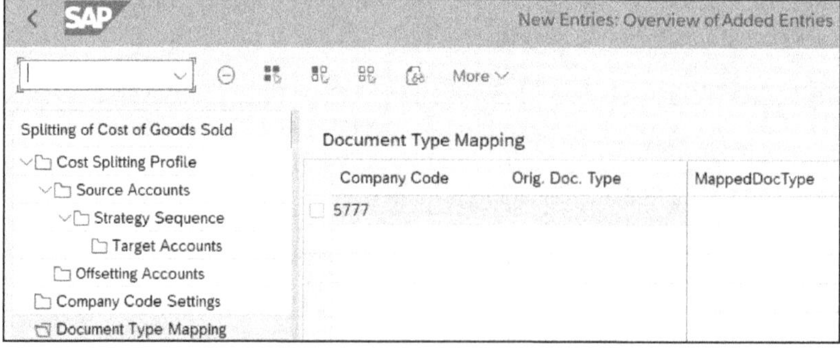

Figure 15.33 Document Type Mapping

For each **Company Code**, you can map a separate **MappedDocType** (mapped document type) to the **Orig. Doc. Type** (original document type).

Now that we've discussed settlement configuration, let's analyze the settlement period-end process. Refer to Chapter 18 for information on event-based processing.

15.3 Settlement Period-End Process

You run period-end settlement with Transaction KK87 (individual) and Transaction CO88 (collective) or by following the menu path **Accounting • Controlling • Product Cost Controlling • Cost Object Controlling • Product Cost by Period • Period-End Closing • Single Functions: Product Cost Collector • Settlement**. A selection screen displays, as shown in Figure 15.34.

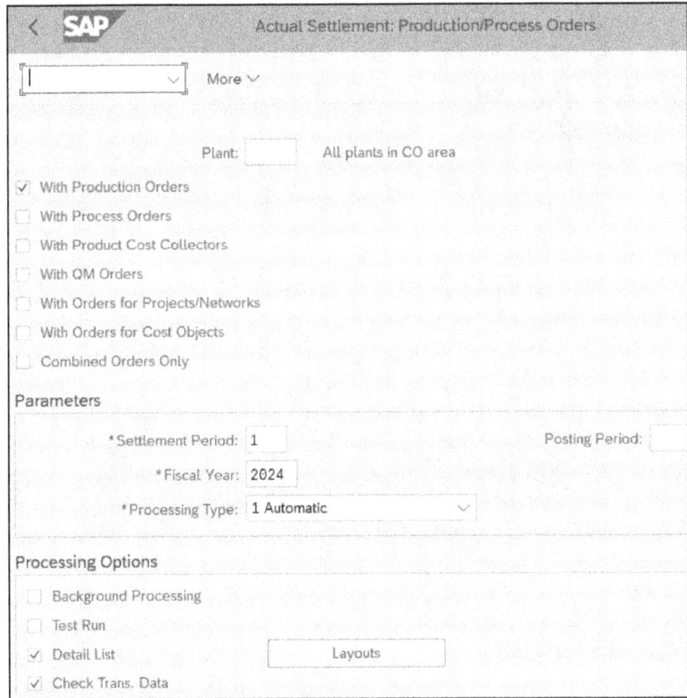

Figure 15.34 Period-End Settlement Selection Screen

> **Production and Process Order Settlement**
>
> You carry out settlement for production and process orders with Transaction KO88 (individual) and Transaction CO88 (collective).

Plant is not mandatory, so you can enter a plant or leave it blank to run settlement for **All plants in the CO area** (all plants in the Controlling area). In this example, we'll settle only production orders by selecting the **With Production Orders** checkbox. You can include other objects in the settlement without running another transaction by selecting the relevant checkboxes.

In this section, we explain the parameters that affect settlement and determine how to process the orders to be settled.

15.3.1 Parameters

Let's discuss the fields in the **Parameters** section of the **Settlement** screen:

- **Settlement Period**
  The **Settlement Period** field determines for which period you *calculate* the settlement.
- **Posting Period**
  Product cost by period requires you to settle each period sequentially, which can lead to difficulties if a prior period needs to be reversed. Using the **Posting Period** field shown to the right, you can reverse, correct, and resettle in prior periods by *posting* to the present or previous period.
- **Fiscal Year**
  The fiscal year can be the same as the current calendar year, but it doesn't have to be.
- **Processing Type**
  Click in the **Processing Type** field in Figure 15.34 and press F4 to display the list of possible entries shown in Figure 15.35.

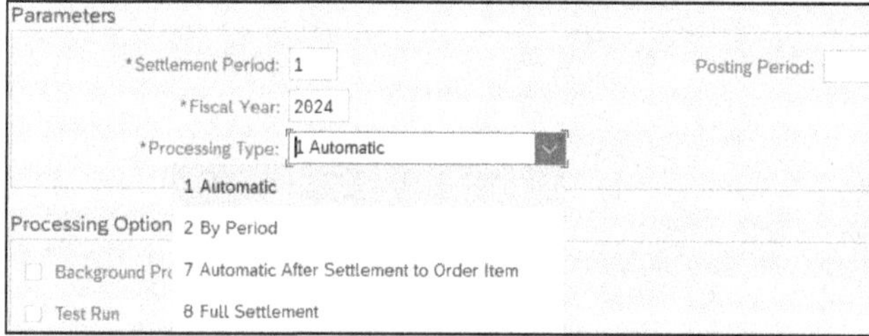

Figure 15.35 Processing Type Possible Entries

  The default **Processing type** field selection is **1 Automatic.** Following is a description of each processing type:

  - **1 Automatic**
    Settles order costs per the settlement type of each order.
  - **2 By Period**
    Selects orders with settlement type PER.
  - **7 Automatic After Settlement to Order Item**
    Choose this processing type if you first settle with Transaction CO8A as part of the *distribution co-product orders* function.

- **8 Full Settlement**
  An error message is generated if costs for previous periods occur on a PER order.

Click the **Processing type** field and press [F1] to access standard detailed documentation on your options.

> **Note**
> You can find more information on processing types in Tip 56 of *100 Things You Should Know About Controlling with SAP* (2nd ed., SAP PRESS, 2015) at *https://www.sap-press.com/3746/*.

15.3.2 Processing Options

Let's discuss the fields in the **Processing Options** section of the **Settlement** screen:

- **Background Processing**
  You can use background processing to avoid overloading the system at busy times by selecting the **Background Processing** checkbox.
- **Test Run**
  Selecting the **Test Run** checkbox has the following outcomes:
  - The program executes normally.
  - The system performs all checks in the program.
  - Results are output in various forms, for example, as a:
    - Single message
    - Collection of messages
    - Log
  - Existing data isn't changed.
- **Detail List**
  Select the **Detail List** checkbox to display a detail list for analysis following settlement (as shown later in Figure 15.37). If you don't select the checkbox, only the basic list is available following settlement, as shown in Figure 15.36. You can also analyze settlement postings by displaying and sorting the settlement general ledger account line-item report in financial accounting with Transaction FBL3N.
- **Check Trans. Data (check transaction data)**
  If you select the **Check Trans. Data** checkbox, the system checks whether you posted any transaction data to the product cost collector or the manufacturing order since the last settlement. If you didn't post any transaction data, sender processing stops, which improves processing time. The system won't issue the error messages that it could have issued during settlement, which improves message analysis by reducing the number of redundant messages.

Complete the selection screen as described in the steps above and click the **Execute** button, or press [F8] to display the settlement basic screen shown in Figure 15.36.

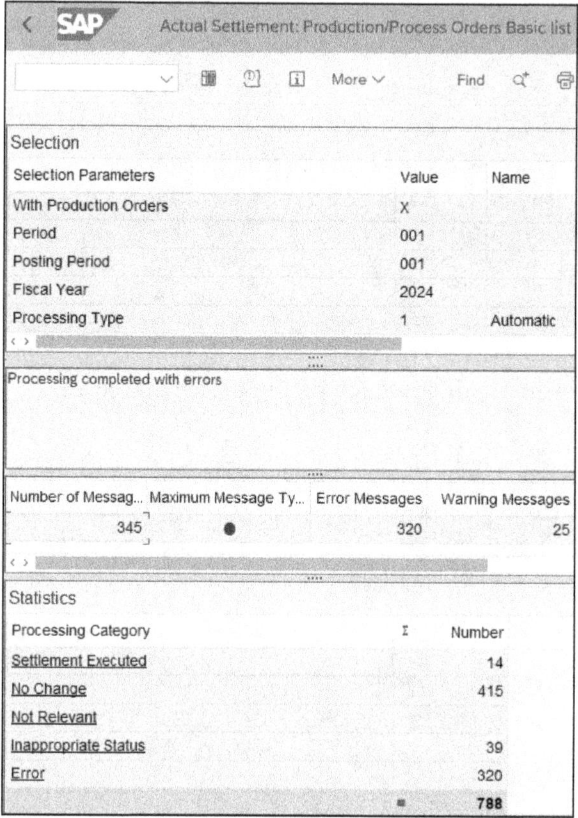

Figure 15.36 Actual Settlement Basic Screen

This screen provides basic information on the parameters you entered in the previous selection screen, such as **Period**, **Fiscal Year**, and **Processing Type**. Click the **Detail Lists** (grid) icon to proceed to a detailed list of settlement values, as shown in Figure 15.37.

In Figure 15.37, the **ValCOArCur** (value in controlling area currency) column is sorted in descending order, which provides visibility of the production orders with the largest settlement amounts. You can reconcile the sum of values settled at the bottom of Figure 15.37 with the sum of the collective variance calculation discussed in Chapter 14. You can also reconcile with postings to settlement financial accounts by sorting the settlement account line-item report with Transaction FBL3N.

When you settle to costing-based profitability analysis, you will also see value fields listed as receivers in Figure 15.37.

Settlement is the last step in period-end closing for product cost collectors and manufacturing orders.

Senders	Text send.	Receiver	ValCOArCur
ORD 1002009	T-HA01	MAT 1710/T-HA01	800.00
ORD 1002055	Bike	MAT 1710/2630	360.00
ORD 1002013	T-HA01	MAT 1710/T-HA01	300.00
ORD 1002053	Bike	MAT 1710/2630	180.00
ORD 1002059	Bike	MAT 1710/2630	180.00
ORD 1002014	T-HA01	MAT 1710/T-HA01	100.00
ORD 1002054	Bike	MAT 1710/2630	100.00
ORD 1002040	Bike	MAT 1710/2630	75.00
ORD 1002042	Bike	MAT 1710/2630	72.00
ORD 1002086	Bike	MAT 1710/2630	60.00
ORD 1002087	Bike	MAT 1710/2630	60.00
ORD 1002090	Bike	MAT 1710/2630	30.00
ORD 1002052	Bike	MAT 1710/2630	10.00
ORD 1002089	Bike	MAT 1710/2630	6.00
			2,333.00

Detail list - Settled values

Figure 15.37 Actual Settlement Detail List

15.4 Summary

In this chapter, we examined settlement configuration transactions, including the settlement profile, which defaults from the order type and includes parameters, such as a requirement for the order to settle in full before you can delete it, and the default values for other settlement transactions. The settlement profile also defines valid receivers such as general ledger accounts and cost centers.

We also examined the allocation structure, which allocates the costs incurred on a sender by cost element or cost element group. An assignment assigns a cost element or cost element group to a settlement cost element.

We examined the source structure, which you define when settling and costing joint products. A source structure contains several source assignments, each containing the individual cost elements or cost element intervals to settle using the same distribution rules.

We also looked at the profitability analysis transfer structure, which assigns source costs to value fields in costing-based profitability analysis.

We looked at how to configure the variance split to map general ledger accounts to variance categories for variance reporting in margin analysis. We also set up the cost of goods sold split to map general ledger accounts to individual cost components.

Finally, we examined the settlement period-end processing step. Later on, in Chapter 18, we'll examine event-based processing. In the next chapter, we'll review actual costing.

Chapter 16
Actual Costing

Actual costing valuates all goods movements within a period at the standard price. At the end of the period, the price and exchange rate differences are used to calculate an actual weighted average price for the period, called the periodic unit price (PUP).

Some accounting standards state that standard costs are acceptable if adjusted at reasonable intervals to reflect current conditions. Actual costing allows the automation of variance capitalization by calculating the actual cost for manufactured goods.

Actual costing valuates all goods movements within a period at the standard price. All price and exchange rate differences for materials are collected as Material Ledger documents. At the end of the period, the differences are used to calculate an actual average price for the PUP.

In this chapter, we'll explore how to configure, set up, and run actual costing.

16.1 Basic Concepts

In SAP S/4HANA, the Material Ledger is always active. You choose within the ledger settings how many currencies will be available (up to three before universal parallel accounting and up to 10 with it—see Chapter 19). In addition, you can carry inventory in two additional valuations to reflect the group and profit center view. Currency amounts are translated into foreign currencies at historical exchange rates directly at the time of posting and then rolled through the quantity structure during the costing run.

> **Note**
> If you use transfer prices when moving materials between internal legal entities, the Material Ledger allows you to view parallel inventory valuations, including transfer pricing for legal reporting purposes and excluding transfer pricing for internal management and consolidated reporting requirements.

Actual costing is not activated by default, and transaction-based price determination is set to **2**. (**The Price Determ.** [price determination] field in the **Accounting 1** view has a value of **2**.) However, the use of actual costing is becoming more common as economic conditions become more volatile. Refer to Chapter 3 for more information on the **Price Determ.** field in the **Accounting 1** view.

Changing Currencies and Actual Costing

Refer to SAP Note 53947 for more information on changing currencies and the simplification list at *http://s-prs.co/v58303* for technical changes in the Material Ledger with actual costing.

Actual costing is based on the Material Ledger documents collected with every goods movement. The documents valuate all goods movements within a period at the standard price (preliminary valuation). At the end of the period, you calculate the actual price for each material based on the actual costs of the period. As mentioned, this actual price is known as the PUP, and you can use it to revaluate the inventory, the work in process (WIP), and the cost of goods sold for the period you are closing. In addition, you can use the actual price as the standard price for the next period.

Actual costing determines what portion of the variance you debit to the next-highest level using material consumption. The actual BOM enables you to roll up variances over multiple production levels all the way to the finished product and to revalue any inventories held for the affected materials. Additionally, you can choose to have variances from cost centers and business processes considered and to make cross-company variance assignments.

If you want the system to calculate a PUP for your materials based on the actual costs incurred in a period, you must activate actual costing. In the next section, we'll discuss the configuration required to activate actual costing.

In addition to standard price control, you must choose price determination **3** to activate actual costing for your materials. As mentioned previously, the **Price Determ.** field is in the **Accounting 1** view. Refer to Chapter 3 for more information on the **Price Determ.** field.

16.2 Configuration

In the following sections, we'll examine the configuration steps required for actual costing.

16.2.1 Activate Valuation Areas for Material Ledger

Material Ledger is activated by default with S/4HANA. You need to determine the price determination setting with Transaction OMX1 or by following the IMG menu path

Controlling • Product Cost Controlling • Actual Costing/Material Ledger • Activate Valuation Areas for Material Ledger. Double-click the **Activate Material Ledger** text to display the screen shown in Figure 16.1.

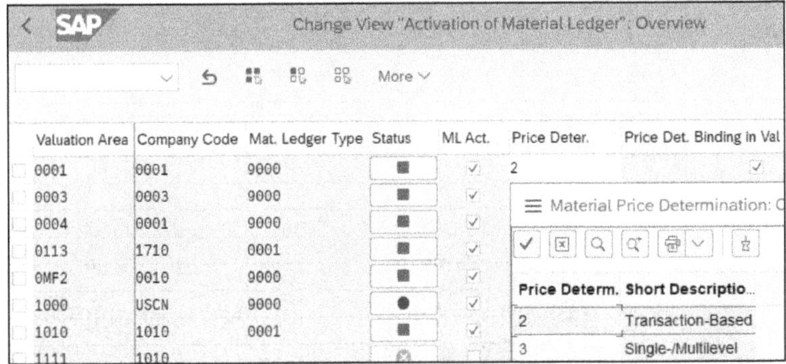

Figure 16.1 Configure Material Ledger and Price Determination

Let's examine the checkboxes:

- **ML. Act.**
  The **ML Act.** (Material Ledger active) checkbox will already be selected in production, indicating the Material Ledger is active for a **Valuation Area** (plant) and **Company Code** combination in SAP S/4HANA.
- **Price Deter.**
  Click the **Price Deter.** (price determination) field and then press F4 to display the list shown in Figure 16.1. This setting determines the proposed material price determination when you create a new material. This entry is ignored at production startup and is automatically set to "2" for all materials. Refer to Chapter 3 for more information on the **Price Determ.** checkbox in the **Accounting 1** view.
- **Price Det. Binding in Val Area**
  Select this checkbox so price determination cannot be changed for the materials in the **Valuation area**.

16.2.2 Assign Currency Types to Material Ledger Type

You assign currency types to the Material Ledger type with Transaction OMX2 or by following the IMG menu path **Controlling • Product Cost Controlling • Actual Costing/ Material Ledger • Assign Currency Types to Material Ledger Type.** The screen in Figure 16.2 is displayed.

You assign currency types (**CrcyTyp**) to Material Ledger types (**ML Type**). In SAP S/4HANA, the **Manual** checkbox is selected by default because you can have up to 10 financial accounting currency types, while you can only have three in the Material Ledger. You choose the three financial accounting currency types for the Material Ledger manually by double-clicking the **Define individual characteristics** folder on the left.

Change View "Define material ledger type": Overview

New Entries More

Dialog Structure
- Define material ledger type
  - Define individual characteristics

ML Type	CT from FI	CO CrcyTyp	Manual	Description
0001	☐	☐	☑	Crcy type/val. 10 30
0002	☐	☐	☑	Test
1710	☐	☐	☑	Material Ledger for 1710

Figure 16.2 Assign Currency Types to Material Ledger Type

With Transaction OB22 in SAP ERP prior to SAP S/4HANA, you defined up to three currency types in financial accounting for each company code.

In SAP S/4HANA, you receive a message that you must edit currency settings with Transaction FINSC_LEDGER. You can still display the **Additional Local Currencies** screen by double-clicking on a company code to proceed to the display-only screen shown in Figure 16.3.

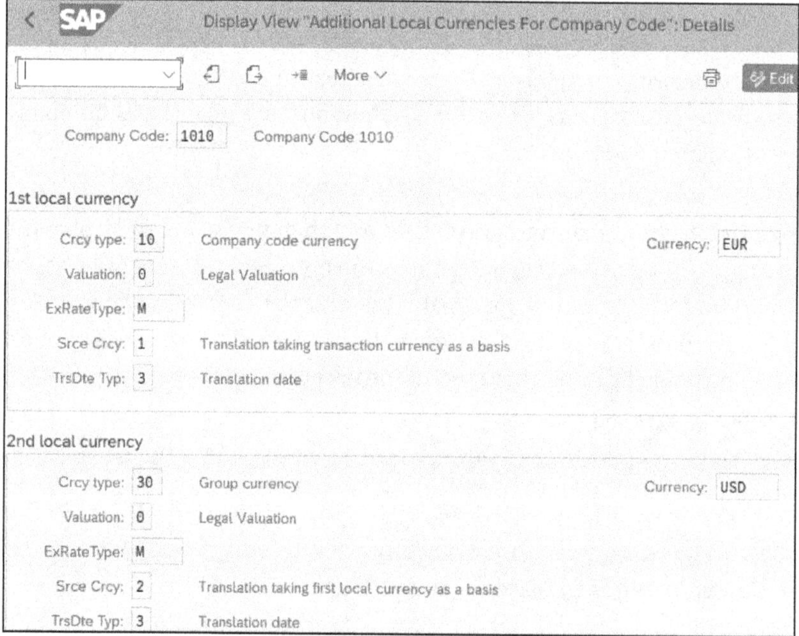

Figure 16.3 Display Additional Local Currencies for Company Code 1010

The system uses the currency types defined in customizing as additional local currencies for a company code. If you define additional local currencies as shown in the **2nd local currency** section of Figure 16.3, every financial accounting document includes the postings in the additional local currencies. You may need to add document layout columns to display the additional currencies in accounting documents.

The Universal Journal allows up to eight user-defined currency types in addition to the company code currency and the global or controlling area currency. This means you have 10 currency fields in financial accounting in table ACDOCA. (You do not have to use all 10 currency fields.) You define the currency types for the Universal Journal with Transaction FINSC_LEDGER or via the IMG menu path **Financial Accounting • Financial Accounting Global Settings • Ledgers • Ledger • Define Settings for Ledgers and Currency Types.** Select a ledger and double-click the **Company Code Settings for the Ledger** folder. The screen in Figure 16.4 is displayed.

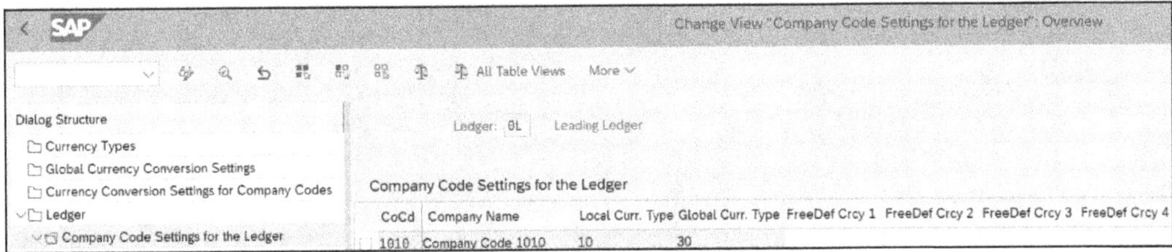

Figure 16.4 Currency Types for Ledger 0L Leading Ledger

Company Code 1010 has been assigned two currency types: **Local 10** and **Global 30**. You can see the first four fields of the eight possible user-defined currency types to the right. Double-click the line to see the valuation views assigned to this ledger, as shown in Figure 16.5.

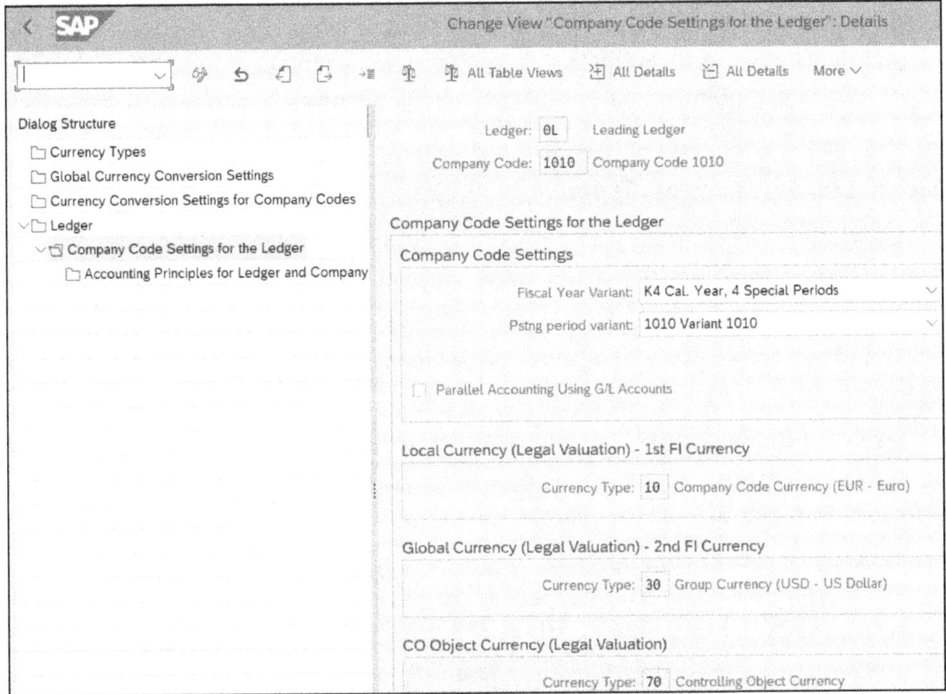

Figure 16.5 Valuation Views of Ledger 0L the Leading Ledger

The **1st FI Currency** and **2nd FI Currency** in Figure 16.5 correspond to those shown earlier in Figure 16.3. Double-click the **Currency Types** folder on the left to display the screen shown in Figure 16.6.

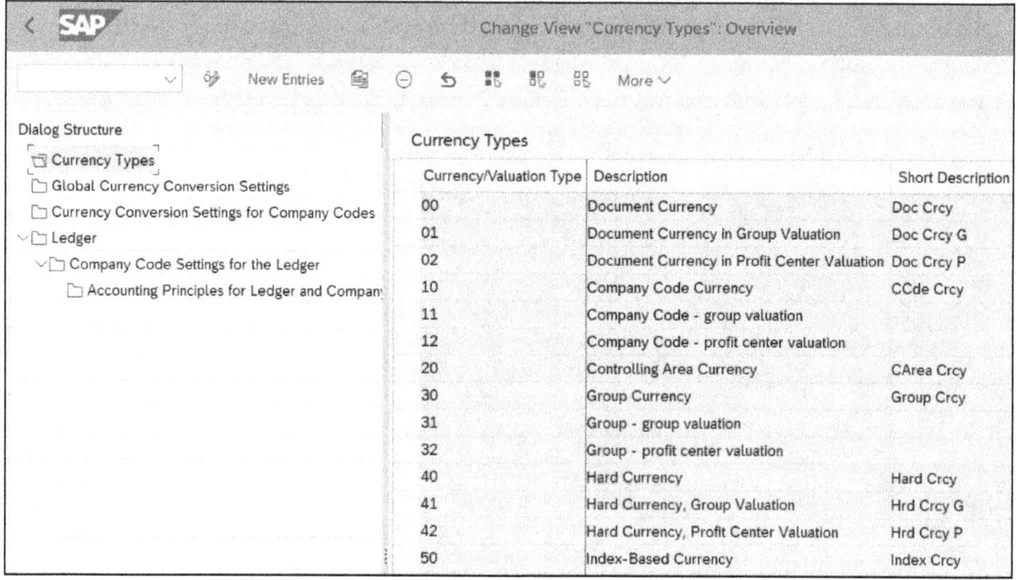

Currency/Valuation Type	Description	Short Description
00	Document Currency	Doc Crcy
01	Document Currency in Group Valuation	Doc Crcy G
02	Document Currency in Profit Center Valuation	Doc Crcy P
10	Company Code Currency	CCde Crcy
11	Company Code - group valuation	
12	Company Code - profit center valuation	
20	Controlling Area Currency	CArea Crcy
30	Group Currency	Group Crcy
31	Group - group valuation	
32	Group - profit center valuation	
40	Hard Currency	Hard Crcy
41	Hard Currency, Group Valuation	Hrd Crcy G
42	Hard Currency, Profit Center Valuation	Hrd Crcy P
50	Index-Based Currency	Index Crcy

Figure 16.6 Currency Types Overview

In the **Currency/Valuation Type** column, the first digit refers to the currency, while the second digit refers to the valuation. Together, the two digits refer to the valuation approach. Let's examine currency types in more detail next.

16.2.3 Standard Currency Types

In this section, we'll examine currency types and valuation views and how to combine them to create valuation approaches in the currency and valuation (C&V) profile.

Currency types specify what the currency is used for. You assign a currency key (e.g., USD, EUR) to a currency type. The following bullet points explain some of the standard SAP currency types in Figure 16.6:

- **Document currency (00)**
  This is the currency in which an accounting document is posted. It can be determined automatically or entered manually.
- **Company code currency (10)**
  This is typically the currency of the company code country. You set the company code currency in the currency settings with Transaction OX02.

- **Controlling area currency (20)**
  The controlling area currency is defined in controlling area configuration with Transaction OKKP, as shown in Figure 16.7. The controlling area **Currency** is **USD** in this example. You also typically set the controlling area **Currency Type** to group currency (**30**), as shown in Figure 16.7.

Controlling Area: US00
*Name: US Controlling Area
Person Responsible:
Assignment Control
*CoCd->CO Area: 2 Cross-company-code cost accounting
Currency Setting
*Currency Type: 30 Group currency
*Currency: USD United States Dollar ☑ Diff. CCode Currency
Curr/Val. Prof.: ☐ Active

Figure 16.7 Controlling Area Currency

- **Group currency (30)**
  The group currency is defined in the settings for the SAP client in Transaction SCC4. The default group currency delivered with the SAP Best Practices content is USD.
- **Hard currency (40)**
  Hard currency is used for external reporting for countries with high inflation. You assign it to the country settings in Transaction OY01 as shown in Figure 16.8. The **Hard Currency**, **USD**, is set for **Venezuela** in this example.

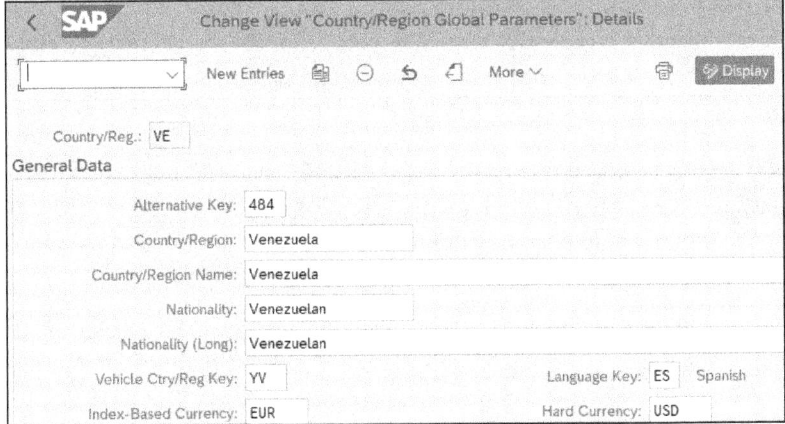

Figure 16.8 Index-Based Currency and Hard Currency

- **Index-based currency (50)**
  This currency is used for external reporting for countries with extra-high inflation. You assign it to the country settings in Transaction OY01 in the **Index-Based Currency** field, as shown in Figure 16.8.
- **Global company currency (60)**
  This currency is assigned to the company (trading partner) configuration in Transaction OX15, as shown in Figure 16.9. In Figure 16.9, **Currency USD** has been assigned to **Company 1710**, which means that USD is the global currency for **Company 1710**.

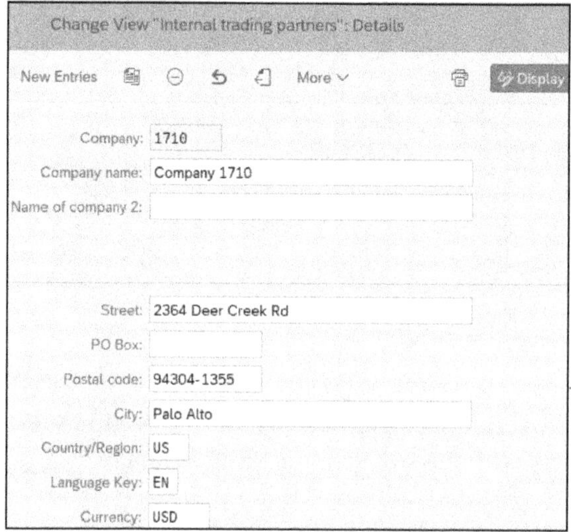

Figure 16.9 Global Currency

- **Controlling object currency (70)**
  This is the currency type of the controlling object, such as a cost center, internal order, production order, or WBS element. The company code currency typically defaults to the controlling object currency. You can change it if the company code currency is the same as the controlling area currency.

We've discussed currency types, so now, let's look at valuation views. These are needed to value inventory in views other than just legal valuation. International companies that use transfer pricing may also need to value inventory in a valuation that excludes markup, such as group valuation.

In the C&V profile, you determine which currency profile to associate with a valuation view, which creates a valuation approach. Let's discuss the most common valuation views:

- **Legal Valuation**
  Every company needs to value inventory in legal compliance within its jurisdiction. The legal valuation approach must be managed in the company code currency (currency type 10) and is fixed.

- **Group Valuation**
  This view eliminates markup from the legal view for group reporting. You can manage the group valuation in the company code currency (currency type 10) or the group currency (currency type 30).
- **Profit Center Valuation**
  This allows companies to report income statements for divisions separately, including internal profits. The inventory held for the buying division will include any profit created by the selling division. You can manage the profit center valuation approach in the company code currency (currency type 10) or the group currency (currency type 30).

The C&V profile combines a currency type and valuation view to create a valuation approach. For example, valuation approach 31 combines 30 for group currency and 1 for group valuation.

To manage multiple valuations in parallel, you must set up a C&V profile with Transaction 8KEM or by following the IMG menu path **Controlling • General Controlling • Multiple Valuation Approaches/Transfer Prices • Basic Settings • Maintain Currency and Valuation Profile.** The screen in Figure 16.10 is displayed.

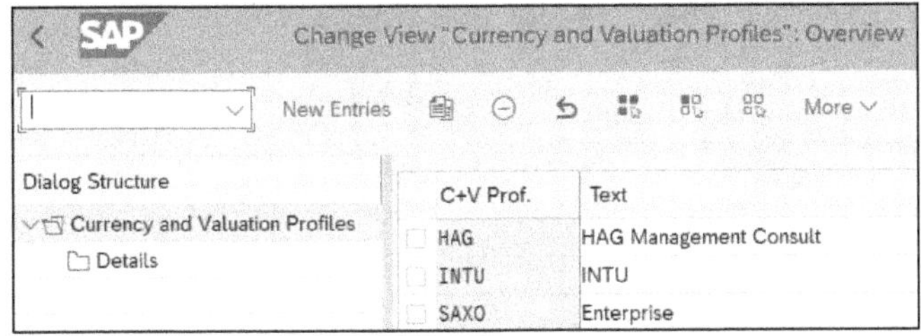

Figure 16.10 Maintain Currency and Valuation Profile Overview

A list of **Currency and Valuation Profiles (C+V Prof.)** is displayed. To maintain the details of a profile, select one and double-click the **Details** folder. The screen in Figure 16.11 is displayed.

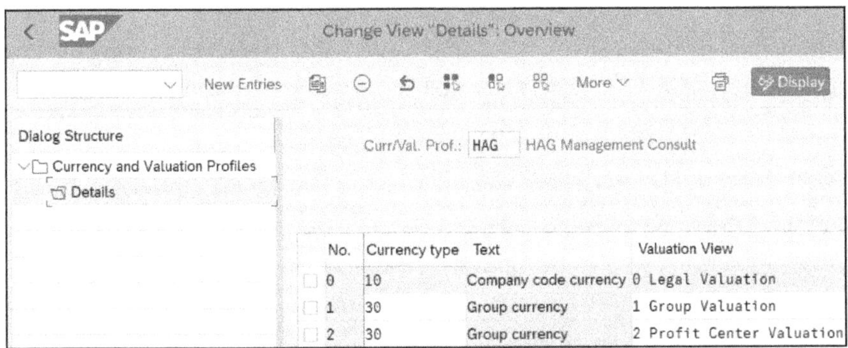

Figure 16.11 Maintain Currency and Valuation Profile Details

You define additional currency types for all posted documents. **Currency and Valuation Profiles** are needed to manage multiple valuations in parallel.

After you create a C&V profile, you assign it to a controlling area with Transaction 8KEQ or by following the IMG menu path **Controlling • General Controlling • Multiple Valuation Approaches/Transfer Prices • Basic Settings • Assign Currency and Valuation Profile to Controlling Area**. You can assign a C&V profile directly to a controlling area with Transaction OKKP.

When you maintain versions with Transaction OKEQ, you can check the C&V profile in a controlling area. Select a version, double-click **Controlling Area Settings**, and click the **Valuation** button. If you work with the best-practice settings, group valuation is delivered as scope item 5W2 and profit center valuation as scope item 6VQ. Profit center valuation is not available in SAP S/4HANA Cloud.

16.2.4 Activate the Currency and Valuation Profile

To use transfer pricing in the controlling area, you must activate the C&V profile with Transaction 8KEP or by following the IMG menu path **Controlling • General Controlling • Multiple Valuation Approaches/Transfer Prices • Activation • Multiple Valuation Approaches: Check/Execute Activation**. The screen in Figure 16.12 is displayed.

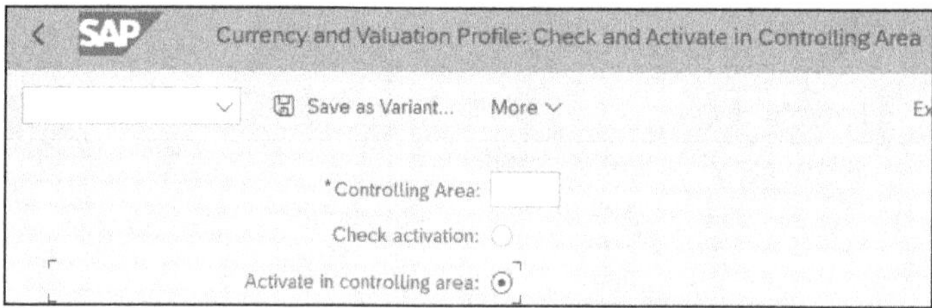

Figure 16.12 Activate the Currency and Valuation Profile for a Controlling Area

Enter your **Controlling Area**, select **Activate in controlling area**, and click the **Execute** button or press F8. The group and profit center valuation views are now available in the controlling area.

In the following section, we'll move on to the next Material Ledger configuration setting. In Chapter 18, we'll explain an alternative approach using universal parallel accounting.

16.2.5 Assign Material Ledger Types to Valuation Area

Now that we've maintained Material Ledger types, you assign them to valuation areas (plants) with Transaction OMX3 or by following the IMG menu path **Controlling • Product Cost Controlling • Actual Costing/Material Ledger • Assign Material Ledger Types to Valuation Area**. The screen in Figure 16.13 is displayed.

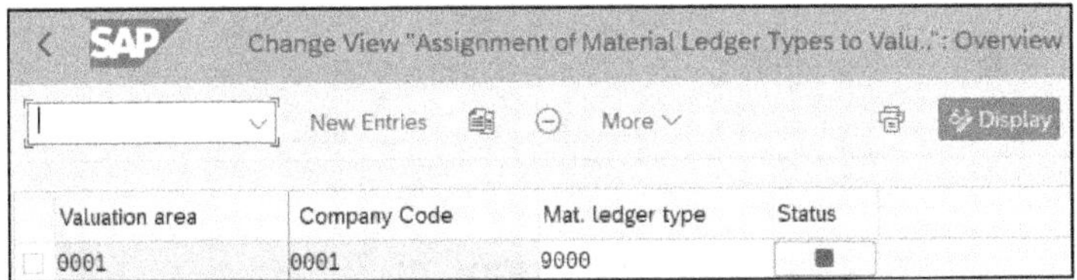

Figure 16.13 Assign Material Ledger Types to Valuation Area

We've now discussed the main configuration transactions required to use the Material Ledger for multiple valuation approaches. Next, we'll consider the configuration required for actual costing.

16.2.6 Activate Actual Costing

You activate actual costing by following the IMG menu path **Controlling • Product Cost Controlling • Actual Costing/Material Ledger • Actual Costing • Activate Actual Costing.** Double-click the text **Activate Actual Costing** to display the screen shown in Figure 16.14.

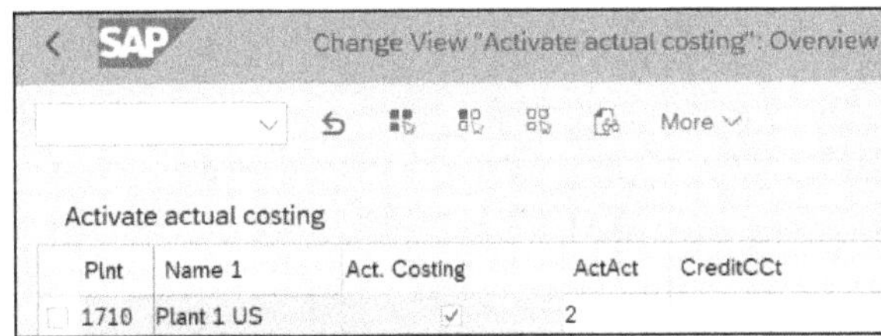

Figure 16.14 Activate Actual Costing

If you work with SAP S/4HANA Cloud or use best practices in SAP S/4HANA, these settings are delivered with scope item 33Q (actual costing). You activate actual costing per plant (**Plnt**) by selecting the **Act. costing** (actual costing) checkbox.

The **ActAct** (update of activity consumption in the quantity structure) field updates the consumption of activities and processes in the actual quantity structure. The following three settings are available:

- **0**
  The update isn't active.
- **1**
  The update is active but not relevant to price determination. You update consumption in the quantity structure, but you don't consider it for price determination.
- **2**
  The update is active and relevant to price determination. Subsequently, you adjust variances between the activity and process prices posted during the period and the actual price at the end of the period.

The **CreditCCt** (credit of cost centers) field is only relevant if you choose option 2 in the update of activity consumption in the quantity structure (**ActAct**) field. You define here how to carry out the credit of the cost centers. Two settings are available:

- **(blank)**
  The cost center is credited during the actual costing run.
- **1**
  The cost center is credited during the alternative valuation run (AVR).

16.2.7 Activate Work in Process at Actual Cost

Work in process (WIP) is the amount posted to the balance sheet before a production order is fully delivered or completed. You valuate WIP at actual via the IMG menu path **Controlling • Product Cost Controlling • Actual Costing/Material Ledger • Actual Costing • Activate WIP at Actual Cost.** Click **New Entries** and activate a plant by selecting the **WIP Active** checkbox in Figure 16.15.

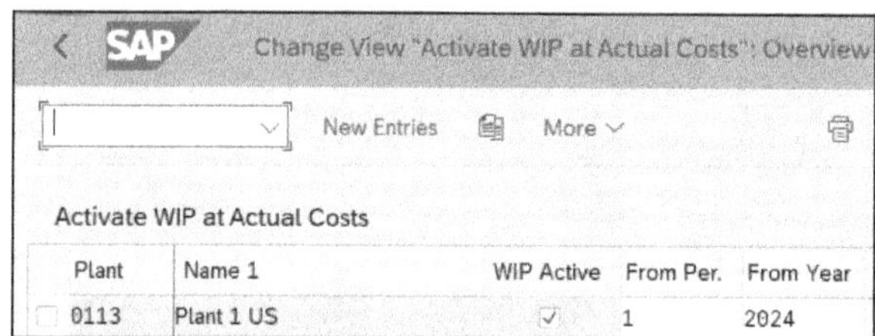

Figure 16.15 Activate WIP at Actual Cost

16.3 Updating the Quantity Structure

Quantity structure generally refers to the assembly BOM and routing, but in actual costing it refers to the goods movements and price documents that impact actual costing and is updated with every goods movement and price change, as we saw in the introduction. It is structured by material and procurement alternative for raw materials and material and production alternative for manufactured goods. Here, we'll discuss single-level or individual materials. An example of a single level is a purchased material with purchase price variance (PPV). These variances stay with the material during single-level price determination. Let's look at single-level price determination versus multi-level price determination:

- **Single-level price determination**
  Single-level price determination refers to an individual material and its procurement process. A simple example of three levels within the Material Ledger includes purchased materials, subassemblies, and finished products. With actual costing, you valuate all materials with a *preliminary* PUP that remains constant for a period. This

price can be, for example, a standard price or an actual price from the previous period determined by the Material Ledger. Review the introduction of this book for an example of a production order goods receipt, settlement, and quantity structure.

Settlement in SAP S/4HANA

In SAP S/4HANA, single- and multilevel price determination and the update of COGS and WIP are combined in the preparation and settlement steps of the actual costing run, which we'll discuss in detail in the next section.

- **Multilevel price determination**

  During *multilevel price determination*, you assign the differences calculated during single-price determination progressively to the next-highest levels of the production process using a multilevel actual quantity structure (the actual BOM). For example, you roll up purchase price differences for raw materials to subassemblies and finished products.

  Periodic unit prices (PUPs) are calculated for the period and updated for information only in all valuation approaches in the material price analysis. The price in the company code currency is updated in the material master for the period.

  During a period, price differences are posted to a price difference account and are updated separately for each material. At the end of a period, single-level price determination assigns the variances for each material. Single-level price determination allows you to assign the cumulative price differences proportionally to the ending inventory quantity or the work in process and to the material consumption of the period (usually the cost of goods sold). If you work with margin analysis, the posting step in actual costing will update the variances for the cost of goods sold to every market segment with goods movements in the period.

Now that we've discussed configuration settings, let's examine the actual costing period-end processing.

16.4 Performing the Actual Costing Run

Typically, the actual costing run is one of the last steps of period-end processing you carry out during the first week of the new period. Prerequisites include:

- All goods movements must be complete for the period
- All price and exchange rate variances must be collected, and all adjustments made
- Activity prices must be calculated
- The new materials management period is open

16.4.1 Getting Started

Material Ledger and actual costing provide inventory reporting that you access with Transaction CKM3 or by following the menu path **Accounting • Controlling • Product Cost Controlling • Actual Cost/Material Ledger • Material Ledger • Material Price Analysis**. The screen in Figure 16.16 is displayed.

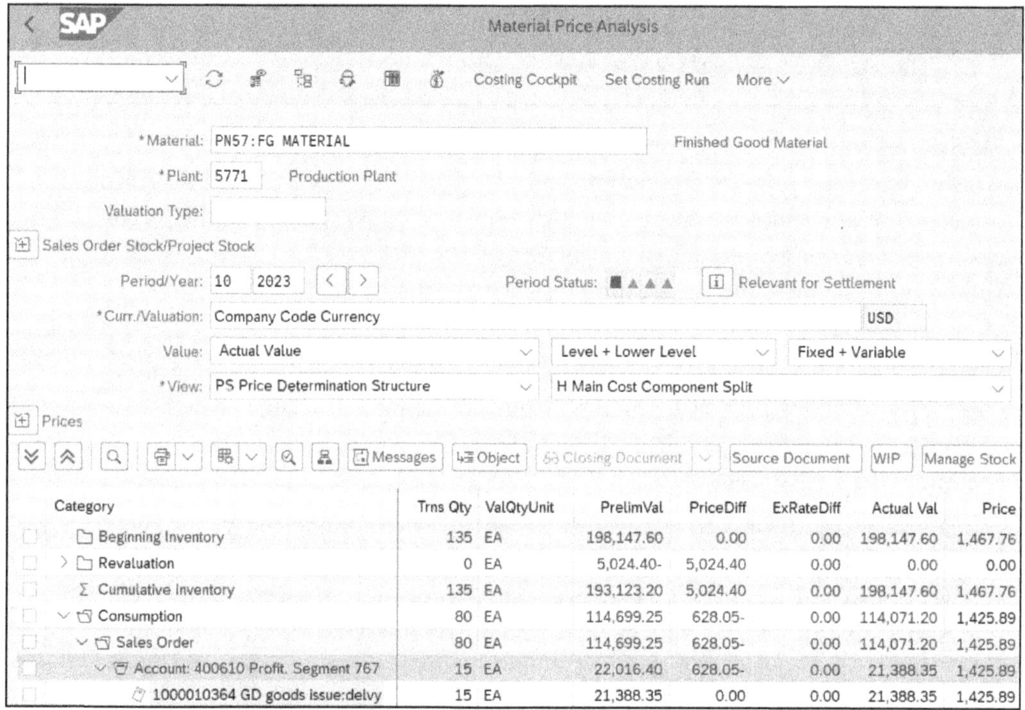

Figure 16.16 Material Price Analysis

This screen provides access to all material transactions for a period. Transactions are grouped by **Category**, which you can progressively expand to display all transaction documents within a category.

16.4.2 Actual Costing Run

Let's look at the first level of period-end processing. To carry out actual costing period-end processing, you create an actual costing run with Transaction CKMLCP or by following the menu path **Accounting • Controlling • Product Cost Controlling • Actual Cost/Material Ledger • Actual Costing • Edit Costing Run**. The screen in Figure 16.17 is displayed.

You first assign plants to the actual costing run and then follow the rows listed in the **Flow step** column of Figure 16.17. In SAP S/4HANA, all plants in a company code must be handled together by the actual costing run. Let's start with the **Selection** step.

Processing													
Flow step	Authorizn	Locked	Parameters	Execute	Log	Mat Stat	Successful	Errors	Still open	Act Stat	Successful	Errors	Still open
Selection							1	0	0		0	0	0
Preparation		X					1	0	0		0	0	0
Settlement		X					1	0	0		0	0	0
Post Closing		X					1	0	0		0	0	0
Mark Prices							1	0	0		0	0	0

Figure 16.17 Edit Actual Costing Run

16.4.3 Selection

All materials with a price determination of 3 in the **Accounting 1** view and all activity types with an activity consumption update (**ActAct**) indicator of 2 are selected for processing. Click the **Parameters** icon in the selection row to display the parameters in Figure 16.18.

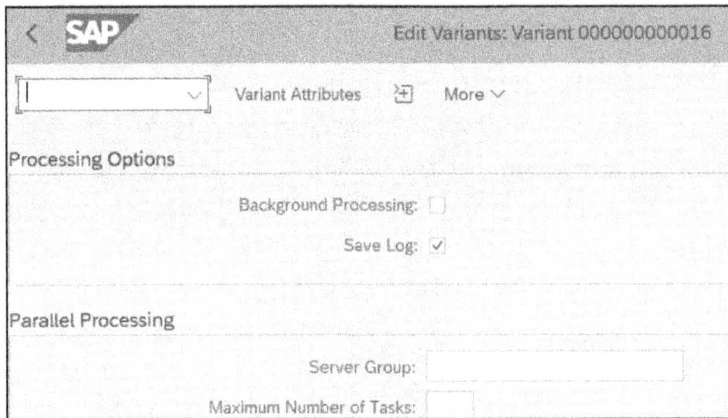

Figure 16.18 Selection Step Parameters

Complete the selection parameters screen with the following steps:

1. Select **Background Processing** only for actual costing runs, with many materials that may time out.
2. Select the **Save Log** checkbox to keep a record of any messages.
3. **Server Group** allows you to spread the processing of the job over several servers.
4. **Maximum Number of Tasks** allows you to limit the number of tasks used by parallel processing.
5. Click the **Save** button to save your parameters.
6. Click the execute icon in the **Selection** step to select the materials and activities.

Now, let's examine the **Preparation** step shown earlier in Figure 16.17.

16.4.4 Preparation

You prepare materials for settlement with the following steps:

1. Determine the single-level variance per material.
   - This assigns the purchase price variances for raw materials and the production variances for assemblies and finished goods.
2. Determine multilevel variance by checking which materials and activities were consumed upward.
3. Prepare a revaluation of consumption—variances that should be transferred to nonmaterial receivers, including the cost of goods sold (COGS) in margin analysis.
   - The posting run in actual costing updates the COGS. In costing-based profitability analysis, you pull the changes due to actual costing using a separate transaction, while in margin analysis, the update is made directly from the costing run.
4. Prepare WIP revaluation by determining how the WIP is impacted by the actual costs.

To proceed with the **Preparation** step, click the padlock icon in the **Authorizn** column shown earlier in Figure 16.17. This allows you to process both the **Preparation** and **Settlement** steps. You can lock goods movements for the period that is being closed by clicking the **X** icon. Be aware that goods movements for the current period won't be locked. The period being processed for actual costing is usually before the current period. Next, click the **Parameters** icon in the **Preparation** row shown earlier in Figure 16.17 to display the screen shown in Figure 16.19.

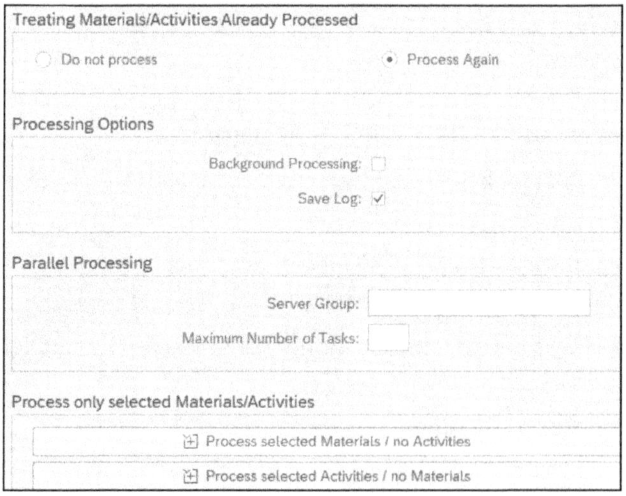

Treating Materials/Activities Already Processed
Do not process
Process Again
Processing Options
Background Processing:
Save Log:
Parallel Processing
Server Group:
Maximum Number of Tasks:
Process only selected Materials/Activities
Process selected Materials / no Activities
Process selected Activities / no Materials

Figure 16.19 Preparation Step Parameters

Let's examine the settings in the four sections of this screen:

- **Treating Materials/Activities Already Processed**
  Select **Do not process** if you want to settle only those materials not settled successfully in the last material settlement. The advantage is that fewer materials are settled. For multilevel material price determination, select **Process Again** to minimize errors.
- **Processing Options**
  For actual costing with many materials, you may need to select **Background Processing** to avoid timing out in the foreground. Select **Save Log** to keep a record of messages.
- **Parallel Processing**
  You can choose a **Server Group** for background processing to spread the processing load over several servers. You can also enter the **Maximum Number of Tasks** that can be processed in parallel.
- **Process only selected Materials/Activities**
  You have the option to select materials for the preparation step. Click the **Process selected Materials/no Activities** button to display the screen shown in Figure 16.20.

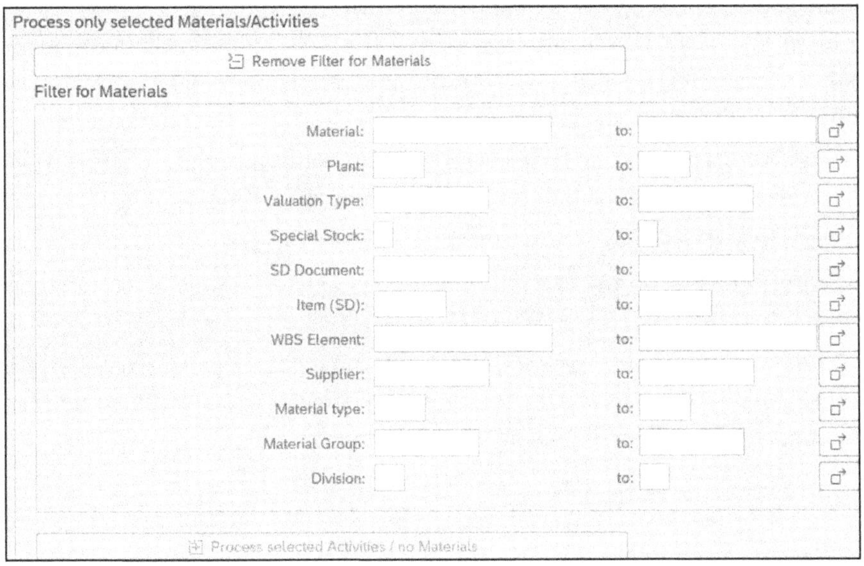

Figure 16.20 Processing Only Selected Materials and No Activities

Enter the **Material** or range of materials you would like to **Filter** for processing for the **Preparation** step. You can also click the **Process selected Activities/no Materials** button to display the **Filter for Activities** fields shown in Figure 16.21.

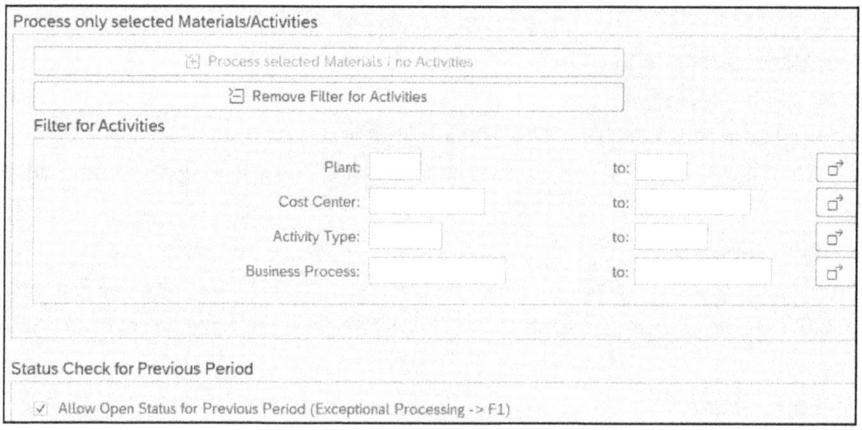

Process only selected Materials/Activities

Process selected Materials / no Activities

Remove Filter for Activities

Filter for Activities

Plant: to:

Cost Center: to:

Activity Type: to:

Business Process: to:

Status Check for Previous Period

Allow Open Status for Previous Period (Exceptional Processing -> F1)

Figure 16.21 Process Only Selected Activities No Materials

Complete the **Cost Center**, **Activity Type**, or **Business Process** fields and press the **Save** button to filter the selection for the **Preparation** step.

- **Status Check for Previous Period**

  Selecting the **Allow Open Status for Previous Period** checkbox allows you to process materials and activities in the **Preparation** step when the previous period hasn't been closed. You receive a warning message that the previous period is still open for the relevant materials/activities. This option should only be used in the following exceptional scenarios:

  - When the actual costing run hasn't been run in previous months in a test system, and you want to test the actual costing run for the current month.
  - For the first Material Ledger close after the plant is set to productive
  - There were errors with the previous actual costing run, and there's no time for reprocessing.

  Press the **Save** button, and then click the execute icon in the **Preparation** step shown earlier in Figure 16.17 to display a results screen that shows the **Status** and number of materials processed per **Costing Level**.

Now, let's proceed to the **Settlement** step shown earlier in Figure 16.17.

16.4.5 Settlement

This step accesses the data set up in the preparation step, including quantity structures, actual activity prices, and costing levels. Variances in the consumption folder in the material price analysis screen shown earlier in Figure 16.16 are considered. Settlement carries out the following four functions:

- **Single-level price determination**
  Calculates the actual price at a single level without considering the multilevel BOM structure.
- **Multilevel price determination**
  Calculates the variances from lower-level materials and activity types.
- **WIP revaluation**
  Calculates variances that relate to WIP revaluation and WIP reduction.
- **Revaluation of consumption**
  Calculates variances to transfer to non-material receivers, including the profitability segments in margin analysis.

To carry out the settlement step, click the parameters icon in the settlement row shown earlier in Figure 16.17 to display the screen shown in Figure 16.22.

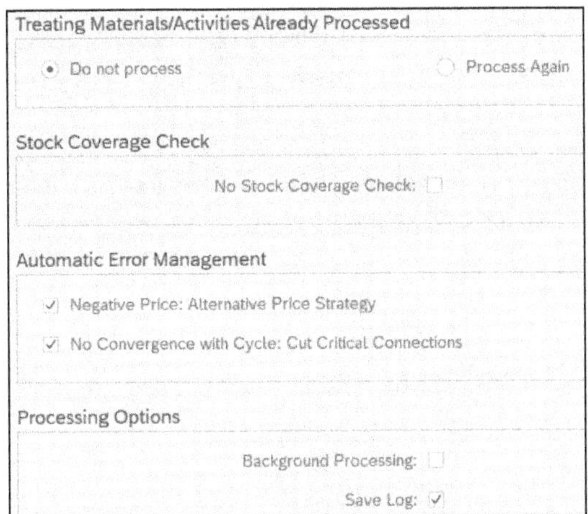

Figure 16.22 Settlement Step Parameters

Let's examine the settings in the four sections of this screen.

- **Treating Materials/Activities Already Processed**
  Select **Do not process,** if you want to settle only those materials not settled successfully in the last material settlement. The advantage is that fewer materials are settled. For multilevel material price determination, select **Process Again** to minimize errors.
- **Stock Coverage Check**
  Single-level price determination calculates the material variances to revalue inventory only if enough inventory is left in stock. This is known as the *coverage check*. If there isn't enough inventory left in stock, only a proportional amount of the variances will be used to revalue inventory to avoid creating a distorted or negative cost. A stock coverage check is applied by default, which you can ignore by selecting the **No Stock Coverage Check** checkbox.

- **Automatic Error Management**
  During price determination, if a negative price is calculated, and you select the **Negative Price: Alternative Price Strategy** checkbox, an alternative price is used to calculate the actual cost. The alternative price is determined via the following sequence until successful:
  - Receipt price of receipts during the period
  - Beginning inventory price
  - Actual cost of the previous period
  - Standard cost of the current period

  The **No Convergence with Cycle: Cut Critical Connections** checkbox allows you to settle variances when a material is a component of itself. Otherwise, the variances may continue in an endless cycle.
- **Processing Options**
  For actual costing with many materials, you may need to select **Background Processing** to avoid timing out in the foreground. Select **Save Log** to keep a record of messages.

Click the **Save** button to save your settlement parameters. Then click the execute icon in the settlement row in Figure 16.17 to display the settlement results screen.

16.4.6 Post Closing

Now, let's move on to the next step in the actual costing cockpit shown in Figure 16.23.

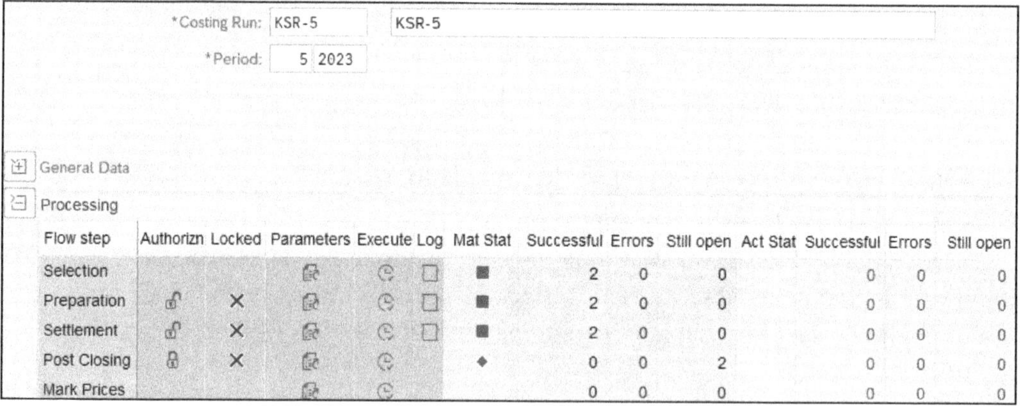

*Costing Run: KSR-5 KSR-5

*Period: 5 2023

General Data

Processing

Flow step	Authorizn	Locked	Parameters	Execute	Log	Mat Stat	Successful	Errors	Still open	Act Stat	Successful	Errors	Still open
Selection							2	0	0		0	0	0
Preparation		X					2	0	0		0	0	0
Settlement		X					2	0	0		0	0	0
Post Closing		X					0	0	2		0	0	0
Mark Prices							0	0	0		0	0	0

Figure 16.23 Actual Costing Cockpit Before Post Closing

The icons in the **Log** column indicate the first three steps have been executed. To proceed with the **Post Closing** step, click the padlock icon in the **Authorizn** row. Then click the icon in the **Parameters** column to display the screen shown in Figure 16.24.

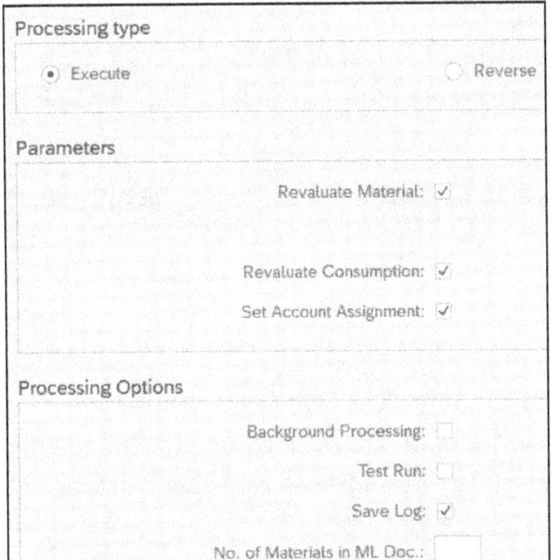

Figure 16.24 Post Closing Step Parameters

Let's examine the settings in the three sections of this screen:

- **Processing type**
  You execute the post-closing step by selecting the **Execute** radio button. Postings are made to Material Ledger accounts, and inventory postings can no longer be made in the period. The status of the materials is also set to **Closing Entries Completed**. When you select the **Reverse** radio button, the postings are reversed, and the material and activity type statuses are reset to **Settlement Completed**.
- **Parameters**
  Select the **Revaluate Material** checkbox to determine whether inventory will be revalued at the actual cost calculated during the settlement step. The material's price control is then changed from S to V for the closed period, while the current period price control remains at S.

  Select the **Revaluate Consumption** checkbox to post the revaluation of consumption variances calculated during the settlement step to corresponding general ledger accounts and profitability segments in margin analysis. Select the **Set Account Assignment** checkbox to update the controlling cost object and margin analysis with the general ledger posting during the revaluation of the consumption posting.
- **Processing Options**
  Select the **Background Processing** checkbox to run the post-closing step in the background. This is a useful option for processing large amounts of data. Select the **Test Run** checkbox to simulate the post-closing step without posting to the database.

Select the **Save Log** checkbox to keep a record of messages. The **No. of Materials in ML Doc.:** field allows you to enter the maximum number of materials one Material Ledger document can contain. This setting can complete the post-closing step if you receive the error message **Maximum number of items reached**. Click the **Save** button to save your post-closing parameters. Then click the execute icon in the **Post Closing** row in Figure 16.23 to display a results screen.

Now let's look at the **Mark Prices** step of the actual costing run, shown earlier in Figure 16.23.

16.4.7 Mark Prices

You can mark the actual cost (PUP) as the standard price for the following period. The actual cost calculated in the settlement step is populated in the **Future price** field in the **Accounting 1** view. Click the **Parameters** icon in the **Mark Prices** step shown earlier in Figure 16.23 to display the selection screen in Figure 16.25.

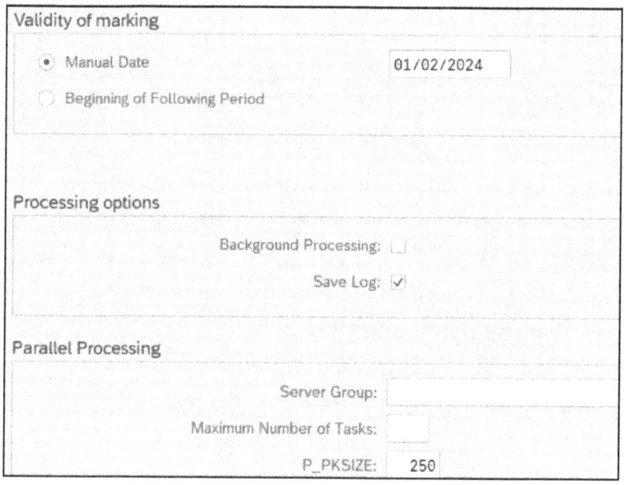

Validity of marking

(•) Manual Date 01/02/2024

( ) Beginning of Following Period

Processing options

Background Processing: [ ]

Save Log: [✓]

Parallel Processing

Server Group:

Maximum Number of Tasks:

P_PKSIZE: 250

Figure 16.25 Mark Prices Step Parameters

You can choose the price marked on a **Manual Date** or at the **Beginning of** the **Following Period**. Save your mark prices parameters, then click the execute icon in the **Mark Prices** step in Figure 16.23. We discussed the **Future Price** field in the **Accounting 1** view in detail in Chapter 3.

The future price set can be released as the standard cost either with Transaction CKME or automatically with *dynamic price release*, which enables the automatic release of a planned price valid on a specified date on the first goods movement in a new period. The planned price that is released is accessed from either of the following fields with the following priority:

1. The **Future Cost Estimate** field in the **Costing 2** view
2. The **Future Price** field in the **Accounting 1** view

You set dynamic price release via the IMG menu path **Controlling • Product Cost Controlling • Actual Costing/Material Ledger • Configure Dynamic Price Changes**. The screen in Figure 16.26 is displayed.

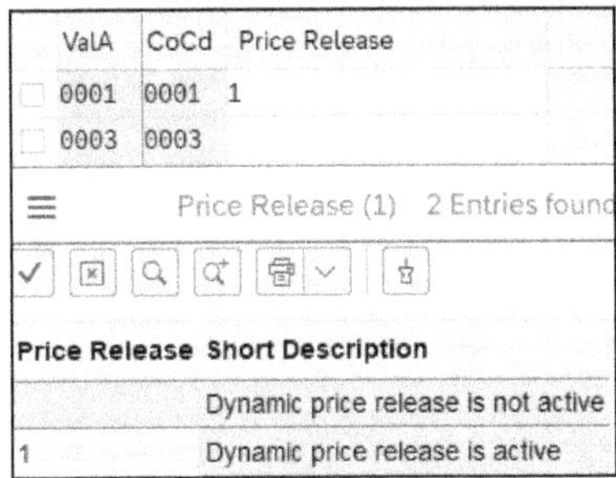

Figure 16.26 Dynamic Release of Planned Prices

Enter 1 in the **Price Release** column to activate dynamic price release for a plant. We've now successfully completed all the steps for the actual costing run. You can display the results of the costing run with the Display Actual Costing Result app (SAP Fiori ID F6073). You can follow the value flow of the materials participating in actual costing with the Display Material Value Chain app (SAP Fiori ID F4095).

16.5 Summary

In this chapter, we discussed the basic concepts of the Material Ledger and actual costing, including the fact that the Material Ledger is always active in SAP S/4HANA. We also examined currency types, valuation approaches, and the configuration steps to activate the C&V profile and actual costing.

We looked at the sequence of steps needed to carry out the actual costing run, which calculates the actual cost for every material with price determination-relevant activity during the period. You analyze the actual costing results with the material price analysis Transaction CKM3.

In the next chapter, we'll discuss sales order controlling, which involves customer-specific sales orders.

Chapter 17
Sales Order Controlling

Sales order costing scenarios involve customer orders that require you to procure components or assemblies manufactured for individual customers or orders.

Because the customer's sales order involves special requirements, you need to treat the costing of sales orders individually. You cannot generally use the standard costs for the material ordered by the customer but must adjust the cost estimate to reflect the customer's specific requirements.

There are three possible options for costing sales order scenarios:

- Sales orders without Controlling and valuated inventory
- Sales orders with Controlling and valuated inventory
- Sales orders with Controlling and non-valuated inventory

We'll first look at the configuration required for these sales order scenarios and then examine postings and period-end processing requirements.

17.1 Configuration

The first indication that sales order Controlling is involved in a process is the requirements type, which you find in the **Procurement** tab of a sales order line item. You display a sales order with Transaction VA03 or via the menu path **Logistics • Sales and Distribution • Sales • Order • Display**. You may find it easier to locate a sales order by displaying a list with Transaction VA05.

You can also manage sales orders with the Sales Order app (SAP Fiori ID F1814).

Click the sales order **Procurement** tab to display the screen shown in Figure 17.1.

Click the **RqTy** (requirements type) field and then press F4 to display a list of requirements types similar to the one shown in Figure 17.1. You typically use requirements type KEK in a make-to-order environment with configurable materials, but KEK is just one of many requirements types that you can use, depending on the planning strategy determined by the production department and the strategy group in the **MRP3** view.

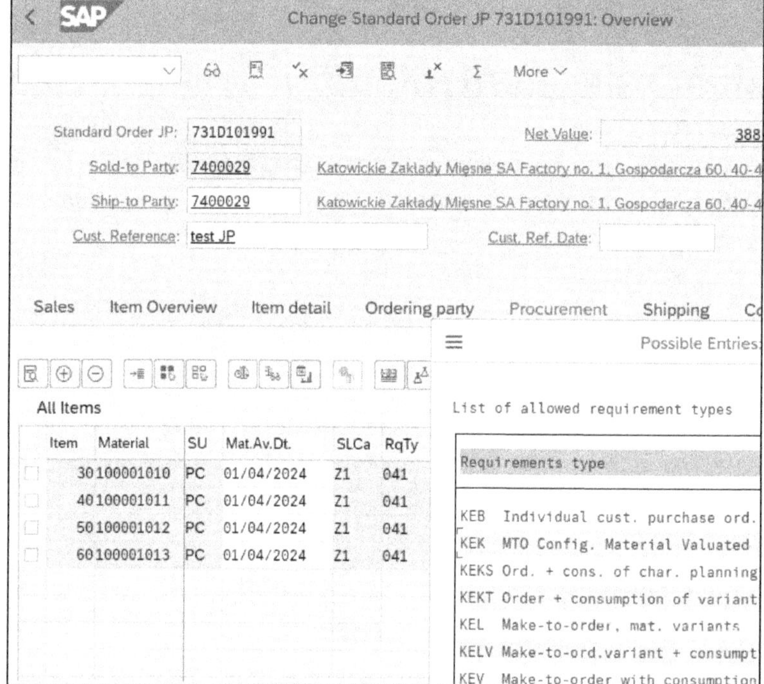

Figure 17.1 Sales Order Requirements Types Possible Entries

You configure the requirements type with Transaction OVZH or by following the IMG menu path **Controlling • Product Cost Controlling • Cost Object Controlling • Product Cost by Sales Order • Control of Sales-Order-Related Production/Product Cost by Sales Order • Check Requirements Types**. The system displays the screen shown in Figure 17.2.

Change View "Requirements Types": Overview

RqTy	Requirements Type	ReqCl	Description
KEB	Individual cust. purchase ord.	KEB	Cust.- individual PO
KEK	Make-to-ord.configurable mat.	046	MMTO config. value.
KEKS	Ord. + cons. of char. planning	043	Mke-ord.cons.charPlg
KEKT	Order + consumption of variant	042	Mke-ord.cons plgVar.
KEL	Make-to-order, mat. variants	047	MkToOrd.-mat.variant
KELV	Make-to-ord.variant + consum...	048	Mke->O.mat.var.cons.
KEV	Make-to-order with consumpti...	045	MTO val. with cons.

Figure 17.2 Mapping of Requirements Type to Requirements Class

This screen maps **Requirements Type** entries to **ReqCl** (requirements class) entries. In this example, requirements type **KEK** maps to requirements class **046**. You can view the configuration settings for a requirements class with Transaction OVZG or by following the IMG menu path **Controlling • Product Cost Controlling • Cost Object Controlling • Product Cost by Sales Order • Control of Sales-Order-Related Production/Product Cost by Sales Order • Check Requirements Classes.** Double-click requirements class **046** to display the screen shown in Figure 17.3.

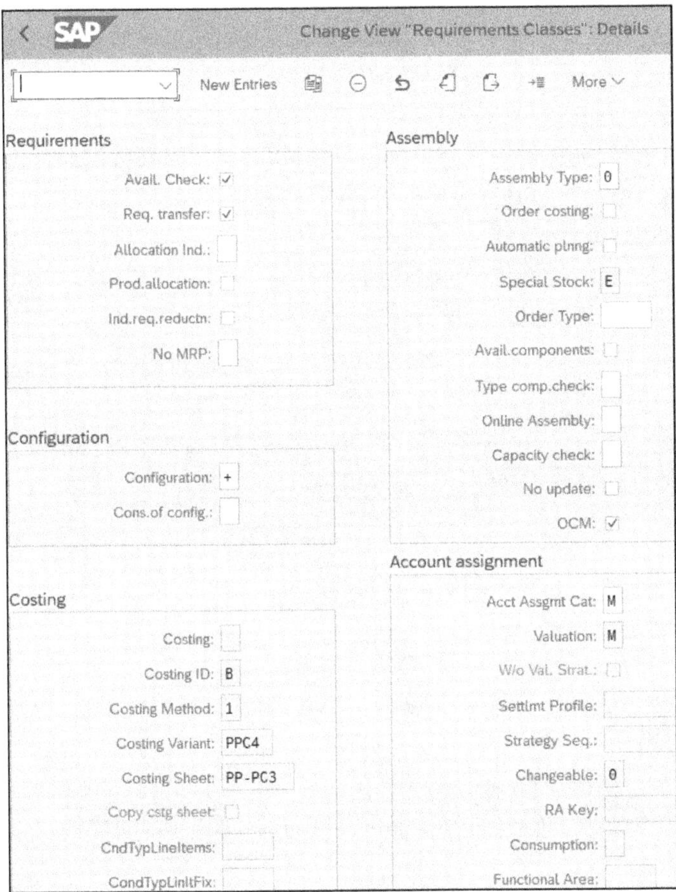

Figure 17.3 Requirements Class Configuration

The settings in this screen determine the sales order costing and inventory valuation scenario. If you work with SAP S/4HANA Cloud or use the best-practice settings, the settings for make-to-order production are in scope item BJE (Make-to-Order Production—Finished Goods Sales and Final Assembly). The use of non-valuated stock is not supported in these scenarios.

Let's examine each relevant field and indicator:

- **Costing**
  The **Costing** field determines if you create a sales order cost estimate. The following entries are possible:

- **X (sales document item to be costed)**
  You must create a sales order cost estimate. Subsequent functions such as delivery and billing are only allowed after you have carried out costing without errors.
- **A (costing to be simulated)**
  The costing of the sales document item is only simulated. You don't save the simulation in the system or update the status. However, you can perform subsequent functions, such as delivery and billing, anytime.
- **(blank) (costing allowed but not necessary)**
  If you do cost the sales document item, then you only allow subsequent functions such as delivery and billing if there were no errors during costing.
- **B (sales document not to be costed)**
  You may use this setting if you don't want the person processing the sales order to see the costing of the line items, for example.

Sales Order Line-item Cost Estimate

To create a sales order line-item cost estimate, select **Extras • Costing** from the menu bar when maintaining a sales order with Transaction VA02.

- **Costing ID**
  The **Costing ID** determines if you create the sales order cost estimate automatically. The following entries are possible:
  - **A**
    You create the cost estimate automatically when saving a sales order.
  - **B**
    You create the cost estimate automatically and mark it when saving a sales order, which is useful for using the marked sales order to evaluate sales order stock.
  - **(blank)**
    You create the cost estimate manually for a sales order line item.
- **Costing Method**
  The **Costing Method** controls whether you cost the items in the sales document automatically in product costing or manually in unit costing. Refer to Chapter 9 for more information on unit cost estimates.
- **Costing Variant**
  The **Costing Variant** determines how the sales order cost estimate determines prices. You can maintain costing variants for sales orders with Transaction OKY9 or via the IMG menu path **Controlling • Product Cost Controlling • Cost Object Controlling • Product Cost by Sales Order • Preliminary Costing and Order BOM Costing • Product Costing for Sales Order Items/Order BOMs • Costing Variants for Product Costing • Check Costing Variants for Product Costing**. Refer to Chapter 7 for more information on costing variants.

- **Costing Sheet**
  The **Costing Sheet** field determines the calculation of overhead. The costing sheet in the sales order line-item defaults from the valuation variant. Refer to Chapter 5 for more information on costing sheets.
- **Copy Costing Sheet**
  The **Copy Costing Sheet** checkbox determines whether you copy the costing sheet to production orders generated by the sales order line-item requirements.
- **CndTypLineItems (condition type for copying costs from line items)**
  If you enter the condition type into the requirements class (**CndTypLineItems** field), you use this condition type for all sales and distribution document items containing a requirements type, which contains this requirements class. The condition type indicates, for example, whether the system applies a price, a discount, a surcharge, or other pricing elements, such as freight costs and sales taxes, during pricing.

  Two standard condition types are available for the cost transfer of line items:
  - **EK01**
    First, transfer the result of the sales order costing to the pricing screen for the item. Then, use the value as the basis for price calculation.
  - **EK02**
    The result of the sales order costing is a statistical value, which you can compare with the price.
- **CondTypLinItFix (condition type for fixed-cost transfer from line items)**
  The fixed-cost part is determined statistically. Transferring the fixed-cost part makes it easier for the system to predict the profit margin. You use standard condition type EK03 in the CondTypLinItFix field for transferring the fixed cost part.

  You can maintain the condition type for determining the fixed-cost part with the condition type for the cost transfer of line items in the previous field.

Now that we've finished looking at the fields in the **Costing** section, let's discuss the fields in the **Assembly** and **Requirements** sections of Figure 17.3:

- **Assembly Type**
  In assembly processing, you generate a manufacturing order together with the sales order. There are two methods of assembly processing:
  - **Static**
    You generate a single manufacturing or planned order for a sales order line item.
  - **Dynamic**
    You generate multiple manufacturing or planned orders for a sales order line item.

  Two static (**1** and **2**) and two dynamic (**3** and **4**) processing options are available for the **Assembly Type** field. If you choose a static option, you can use the preliminary cost estimate of the manufacturing order to set the standard price of the sales order line item.

17

- **Sales Order Costing**
  You select the **Sales order costing** checkbox to generate a sales order cost estimate for setting the standard price of the sales order line item. You can only select this indicator if you select option **2** for the previous **Assembly type** field.

Preliminary Cost Estimates

A consideration when deciding whether to use the preliminary cost estimate of the manufacturing order or a sales order cost estimate is the timing of changes to the sales order item and the manufacturing order. This depends on how long after the sales order is placed you allow the customer to make changes to the sales order item, and how long it takes after the manufacturing order is created for production to take place.

The standard price of the sales order stock is determined by the relevant cost estimate price at the time of the first goods receipt of sales order stock into inventory.

- **Automatic Plnng (automatic planning)**
  If you select the **Automatic Plnng** checkbox, the system automatically carries out single-item, multilevel planning for a sales order if you create or change a sales order item. This function can only be used for make-to-order production.

Now that we've discussed the **Requirements** and **Assembly** fields, let's look at the fields in the **Account assignment** section in the requirements class in Figure 17.3:

- **AcctAssignment Cat. (account assignment category)**
  The **AcctAssignment Cat.** checkbox controls if costs are maintained on the sales order line item with sales order controlling as follows:
  - **E**
    Sales order controlling is active. Therefore, you maintain revenues and costs on the sales order line item, and an associated settlement process occurs at period end.
  - **M**
    Sales order controlling isn't active. You don't maintain costs on the sales order, and no associated settlement process occurs at the period end. You normally choose this setting for sales order costing since it's the simplest, and no period-end process is involved.

Account Assignment Categories

Two popular account assignment categories for costing for sales orders are in the preceding list. You maintain existing account assignment categories or create your own with Transaction OME9 or by following the IMG menu path **Controlling • Product Cost Controlling • Cost Object Controlling • Product Cost by Sales Order • Control of Sales-Order-Related Production/Product Cost by Sales Order • Check Account Assignment Categories**.

- **Valuation**
  The **Valuation** field determines if sales order stock is managed on a valuated or non-valuated basis as follows:
  - **(blank)**
    Sales order stock isn't valuated. You must choose sales order controlling as active in the previous field. You also need to run period-end results analysis to determine revenue and cost of sales (COS) and then settle the values to general ledger accounts and profitability analysis.
  - **M**
    You valuate sales order stock based on:
    - The sales order line-item cost estimate or:
    - The production order preliminary cost estimate during the first goods receipt into inventory

    You display valuated special stocks using report RM07MBWS with Transaction SA38.
  - **A**
    You valuate sales order stock based on the standard price of the material and manage the stock with anonymous warehouse stock.

Valuated Sales Order Stock

Valuated sales order stock production is similar to make-to-stock. Account postings occur at the time of goods movements of the sales order inventory based on the standard price of the customer segment stock.

Valuated sales order stock became available as of SAP R/3 release 4.0 and is the preferred way to manage sales order stock. It eliminates several period-end processing steps with non-valuated sales order stock.

- **Without Val. Strategy (without valuation strategy)**
  Set the **Without Val. Strategy** checkbox to valuate the valuated sales order stock with the standard price of the unallocated warehouse stock. You can use this, for example, if you are producing the same material as an unallocated stock item as well as for mass production related to sales orders.
- **Settlmnt Profile (settlement profile)**
  A settlement profile contains the parameters necessary to create a settlement rule that determines which portions of a sender's costs you allocate to which receivers. You only need a settlement profile to settle the sales order line items.
- **Strategy Sequence**
  During the automatic generation of a settlement rule, the system creates a settlement rule as specified by the strategies in the **Strategy Sequence** field. You can prioritize the strategies in this sequence and have multiple strategies with the same

priority. Evaluation of more than one strategy with the same priority results in a settlement rule with more than one distribution rule.

- **Changeable**
  The **Changeable** checkbox determines the behavior of a settlement rule after you change the master data. While generating settlement rules, another settlement rule could already exist for an object. The checkbox determines how the system should continue if this is the case. The system can differentiate between the following:
  - Manually created or changed distribution rules
  - Automatically generated distribution rules

  Normally, the system doesn't modify manually created or changed rules. If the system originally created a rule, but then you modified it manually, it qualifies as a manually changed rule.
- **RA Key (results analysis key)**
  You normally enter a results analysis key for non-valuated inventory to determine the revenue and cost of sales values at period end.
- **Functional Area**
  A functional area is required to create a profit and loss account in financial accounting using COS accounting.

Now that we've considered configuration for sales order costing, let's look at period-end processing and posting.

17.2 Period-End Processing

There are three possible scenarios for sales order costing as follows:

17.2.1 Sales Orders without Controlling with Valuated Inventory

Valuated sales order inventory has been available since SAP R/3 release 4.0 and is the preferred scenario because it's the simplest. COS and revenue postings occur in real time because sales order stock is valuated. Either you base the sales order stock standard price on the sales order line item cost estimate or the manufacturing order preliminary cost estimate based on the **Sales order costing** checkbox, as discussed in Section 17.1. You base period-end processing, including work in process (WIP) and variance analysis, on manufacturing orders and not the sales orders, as discussed in Chapter 10, Chapter 11, Chapter 12, Chapter 13, Chapter 14, and Chapter 15.

This scenario is applicable if you don't require detailed margin analysis on each sales order or deferred postings to costing-based profitability analysis. If a more detailed margin analysis of each sales order line item is required, you can consider using the scenario described in the next section.

17.2.2 Sales Orders with Controlling and with Valuated Inventory

This option may be useful if you need to monitor the profit margin for each sales order; for example, if you have high-value individual sales orders. You should also consider this option if you need to defer billing until the customer has received and accepted the product, as determined in customer contracts, as discussed in Chapter 20.

In this scenario, postings to general ledger financial accounts occur during goods movement because you valuate the sales order inventory. Postings to costing-based profitability analysis, however, occur during period-end settlement. This process has two advantages:

- Postings to general ledger financial accounts during goods movements and billing are in real time.
- Detailed reporting analysis available per sales order line item in profitability analysis.

One disadvantage is that you must run the two additional period-end processing steps of results analysis and settlement described in the following scenario. We'll follow seven steps in an example scenario shown in Figure 17.4.

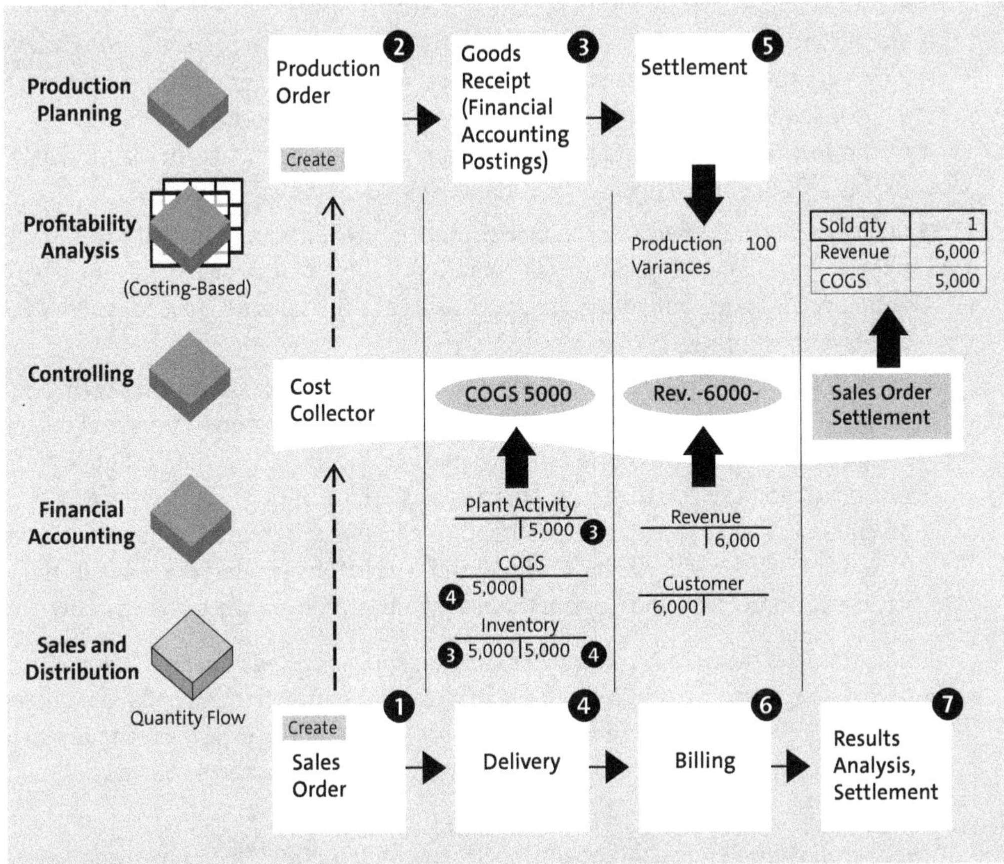

Figure 17.4 Sales Order without Controlling with Valuated Inventory

The following is a description of each of the seven steps in this scenario:

❶ You can use the *sales order cost estimate* as the basis for the standard price of the sales order stock when produced. For configurable products, you save customer choices (e.g., red or blue color for a car) with the sales order.

❷ You can also use the *production order preliminary cost estimate* as the basis for the standard price of the finished good when produced. (Note that this is the only option when using assembly processing, i.e., when a production order is created automatically when the sales order is saved.)

❸ Financial account postings occur when you receive finished goods into inventory. You debit a finished goods inventory account, credit a plant activity profit and loss (P&L) account, and postings to margin analysis occur. You credit the production order with the quantity produced multiplied by the standard price. You determine this price by the sales order or production order preliminary cost estimate. After the first goods receipt occurs, you cannot change the standard price. Nothing is posted to the sales order at this stage.

❹ When the sales order stock is removed from inventory for delivery to the customer, a posting to a COS account is made. The sales order is debited with the COS amount.

❺ During production order settlement, the production variances settle to costing-based profitability analysis, where individual variance categories settle to different value fields. Settling the total production variance to two P&L accounts, such as production variance and plant activity accounts, is standard practice. In margin analysis, the settlement will update the variance split as described in Chapter 20.

❻ Billing results in immediate postings to financial accounts, margin analysis, and the sales order. If there are no postings to deferred accounts (described in the next step), you normally carry out billing on the same day as the sales delivery to match COS and revenue postings.

❼ Period-end results analysis and sales order settlement don't result in postings to financial accounts in this scenario because all required postings to general ledger accounts have already occurred during steps ❹ and ❻. If, however, billing occurs much later than sales delivery, results analysis can be used to ensure that COS and revenue post to costing-based profitability analysis at the same time, usually at period end. Although results analysis and sales order settlement are normally period-end processes, you can run them as batch jobs once a week or even daily.

Because COGS is posted with the delivery and revenue is posted with the invoice when you work with margin analysis, there can be a time lag between them, making it necessary to either defer the COGS until the revenue is posted or to accrue the revenue when the delivery is made. In that case, you use event-based revenue recognition (scope item 1K2).

You can run period-end results analysis with Transaction KKA3 (individual) and Transaction KKAK (collective), and you can run period-end settlement with Transaction

VA88 or by following the menu path **Accounting • Controlling • Product Cost Controlling • Cost Object Controlling • Product Cost by Sales Order • Period-End Closing • Single Functions.**

Now that we've discussed and followed an example of sales order controlling with valuated inventory, let's examine the third and final possible sales order costing scenario.

17.2.3 Sales Orders with Controlling and without Valuated Inventory

If you started your implementation before SAP R/3 release 4.0, your company might still be using non-valuated inventory, as it was the only option then. You could use it after SAP R/3 release 4.0 if non-valuated inventory is necessary—for example, to track asset quantities in inventory. The asset already has its own balance sheet valuation, so valuated inventory would lead to double counting on the balance sheet. This scenario can occur if your company manufactures its own assets, which you then lease or loan to customers. We'll follow the seven steps in the example scenario shown in Figure 17.5.

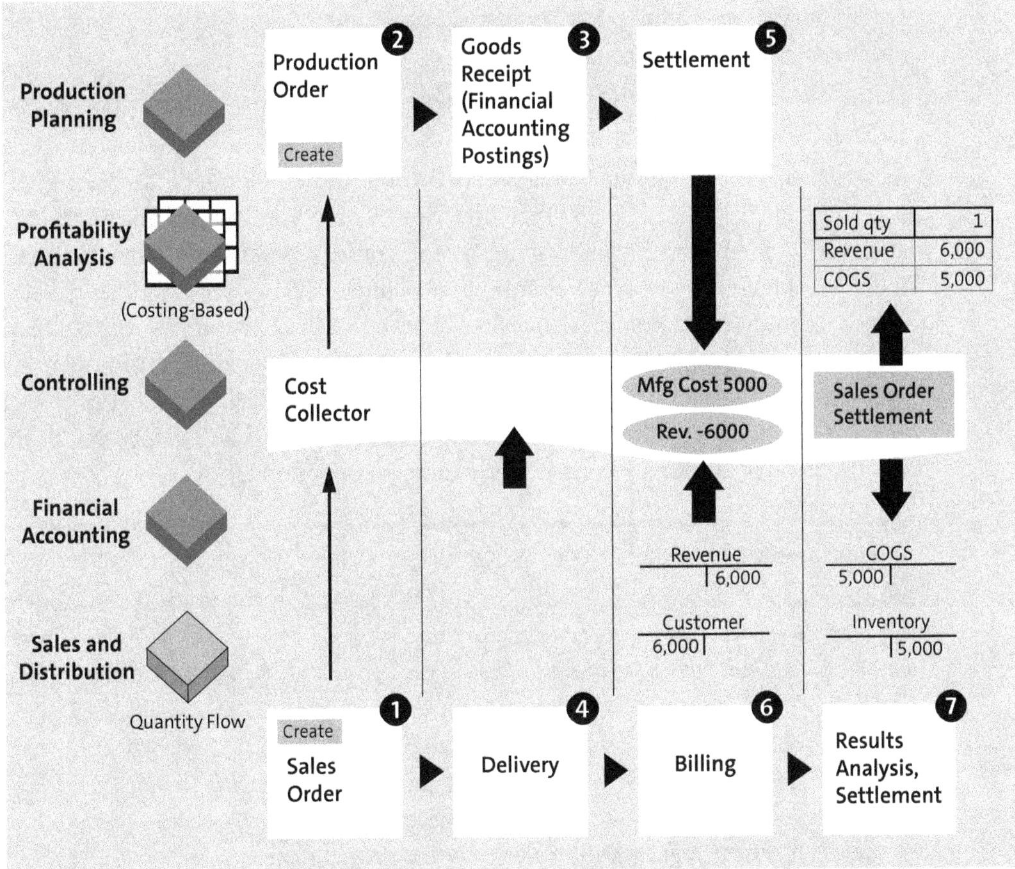

Figure 17.5 Sales Order Controlling with Non-valuated Sales Order Stock

The following is an explanation of each of the seven steps that occur during this scenario:

❶ The process starts with creating a sales order line item that acts as a cost collector in sales order controlling. For configurable products, customer choices (e.g., red or blue color for a car) are saved with the sales order.

❷ You create the production order automatically when you create the sales order (assembly processing) or manually later. The system creates the production order preliminary cost estimate from a bill of materials (BOM) and routing information from the sales order.

❸ During finished goods receipt into inventory from the production order, no accounting entries are made because the sales order stock is non-valuated.

❹ Further, no accounting entries are made when the finished goods are removed from inventory and issued to the customer during sales order delivery.

❺ During production order settlement, you transfer all production costs from the production order to the sales order. This movement of costs only occurs within the sales order controlling module. No financial accounting entries result from production order settlement in this scenario.

❻ Billing results in immediate postings to financial accounts and margin analysis. You also post revenue to the sales order.

❼ At period end, the results analysis transaction determines the sales order billing status. The financial general ledger accounts for COS and WIP postings during the following settlement step are determined. If a sales order has a status of "finally billed," the full production costs received from the production order post to a COS general ledger account. A status of "partially billed" may result in a portion of the production costs being posted to COS and the remainder to WIP. Many possible results analysis methods can be used to determine the portions posted. Two of the most common are revenue-based results analysis and cost-based percentage of completion (POC). We discussed results analysis methods in detail in Chapter 13.

Variance Analysis and Sales Order Controlling

Detailed variance analysis isn't possible with the sales order controlling and non-valuated inventory scenario because no standard price exists to compare with the actual costs of purchase or manufacture. However, you can create a preliminary price based on an initial set of assumptions to calculate variances against. You post all costs to the sales order and compare them with revenue to determine profitability.

Using valuated sales order stock has the advantage of detailed variance analysis. You can carry out variance analysis on the cost of manufacturing the sales order stock because you can compare it with the sales order cost estimate or the manufacturing order preliminary cost estimate.

17.3 Summary

In this chapter, we've explained how sales order controlling differs from controlling for standard products and walked through the sales order controlling configuration and the period-end closing steps. In the next chapter, we'll cover further special topics, including subcontracting, delivery costs, and stock in transit.

Chapter 18
Special Topics

Special topics cover specialized areas within product cost controlling in addition to the master data, configuration, and period-end topics discussed in previous chapters.

Product cost controlling covers a range of topics that are continuously being developed and improved. Previous chapters covered the foundational subject areas of initial planning, master data, configuration, cost estimates, and period-end and event-based processing. There are too many specialized topics to cover in detail in one book.

This chapter covers additional useful topics in more specialized areas under product cost controlling. These topics include:

- Subcontracting
- Delivery costs
- Stock in transit

Let's examine these three topics in detail, starting with subcontracting.

18.1 Subcontracting

Subcontracting involves sending components to a supplier who manufactures an assembly. The supplier then returns the completed assembly, and during goods receipt, the components are issued from the subcontract inventory.

Subcontracting Is a Purchasing Process

One key point of interest in subcontracting is that even though it involves manufacturing by the supplier, it's a purchasing process in SAP because the organization purchases the manufacturing activities from the supplier.

We'll examine the main processes involved in subcontracting in the following sections.

18.1.1 MRP 2 View

The **Special Procurement** type field in the material master **MRP 2** view determines if you manufacture a material with a subcontracting process. You maintain material master views with the Material Master app (SAP Fiori ID F0338A), with Transaction MM02 or by following the menu path **Logistics • Production • Master Data • Material Master • Material • Change • Immediately**. Select the **MRP 2** tab to display the screen shown in Figure 18.1.

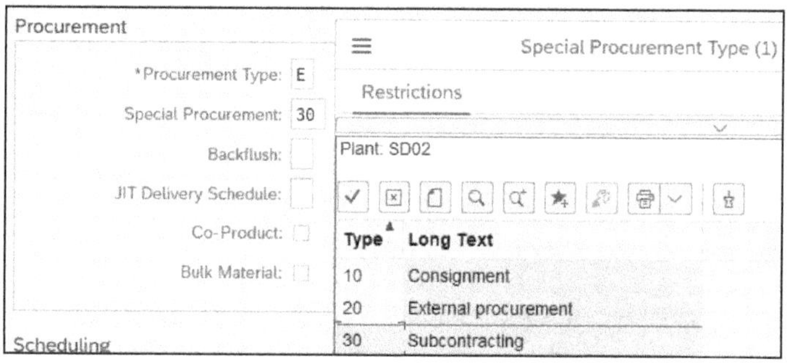

Figure 18.1 MRP 2 View Subcontracting Special Procurement Type

Click in the **Special procurement** field and press F4 to display a list of special procurement types as shown in Figure 18.1. Option **30 Subcontracting** instructs material requirements planning (MRP) to create a purchase requisition or order to issue to the subcontract supplier. The system first searches for a subcontract purchasing info record.

18.1.2 Purchase Order

You create a subcontract purchase order with Transaction ME21N or by following the menu path **Logistics • Materials Management • Purchasing • Purchase Order • Create**. Enter the subcontract material purchasing information to display the screen shown in Figure 18.2.

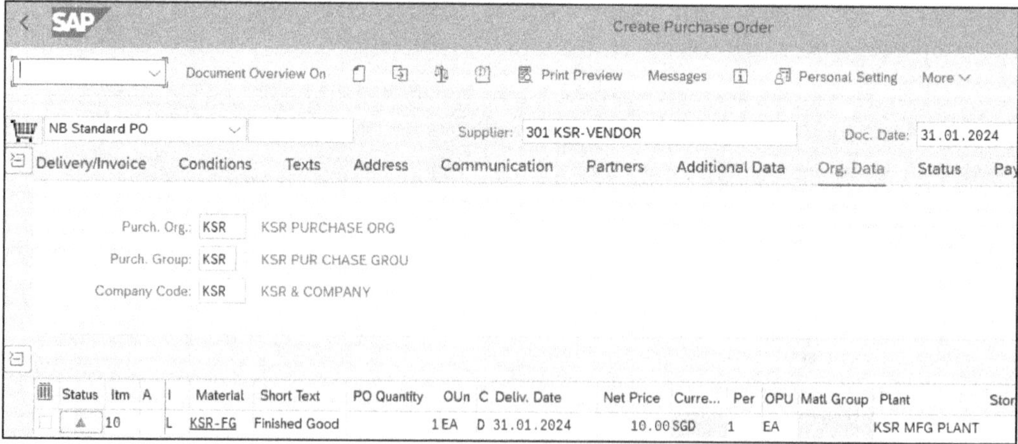

Figure 18.2 Subcontract Purchase Order Line Item

You indicate that this material is to be procured by subcontracting by entering item category **L** (third column) when creating the purchase order line item. When you press [Enter] after entering a subcontract item category, two new icons appear in the **Material Data** tab of the item details, as shown in Figure 18.3.

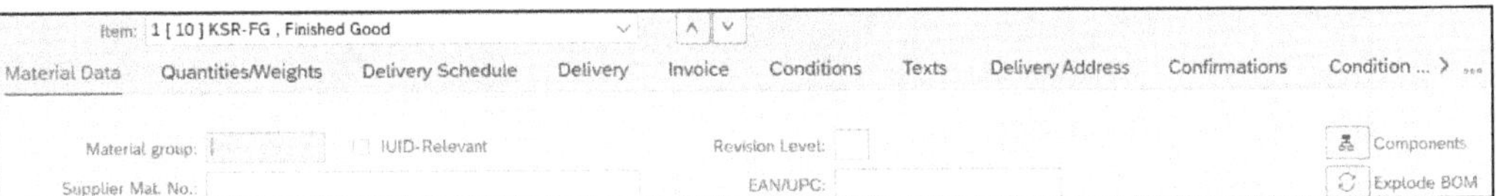

Figure 18.3 Components and Explode BOM Icons in Material Data Tab

Click the **Components** icon to display the screen shown in Figure 18.4. These components will be sent to the subcontractor for processing.

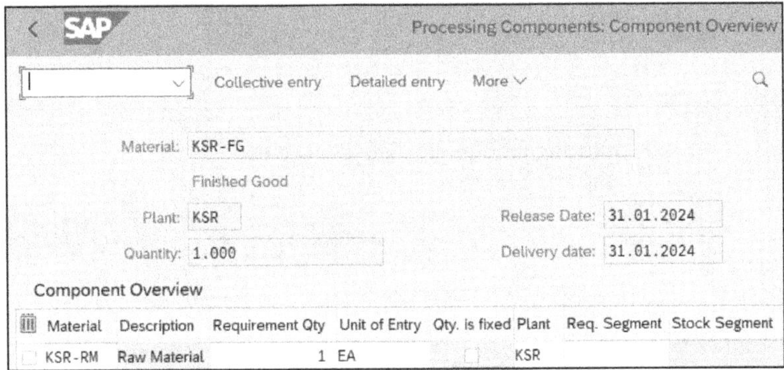

Figure 18.4 Subcontract Components Entry Screen

Subcontract components appear automatically because of the subcontract BOM. You can manually populate the subcontract components if you can't locate a BOM.

18.1.3 Monitoring Stocks Provided to Supplier

You can monitor stocks issued to a subcontract supplier with Transaction ME2ON or by following the menu path **Logistics • Materials Management • Purchasing • Purchase Order • Reporting • Subcontracting Cockpit**. The system displays the selection screen shown in Figure 18.5.

Complete the fields you need information on and click the **Execute** button or press [F8] to display the screen shown in Figure 18.6.

From this screen, you determine the status of subcontract stock. You manage the components provided to the subcontractor as stock provided to the supplier. You can post a transfer posting from unrestricted-use stock to the stock of material provided to the supplier. You can only debit components consumed in a subcontract order from the stock of the material provided to the particular supplier.

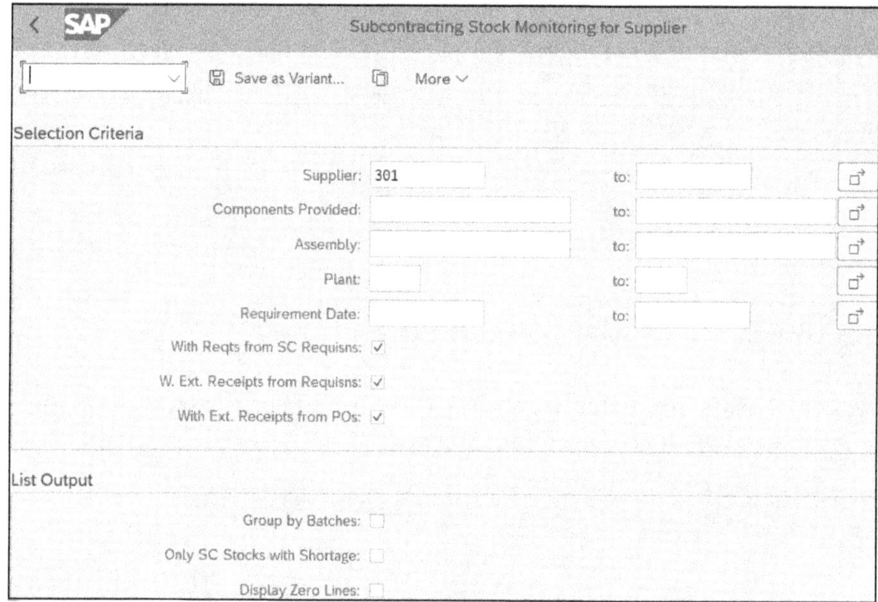

Figure 18.5 Subcontract Stock Monitoring for Supplier Selection Screen

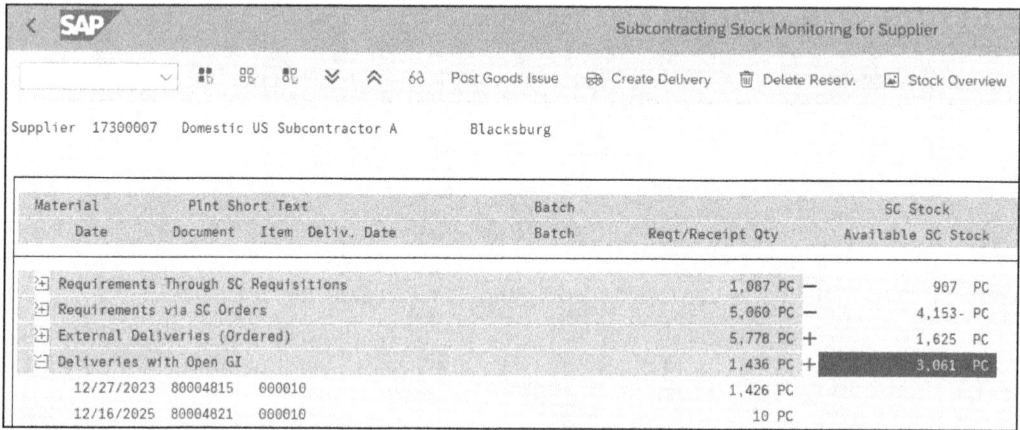

Figure 18.6 Subcontract Stock Monitoring for Supplier

18.1.4 Goods Receipt of End Product

You post the goods receipt for the product with reference to the subcontract order item. At goods receipt, you also make a consumption posting for the components from the stock of material provided to the supplier. For each goods receipt item, the system copies the components with their quantities as goods issue items. If the supplier (subcontractor) consumed a greater or smaller quantity than was planned in the purchase order, you can adjust the component quantity at goods receipt.

18.1.5 Supplier Invoice for Subcontracting Work

The supplier will issue an invoice for the subcontracting work performed with reference to the purchase order; this is considered part of the product costs.

Subcontracting and External Processing

Here is the difference between subcontracting and external processing:

- Subcontracting involves supplying components. The supplier does all the manufacturing, and you place the finished goods into inventory.
- External processing involves single operations in a manufacturing order that an external supplier carries out. Selecting a subcontracting indicator available in the external operation allows you to send components for assembly by the supplier during the external operation.

Put simply: Subcontracting involves only purchasing; external processing involves both operations and purchasing.

18.1.6 Business Function LOG_MM_OM_1 (Outsourced Manufacturing)

Business function `LOG_MM_OM_1` triggers the creation of a Controlling production order to mirror the purchase order item, giving you more transparency into subcontracting costs. Controlling production orders are like production orders but without a BOM or a routing. They are for cost purposes only. They are created automatically with reference to the subcontracting purchase order, include an extra group of fields that include information from the purchase order, and can be displayed using Transaction KKF3 or the menu path **Accounting • Controlling • Product Cost Controlling • Cost Object Controlling • Product Cost by Order • Order • CO Production Order • Display**. You can select Controlling production orders created for the purposes of outsourced manufacturing using [F4] help on the order field and matchcode **W** (Controlling production orders created from outsourced manufacturing).

This business function allows you to capture the costs of the issued materials and display WIP for the process. See the following documentation URL for more information: *http://s-prs.co/v58304*.

Now that we've reviewed the subcontracting process, let's examine delivery costs.

18.2 Delivery Costs

Cost estimates can automatically calculate and report purchased material delivery costs, such as freight and duty, as cost components. To set this up, create a link between purchasing condition types, origin groups, and cost components with the following steps:

1. Identify the purchased material consumption accounts.
2. Configure a costing variant to analyze the delivery costs table.
3. Assign condition types to origin groups.
4. Assign condition types to purchasing info records.
5. Assign origin groups to cost component structures.

We'll review each of these steps in detail in the following sections.

18.2.1 Purchased Material Consumption Account

You need to determine purchased material consumption accounts because you'll assign an origin group together with the consumption accounts to basic material costs in a later step.

You identify purchased material consumption accounts by first determining the valuation class of the purchased materials. Run Transaction MM03 and navigate to the **Costing 2** view of a purchased material. Click in the **valuation class** field and press F4 to display a list of purchased material valuation classes.

You can analyze general ledger accounts assigned to each valuation class with Transaction SE16N (Data Browser). Type in "Table T030," press Enter, and in the selection screen, enter your chart of accounts in the **Chrt/Accts** (chart of accounts) column, enter "GBB" in the **Trans.** (transaction) column, enter "VBR" in the **AM** (account modifier) column, enter your valuation class in the **ValCl** (valuation class) column, and click **Execute** to display the screen shown in Figure 18.7.

Chrt/Accts	Trans.	VGCd	AM	ValCl	G/L Acct	G/L Acct
KSR	GBB	0001	VBR	KSR1	400100	400100
KSR	GBB	0001	VBR	KSR3	400130	400130

Figure 18.7 Listing of General Ledger Accounts in Table T030

Now that we've determined the purchased material consumption account, the next step is configuring the costing variant.

18.2.2 Configure Costing Variant

You configure a costing variant by following the IMG menu path **Controlling • Product Cost Controlling • Product Cost Planning • Material Cost Estimate with Quantity Structure • Define Costing Variants**. Double-click a costing variant and click the **Valuation Variant** button to display the screen shown in Figure 18.8.

Because the first **Priority** in the **Strategy Sequence** searches for **L Price from Purchasing Info Record**, the system must then search the **Sub-Strategy Sequence** to determine the purchasing info record price. The quotation price represents the material supplier quotation stored in the purchasing info record. In Figure 18.8, the cost estimate first searches the purchasing info record for condition types and then accesses the condition table to determine the origin group.

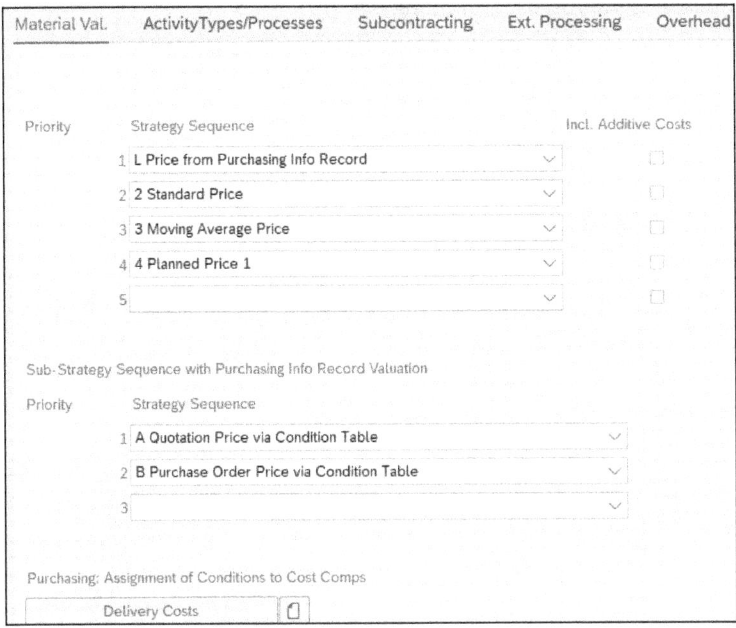

Figure 18.8 Valuation Variant Material Strategy Sequence

> **Valuation Variant Purchasing Info Record Strategies**
>
> You can find more information on valuation variants with purchasing info record strategies in SAP Note 351835.

Next, you create the link between condition types and origin groups.

18.2.3 Assign Conditions to Origin Groups

You create origin groups with Transaction OKZ1. To assign condition types to origin groups, click the **Delivery Costs** button in Figure 18.8 to display the screen shown in Figure 18.9.

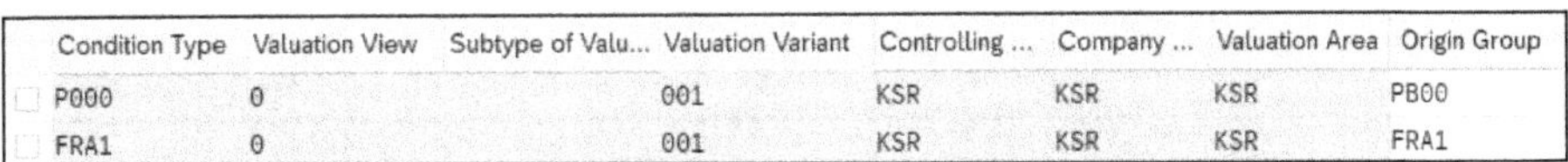

Condition Type	Valuation View	Subtype of Valu...	Valuation Variant	Controlling ...	Company ...	Valuation Area	Origin Group
P000	0		001	KSR	KSR	KSR	PB00
FRA1	0		001	KSR	KSR	KSR	FRA1

Figure 18.9 Assign Condition Types to Origin Group

In this screen, you assign the purchasing **Condition type** to an **Origin group**. If your entry is for all company codes or valuation areas, you can leave the **Company Code** and **Valuation Area** fields blank. You can also leave **Origin group** blank to exclude all cost components with an origin group assigned.

Now let's assign condition types to purchasing info records.

18.2.4 Assign Conditions to Purchasing Info Records

In Figure 18.8, we instructed the cost estimate to search for a purchasing info record. You can display a purchasing info record with Transaction ME1M or by following the menu path **Logistics • Materials Management • Purchasing • Master Data • Info Record • List Displays • By Material**.

In the selection screen, type in the material, purchasing organization, and plant and click the **Execute** icon or press F8. Double-click a purchasing info record and click the **Conditions** button to display the conditions. Ensure condition type FRA1 is added as a supplemental condition type as a percentage of gross price.

Origin Groups and Condition Types

For more on assigning origin groups to condition types and the material master, see SAP Note 1445940.

Next, it's time to assign origin groups to cost components and create a cost estimate.

18.2.5 Assign Origin Groups to Cost Components

To determine which main cost component structure to access, follow these steps:

1. Display a cost estimate.
2. Click the **Costing Data** tab and then double-click the underlined **Costing Variant**.
3. Select the **Assignments** tab and click the **Cost Component Structure** button to display the screen shown in Figure 18.10.

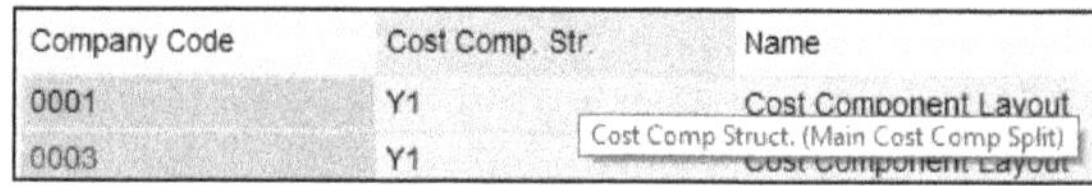

Company Code	Cost Comp. Str.	Name
0001	Y1	Cost Component Layout
0003	Y1	Cost Component Layout

Figure 18.10 Assignment of Cost Component Structures to Company Codes

This screen displays the assignment of the main and auxiliary cost component splits to company codes. Position your mouse over the column headings to display more details of each column, as shown in Figure 18.10.

You assign origin groups to cost components with Transaction OKTZ or by following the IMG menu path **Controlling • Product Cost Controlling • Product Cost Planning • Basic Settings for Material Costing • Define Cost Component Structure**.

Select your cost component structure: Double-click **Cost Components with Attributes** at the left, create a new **Freight** cost component in the **Name of Cost Comp.** column, and assign the freight origin group to the purchased material consumption accounts, as shown in Figure 18.11.

Cost Comp. Str.	Chart of Accts	From cost el.	Origin Group	To cost elem.	Cost Component
01	INT	452000	FRA1	452000	009

Figure 18.11 Assign Freight Origin Group to Freight Cost Component

After you save the new **Freight** cost component, new cost estimates will list **Freight** as a separate cost component.

18.3 Stock in Transit

Stock in transit is where inventory is transferred between two entities but hasn't yet reached its destination. Depending on the legal agreement between the entities, ownership of the inventory can belong to either of the entities while the inventory is in transit. You access advanced stock in transit functionality by activating business function LOG_MM_SIT (Cross-Company-Code Stock/Actual Costing) with Transaction SFW5.

A new stock segment is available to keep the material quantity and value visible during transfer from one company code to another. We'll examine the details in the following two sections.

18.3.1 Cross-Company-Code Stock in Transit

In sales and cross-company transfer processes, modeling stock in transit became possible with SAP ERP EHP 6. Before this, the ownership of the material was not reflected correctly in SAP during the time when the goods were in transit. When selling materials from one company code to another using the two-step approach, the stock value did not automatically appear in the balance sheet after the first step.

This business function enables companies to track and manage ownership changes more effectively when selling materials to a customer or an affiliated company in the same group. Companies can more easily fulfill the legal requirement of showing all the stock values in the group. Figure 18.12 displays the three scenarios enabled by LOG_MM_SIT and the different points in time when ownership is transferred.

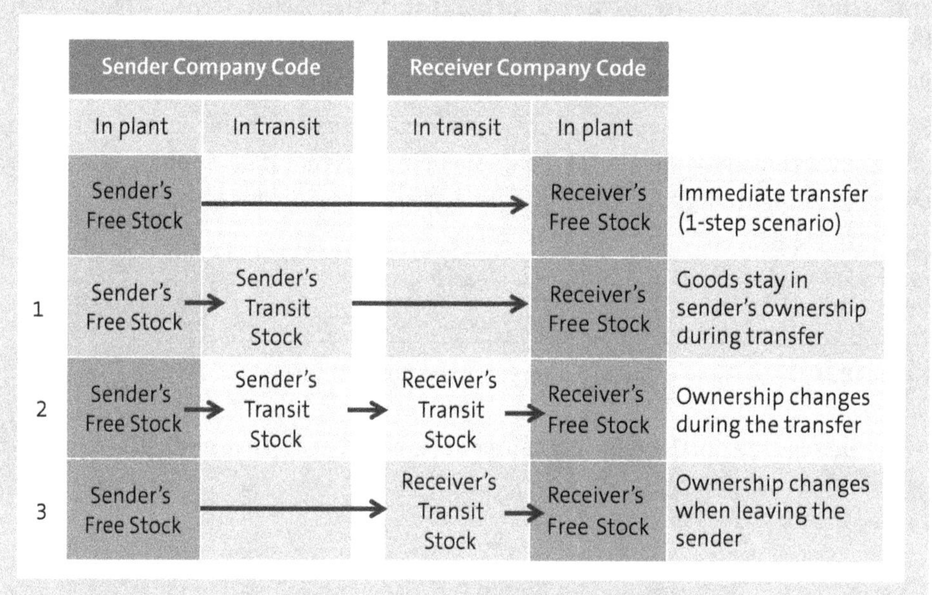

Figure 18.12 Material Transfer Across Company Codes with Stock in Transit

Let's discuss each of these two-step scenarios:

- Option 1: During the transportation phase, the sender company keeps ownership of the materials sent to a customer, or another company code, and the inventory remains visible as stock in transit in the sender company.
- Option 2: You also have the option to move materials from the stock in transit associated with one plant to the stock in transit for another plant. Ownership changes during transit; for example, when the goods arrive at the destination port.
- Option 3: During the transportation phase, ownership of the goods is transferred directly to the receiving company code and becomes visible as stock in transit in the receiving company code.

Now, let's examine the other scenarios available for this business function.

18.3.2 Cross-Company-Code Actual Costing

The other use of stock in transit is to establish the link between the affiliated companies for the purposes of actual costing. Customers can transfer actual cost component split information from one company code to another, so markups can be used in a cross-company-code sales scenario. Table 18.1 displays an example of the cost components in the legal view, where the detailed cost component split in Company A is rolled up to become a single direct material purchase in Company B. This process is known as *arm's length trading* because Company A and Company B are trading with one another as if they were external business partners even though they are affiliated companies within the same group. In this case, actual costing does not provide visibility into the cost structure after intercompany sales

Group				
Company A		Sales Process	Company B	
Costs			Costs	
Raw/Material	20		Raw/Material	100
Energy	25		Energy	0
Labor	15		Labor	0
Production	40		Production	0

Table 18.1 Cost Structure Does Not Display in Receiving Company

By contrast, the desired solution should ensure cost transparency during intercompany sales by transferring the individual cost components in detail and by adding freight costs and intercompany profits to the costs incurred in the selling company to provide a correct valuation from a group perspective, as shown in Table 18.2. This is achieved by using an additional price condition (**KW00**) that ensures that the cost component is selected in the selling company and transferred to the receiving company "at cost" during the stock transfer. The same structure is updated during actual costing to reflect any variances incurred in the selling company.

Group				
Production		Sales Process	Regional Sales	
Total cost	1000		Total cost	1200
Intercompany profit	0		Intercompany profit	100
Materials	500		Materials	500
Labor	400		Labor	400
Overhead	100		Overhead	100
Freight	0		Freight	100

Table 18.2 Detailed Outcome of Intercompany Sales

With this business function, you extend material and actual costing to run across company codes without losing actual costs or losing actual cost component split at the company code border. The intercompany sales process allows the transfer of the costs and cost component split information from the sending company code to the receiving company code. You can also calculate markup or intercompany profit in the value chain. Let's look at an example of how this works in Table 18.3 and then discuss the three possible valuations.

Group							
Plant 1		**Sales Process**	**Plant 2**	**Legal View**	**Transparent Legal View**	**Group View**
Total cost	1000		Total cost	1175	1175	1075
Materials	0		Materials	1100	600	500
Labor	500		Labor	0	400	400
Overhead	400		Overhead	0	100	100
Freight	100		Freight	75	75	75
CC markup	0		CC markup	100	100	0

Table 18.3 Valuation Example in Three Views

Let's discuss each of the three views in detail:

- **Legal view**
  The legal view contains the cross-company (CC) markup but does not show cost component information from the supply chain outside the company. Materials purchased from affiliate companies will include only the material costs component and the markup. The default currency type is 10.
- **Transparent view**
  This view contains the markup and preserves the cost component information from the affiliate companies if the logistical process allows. You can activate this by a BAdI in currency type 10.
- **Group view**
  You can use group view in parallel to the other two valuations. This view does not show a markup since you treat all processes as if the plants were part of the same organization without considering company code borders. You can use either currency type 11 or 31.

An implementation can use either a legal view or a transparent legal view. You can add a group view to both. The two typical scenarios are using a transparent legal view or a legal view with a group view.

Advanced Intercompany Sales

Stock in transit is the key to advanced intercompany sales and stock transfer. We'll examine this when we discuss universal parallel accounting in Chapter 19.

18.4 Summary

In this chapter, we discussed product cost controlling special topics. First, we examined subcontracting and how purchasing controls this process. You issue components to a supplier for assembly and then receive the assembled product into inventory. All financial postings occur at the time of the assembly goods receipt.

Next, we discussed delivery costs for purchased materials and how you can report inbound freight costs as a separate cost component. We also discussed the stock-in-transit functionality available with the business function LOG_MM_SIT. In the next chapter, we'll discuss event-based product costing and universal parallel accounting.

Chapter 19
Event-Based Product Costing

Product cost controlling—the calculation of overhead, work in process (WIP), production variances, and settlement—occurs at period close. However, with the advent of SAP S/4HANA and universal parallel accounting, a significant shift has occurred. You can now transition tasks from the period close to real-time production accounting. This chapter investigates universal parallel accounting and event-based product costing, highlighting the benefits of this shift.

In the previous chapters, we saw how to:

- Assign overhead at period close
- Calculate WIP for unfinished orders and production variances for completed orders
- Settle and create journal entries for WIP and production variances

SAP S/4HANA introduces a new approach, moving journal entries for overhead, WIP, and production variances out of the period close to real time. The entries are created when the underlying business transactions occur during goods issue, activity confirmation, and finished goods receipt into inventory.

Real-time accounting involves:

- Reference documents for the goods issues, confirmations, and goods receipts
- Follow-on documents for overhead, WIP, and variances triggered by configuration settings.

Rest assured; the introduction of SAP S/4HANA doesn't mean everything you've learned changes. Your existing knowledge remains invaluable and will continue to serve you well.

In Chapter 12, we discussed how overhead calculation is defined with costing sheets, a principle that remains unchanged with SAP S/4HANA. The system now charges overhead as a follow-on document in real time, triggered by the issue of raw materials or the confirmation of production activities.

In Chapter 13, you learned about WIP calculation with a results analysis (RA) key, which remains the same. Instead of waiting for the period to close, a follow-on WIP document is now created immediately, based on the combination of the raw material costs and material overhead or the production costs and overhead. This new approach improves the timeliness and accuracy of your reporting.

Results Analysis Key

The RA key for WIP calculation now also controls production variances in event-based processing.

As you saw in Chapter 14, instead of using a variance key, the results analysis key for the WIP calculation also controls the production variances when the finished goods are delivered into stock, and any remaining work in process is reversed.

There is no need for the settlement step we saw in Chapter 15 because the journal entries for WIP and production variances are automatically linked with the market segments in margin analysis. Production controllers can report on the WIP and production variances per production order or product cost collector, while sales controllers can report on the same WIP and production variances per market segment.

Real-time accounting is currently available with the business function universal parallel accounting (FINS_PARALLEL_ACCOUNTING_BF) for pilot customers only. There are still many gaps in the approach, such as the calculation of scrap variances and mixed price variances. Let's discuss the ideas behind universal parallel accounting to understand SAP's future strategy for product cost controlling and changes to the period close activities. We'll discuss these ideas in this chapter.

19.1 Universal Parallel Accounting

Universal parallel accounting is the most significant innovation since introducing the Universal Journal with SAP S/4HANA. The business requirements behind the approach will be familiar to anyone who works in a multinational organization; the innovation is that the ledger and currency settings are used consistently across all financial applications.

In this section, we'll provide an overview of the main use cases for parallel accounting and explain how these scenarios were realized before the introduction of universal parallel accounting. We'll then look at how the ledger and currency settings will change and their impact on activity and material prices.

19.1.1 Overview

For legal purposes, parallel accounting is delivered with two *ledgers*:

- One ledger represents the *common accounting principle* for all company codes in the group, usually US-GAAP for the USA and IFRS for Europe.

- The other ledger supports local accounting principles for each company code in the group. This was supported in SAP ERP by activating the business function Multiple Valuation of Cost of Goods Manufactured (FIN_CO_COGM). You can now carry the same material with multiple legal and standard prices and calculate production variances and contribution margins in each ledger, providing an end-to-end view of product profitability depending on the underlying accounting principle.

With universal parallel accounting:

- The ledger with the common accounting principle also contains the common fiscal year variant.
- The ledger with the local accounting principles can contain different fiscal year variants to meet the local reporting requirements.
- The periods within the fiscal year variants must have the same structure as the common fiscal year variant to support a calendar year for headquarters' fiscal years starting in April for the UK and India, July for Australia, and October for the U.S. federal.

Universal parallel accounting also extends the number of currencies available in financial applications. In SAP ERP, the controlling applications support two currencies:

- Controlling area currency
- Object currency

Financial accounting, asset accounting, and actual costing support three currencies.

The Universal Journal supports 10 currencies. With universal parallel accounting, the currencies are consistent across all financial applications, with no currency conversion. Without universal parallel accounting, the results of an allocation calculated in two currencies are transferred to the general ledger that was configured to receive three currencies, and the third currency must be converted.

Previously, parallel accounting did not support multiple accounting principles from a legal perspective. Instead, working with group valuation and profit center valuation was distinct from the legal valuation we saw in the settings for the costing variant in Chapter 7.

With the group valuation approach, universal parallel accounting delivers a consolidated view of the intercompany value flows separately from the transfer prices used by affiliated companies as they trade at arm's length. This is achieved with the consolidation-like approach in which the intercompany revenues and cost of goods sold (COGS) are eliminated in an additional group valuation ledger whenever an intercompany boundary is crossed, like when a manufacturing company sells to a distribution center or a distribution center sells to the selling company that handles the business with the final customer. Separate material prices for group valuation ensure that intercompany goods movements are valuated at cost so that profit in inventory is excluded in the group view.

Finally, universal parallel accounting supports *profit center valuation*, where companies agree on transfer prices between profit centers with a management view of their operations in a separate profit center valuation ledger. The agreed transfer prices can be either intercompany or *intracompany* (plant-to-plant). Intercompany goods movements will also trigger elimination postings in the profit center valuation ledger following the same logic as in the group valuation ledger.

19.1.2 FIN_CO_COGM: Multiple Valuation of Cost of Goods Manufactured

Let's first look at what's possible in SAP ERP with the business function Parallel Cost of Goods Manufactured (FIN_CO_COGM). This allows multiple legal valuations with ledgers for the common accounting principle and the local accounting principles. The differences are in the other finance applications since asset accounting continues to use different depreciation areas to represent the various valuation approaches, and Controlling continues to use different versions and actual costing currency types. You must map between the different entities in each application.

To use multiple valuation of cost of goods manufactured (COGM), activate the business function Parallel Cost of Goods Manufactured (FIN_CO_COGM). The activation process is described in SAP Note 1852519. You can display business functions activated with Transaction SFW5.

First, you define a *currency and valuation profile*, which you can use for group and profit center valuation. To maintain a Currency and Valuation Profile, follow the IMG path **Controlling • General Controlling • Multiple Valuation Approaches/Transfer Prices • Basic Settings • Define Currency and Valuation Profile**. Select a profile and double-click **Details** on the left to display the screen in Figure 19.1.

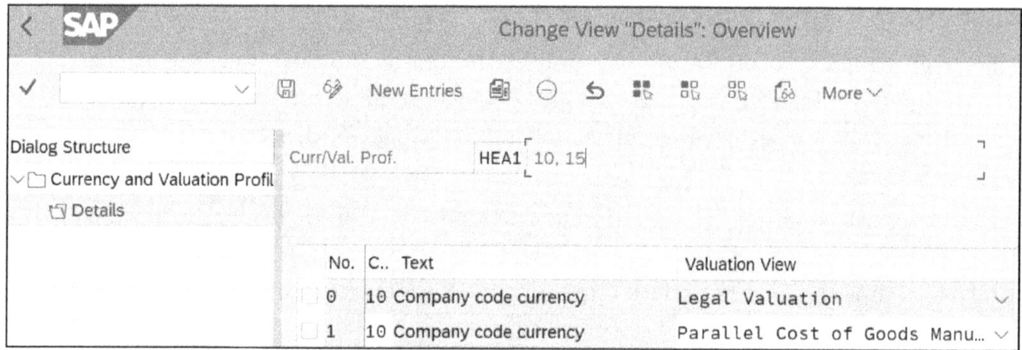

Figure 19.1 Currency and Valuation Profile for Parallel Cost of Goods Manufactured

The **Legal Valuation** is in **Company code currency 10**, and the **Parallel Cost of Goods Manu...** (Parallel cost of goods manufactured), is also in **Company code currency 10**. You can add up to three valuations in addition to the **Legal Valuation**. You can activate the parallel cost of goods manufactured retroactively because, unlike group and profit center valuation, you don't have to add a new currency to the Universal Journal to create an additional legal valuation. Once you define a currency and valuation profile, you assign it to a controlling area via the IMG path **Controlling • General Controlling • Multiple Valuation Approaches/Transfer Prices • Basic Settings • Assign Currency and Valuation Profile.**

Next, you set up Controlling versions to represent each parallel valuation. Follow the IMG menu path **Controlling • General Controlling • Organization • Maintain Versions** and double-click **Controlling Area Settings** for a version as shown in Figure 19.2.

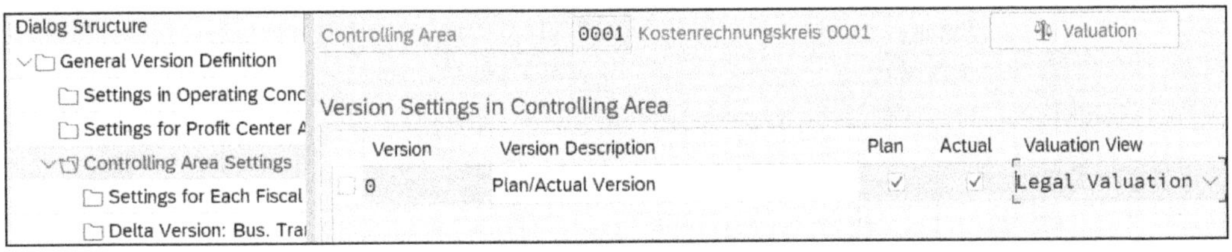

Figure 19.2 Valuation Views for the Controlling Version

The next step is to set up Controlling versions to represent each of the parallel valuations you need. To do this, choose **Controlling • General Controlling • Organization • Maintain Versions** from the menu bar and choose the **Controlling Area Settings** for your version, as shown in Figure 19.3. All systems include a version **0**, which is automatically flagged for **Legal Valuation**. You then need to create as many additional versions as you need and flag them with the **Valuation View** for **Parallel Cost of Goods Manufactured.**

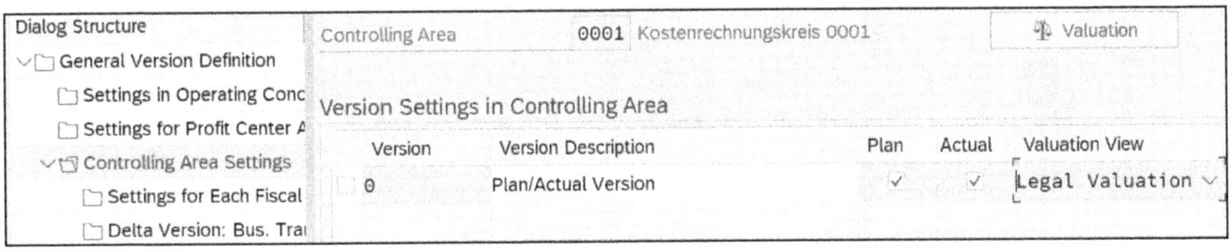

Figure 19.3 Valuation Views for the Controlling Version

When the business function FIN_CO_COGM is active, you'll find the remaining IMG steps by following the menu path **Controlling • Product Cost Controlling • Actual Costing/Material Ledger • Actual Costing • Parallel Valuation of Goods Manufactured.** Unlike group and profit center valuation, which are activated at the level of the Controlling area, you can decide company code by company code whether there is a need for a different legal valuation for the purposes of actual costing. To define the company codes that require an additional valuation, follow the menu path **Controlling • Product Cost Controlling • Actual Costing/Material Ledger • Actual Costing • Parallel Valuation of**

Goods Manufactured • Set Up Parallel Cost of Goods Manufactured for Company Codes. In Figure 19.4, we see that in company code **1000** for Accounting Principle **IAS**, there is no requirement for actual costing (so only standard costs are used), but actual costing is used according to the accounting principles **IASA**, **LOGA**, and **LOTA**. In the case of accounting principle **IASA**, actual costing is performed using a normal costing run, and for accounting principles **LOGA** and **LOTA**, it uses an alternative valuation run. In the context of parallel cost of goods manufactured, the leading version in Figure 19.2 is version **0** in Controlling, while valuations 5, 6, and 7 are the additional Controlling versions added for the purpose of parallel valuation in Figure 19.3.

Transfer ML Postings to CO

CoCd	AccP	Excl. Updt	ML Application
1000	IAS	Exclusively to Leading Versic	
1000	IASA	Exclusively to Valuation 7	Actual Costing
1000	LOGA	Exclusively to Valuation 5	Alternative Valuation
1000	LOTA	Exclusively to Valuation 6	Alternative Valuation

Figure 19.4 Activate Parallel Valuation by Company Code

This same mapping is used to access the depreciation values from asset accounting and link the charts of depreciation with the versions in Controlling. You'll find the settings for this mapping by following the menu path **Controlling • Product Cost Controlling • Actual Costing/Material Ledger • Actual Costing • Parallel Valuation of Goods Manufactured • Transfer Depreciation from Asset Accounting to Controlling**, as shown in Figure 19.5. Here you can see that for the chart of depreciation **PCGM**, many different depreciation areas are being mapped to the leading version (**0**) and valuations 5, 6, and 7.

Transfer Depreciation to CO

ChDep	Ar.	Excl. Updt
PCGM	1	Exclusively to Leading Versic
PCGM	2	Exclusively to Valuation 6
PCGM	3	Exclusively to Valuation 6
PCGM	10	Exclusively to Valuation 6
PCGM	15	Exclusively to Valuation 6
PCGM	60	Exclusively to Valuation 5
PCGM	65	Exclusively to Valuation 6
PCGM	68	Exclusively to Valuation 7
PCGM	71	Exclusively to Leading Vers...
PCGM	72	Exclusively to Leading Vers...
PCGM	81	Exclusively to Valuation 7
PCGM	83	Exclusively to Valuation 7

Figure 19.5 Transfer Depreciation to Controlling

While this setting controls the flow of costs into Controlling from asset accounting, there is also a configuration to control how the values shown in Figure 19.4 flow back to the various ledgers in financial accounting. To access the settings shown in Figure 19.6, choose **Controlling • Product Cost Controlling • Actual Costing/Material Ledger • Actual Costing • Parallel Valuation of Goods Manufactured • Set Up Real-Time Integration for Parallel Cost of Goods Manufactured** from the menu bar.

Transfer CO Postings to FI for Parallel Valuations

Company Co...	Accounting ...	Excl. Update	Real-Time Integ. for Va...	General Ledger Update	Ledger Group (FI)
1000	IAS	Exclusively to Leading Versic	✓	Accounting Principle FI	0L
1000	IASA	Exclusively to Valuation 7	✓	Accounting Principle FI	L7
1000	LOGA	Exclusively to Valuation 5	✓	Accounting Principle FI	L5
1000	LOTA	Exclusively to Valuation 6	✓	Accounting Principle FI	L6

Figure 19.6 Transfer Controlling Postings to Financial Accounting for Parallel Valuations

19.1.3 Multiple Valuations with Universal Parallel Accounting

Universal parallel accounting is useful for all organizations operating in multiple legal entities with their different local accounting requirements, who also wish to achieve a common view of their activities according to a single accounting principle. The change compared to the previous section, is that there is no mapping between the applications as we saw earlier in Figure 19.1 through Figure 19.6. With Universal parallel accounting, the settings are much simpler, with many being made directly in the ledger. To view the ledger settings in the IMG, choose Transaction FINSC_LEDGER or follow menu path **Financial Accounting • Financial Accounting Global Settings • Ledger • Define Settings for Ledgers and Currency Types.** Figure 19.7 shows the ledger settings for three standard ledgers, two of which are for the purposes of **Legal Valuation** and the third for **Group Valuation**. The idea of the **Valuation View** is brought over from the classic solution (see Figure 19.3), but with universal parallel accounting, the ledger settings are central, and the settings for asset accounting (see Figure 19.5) and actual costing (see Figure 19.4) only include those settings over and above the ledger settings. The costing runs in Controlling and the settlement of values to asset under construction and inventory are based entirely on the ledger. The downside of these central settings is that you cannot activate universal parallel accounting company code by company code but only at system level by activating the business function `FINS_PARALLEL_ACCOUNTING_BF` (universal parallel accounting). A full description of all the configuration settings is beyond the scope of this book, but you will find a complete implementation guide for OP2022 and OP2023 for reference purposes in SAP Note 3207221.

Dialog Structure
- Currency Types
- Global Currency Conversion Se
- Currency Conversion Settings f
- Ledger
  - Company Code Settings for

Ledger

Ld	Ledger Name	Leading	Ldgr Type	Ex...	Underlyin...	Valuation View
0L	Ledger 0L	✓	Standard Ledger			Legal Valuation
2L	Ledger 2L		Standard Ledger			Legal Valuation
4G	Ledger 4G		Standard Ledger			Group Valuation

Figure 19.7 Ledger Settings for Universal Parallel Accounting with Multiple Valuations

If we then select the company code settings for the first ledger via the dialog structure, we see the default settings, currency codes **10** (local) and **30** (global), and up to eight additional currencies as required within the relevant company codes, as shown in Figure 19.8. While most organizations can make a case for a local and a global currency, it is becoming increasingly common to keep a third *functional currency* together with a hard currency or index-based currency for countries with high inflation. Notice that you can also create new currency types beginning with Z* to meet your specific currency needs. What makes universal parallel accounting special is that all the currencies in this table are used consistently in asset accounting, actual costing, Controlling, and so on, preventing the accidental inconsistencies caused by currency conversions where an application that only supported two or three currencies writes a document to the many currency fields in the universal journal.

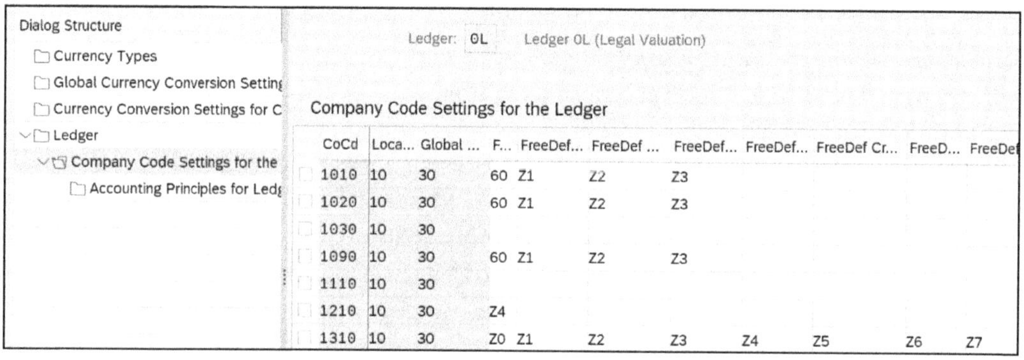

Dialog Structure
- Currency Types
- Global Currency Conversion Settin
- Currency Conversion Settings for C
- Ledger
  - Company Code Settings for the
    - Accounting Principles for Led

Ledger: 0L Ledger 0L (Legal Valuation)

Company Code Settings for the Ledger

CoCd	Loca...	Global ...	F...	FreeDef...	FreeDef ...	FreeDef...	FreeDef...	FreeDef Cr...	FreeD...	FreeDe
1010	10	30	60	Z1	Z2	Z3				
1020	10	30	60	Z1	Z2	Z3				
1030	10	30								
1090	10	30	60	Z1	Z2	Z3				
1110	10	30								
1210	10	30	Z4							
1310	10	30	Z0	Z1	Z2	Z3	Z4	Z5	Z6	Z7

Figure 19.8 Company Code Settings for the Ledger with Multiple Currencies

Use of Functional Currency prior to Universal Parallel Accounting

If you need to activate a functional currency but are not yet ready to implement the business function for universal parallel accounting and are using SAP S/4HANA 1909 or later, refer to the details in SAP Note 3200089. The default setting is that the local currency will be treated as the functional currency, but if you have a subsidiary that needs a different functional currency because of the primary economic environment in which the entity operates, you can include this additional currency in the ledger settings and run program FINS_MIG_FUNCTIONAL_CURCY to include figures in the third currency on the database.

If you switch from ledger **0L** to ledger **2L** and scroll to the right, in Figure 19.9 you can see the multiple fiscal-year variants and multiple accounting principles used in the second ledger. In ledger **0L**, the same calendar and accounting principle are used in all company codes. Notice that compared to the accounting principles used in the previous section (see Figure 19.4), with universal parallel accounting a different accounting principle is used for every country (**DEAP** for Germany, **GBAP** for Great Britain, **FRAP** for

France, and so on). This is because of the settings in asset accounting that are now linked directly with the accounting principle.

Ledger: 2L Ledger 2L (Legal Valuation)

Company Code Settings for the Ledger

CoCd	Fisc.Year Variant	Posting Variant	Accounting Princip...
1010	K4	1010	DEAP
1020	V3	1010	DEAP
1030	K4	1010	DEAP
1110	K4	1110	GBAP
1210	K4	1210	FRAP
1310	K4	1310	CNAP
1510	K4	1510	JPAP
1710	V6	USV6	USAP
1810	V3	INV3	INAP
1910	K4	1910	SEAP
2110	K4	2110	HUAP

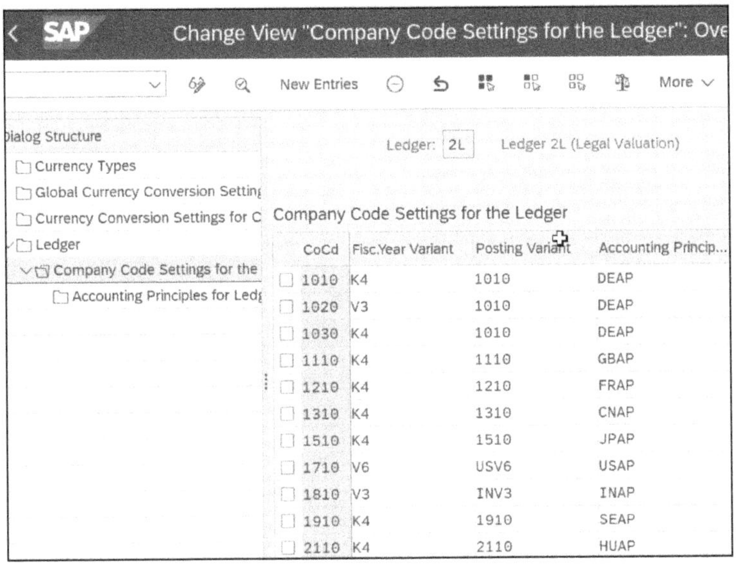

Figure 19.9 Company Code Settings for the Ledger with Multiple Accounting Principles

19.1.4 Activity Prices in Multiple Ledgers

We introduced the master data for the activity type in Chapter 2 and explained how to perform activity price planning for the manufacturing activities in Chapter 1. This process changes with universal parallel accounting because activity prices can be ledger-specific. You can, of course, continue to use the same activity price in all ledgers as before, but behind the scenes SAP has swapped table COST for table ACCOSTRATE to accommodate the changes. This means that you won't be able to use Transaction KP26 to enter activity rates but must instead log on with the Overhead Accountant role and use the SAP Fiori app Manage Cost Rates—Plan (SAP Fiori ID F3162) to enter the activity rates manually or to show the result of the data transfer from SAP Analytics Cloud for planning.

Figure 19.10 shows the Manage Cost Rates—Plan app, where we've selected the cost rates for **Cost Center** 10101301 and **Activity Type** 1 in the ledgers **0L**, **2L**, and **3L**. To display the planned rates, we've chosen the values in ledger **2L** for January 2023. These values are used in the standard cost estimate provided the valuation variant is linked with the appropriate ledger to calculate standard costs that use this activity. When you enter a confirmation for an operation running at a work center assigned to **Cost Center** 10101301 and **Activity Type** 1, the hours confirmed will potentially result in different values being recorded in each of the ledgers listed. Similarly, these activity rates will be used as the preliminary valuation in actual costing and may be updated to reflect the actual costs if the underlying costs per ledger change during the period.

Manage Cost Rates - Plan

Standard*

Edit Cost Rates

Cost Rates (1)

Cost Center	Activity Type	From Year*	From Period*	Fixed Rate	Variable Rate	Currency*	Per*
10101301	1	2023	001	99.00	0.00	USD	1

Save Cancel

Cost Rates (5) Standard

Add Edit Copy Timeline Upload Analysis Delete

Cost Center	Activity Type	Ledger	From Year	From Period
10101301 (Manufacturing 1 (DE))	1 (Machine hours 1)	2L (Ledger 2L)	2023	001
10101301 (Manufacturing 1 (DE))	1 (Machine hours 1)	3L (Ledger 3L)	2023	001
10101301 (Manufacturing 1 (DE))	1 (Machine hours 1)	3L (Ledger 3L)	2023	007
10101301 (Manufacturing 1 (DE))	1 (Machine hours 1)	0L (Ledger 0L)	2023	007
10101301 (Manufacturing 1 (DE))	1 (Machine hours 1)	3L (Ledger 3L)	2023	008

Figure 19.10 Manage Cost Rates - Plan, Showing Activity Rates for Multiple Ledgers

19.1.5 Material Prices in Multiple Ledgers

We introduced the material master data in Chapter 3 and explained how to enter raw material prices. This process changes because material prices can also be ledger-specific with universal parallel accounting. You can, of course, continue to use the same material price in all ledgers as before, but behind the scenes, SAP has swapped table MBEW for table FMLT_PRICE to accommodate the changes. The material master transaction (Transaction MM03) has changed slightly so that you can see the various prices in the **Accounting** view. To scroll between the various valuations in Figure 19.11, use the **Currency Type** field to scroll between the ledgers, currencies, and valuations. The number of combinations available is determined by the number of ledgers and the number of currency types (the combination of currency and valuation) activated in each ledger (refer back to Figure 19.8), giving you much greater flexibility than we had when we looked at the material prices in Chapter 3.

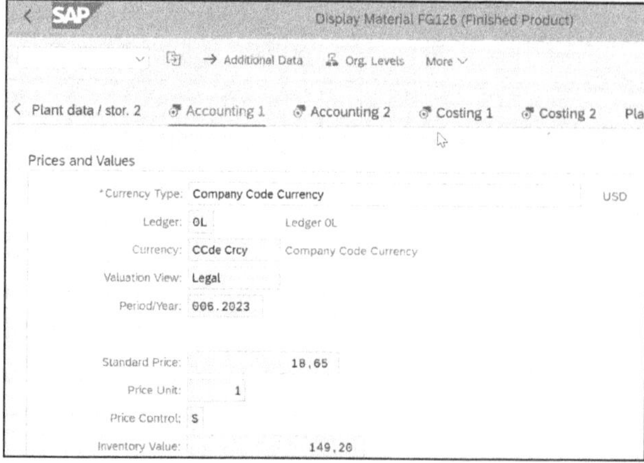

Figure 19.11 Accounting View of Material Master Showing Multiple Valuations

Alternatively, you can log on in the role of an inventory accountant and use the SAP Fiori app Manage Material Valuations (SAP Fiori ID F2680) to view the material prices in the various ledgers. Figure 19.12 shows the current material valuation for Material **FG126** within the Manage Material Valuations app, where we've selected the values for **Legal Valuation** in ledger **0L** and currency type **Company Code Currency**.

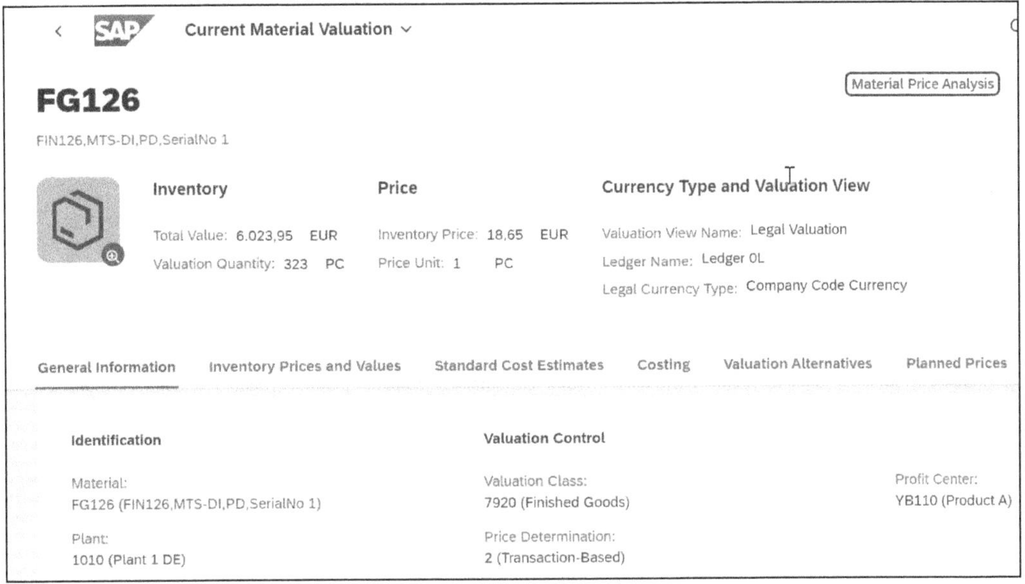

Figure 19.12 Manage Material Valuations App, Showing Current Material Valuation for Material FG126

19

The introduction of the ledger for the material prices also affects the way you create standard costs. You can choose either to continue to use your existing costing variants while accepting that they have no connection to a ledger and so all legal valuation ledgers will be updated with the same value when you release the standard cost estimate (see Chapter 9), or to use new costing variants that link to the ledger via the costing type. This will allow you to create different cost estimates for the purpose of setting the standard costs in each ledger and include different activity prices and material prices in these valuations via the valuation variant. If you work with best practices, the settings for product cost planning are delivered via scope item 6DF (Universal Parallel Accounting). Figure 19.13 shows the costing variants to update ledgers **0L**, **2L**, and **3L** in the context of universal parallel accounting.

To understand the settings associated with these costing variants, choose costing variant **POOL**. Figure 19.14 shows that both the costing type and valuation variant associated with this variant are ledger-specific. This means that you will only be able to use a cost estimate using costing variant **POOL** to update inventory in ledger **0L**, and the material prices and activity prices that support this valuation will be selected from ledger **0L**.

Costing Variants

Costing Variant	Name
P00L	Legal Mat Cst Est Ldgr-0L
P02L	Legal Mat Cst Est Ldgr-2L
P03L	Legal Mat Cst Est Ldgr-3L

Figure 19.13 Costing Variants for Ledger-specific Update of Standard Costs

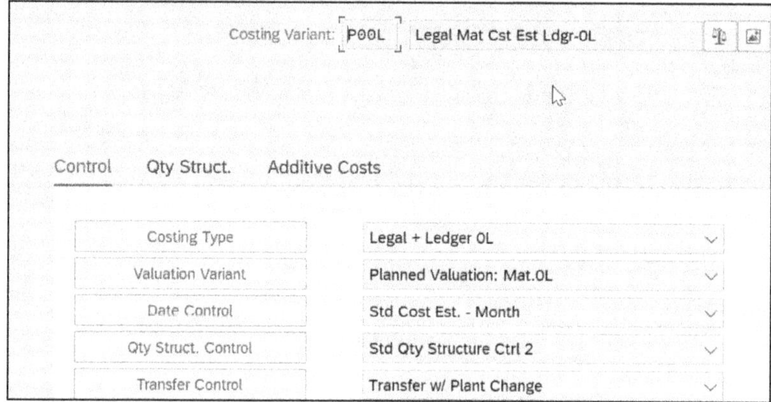

Figure 19.14 Ledger-specific Costing Type and Valuation Variant

19.2 Event-Based Production Accounting

At the time of writing, event-based production accounting is only supported in combination with universal parallel accounting. In this section, we'll look at the configuration settings that control the posting logic for the calculation of overhead, work in process, and production variances. We'll then explain how the system calculates overhead, WIP, and production variances as the underlying business transactions are captured.

19.2.1 Settings for Event-Based Production Accounting

When you make the move to event-based production accounting, you can keep the rules for the calculation of overhead that we saw in Chapter 5, but you must flag your costing sheet as **Evt. based**, as shown for **Costing Sheet 1010EP** in Figure 19.15. It's not possible to combine non-event-based costing sheets with event-based WIP, because the assumption is that the creation of the reference document will trigger a first follow-on document for the overhead that will then be followed by a follow-on document for either WIP or production variances, depending on the status of the order.

Dialog Structure
- Costing Sheets
  - Costing Sheet Rows
    - Base
    - Overhead Rate
    - Credit

Costing Sheets

Costing Sheet	Description	Evt. based
1010EP	Event based costing sheet (DE)	✓
1010PC	Costing Sheet Production-DE	
1010PI	PS Projects DE	

Figure 19.15 Costing Sheet for Event-Based Production Accounting

You determine that a production order, a process order, or product cost collector should generate journal entries for WIP and production variances by assigning the appropriate **Event-Based Processing Key** to the order. As we saw for WIP calculation in Chapter 13, the default key is selected using the combination of the order type and plant. Figure 19.16 shows the event-based processing key for production orders, which combines the settings for WIP calculation, variance calculation, and potential settlement receivers. Notice that if you use this key, WIP is calculated at actual costs, variances (if calculated) can be split into the categories **Input Price Variance**, **Input Quantity Variance**, **Resource Usage Variance**, **Lot Size Variance**, and **Remaining Variance**, and it's possible to assign some costs to cost centers or WBS elements rather than assigning all costs to **Material**. To access these settings, chose **Controlling • Product Cost Controlling • Product Cost by Order • Period-End Closing • Event-Based WIP and Variance Posting** from the menu bar and select the appropriate **Event-Based Processing Key**. If you work with best practices, these settings are delivered using scope item 3FO and have been available since SAP S/4HANA 2022.

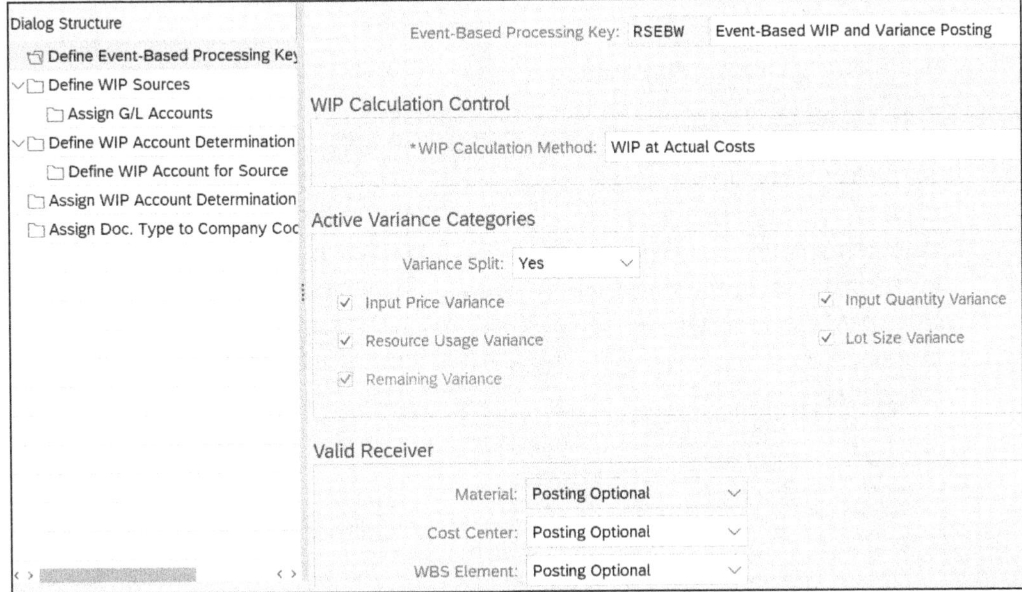

Figure 19.16 Event-Based Processing Key for Product Cost by Order

By comparison, the event-based processing key for product cost by period shown in Figure 19.17 calculates **WIP at Target Costs**—in other words, using the target costs for the operation recorded at the reporting point reached. Production variances are only split to the categories **Input Price Variance**, **Input Quantity Variance**, **Resource Usage Variance**, and **Remaining Variance**, and all costs must be settled to the material. If you don't create any reporting points and record all goods issues and confirmations as a backflush on completion of the finished product, no WIP will be calculated, but only variances if changes are made to the standard BOM or routing quantities. These settings are also delivered with scope item 3F0 and have been available since SAP S/4HANA 2023.

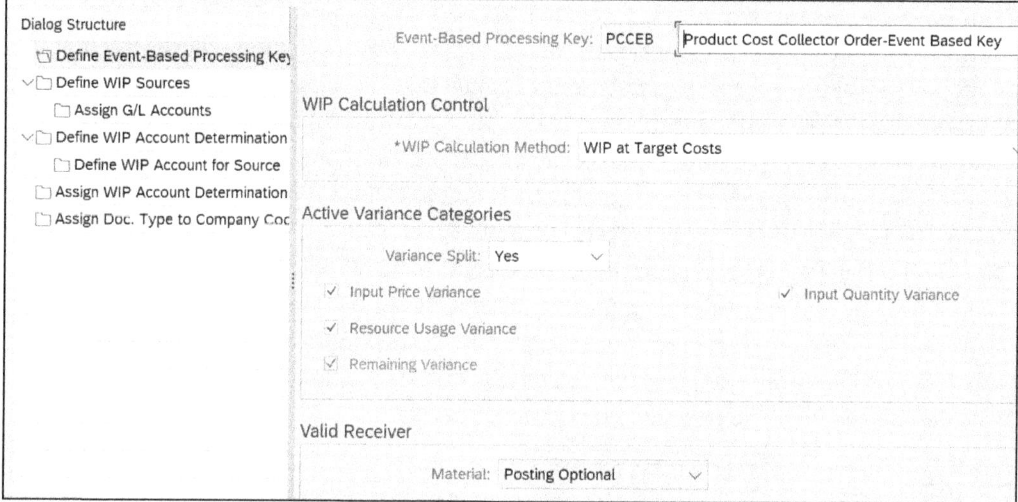

Figure 19.17 Event-Based Processing Key for Product Cost by Period

Unless you are working with the best-practice settings, you should check that all general ledger accounts for goods issues and confirmations to the orders and goods receipts on completion, together with the follow-on costs for overhead, have been assigned to the delivered **WIP sources**, as shown in Figure 19.18. The system will increase work in process for the values posted under the general ledger accounts for material cost (**0MAT**), labor cost (**0LBR**), and overhead cost (**0OVH**) and decrease work in process in proportion to the values posted under the general ledger accounts for delivered costs (**0DLV**).

Dialog Structure
Define Event-Based Processing Key
Define WIP Sources
Assign G/L Accounts
Define WIP Account Determination
Define WIP Account for Source
Assign WIP Account Determination to
Assign Doc. Type to Company Codes

Define WIP Sources

Source	Description	Source Type
0DLV	Delivered Cost	Delivered Costs
0LBR	Labor Cost	Production Costs
0MAT	Material Cost	Production Costs
0OVH	Overhead Cost	Production Costs

Figure 19.18 WIP Sources for Event-Based Processing

You should also check the general ledger accounts to be updated to account for work in process and reserves for unrealized costs for each of the WIP sources, as shown in Figure 19.19.

Dialog Structure
- Define Event-Based Processing Key
- Define WIP Sources
  - Assign G/L Accounts
- Define WIP Account Determination
  - Define WIP Account for Source
- Assign WIP Account Determination to
- Assign Doc. Type to Company Codes

WIP Account Determination Rule: SEBW

Define WIP Account for Source

Chrt/Accts	Item	Source	WIP (B/S Acct)	WIP Offset (P/L)	Reserves (B/S Acct)	Rsrvs Offset (P/L)
YCOA	1	Material Cost	13200000	54200100	24098600	54200100
YCOA	2	Labor Cost	13200000	54200100	24098600	54200100
YCOA	3	Overhead Co...	13200000	54200100	24098600	54200100
YCOA	4	Delivered C...				

Figure 19.19 WIP Account Determination Rules

19.2.2 Event-Based Overhead

The rules for the application of overhead are the same as we saw in Chapter 5, the only difference being that the overheads are applied at a different point in time, so the example in Figure 19.20 shows the material overhead being calculated when the raw materials were issued to the production order. What's important here is the link to the **Reference document** because it provides the link to the goods issue that resulted in the line item for material overhead. If the goods issue is cancelled for any reason, this link is needed to find and reverse the associated overhead and WIP. Notice also that these documents have a new business transaction type, **KZPI**. If you are involved in discussions about system sizing, bear in mind that an overhead document is created for each reference document, rather than for the whole order during the period close, and that this will result in extra documents, especially if you have many different confirmations and goods issues on a single order.

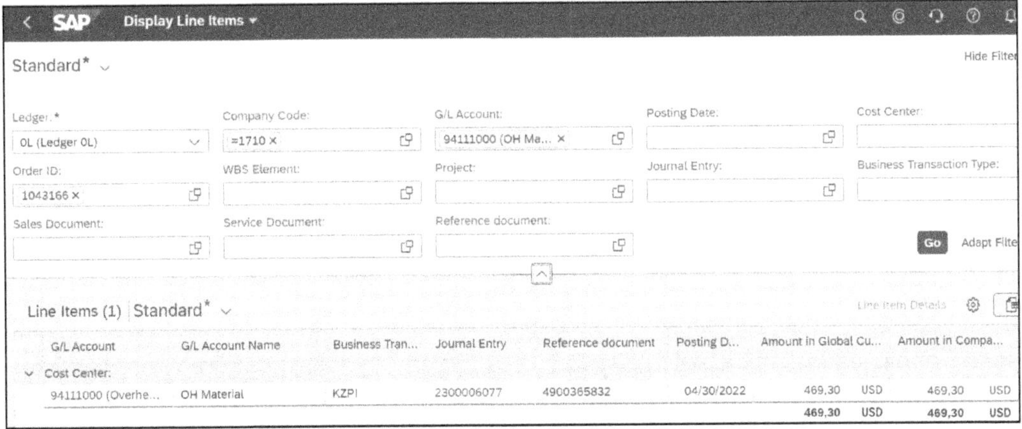

Figure 19.20 Line Item for Event-Based Overhead

In the context of universal parallel accounting, notice that in this example we are looking at the journal entry in ledger **0L**, but that depending on configuration, further ledgers may have been updated. The rules in the costing sheet are not yet ledger-specific, so a material overhead of 10% is applied regardless of ledger, but the underlying raw material costs could be different in each ledger.

19.2.3 Event-Based Work in Process

Just like the overhead calculation, WIP is also created in association with a reference document, so in Figure 19.21 we see two reference documents: one for the confirmation and one for the goods issue, and the associated WIP posting with business transaction type EBWP. The WIP in the first line includes the activity costs for the operation confirmed and the production overhead, and the WIP in the second line includes the raw material costs for the goods issued at this operation and the material overhead (see Figure 19.20). In neither case was WIP calculation triggered by the user. It was calculated automatically when the underlying reference document was posted. At this point, all the costs on order **1043166** are considered work in process because there has not yet been a delivery to stock.

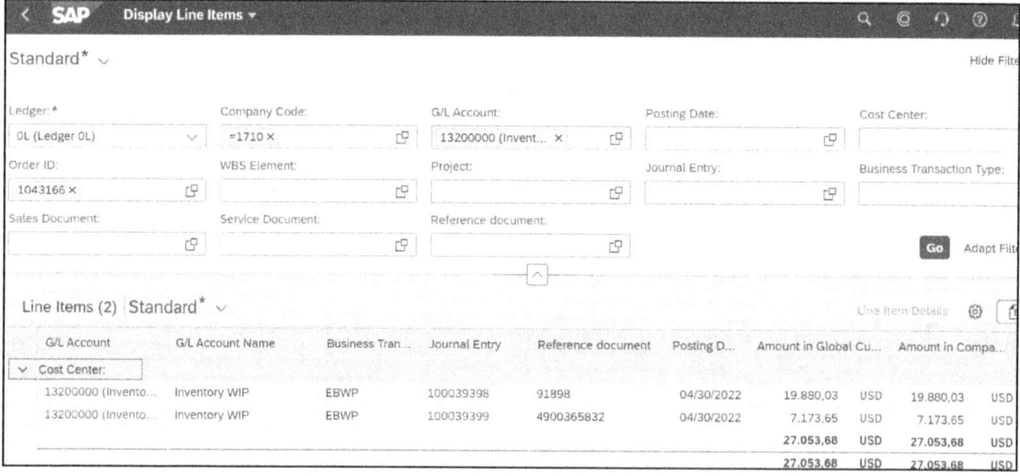

Figure 19.21 Line Items for Event-Based WIP (Goods Issue and Confirmation)

By contrast, in Figure 19.22, we see two additional reference documents for an additional confirmation (production activities) and then for the delivery of the finished goods to stock, which have resulted in the WIP being first increased to cover the additional production activity costs and then the total WIP having been reversed. In this example, the lot size of the order was delivered in a single goods movement, but it's also possible to make partial deliveries, which results in a partial reversal of the WIP captured on the order. If you don't make a final delivery, any remaining work in process is reversed when you set the status to **Technically Closed** for the order.

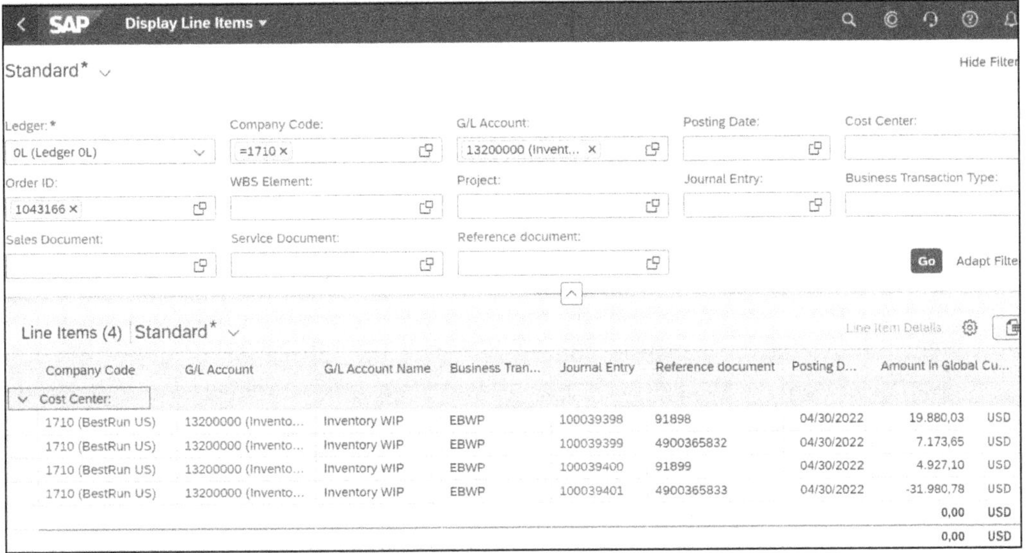

Figure 19.22 Line Items for Event-Based WIP (Goods Issue, Confirmations, and Reversal)

While the line-item reports are useful during testing, the easier way to monitor the work in process for the orders in a plant is to use the Event-Based Work in Process app shown in Figure 19.23. This shows the work in process by plant, product, order, and so on, together with the estimated cost to complete in accordance with the planned costs for the order. The journal entries for the WIP can also be seen in all financial reports (financial statements, trial balance, and so on) without the need for settlement.

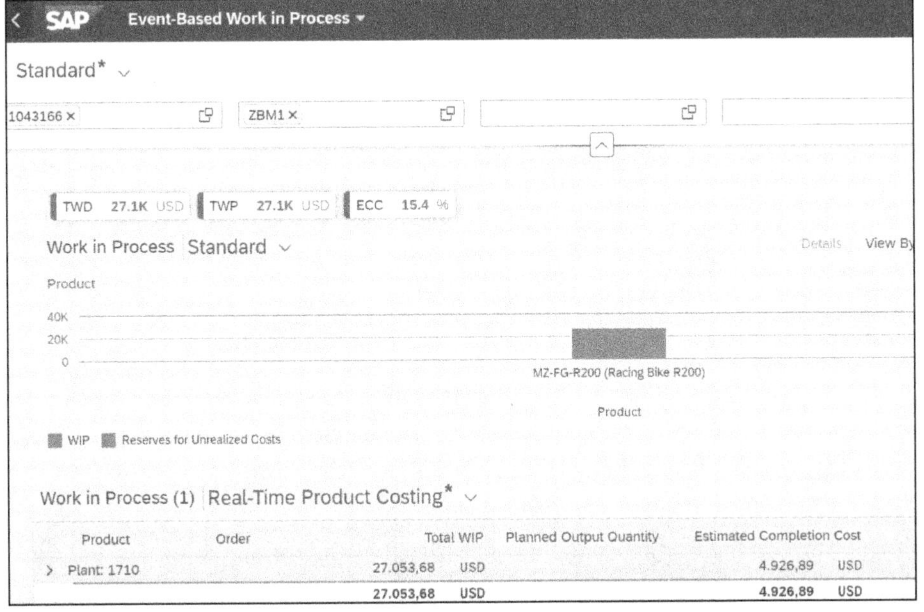

Figure 19.23 Event-Based Work in Process App

19.2.4 Event-Based Variances

Whereas the calculation of overhead and WIP is essentially the same for production/ process orders and product cost collectors, the calculation of production variances is different depending on whether the process is *status-driven* (production or process orders) or *period-driven* (product cost collectors). In the case of the production or process order, the process is status-driven and relies on an event, usually the status **Final Delivery**, but occasionally the status **Technically Closed**, to trigger the calculation of the production variances. In the case of the product cost collector, production is continuous, and there is no event to signal order completion and trigger the calculation of production variances, so instead you must trigger the calculation of the production variances from the Manage Event-Based Processing Errors—Product Costing app using the **Post Processing—Product Costing** button at period close.

Figure 19.24 shows the line items for the production variances on order **1043166**. These have been triggered with reference to the goods receipt document **400036870**. Notice that they have all been assigned to the same general ledger account: **52070000** (loss from production). To see the variance categories, use the wheel icon ⚙ to add the field **SLA Line Item Type** to the report layout.

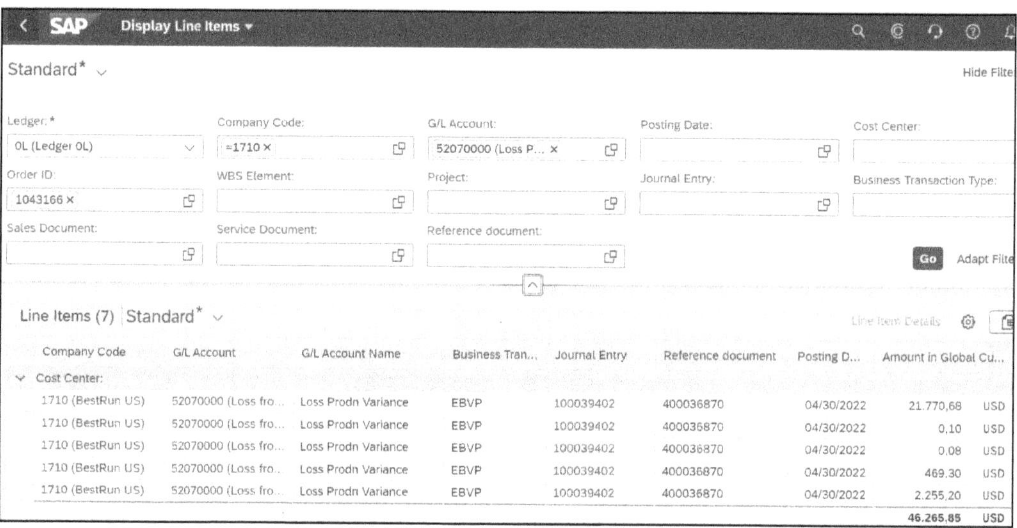

Figure 19.24 Line Items for Event-Based Production Variances

Figure 19.25 shows a different set of line items with the field **SLA Line Item Type** in the report layout, so that we see the various items for **Resource Usage Variances**, **Lot Size Variances** and **Remaining Variances**. The formulas for calculating the variances are the same as we described in Chapter 14. With release OP2023, it is not yet possible to calculate scrap variances or mixed-price variances, but there are plans to add these in subsequent releases.

Another significant change in the new approach to variance calculation is that there is no need to run settlement to move the production variances to margin analysis. The journal entry for the production variances already includes not just the production order, but also the market segments derived for that order, such as material, product group, and so on. This uses the derivation logic that used to take place to derive the profitability segment during settlement but now takes place when the production variances are calculated.

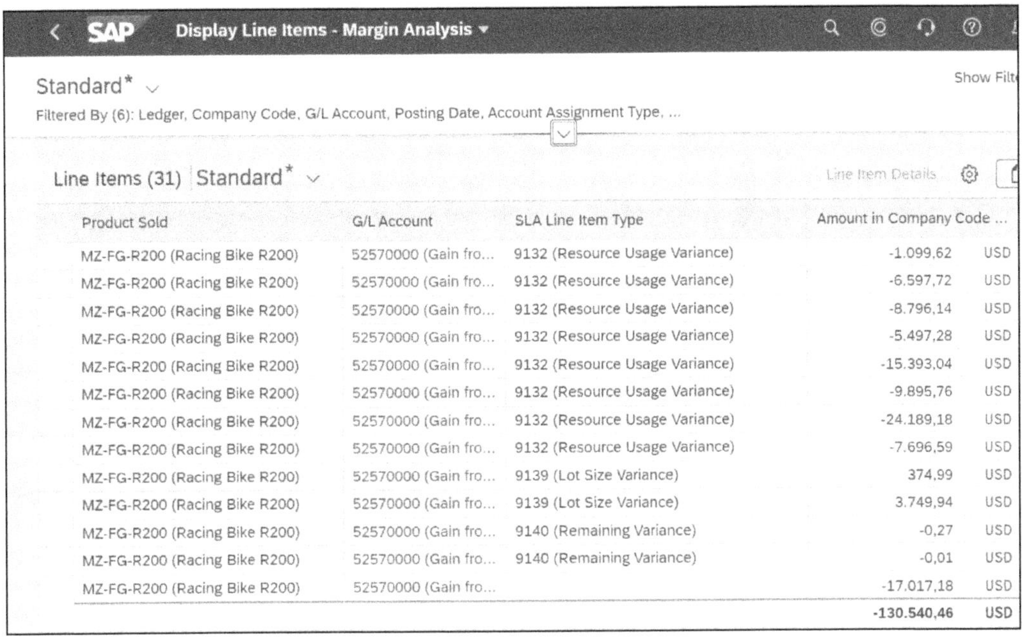

SAP Display Line Items - Margin Analysis

Standard*

Filtered By (6): Ledger, Company Code, G/L Account, Posting Date, Account Assignment Type, ...

Line Items (31) Standard*

Product Sold	G/L Account	SLA Line Item Type	Amount in Company Code ...	
MZ-FG-R200 (Racing Bike R200)	52570000 (Gain fro...	9132 (Resource Usage Variance)	-1.099,62	USD
MZ-FG-R200 (Racing Bike R200)	52570000 (Gain fro...	9132 (Resource Usage Variance)	-6.597,72	USD
MZ-FG-R200 (Racing Bike R200)	52570000 (Gain fro...	9132 (Resource Usage Variance)	-8.796,14	USD
MZ-FG-R200 (Racing Bike R200)	52570000 (Gain fro...	9132 (Resource Usage Variance)	-5.497,28	USD
MZ-FG-R200 (Racing Bike R200)	52570000 (Gain fro...	9132 (Resource Usage Variance)	-15.393,04	USD
MZ-FG-R200 (Racing Bike R200)	52570000 (Gain fro...	9132 (Resource Usage Variance)	-9.895,76	USD
MZ-FG-R200 (Racing Bike R200)	52570000 (Gain fro...	9132 (Resource Usage Variance)	-24.189,18	USD
MZ-FG-R200 (Racing Bike R200)	52570000 (Gain fro...	9132 (Resource Usage Variance)	-7.696,59	USD
MZ-FG-R200 (Racing Bike R200)	52570000 (Gain fro...	9139 (Lot Size Variance)	374,99	USD
MZ-FG-R200 (Racing Bike R200)	52570000 (Gain fro...	9139 (Lot Size Variance)	3.749,94	USD
MZ-FG-R200 (Racing Bike R200)	52570000 (Gain fro...	9140 (Remaining Variance)	-0,27	USD
MZ-FG-R200 (Racing Bike R200)	52570000 (Gain fro...	9140 (Remaining Variance)	-0,01	USD
MZ-FG-R200 (Racing Bike R200)	52570000 (Gain fro...		-17.017,18	USD
			-130.540,46	**USD**

Figure 19.25 Line Items for Production Variances, Showing SLA Line-item Type

Figure 19.26 shows the same journal entries for the production variances from the perspective of the sales accountant working with margin analysis. This time, we've selected the account assignment type **EO** (for market segment) and can see the variances for product **MZ-FG-R200** (racing bike) without the need for settlement to move the values between applications.

While there is no longer a need to run overhead calculation, WIP calculation or variance calculation manually, you should ensure that key users have access to the Manage Event-Based Processing Errors—Product Costing app (SAP Fiori ID F5132) shown in Figure 19.27. This will show you any orders for which configuration errors prevented event-based processing. When the error has been corrected, you can trigger the relevant journal entries using the buttons **Post Process Event-Based Postings** for the relevant process. You can use the same buttons to trigger the calculation of production variances for the product cost collectors and to trigger event-based postings for manual journal entries and correction postings to the orders.

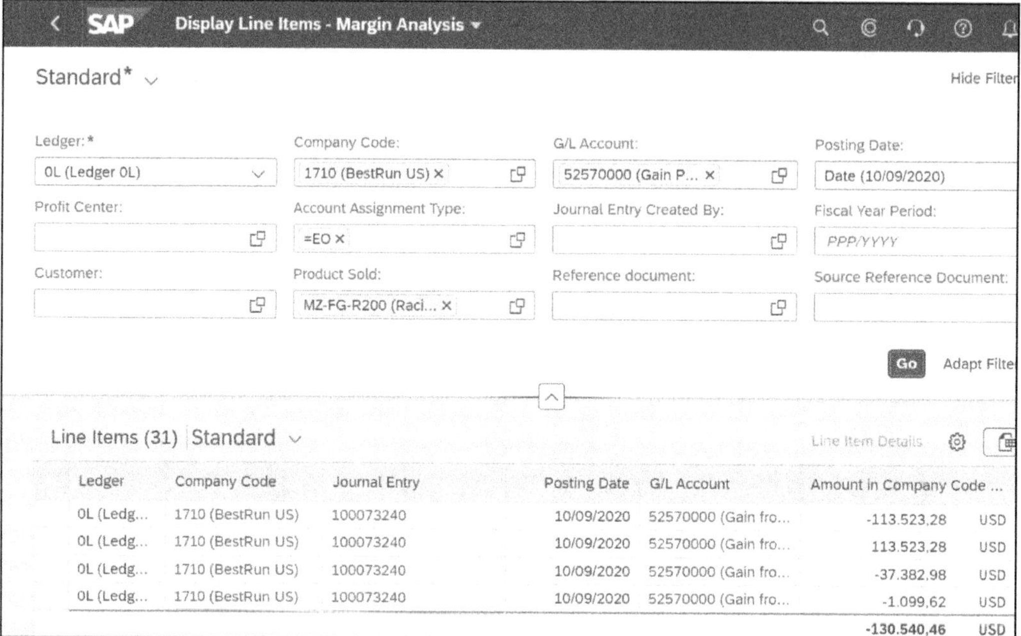

Figure 19.26 Line Items for Production Variances in Margin Analysis

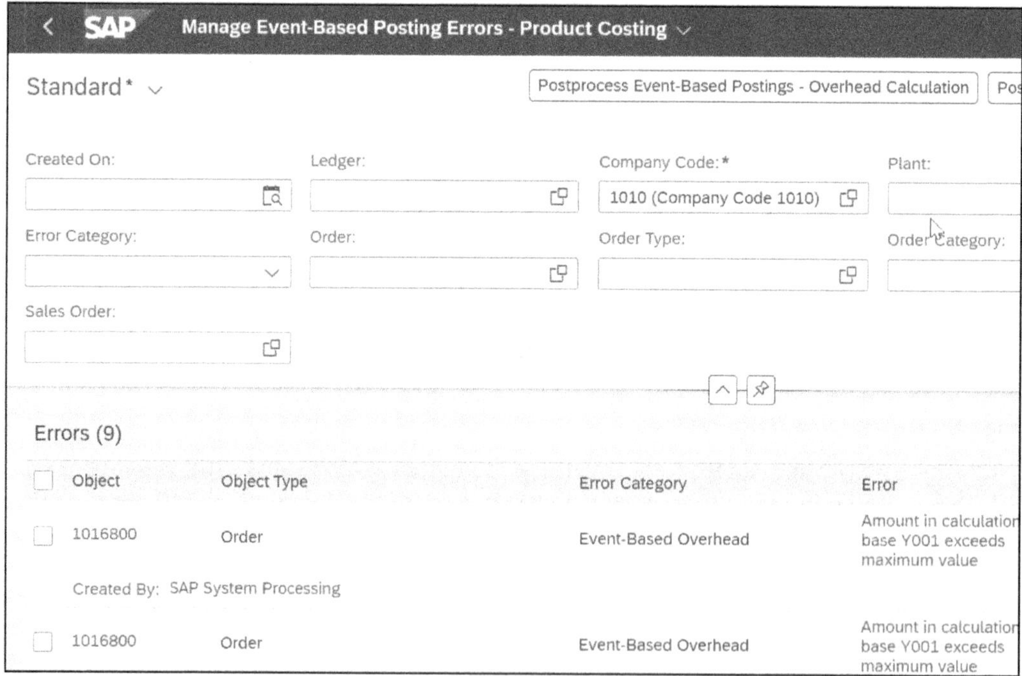

Figure 19.27 Manage Event-Based Processing Posting Errors - Product Costing

19.2.5 Margin Analysis

The production variances are used in margin analysis to adjust the contribution margin to reflect the difference between the standard costs that were used to set the cost of goods sold (COGS) and the actual costs for each production order. The variances are displayed to the sales accountant in apps such as Display Line Items—Margin Analysis (see Figure 19.26), and Product Profitability with Production Variances, shown in Figure 19.28. This report includes a line for each of the supported variance categories (**Input Price Variance**, **Input Quantity Variance**, **Resource-Usage Variance**, **Lot-Size Variance**, and **Input Remaining Variance**).

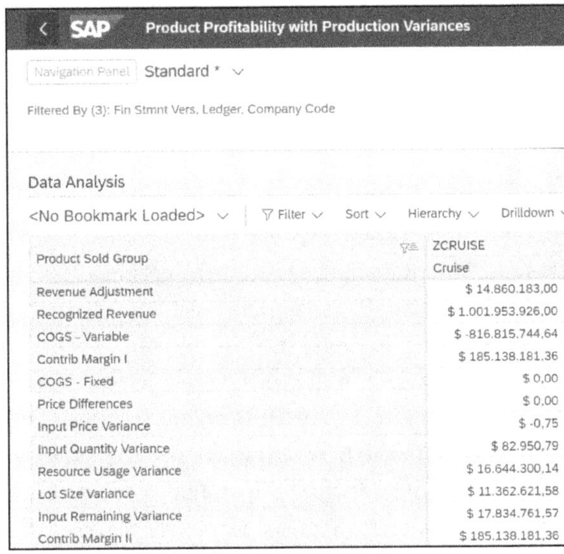

Product Profitability with Production Variances

Navigation Panel Standard *

Filtered By (3): Fin Stmnt Vers, Ledger, Company Code

Data Analysis

<No Bookmark Loaded> | Filter | Sort | Hierarchy | Drilldown

Product Sold Group	ZCRUISE Cruise
Revenue Adjustment	$ 14.860.183,00
Recognized Revenue	$ 1.001.953.926,00
COGS - Variable	$ -816.815.744,64
Contrib Margin I	$ 185.138.181,36
COGS - Fixed	$ 0,00
Price Differences	$ 0,00
Input Price Variance	$ -0,75
Input Quantity Variance	$ 82.950,79
Resource Usage Variance	$ 16.644.300,14
Lot Size Variance	$ 11.362.621,58
Input Remaining Variance	$ 17.834.761,57
Contrib Margin II	$ 185.138.181,36

Figure 19.28 Manage Product Profitability with Production Variances App

To see the production variances in the Display Line Items—Margin Analysis app, adjust the fields shown in the result list by choosing the settings wheel ⚙ and adding the **SLA Line Item Type** field, as shown in Figure 19.29.

Display Line Items - Margin Analysis

Standard*

Filtered By (6): Ledger, Company Code, G/L Account, Posting Date, Account Assignment Type, ...

Line Items (12.066) Standard*

Line Item Details

Ledger	Company Code	Journal Entry	G/L Account	SLA Line Item Type	Amount in Company Co...	
0L (Ledg...	1710 (BestRun US)	100070289	52570000 (Gain fro...	9132 (Resource Usage Variance)	-1.387,57	USD
0L (Ledg...	1710 (BestRun US)	100060116	52570000 (Gain fro...	9139 (Lot Size Variance)	-1.384,88	USD
0L (Ledg...	1710 (BestRun US)	100066022	52570000 (Gain fro...	9132 (Resource Usage Variance)	-1.383,30	USD
0L (Ledg...	1710 (BestRun US)	100066022	52570000 (Gain fro...	9132 (Resource Usage Variance)	-1.383,30	USD
0L (Ledg...	1710 (BestRun US)	100067967	52570000 (Gain fro...	9139 (Lot Size Variance)	-1.382,25	USD
0L (Ledg...	1710 (BestRun US)	100059946	52570000 (Gain fro...	9139 (Lot Size Variance)	-1.376,98	USD

Figure 19.29 Display Line Items – Margin Analysis App, with Variance Categories

If you build your own reports, you can build the relevant key figures by combining the general ledger account used to capture the price differences with the **SLA Line Item Type** delivered by SAP. These are shown in Table 19.1.

SLA Line Item Type	Variance Category
9130	Input Price Variance
9131	Input Quantity Variance
9132	Resource Usage Variance
9139	Lot Size Variance
9140	Input Remaining Variance

Table 19.1 Key Figure Definition for Production Variances

19.3 Actual Costing

Since by its very nature, actual costing always takes place either at period close or cumulatively, if you calculate your actual costs on a year-to-date basis, there are less dramatic changes compared to what we described in Chapter 16 with the introduction of universal parallel accounting. Again, only those materials are included in actual costing that are flagged as using **Price Determination 3**. In Figure 19.30, we see that the sample material is classified as **3 (Single/Multilevel)** in the **Accounting** view of the material master (Transaction MM03), which means it will be included in actual costing. From here, you can use the **Material Price Analysis** button (or Transaction CKM3N) to see the various goods movements that have been collected for the material during the period.

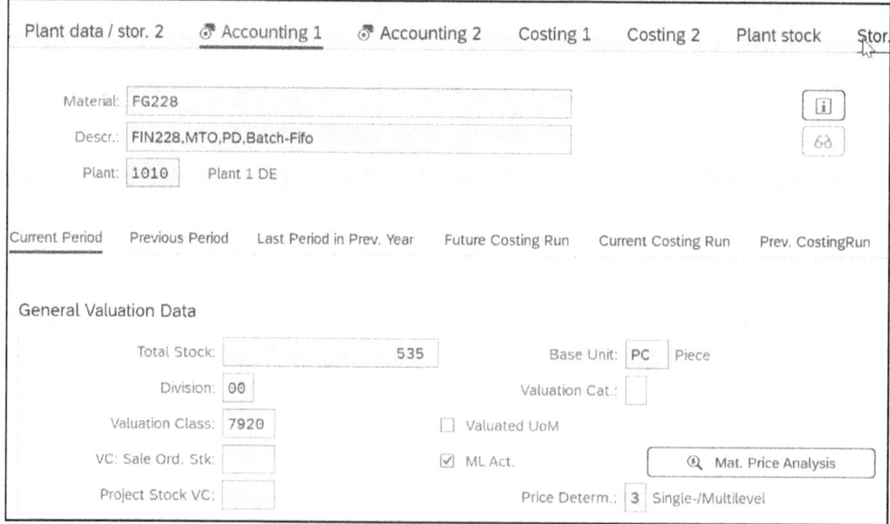

Figure 19.30 Material Master, for a Material Using Actual Costing

Figure 19.31 shows the **Material Price Analysis** view of a produced material prior to actual costing. Notice that there have been two goods receipts with a preliminary valuation at the standard price for the finished goods and two order settlements, resulting in price differences that must be assigned to the goods in inventory using actual costing. The status **Relevant for Settlement** is set because the material is classified as **3 (Single/Multilevel)** and means that this material will be included in the costing run at period close. To see the costs in the various currencies and valuations, you can scroll through the entries in the **Curr./Valuation:** field to see all ledger/currency/valuation combinations shown earlier in Figure 19.8.

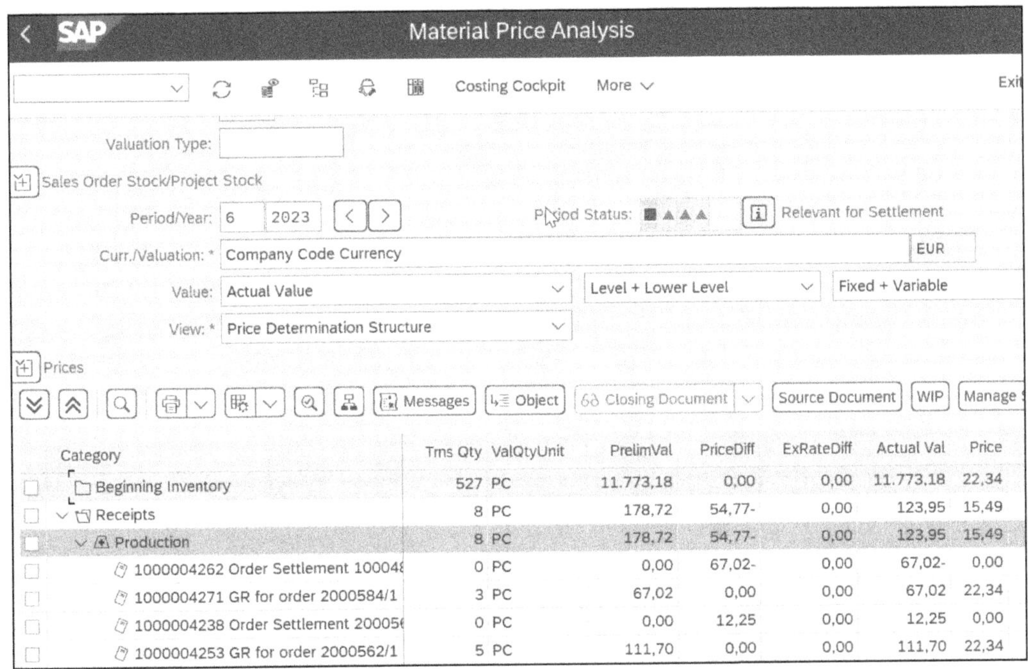

Category	Trns Qty	ValQtyUnit	PrelimVal	PriceDiff	ExRateDiff	Actual Val	Price
Beginning Inventory	527	PC	11.773,18	0,00	0,00	11.773,18	22,34
Receipts	8	PC	178,72	54,77-	0,00	123,95	15,49
Production	8	PC	178,72	54,77-	0,00	123,95	15,49
1000004262 Order Settlement 10004	0	PC	0,00	67,02-	0,00	67,02-	0,00
1000004271 GR for order 2000584/1	3	PC	67,02	0,00	0,00	67,02	22,34
1000004238 Order Settlement 20005	0	PC	0,00	12,25	0,00	12,25	0,00
1000004253 GR for order 2000562/1	5	PC	111,70	0,00	0,00	111,70	22,34

Figure 19.31 Material Price Analysis, Showing Goods Receipts and Price Differences

What changes with universal parallel accounting is that you will have to create your costing run with reference to a *run template*. The idea behind the run template is that it contains the settings for your costing run(s), so you don't need to reenter them in each new period. Figure 19.32 shows a sample run template to update the actual costs in ledger **0L** period-by-period (rather than year-to-date). To create a template, choose Transaction CKMLCP and the icon to **Create an Actual Costing Ledger Run**. Then choose F5 to activate **Create/Edit Run Template**. Since the templates are ledger-specific, you might only create a template for those countries requiring actual costing according to their local GAAP, or you could create templates for all ledgers. The entries in the **Company Code Assignment** tab ensure that the costing run covers all plants in the selected company code; it is not possible to deselect plants in a company code in SAP S/4HANA.

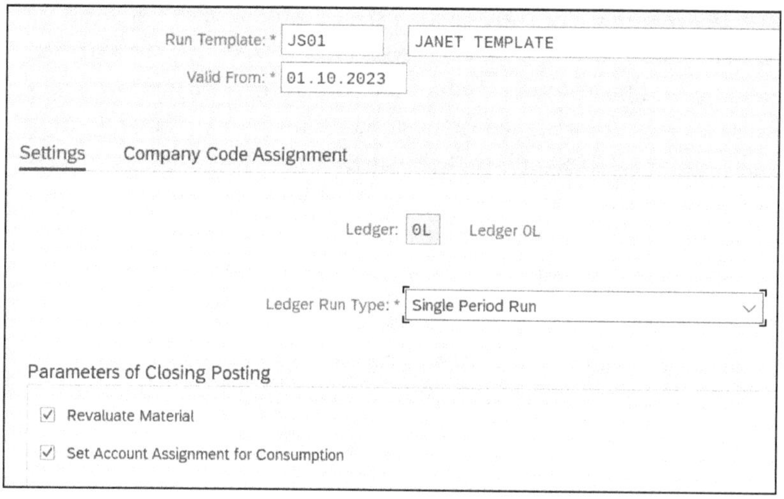

Figure 19.32 Run Template for Actual Costing

You can create the costing run with reference to this run template immediately, as shown in Figure 19.33. Notice that the costing run has an initial period (**10.2023**), and each subsequent costing run is also created with reference to the run template. You can then schedule the various jobs to select the materials for costing, calculate the new prices, update the inventory values and cost of goods sold, and mark the future prices at period close, as described in Chapter 16.

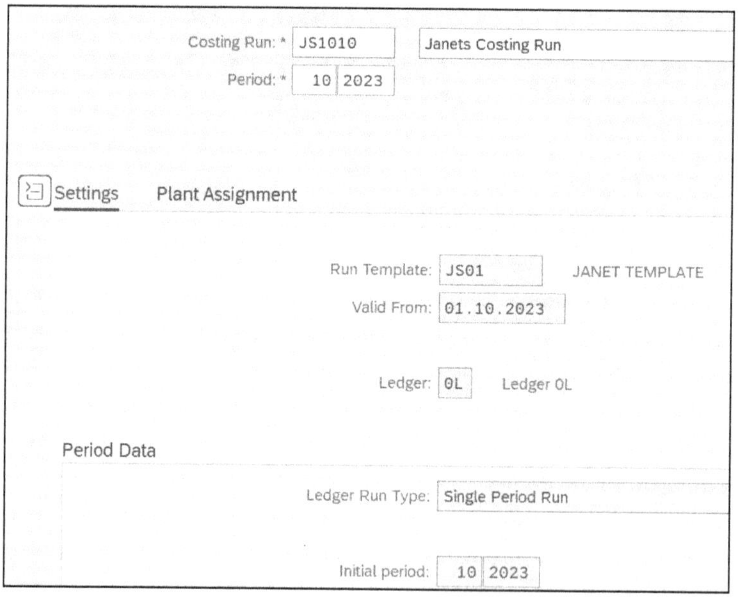

Figure 19.33 Costing Run for Actual Costing

19.4 Intercompany Processes

So far, we have looked at the costs associated with manufacturing within a single plant, but business processes are rarely so simple. Most companies split their activities so that a selling company handles trade with the final customer and requests their manufacturing plant to supply them with the relevant products. Others will make the value chain longer by having the manufacturing plant deliver to a distribution center that will in turn supply goods to the selling company for final sale. Then there are complex manufacturing processes, where different components are manufactured in different parts of the world and brought together for final assembly before distribution. Within each of these plants there will be a *legal cost estimate* used to set the inventory values in each plant, as described in the previous chapters. In some cases, the movement of the finished good to the distribution center takes place within the same company code, so this standard cost can also be used to value the goods movement from manufacturing plant to distribution center. More commonly, the manufacturing plant and the distribution center belong to different company codes and the goods movement will involve a sale by the manufacturing plant to the distribution center at an agreed *transfer price*. This transfer price is captured in sales conditions just as if the goods were being sold to an external customer. You'll often hear this process referred to as *arm's length trading*, where the manufacturing plant and the distribution center are trading with one another as if they were external business partners even though both entities belong to the same corporate group.

The process of applying transfer prices for intercompany goods movements gives rise to a requirement for an alternative view that simply looks at the product costs for the whole value chain from a group perspective, as if there were no company code barriers with agreed transfer prices but only a BOM structured to include multiple plants. You can create a *group cost estimate* using a different costing variant from that used for the legal cost estimate. This can either explode the BOM to bring together semi-finished products from multiple plants and roll the costs up to deliver a group view of the finished product or work using a *reference cost estimate* so that the BOMs are not exploded twice but the itemization for the legal cost estimate is simply referenced during the group cost estimate as described in Chapter 7.

Finally, some organizations define transfer prices both between company codes and between profit centers in a *profit center valuation*: The profit centers can either be above the level of the company code and represent large divisions within the group or below the level of the company code and represent plants where there is a management requirement to see an internal profit by establishing transfer prices at profit-center level. We'll look at both group valuation and profit center valuation in this section.

19.4.1 Intercompany Sales and Stock Transfers with Group Valuation

When you work with universal parallel accounting, the group valuation is performed with reference to a ledger that has been flagged for **Group Valuation**, as we saw earlier in Figure 19.7. In our example, this means that the costing type chosen must reference ledger **4G** if you want to use the cost estimate to update inventory in ledger **4G**. It also means that all input values for the cost estimate material prices (refer back to Figure 19.11), activity prices (refer back to Figure 19.10), and so on will be selected from ledger **4G**. The biggest challenge for **Group Valuation** is how to establish the link between the cost estimates in the various sites, and this takes place by referencing the **Special Procurement Key** that we met when we looked at the material master fields in Chapter 3. It is the task of the special procurement key to build the bridges between the various manufacturing companies, the distribution centers, and the selling company so that when the system explodes the bill of material (BOM; see Chapter 4) it builds a cost estimate by starting in the selling company and following the links in the bill of material to create new cost estimates in each of the associated plants or copies the existing cost estimates in these plants in accordance with the settings for transfer control we looked at in Chapter 7.

Figure 19.34 shows a sample cost estimate in which the costing structure on the left shows the multilevel BOM with the same material being moved between three different plants, where **LT03** is the selling company, **LT02** is the distribution center, and **LT01** is the manufacturing company (where we also see the raw materials consumed during the production process).

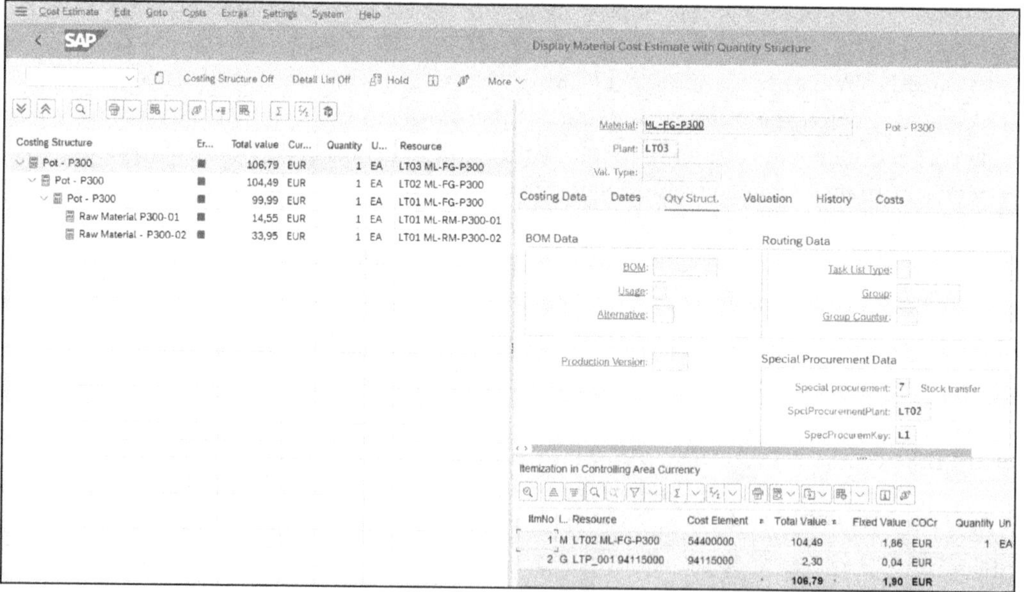

Figure 19.34 Group Cost Estimate with Quantity Structure Showing Special Procurement Data

In this example, the manufacturing costs are calculated in plant **LT01** and then rolled up to plants **LT02** and **LT03**. On the right, we've selected the quantity structure details to show how the BOM and routing in plant **LT03** are replaced by the special procurement key that establishes a link between plants **LT03** and **LT02**. If we selected the quantity structure details for plant **LT02**, we would see a similar special procurement key being used to establish the link between plants **LT02** and **LT01**, where production takes place.

This multilevel rollup across multiple plants and company codes was possible in SAP ERP, but what changes with universal parallel accounting is that the impact of the sale of the material in **LT01** to plant **LT02** and from plant **LT02** to **LT01** is eliminated to a group valuation clearing account when the goods movements are posted so that only the sale of the material in **LT01** to the final customer is visible in ledger **4G**. This is done by moving the intercompany revenue and cost of goods sold in plants **LT02** and **LT01** to a *valuation clearing account*. You can maintain a valuation clearing account for each company code in the IMG by choosing **Controlling • General Controlling • Manage Multiple Valuation Approaches/Transfer Prices • Level of Detail • Define Valuation Clearing Account** from the menu bar or using Transaction 8KEN. This will result in a transactional consolidation under Transaction TCV (Transactional Consolidation). Figure 19.35 shows a sample posting to the group valuation ledger, where inventory has been posted against cost of goods sold (**GBB**) and elimination postings created with Transaction TCV to remove the impact of these intercompany postings in the group view. Note that these elimination postings are only created if you perform group valuation in combination with the advanced intercompany sales process or the advanced intercompany stock transfer process. If you continue to work with classic stock transfer orders, no elimination postings will be triggered.

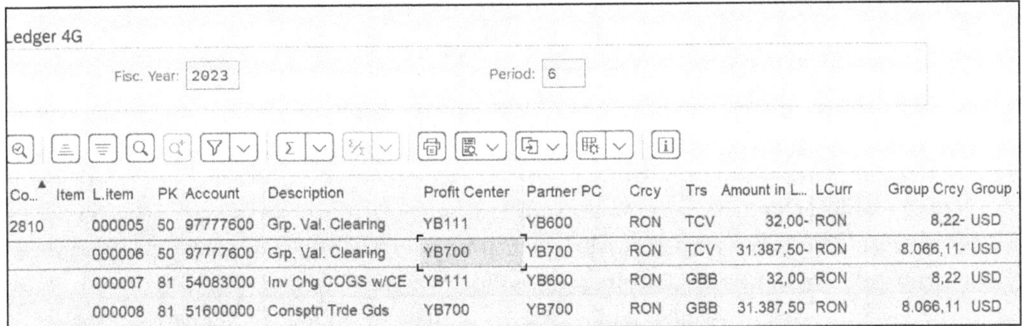

Ledger 4G

Fisc. Year: 2023 Period: 6

Co...	Item	L.item	PK	Account	Description	Profit Center	Partner PC	Crcy	Trs	Amount in L...	LCurr	Group Crcy	Group ...
2810		000005	50	97777600	Grp. Val. Clearing	YB111	YB600	RON	TCV	32,00-	RON	8,22-	USD
		000006	50	97777600	Grp. Val. Clearing	YB700	YB700	RON	TCV	31.387,50-	RON	8.066,11-	USD
		000007	81	54083000	Inv Chg COGS w/CE	YB111	YB600	RON	GBB	32,00	RON	8,22	USD
		000008	81	51600000	Consptn Trde Gds	YB700	YB700	RON	GBB	31.387,50	RON	8.066,11	USD

Figure 19.35 Elimination Posting to Group Valuation Clearing Account in Ledger 4G

The material price for these goods movements is selected using the group valuation price in the material master (see Figure 19.12). This means that when the goods are moved to the consuming company code, they are transferred *at cost*, without the markup that is applied to these values in the legal view.

19.4.2 Profit Center Valuation

Just like group valuation, profit center valuation with universal parallel accounting requires the creation of an additional ledger that is assigned to the valuation view **Profit Center Valuation**. As we discussed for group valuation, the material prices are also ledger-specific, so you can define the material prices to be used for profit center valuation in the material master (Transaction MM03). The additional settings for pricing are made in the IMG by following the menu path **Controlling • Profit Center Accounting • Transfer Prices • Basic Settings for Pricing**. Figure 19.36 shows the sample settings for transfer pricing, where you can select a fixed price from the material master or apply a percentage surcharge to this material price.

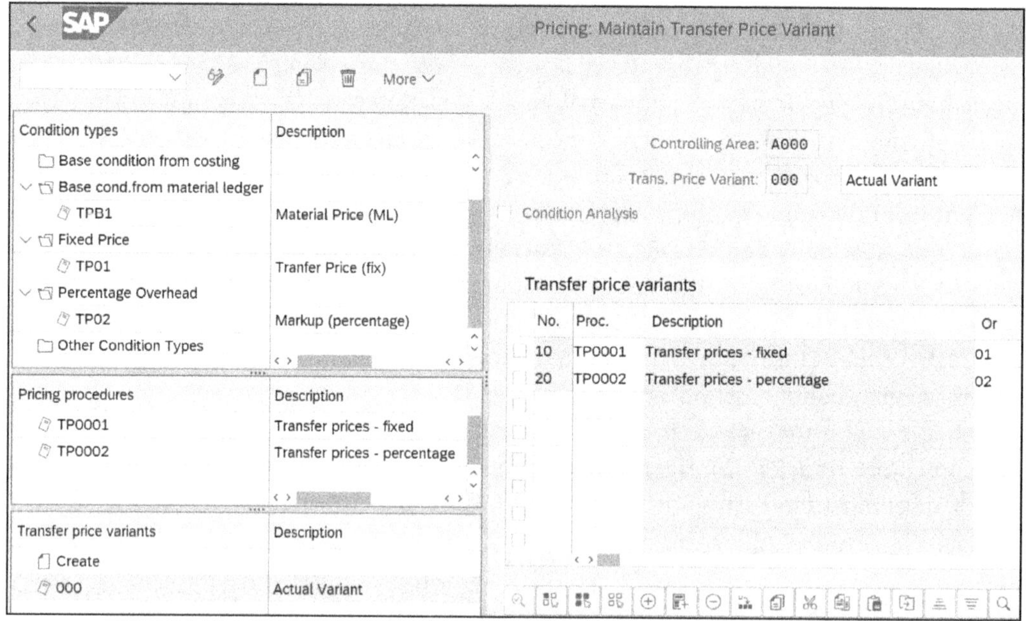

Figure 19.36 Transfer Price Settings for Profit Center Valuation

Transfer prices can be applied to goods movements between plants in the same company code or between different company codes. Transfer prices are used between plants in the same company code when there is a requirement to handle the plants as true profit centers with a profit and loss for a management view of the organization. Transfer prices can also be applied between divisions at a level higher than the company code. Figure 19.37 shows a revenue posting between profit centers at divisional level that has resulted in intercompany clearing postings under Transaction TCV.

Profit center valuation was first enabled for universal parallel accounting with SAP S/4HANA 2023.

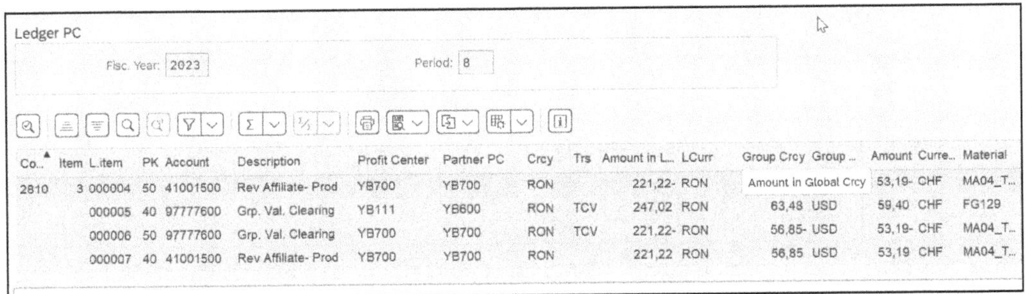

Figure 19.37 Elimination Postings in Profit Center Valuation

19.5 Further Topics

The other topic to change with universal parallel accounting is *balance sheet valuation*, in which companies reduce the value of their inventory according to various methods at period close. We introduced the idea of valuation alternatives in Chapter 3, and here we'll explain the changes. Finally, we'll introduce the transition tools that allow customers to make the move to universal parallel accounting if they are already running SAP S/4HANA.

19.5.1 Balance Sheet Valuation

We discussed the topic of balance sheet valuation in Chapter 3. Universal parallel accounting currently supports the following valuation alternatives for the purposes of balance sheet valuation:

- BSV: Balance Sheet Value
- FIF: FIFO (first in first out) value from Material Ledger
- LMP: Lowest Value by Market Price
- LMR: Lowest Value by Movement Rate
- ROC: Lowest Value by Range of Coverage

In many cases, the classic balance sheet valuation transactions (Transactions MRN1, MRN2, and MRN9) have gained an N to indicate that they support the ledger, becoming Transactions MRN1N, MRN2N, and MRN9N, respectively. Figure 19.38 shows the selection screen for Transaction MRN2N (Determine Lowest Value: Movement Rate) with the link to the ledger. This makes it possible to support different valuation alternatives for each ledger/company code combination depending on the requirements of the accounting principle to be supported. If you work with best practices, the settings for balance sheet valuation are delivered via scope item 6DF (Universal Parallel Accounting).

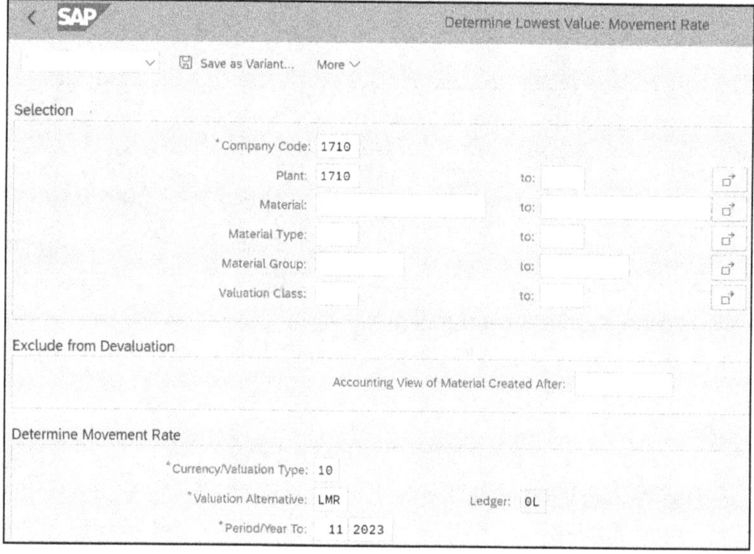

Figure 19.38 Determine Lowest Value by Movement Rate, Showing Link to Ledger 0L

19.5.2 Transition Tools

If you work in the public cloud, universal parallel accounting is automatically active if you work with multiple ledgers. You can choose whether to activate group valuation (scope item 5W2) or profit center valuation (scope item 6VQ) as part of your initial scoping, but you cannot activate these scope items retrospectively because that would require creating an additional ledger and filling it with historical data.

If you work in the private cloud or on-premise, the first tools to migrate to universal parallel accounting are delivered with release OP2023, but these focus on using multiple ledgers for legal valuation (parallel cost of goods manufactured). It is not yet possible to add a ledger for group valuation or profit center valuation and fill it with historical data. Even if you don't use universal parallel accounting, you cannot add group valuation or profit center valuation retroactively because it would involve adding a new currency to the Universal Journal.

19.6 Summary

In this chapter, we looked at the topic of universal parallel accounting and its impact on product cost controlling by allowing organizations to support the requirements of different accounting principles when calculating product costs and determining production variances. We also examined its impact on intercompany processes and balance sheet valuation and introduced the first transition tools. The final topic we discuss in our examination of product cost controlling is reporting, which we'll look at next in Chapter 20.

Chapter 20
Reporting

SAP S/4HANA offers improved reporting with SAP GUI reports and SAP Fiori apps, which access the Universal Journal. In this chapter, we'll examine what's available for product costing and margin analysis.

Product cost reporting in SAP S/4HANA is based on creating standard cost estimates from master data, including material masters, which we discussed in Chapter 3, and BOMs and routings, as we discussed in Chapter 4.

As discussed in Chapter 9, you create a standard cost estimate that displays costs divided into material, activity, and overhead cost components. You then create, mark, and release a standard cost.

As you manufacture assemblies, production orders are debited with actual costs, compared with the standard cost, which credits the production order at the time of goods receipt. The difference between debits and credits is the *production variance*, which is divided into categories during variance analysis to assist you in determining the reason for the variance.

In addition to the information available in an individual cost estimate, reports list cost estimates according to multiple criteria. Reports are also available to analyze costing run data, like the analysis section of the costing run step discussed in Chapter 9. We'll look at these reports in detail in this chapter.

20.1 Analyze Costing Runs

Summarized reporting involves analyzing costing runs, which mass-produce cost estimates.

You can analyze costing runs from several locations:

- Manage Costing Runs—Estimated Costs app (SAP Fiori ID F1865)
- Transaction CK40N
- The information system menu path

We discussed analyzing a costing run with the Manage Costing Runs—Estimated Costs app and Transaction CK40N in Chapter 9. Let's now look at analysis with the information system. You can analyze a costing run with Transaction S_ALR_87099930 or by following the menu path **Accounting • Controlling • Product Cost Controlling • Product Cost Planning • Information System • Summarized Analysis • Analyze Costing Run**. The selection screen in Figure 20.1 is displayed.

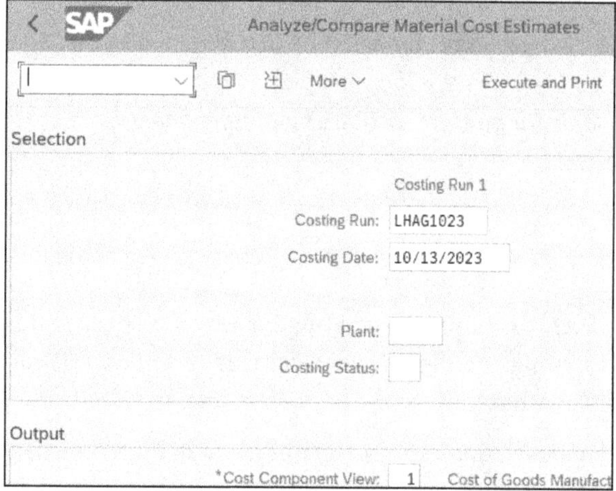

Figure 20.1 Analyze Costing Run Selection Screen

This selection screen is useful for analyzing one costing run with limited selection criteria. Enter a **Costing Run** and **Costing Date** and click the **Execute** button or press [F8] to display the results screen shown in Figure 20.2.

Analyze/Compare Material Cost Estimates

Costing Run LHAG1023 10/13/2023
Currency USD United States Dollar
Base Values Based On Costing Lot Size
Cost Component View 01(Cost of Goods Manufactured)

Material	Material Description	Plant	Status	Costing Result	Lot Size	Base Unit
E-ENGINE_BLOCK	Enginge Block	US00	FR	500.00	1	PC
OPF	Otto particle filter	US00	FR	60.00	1	PC
STEEL	Steel	US00	FR	50.00	1	PC
T-GEAR1	Transmission - Gear Set 1	US00	FR	56.00	1	PC

Figure 20.2 Analyze Material Cost Estimates

This screen displays a list of cost estimates generated during the costing run, depending on your selection criteria. Double-click any line to display the corresponding individual cost estimate details. You can also display different columns on this screen by clicking the change layout (grid) icon on the right.

To view more selection fields and compare two costing runs, click the plus sign icon shown earlier in Figure 20.1. The selection screen shown in Figure 20.3 is displayed.

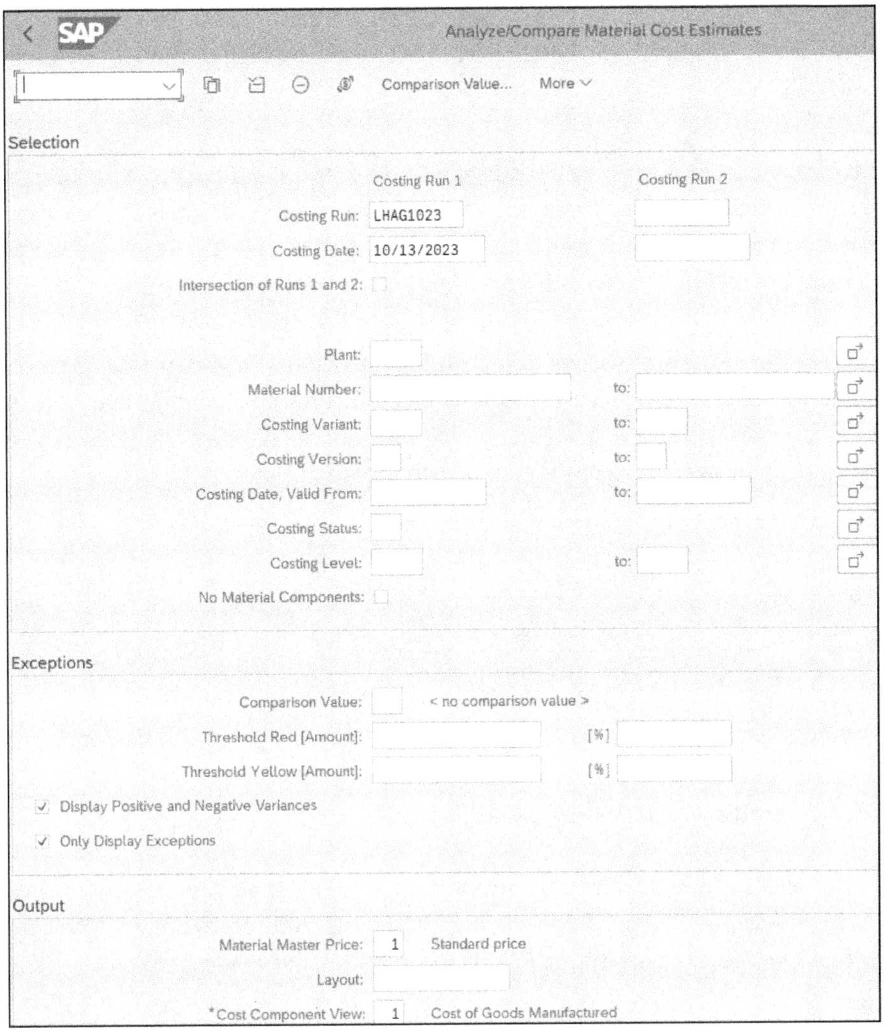

Analyze/Compare Material Cost Estimates

Comparison Value... More

Selection

	Costing Run 1		Costing Run 2
Costing Run:	LHAG1023		
Costing Date:	10/13/2023		
Intersection of Runs 1 and 2:			
Plant:			
Material Number:		to:	
Costing Variant:		to:	
Costing Version:		to:	
Costing Date, Valid From:		to:	
Costing Status:			
Costing Level:		to:	
No Material Components:			

Exceptions

Comparison Value: < no comparison value >

Threshold Red [Amount]: [%]

Threshold Yellow [Amount]: [%]

Display Positive and Negative Variances

Only Display Exceptions

Output

Material Master Price: 1 Standard price

Layout:

*Cost Component View: 1 Cost of Goods Manufactured

Figure 20.3 Analyze Costing Runs with Expanded Selection Fields

This expanded selection field screen allows you to choose cost estimates from a costing run or compare costing runs. There are four types of comparisons:

- Compare the value calculated for cost estimates for two **Costing Run** field entries in the **Selection** section.
- Compare the value calculated for cost estimates with a **Comparison Value** entry in the **Exceptions** section.
- Compare the calculated values with **Threshold Amounts** entries in the **Exceptions** section.
- Compare the calculated values with a **Material Master Price** entry in the **Output** section.

The reason these comparisons are important is that you are about to revalue your inventory with the results of the costing run. We'll examine each of these screen sections in detail:

- **Selection**
  You typically enter two costing runs in the **Costing Run 1** and **Costing Run 2** fields with corresponding **Costing Dates** and click the **Execute** icon or press F8 to see a comparison of the cost estimates in the two costing runs. Select the **Intersection of Runs 1 and 2** checkbox to view only cost estimates in both costing runs.
- **Exceptions**
  The **Exceptions** section allows you to decide which values to compare with the cost estimate values and to display exceptions beyond certain percentages you set in threshold values. Let's examine these fields in more detail. Click the **Comparison Value** field and press F4 to display the screen shown in Figure 20.4.

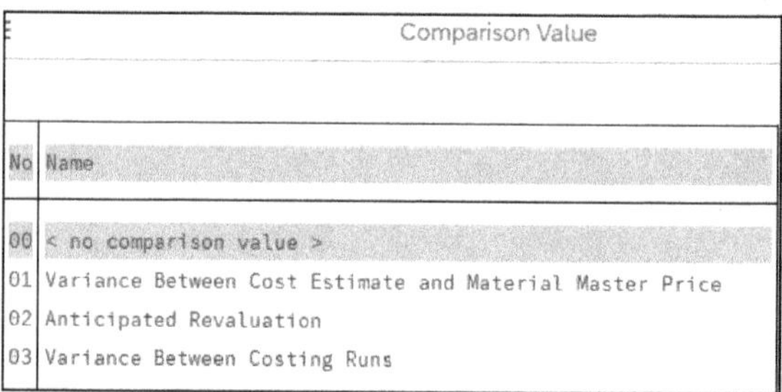

Figure 20.4 Comparison Value Possible Entries

  After you select a **Comparison Value** option, you set a **Threshold Amount** or percentage in the subsequent fields, which will highlight value differences greater than you set with either red or yellow traffic light icons in the output report.
- **Output**
  Click the **Material Master Price** field in Figure 20.3 and press F4 to display the list shown in Figure 20.5.

  The selection you make determines the **Material Master Price** column that appears in the report results screen. You can compare the **Material Master Price** field you choose with the cost estimate proposed price. You can also review the exceptions that you defined in the **Exceptions** section.

  The default selection for this field is the **Standard Price**. You can see from the list of possible entries that you can also choose any of the 15 **Material Master Price** fields to compare with the cost estimate values.
- **Base**
  The **Base** section is in the **Output** section, as shown in Figure 20.6.

  The default selection for the **Base** section is **Costing Lot Size**. For example, if the costing lot size is 1,000 for a material, you'll see the prices in the results screen displayed in units of 1,000. Depending on your costing lot size settings, displaying the results screen with the **Price Unit in Material Master** setting is often useful since it is typically in units of 10.

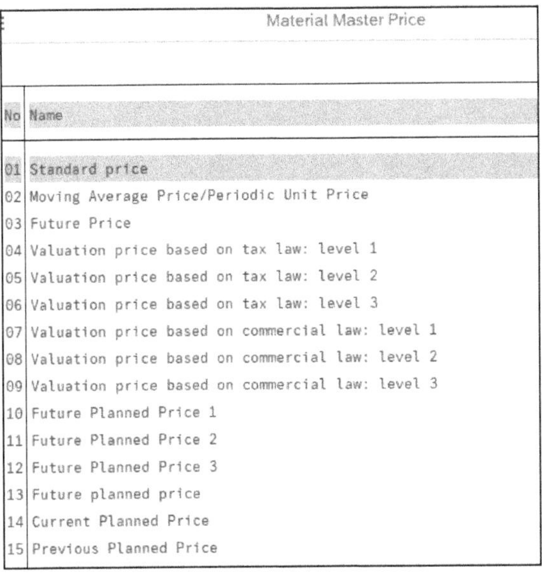

Material Master Price

No	Name
01	Standard price
02	Moving Average Price/Periodic Unit Price
03	Future Price
04	Valuation price based on tax law: level 1
05	Valuation price based on tax law: level 2
06	Valuation price based on tax law: level 3
07	Valuation price based on commercial law: level 1
08	Valuation price based on commercial law: level 2
09	Valuation price based on commercial law: level 3
10	Future Planned Price 1
11	Future Planned Price 2
12	Future Planned Price 3
13	Future planned price
14	Current Planned Price
15	Previous Planned Price

Figure 20.5 Material Master Price Possible Entries

Output
Material Master Price: 1 Standard price
Layout:
*Cost Component View: 1 Cost of Goods Manufactured
Base
Costing Lot Size: ●
Price Unit in Material Master: ○

Figure 20.6 Base Section in Output Section

Now that we've analyzed costing runs, let's analyze cost estimate lists in the next report.

20.2 List Material Cost Estimates

For each material in a plant, there can be many cost estimates created with different costing variants, versions, or dates. You can display a list of cost estimates for materials with the Manage Material Cost Estimates app (SAP Fiori ID F7650), with Transaction S_P99_41000111 or via the menu path **Accounting • Controlling • Product Cost Controlling • Product Cost Planning • Information System • Object List • For Material**. The selection screen in Figure 20.7 is displayed.

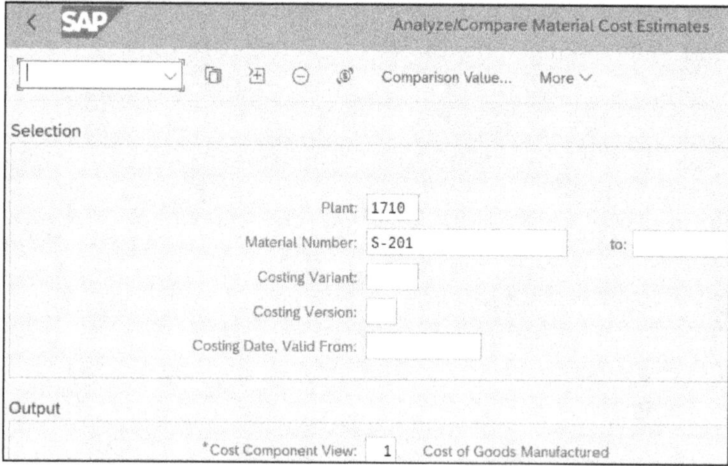

Figure 20.7 Display Cost Estimate List Selection

Complete the **Plant** and **Material Number** fields and click the **Execute** button or press [F8] to display the results screen shown in Figure 20.8.

Analyze/Compare Material Cost Estimates

More

Plant	1710
Material Number	S-201
Currency	USD United States Dollar
Base	Values Based On Costing Lot Size
Cost Component View	01(Cost of Goods Manufactured)

Material	Material Description	Plant	Status	Costing Result	Lot Size	BUn
S-201	SEMI201,MTS,D1,Subassembly	1710	KF	141.56	100	PC
S-201	SEMI201,MTS,D1,Subassembly	1710	KF	148.90	100	PC
S-201	SEMI201,MTS,D1,Subassembly	1710	KA	128.78	100	PC

Figure 20.8 Material Cost Estimate List

Double-click any line to display the corresponding cost estimate details. There can be many cost estimates in this list. The quickest method to locate the latest released standard cost estimate is to click the **Current Cost Estimate** button in the **Costing 2** view, which we discussed in Chapter 3. Displaying and sorting on the **Valid from** date column can help identify individual cost estimates.

Now that we've examined cost estimate list reports, let's examine summarized analysis reports and line-item reports in the following sections. We'll start with some new reports and margin analysis.

The Display Material Value Chain—Estimated Cost app (SAP Fiori ID F4898) allows you to display the progression of planned quantities and planned values for a material along a value chain using estimated costs from a specified validity date. You can see a visual flow of the valuated transactions for a material, its procurement alternatives, and the activities related to that material. Color-coded nodes display material values, activities, and subcontracting information, while the connectors between the nodes display the valuated transactions.

> **Classic Reports versus SAP Fiori Apps**
>
> In this chapter, we discuss reports and SAP Fiori apps. For future direction, refer to the strategy described in SAP Note 2579584. SAP will gradually replace classic reports with SAP Fiori apps, which use core data services (CDS) views to select and aggregate the data for reporting and are more dynamic.

20.3 Margin Analysis

Margin analysis reports on the profitability of the products you manufacture and sell by market segment. While the first contribution margin is calculated by comparing the revenue with the standard cost of goods sold (COGS), you must include production variances in the second contribution margin to see the full manufacturing costs and the individual variance categories.

In the next section, we'll set up the variance split to include variance category details in margin analysis.

20.3.1 Variance Split

20

For profitability segments to contain the full manufacturing cost, you must include the production order's actual production cost variance from the standard. You do this by mapping variance categories to general ledger accounts in the variance split. You set up the variance split via the IMG menu path **Financial Accounting • General Ledger Accounting • Periodic Processing • Integration • Materials Management • Define Accounts for Splitting Price Differences**. You'll see the screen displayed in Figure 20.9.

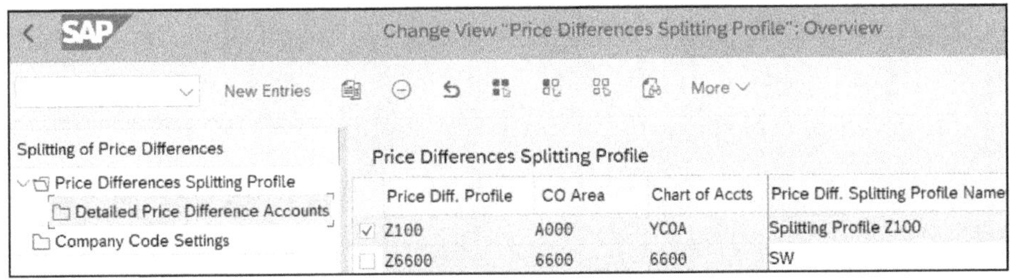

Figure 20.9 Define Splitting Profile

Select a **Splitting Profile**, double-click **Detailed Price Difference Accounts**, and click **New Entries** to display the screen shown in Figure 20.10.

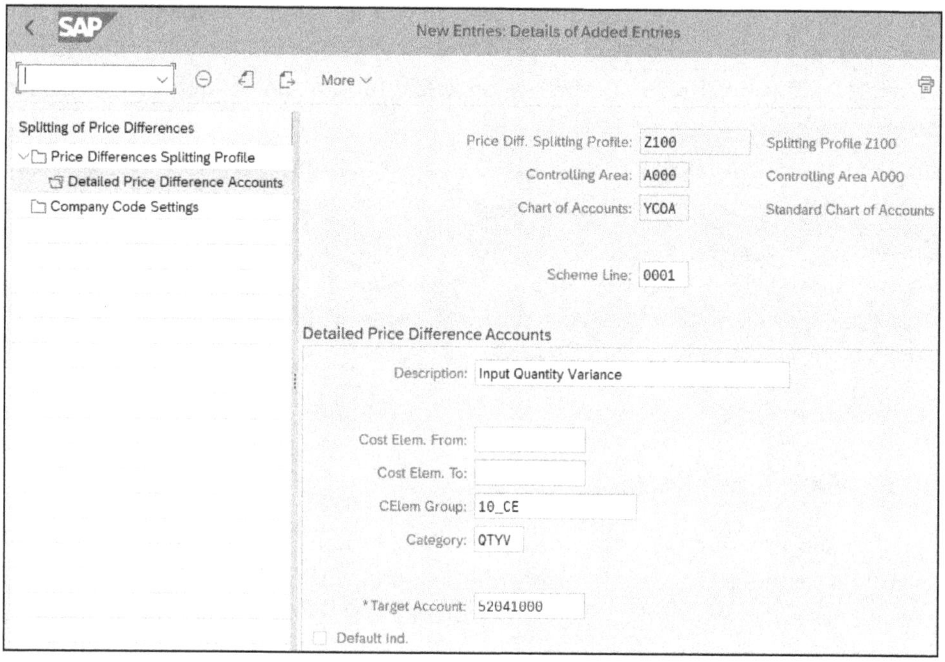

Figure 20.10 Assignment of Variance Categories to Target Accounts for Splitting

Complete the screen as follows:

1. Enter **0001** for the **Scheme Line**.
2. Type in the variance **Description**, typically the name of the variance category.
3. Enter a single **Cost Elem.** (cost element), range, or group as the source cost elements for the variance calculation.
4. Enter the variance **Category** to split with the **Target Account**.
5. Enter the **Target Account**, which is the general ledger account the variance category will post to when the production order is settled.
6. Select the **Default Ind.** checkbox for one variance category. If a target account isn't specified for a cost element/variance category, the system automatically posts the amounts to the default general ledger account.
7. Save your work.
8. Add a **Scheme Line** for every variance category to the **Splitting Profile**.

Now that we've set up the variance split, we'll discuss setting up the COGS split in the next section.

20.3.2 Cost of Goods Sold Split

The system determines the cost component split from the cost estimate, breaks out the COGS amount into cost components, and posts these to general ledger accounts.

You configure the COGS split by following the IMG menu path **Financial Accounting • General Ledger Accounting • Periodic Processing • Integration • Materials Management • Define Accounts • Define Accounts for Splitting the Cost of Goods Sold**. Click **New Entries** to display the screen shown in Figure 20.11.

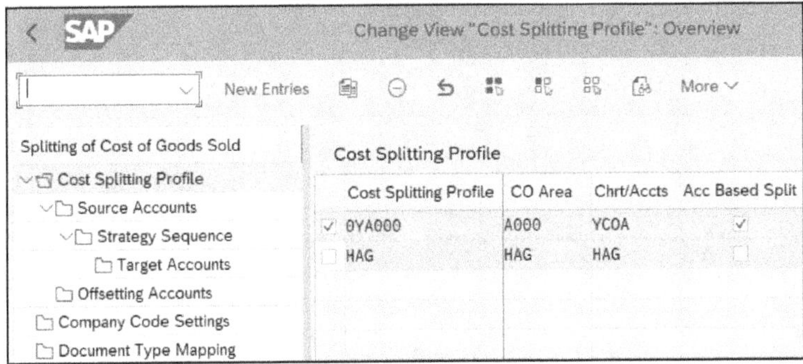

Figure 20.11 Configuring the Cost Splitting Profile

Create **Cost Splitting Profile 0YA000** and assign it to controlling area (**CO Area**) **A000** and chart of accounts (**Chrt/Accts**) **YCOA**. Also, select the **Acc Based Split** checkbox so that the COGS account will always be split when the source account receives a posting.

You can specify the accounts and valuation views for which the split will be performed by selecting the **Cost Splitting Profile 0YA000** and then double-clicking the **Source Accounts** folder to reveal the view in Figure 20.12.

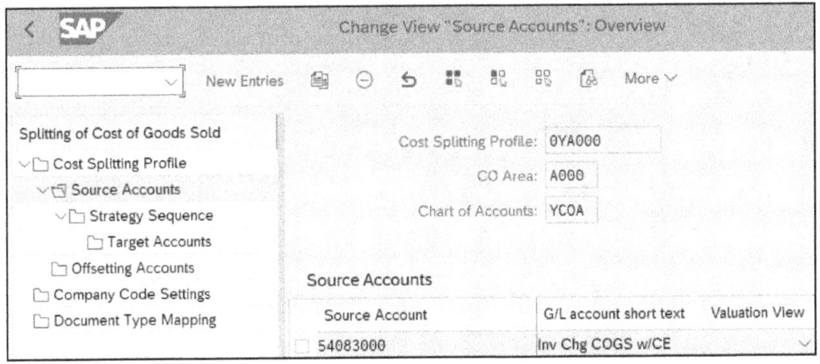

Figure 20.12 Source Account and Valuation View for the COGS Split

Enter the **Source Account 54083000** as the COGS accounts to split. Don't enter a **Valuation View** because you want all target accounts to be split regardless of the valuation view.

Next, define the strategy sequence that will be used as a basis for the COGS account split by selecting the COGS account you defined and double-clicking the **Strategy Sequence** folder to display the screen in Figure 20.13.

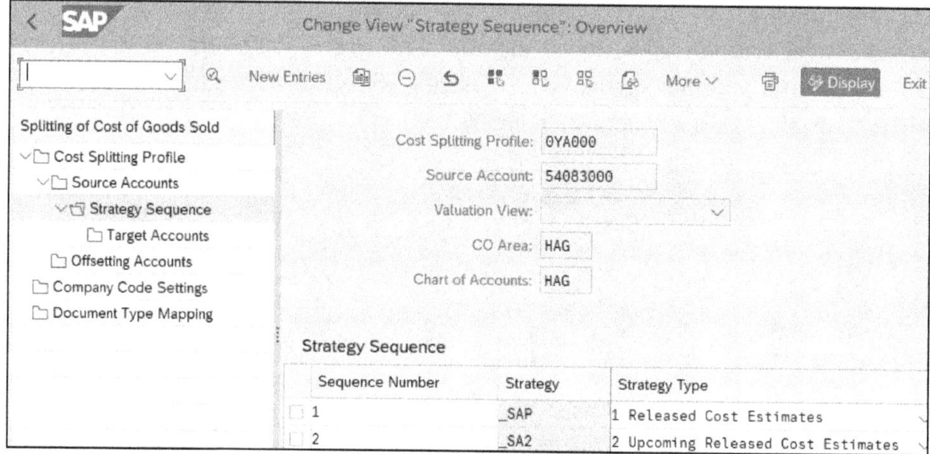

Figure 20.13 Strategy Sequence for Splitting the COGS Account

Enter **Sequence Number 1** and **Strategy _SAP** and choose **1 Released Cost Estimates** for the **Strategy Type** column. Then, enter **Sequence Number 2** and **Strategy _SA2** and choose **2 Upcoming Released Cost Estimates** as the **Strategy Type**. This means the COGS split can be based on a released cost estimate or, if none yet exists, on a future cost estimate in the COGS posting period. Double-click the first **Sequence Number** to display the screen shown in Figure 20.14.

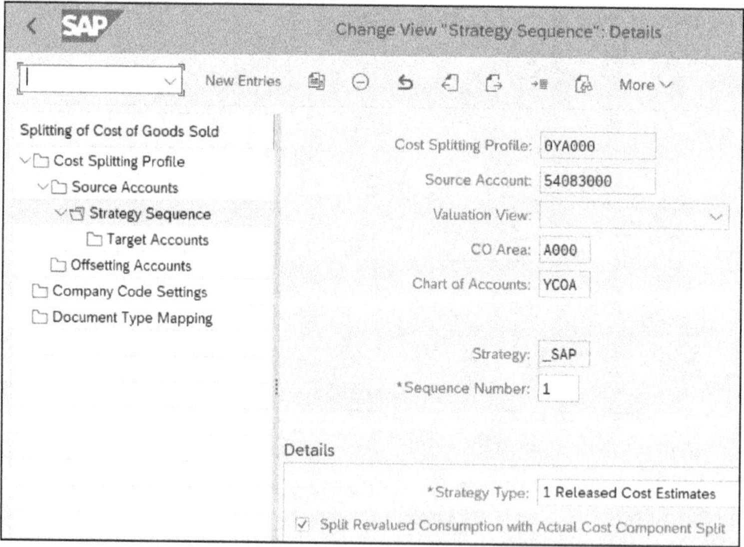

Figure 20.14 Split Revaluation of Consumption with Actual Cost Component Split

Select the **Split Revalued Consumption with Actual Cost Component Split** checkbox so that the costs related to the revaluation of consumption in the actual costing cockpit are split based on the actual cost component split. Next, define the accounts for each cost component in the cost component structure by double-clicking the **Target Accounts** folder on the left. The screen in Figure 20.15 is displayed.

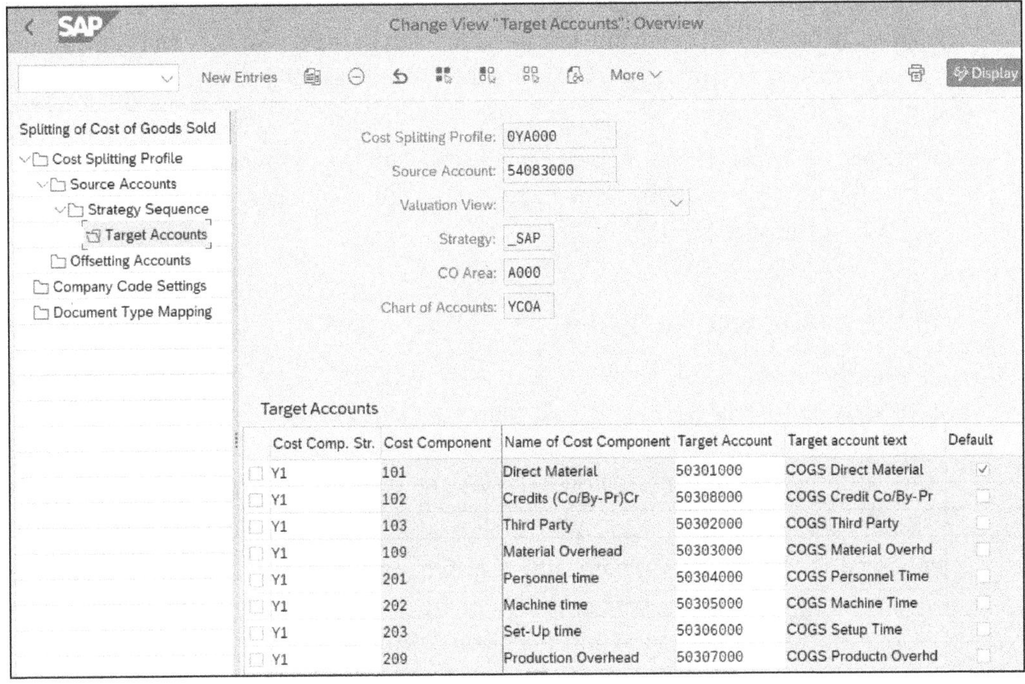

Cost Comp. Str.	Cost Component	Name of Cost Component	Target Account	Target account text	Default
Y1	101	Direct Material	50301000	COGS Direct Material	✓
Y1	102	Credits (Co/By-Pr)Cr	50308000	COGS Credit Co/By-Pr	
Y1	103	Third Party	50302000	COGS Third Party	
Y1	109	Material Overhead	50303000	COGS Material Overhd	
Y1	201	Personnel time	50304000	COGS Personnel Time	
Y1	202	Machine time	50305000	COGS Machine Time	
Y1	203	Set-Up time	50306000	COGS Setup Time	
Y1	209	Production Overhead	50307000	COGS Productn Overhd	

Figure 20.15 Target Accounts for the COGS Split

Assign a **Target Account** to each **Cost Component** in the cost component structure (**Cost Comp. Str.**). Check the **Default** checkbox for one cost component—the **COGS Direct Material** cost component in this example. If any cost components don't have an account assigned, the amounts for those cost components post to the **Default** account.

Now that we've set up the variance and COGS splits for margin analysis, let's discuss SAP Fiori apps and then summarized reports in the next section.

Refer to the introduction of this book to see more SAP Fiori apps for reporting. There, we reviewed product profitability with the Production Variances app (SAP Fiori ID W0182), which displays profitability-related information in the Universal Journal.

20.4 Summarized Analysis Reports

One of the most useful features of standard reporting is the drilldown functionality. From high-level summarized reports, you can drill down through detailed and line-item reports to source documents.

The highest-level report you can drill down from is a summarized report, which allows you to display variances at a highly summarized level—for example, manufacturing variances for a plant.

Summarization hierarchies allow you flexibility when setting up and displaying summarized data. They group manufacturing orders or product cost collectors at the lowest-level summarization nodes, which are then grouped at higher-level nodes, creating a pyramid structure.

There are two types of summarization reports available: product drilldown and summarization hierarchy reports, which we'll cover in this section. Later on in this section, we'll discuss detailed reports.

20.4.1 Product Drilldown Reports

Product drilldown reports allow you to slice and dice data based on characteristics such as product group, material, plant, cost component, and period. You can easily navigate between characteristics and drill down to lower-level characteristics. The navigation methods used for product drilldown reports are like those used with margin analysis reports.

Product drilldown reports use predefined summarization levels, which means they are already set up for you. The predefined summarization levels are suitable for most scenarios. However, if they do not fully meet your requirements, you can create your hierarchies with summarization hierarchy reports, which we discuss in detail in Section 20.4.2.

Now that we've briefly discussed product drilldown reports, we'll examine the three steps necessary to display them (configuration, data collection, and running the report) in the following sections.

Configuration

You can run product drilldown reports without this configuration step. Its only purpose is to add a product group, such as a material group, to drilldown reports. You can try running the reports first, to see if you have a requirement to add a product group. You should not need to change the product drilldown configuration after making the initial settings with the following procedure.

You can change the configuration setting with Transaction OKNO or via IMG menu path **Controlling • Product Cost Controlling • Information System • Control Parameters**. Click on the **Data Extraction/Product Drilldown** tab to display the screen shown in Figure 20.16.

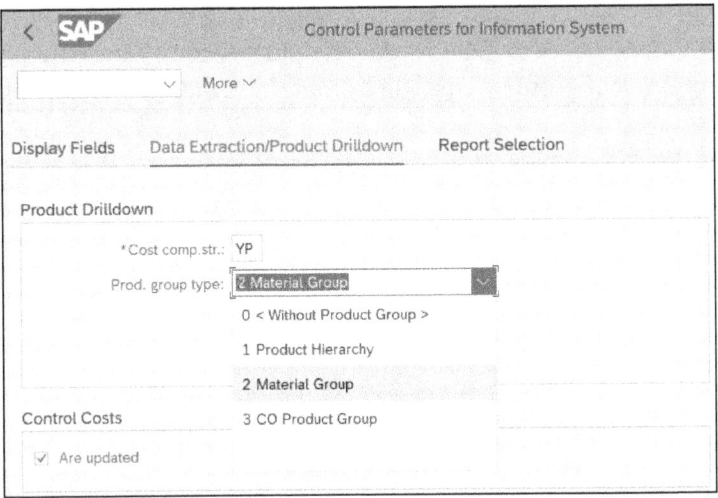

Figure 20.16 Product Drilldown Setting in Control Parameters

Click in the **Prod. Group type** field to display a list of possible product groups. You select a line to assign the corresponding product group to product drilldown reports with associated summarized data. The two product groups you can assign are explained as follows:

- **Product Hierarchy**
  You can assign the **Product Hierarchy** in the following two material master views:
  - **Basic data 1**
  - **Sales: sales org. 2**

  A *product hierarchy* is recorded by the sequence of digits within a hierarchy number. The hierarchy number can have a maximum of 24 digits with an unlimited number of levels. The product hierarchy can be used to classify a material for use in the following:
  - Pricing, with each level being used as a field in the condition technique
  - Margin analysis reports
- **Material Group**
  You can assign a **Material Group** in the following two material master views:
  - **Basic data 1**
  - **Purchasing**

  Material groups allow you to group materials or services that have similar attributes. Material groups are important for the following:
  - Searching for materials
  - Purchasing info records used in production orders, where material is sent to suppliers for processing

As we noted when discussing the product hierarchy, you do not need to understand how material groups are organized initially. In a test system, you can assign **Material Group** to the **Control Parameters** (shown in Figure 20.16) and examine the resulting product drilldown report.

You do not need to fully understand, at least initially, the structure of the product hierarchy. In a test system, you can assign **Product Hierarchy** to the **Control Parameters,** as shown in Figure 20.16 and examine the resulting product drilldown report. You'll quickly understand the logic behind creating the product hierarchy.

Data Collection for Product Drilldown

Now that we've assigned a product group to the product drilldown report in configuration, the next step is to populate the predefined summarization hierarchy with data collection. You run data collection for product drilldown reports with Transaction KKRV or via menu path **Accounting • Controlling • Product Cost Controlling • Cost Object Controlling • Product Cost by Period • Information System • Tools • Data Collection • For Product Drilldown.** The screen in Figure 20.17 is displayed.

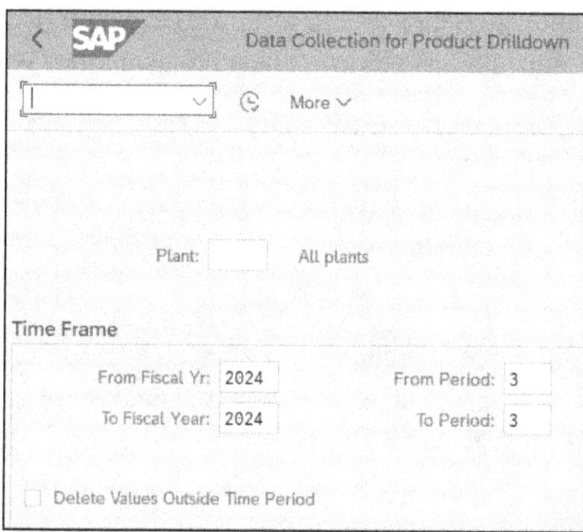

Figure 20.17 Data Collection for Product Drilldown Reports

You can run data collection for product drilldown for an individual **Plant**, or run it for **All plants** by leaving the **Plant** field blank, as shown.

You typically run data collection following period-end closing for the current and previous periods because order data can change within open financial periods. If data collection has already been run, the system resets and recalculates all data within the summarization **Time Frame** of the new data collection run. Data outside the time frame is retained unless you select the **Delete Values Outside Time Period** checkbox.

Complete the fields as described above and save your work. A data collection results screen indicates the number of records read during the run.

Now that we've configured and populated the hierarchy, the next step is to run the report, as described in the next section.

Run Reports

Following data collection, you run product drilldown reports using Transaction S_ALR_87013139 or via the menu path **Accounting • Controlling • Product Cost Controlling • Cost Object Controlling • Product Cost by Period • Information System • Reports for Product Cost by Period • Summarized Analysis • With Product Drilldown • Variance Analysis.** The screen in Figure 20.18 is displayed.

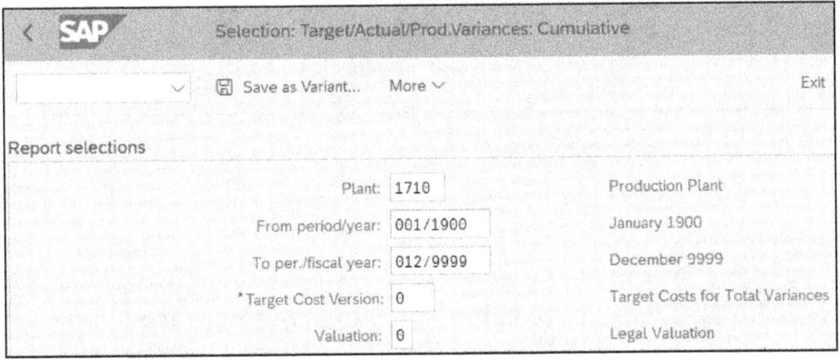

Figure 20.18 Product Drilldown Report Selection Screen

You can run the report for a **Plant** with a wide **period** range, as shown in Figure 20.18. You can also restrict the period range to improve performance if you only need to report over several periods.

Complete the fields as follows:

1. Enter the **Plant**
2. Complete the **From period/year** and **To per./fiscal year** fields described above.
3. You typically complete the **Target Cost Version** field with **0** for **Target Costs for Total Variances.**
4. You enter **Valuation 0** for **Legal Valuation, 1** for **Group Valuation**, or **2** for **Profit Center Valuation.**
5. Click the **Execute** button or press F8 to run and display the product drilldown report.

Each row represents a material, while variance appears in columns. This provides visibility to materials with the largest accumulated variance. Drill down on any row to find more details by period, year, and cost component. You can further analyze variance per material by displaying a detailed report of the product cost collector with Transaction KKBC_PKO or Transaction KKBC_ORD for manufacturing orders, as described in Section 20.5.

Cumulative versus Periodic Drilldown Reports

In addition to the cumulative drilldown report we discussed previously, a periodic drilldown report is available in the same menu path. These reports allow you to display period data on an individual or summarized basis:

- **Periodic**
  These drilldown reports display results by period, enabling fast navigation across multiple periods.
- **Cumulative**
  These drilldown reports display results summarized across periods—for example, by fiscal quarter or year.

20.4.2 Summarization Hierarchy Reports

Now that we've examined how product drilldown reports with a predefined hierarchy work, let's look at how you create and report on your own hierarchy with summarization hierarchy reports. Summarization hierarchies allow you flexibility in setting up and displaying summarized data. A *summarization hierarchy* groups manufacturing orders or product cost collectors at the lowest-level nodes, which are grouped at higher-level nodes to create a pyramid structure.

You set up each summarization level in configuration. The hierarchy levels are presented as expandable and collapsible rows in the reports. Here is an example of hierarchy levels, starting at the top:

- Controlling area
- Profit center
- Plant
- Material group
- Material number

Because you can display summarized data at each node, you may decide to define hierarchy levels based on responsibility at lower management levels—for example, profit centers, material groups, or order types. You may also create an alternate summarization hierarchy with levels based on profitability reporting requirements, such as profit center, division, and sales document.

After you define the summarization hierarchy in configuration, you run a data collection transaction, which populates the hierarchy with current data. We'll discuss these topics in the following sections.

During a data collection run, the following cost data is collected for all orders included in the run and rolled upward through the hierarchy nodes and levels:

- Plan
- Target
- Actual
- Variances
- Work-in-process (WIP)

Following the data collection run, you can add the preceding cost data as columns to the summarization report. You then run a summarization report, which displays the data summarized at each of the hierarchy levels. You may display the following cost data as columns at each level: **Target**, **Actual**, and **Variance**. This summarization report allows you to display the total variance for a plant and then at each of the lower hierarchy levels by expanding the levels as required. At the lowest level, you can display variance details of individual orders. You can then display a detailed report with cost element rows to drill down to line-item reports and source documents if necessary.

Now that we've described how summarization hierarchy reports work, we'll examine in detail the three steps necessary to display the reports:

- Configuration
- Variance calculation (see Chapter 14)
- Running reports with data collection

Configuration

This configuration step is required only once. You only need to change the configuration if you need an additional hierarchy or if you need to change an existing hierarchy.

> **Configuration**
>
> You typically don't need to make changes to hierarchies, provided they are tested and reviewed by users before implementation in the production system. Testing and user review procedures for configuration implementation are good practices for configuration changes, not just summarization hierarchies.

You maintain summarization hierarchies with Transaction KKRO or via the menu path **Accounting • Controlling • Product Cost Controlling • Cost Object Controlling • Product Cost by Period • Information System • Tools • Summarized Analysis: Preparation • Create Summarization Hierarchy**. The screen in Figure 20.19 is displayed.

This screen displays existing hierarchies and allows you to define new ones. Select a **Hierarchy**, **YBMF0001** in this example, and double-click on **Data scope (object types)** to display details shown in Figure 20.20.

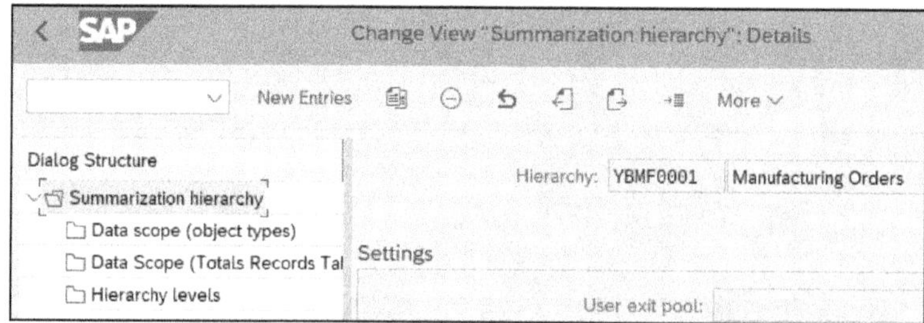

Figure 20.19 Change Summarization Hierarchy: Initial Screen

Change View "Data scope (object types)": Overview

More Display

Dialog Structure
- Summarization hierarchy
  - Data scope (object types)
  - Data Scope (Totals Records Tal
  - Hierarchy levels

Hierarchy	Description	Summarize	Prio.	Status sel.	Sel. variant
YBMF0001	Internal Orders	☐	0		
YBMF0001	Maintenance/Service Orders	☐	0		
YBMF0001	Prod. Orders, QM Orders, Prod. Cost Coll	☑	1		
YBMF0001	Projects	☐	0		
YBMF0001	Sales Orders Without Dependent Orders	☐	0		
YBMF0001	Sales Orders with Dependent Orders	☐	0		

Figure 20.20 Change Data Scope Object Types Overview

In this screen, you specify which object types you want to summarize in the hierarchy. Read the **Description** in the row with the **Summarize** checkbox selected to understand the hierarchy's purpose.

You can also specify a status selection profile in the **Status sel.** column. Enter a status selection profile if you want to select only objects with a status that matches the selection criteria in the profile. For example, you may be interested in summarizing production orders with a status of fully delivered (DLV) or technically complete (TECO).

To display further hierarchy details, double-click on the **Hierarchy levels** folder on the left. The screen in Figure 20.21 is displayed.

Change View "Hierarchy levels": Overview

New Entries More

Dialog Structure
- Summarization hierarchy
  - Data scope (object types)
  - Data Scope (Totals Records Tal
  - Hierarchy levels

Hierarchy	Level	Hierarchy Field	Name	Tot. length
YBMF0001	1	KOKRS	Controlling Area	4
YBMF0001	2	BUKRS	Company Code	4
YBMF0001	3	WERKS	Plant	4
YBMF0001	4	AUART	Order Type	4

Figure 20.21 Change Hierarchy Levels

In this screen, you specify the **Hierarchy levels** of the **Summarization hierarchy**. You can include as many hierarchy levels as you like; however, you should minimize the number of levels for ease of navigating and analyzing the summarization report.

> **Controlling Area Level**
>
> The system requires the **Controlling Area** as the **Level 1** hierarchy level. The field is not changeable.

Click in any row in the **Hierarchy Field** column and press F4 to display a list of possible entries, as shown in Figure 20.22.

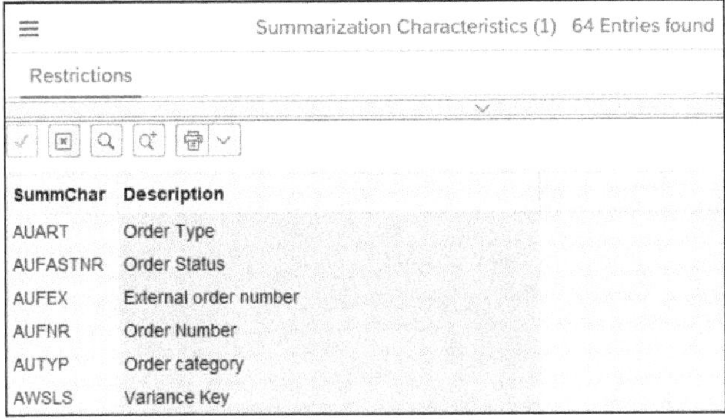

Figure 20.22 Summarization Characteristic Possible Entries

Some typical hierarchy levels are order type, plant, profit center, and material.

Long runtimes can result if the order number is defined as a hierarchy level. It is also quicker and easier to navigate summarization reports without the order number included. Instead, you can display a separate list of individual production orders directly from a summarization report. You can test different summarization hierarchies in a test system without affecting any configuration or data in your production system.

Status Selection Profile

Now that we've maintained a summarization hierarchy, let's examine how to narrow the selection of summarized orders with a status selection profile. You maintain status selection profiles with Transaction BS42 or via the IMG menu path **Controlling • Product Cost Controlling • Information System • Cost Object Controlling • Settings for Summarized Analysis/Order Selection • Define Status Selection Profiles**. The screen in Figure 20.23 is displayed.

Change View "Status selection schema": Overview

SelProf	Text	StatProf	Text
SAP001	Operations Dispatched		
SAP002	Operations Not Dispatched		
SAP003	Split Planned		
SAP004	Split Not Planned		

Figure 20.23 Status Selection Profile Overview

This screen shows the existing **SelProf** (selection profile) list and allows you to define new profiles. Select an existing selection profile and double-click the **Selection conditions** folder on the left to display selection profile details, as shown in Figure 20.24.

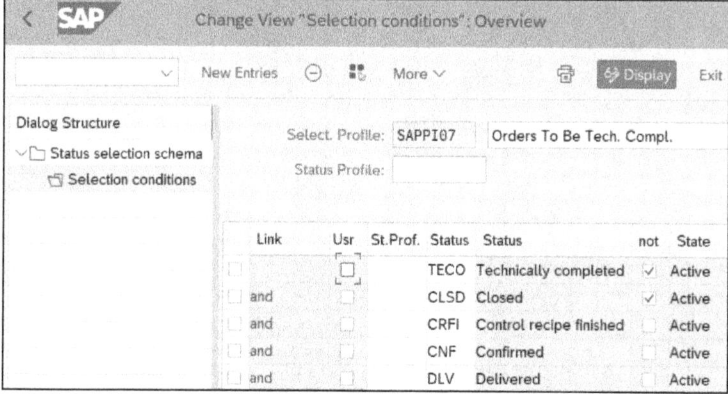

Figure 20.24 Selection Profile Selection Conditions

The selection conditions displayed in Figure 20.24 result in the selection of orders with a matching **Status**, in this example: **Technically completed, Closed, Control recipe finished, Confirmed, and Delivered.**

Variance Calculation

Now that we've defined the summarization hierarchy and examined creating a status selection profile, the next step is to calculate target costs and run variance calculations. You run variance calculations with Transaction KKS1H, as explained in Chapter 14. The selection screen in Figure 20.25 is displayed.

You run variance calculation by **Plant**, which is a mandatory field. You normally select **With Production Orders**, **W/Product Cost Collectors**, and **With Process Orders**. You typically run it for the current **Period** and **Fiscal Year** and **All Tgt Cost Vsns**. You can change the **Selected Target Cost Vsns** (Versions) by selecting **Settings • Target Cost Version** from the menu bar.

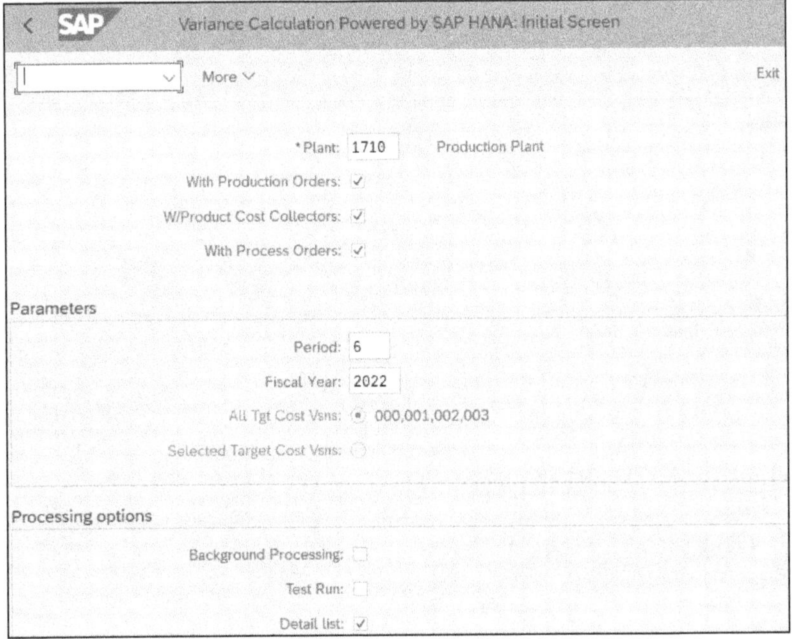

Figure 20.25 Variance Calculation for Summarization Report Selection Screen

A *target cost version* determines the basis for calculating target costs. Target cost version 0 calculates total variance and is used to explain the difference between actual debits and credits on an order. You typically select all target cost versions to provide more reporting data.

When calculating variances, select **Test Run** and **Detail list**, deselect **Test Run**, and run the transaction after everything looks OK.

Run Reports with Data Collection

After you run variance calculation, you run summarization reports that gather data and run the report simultaneously (also known as aggregating on the fly). This is done using Transaction KKBC_HOE_H or via menu path **Accounting • Controlling • Product Cost Controlling • Cost Object Controlling • Product Cost by Period • Information System • Reports for Product Cost by Period • Summarized Analysis • On-the-fly Summarization.** The selection screen in Figure 20.26 is displayed.

Click the hierarchy icon to display the summarization hierarchy shown in Figure 20.27.

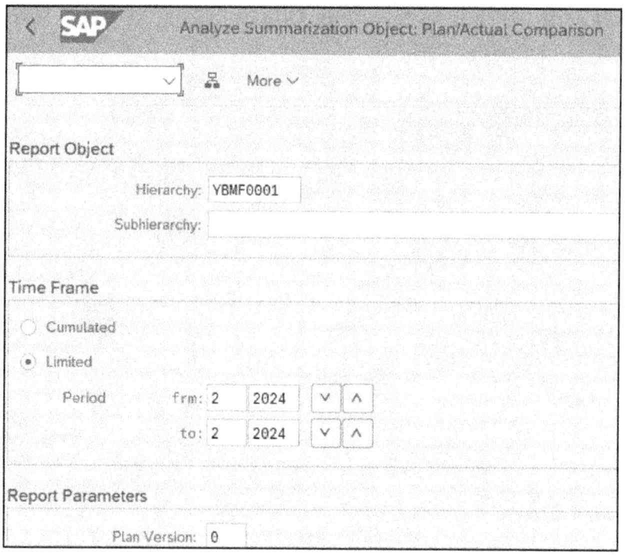

Analyze Summarization Object: Plan/Actual Comparison

More

Report Object

Hierarchy: YBMF0001

Subhierarchy:

Time Frame

Cumulated

Limited

Period frm: 2 2024

to: 2 2024

Report Parameters

Plan Version: 0

Figure 20.26 Summarization Report Selection Screen

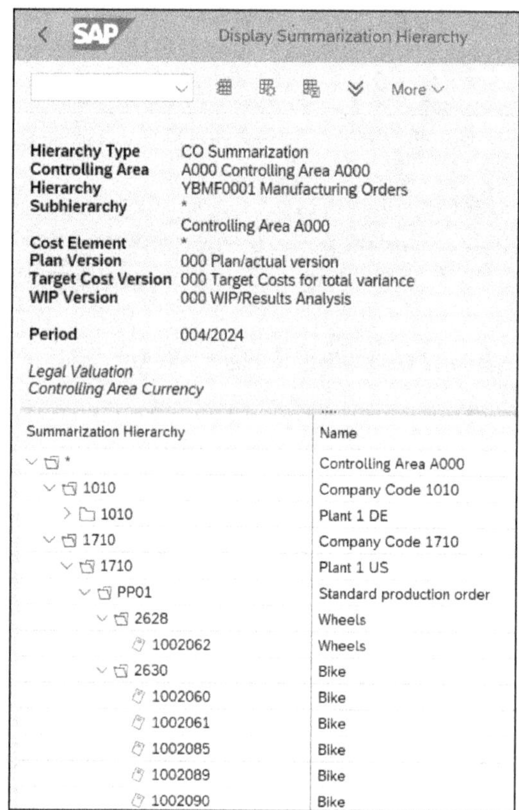

Figure 20.27 Display Summarization Hierarchy

Expand the **Summarization Hierarchy** to display the objects **Wheels** or **Bike** on the right in this example. Double-click on any node or object to display a plan-versus-actual cost report as shown in Figure 20.28.

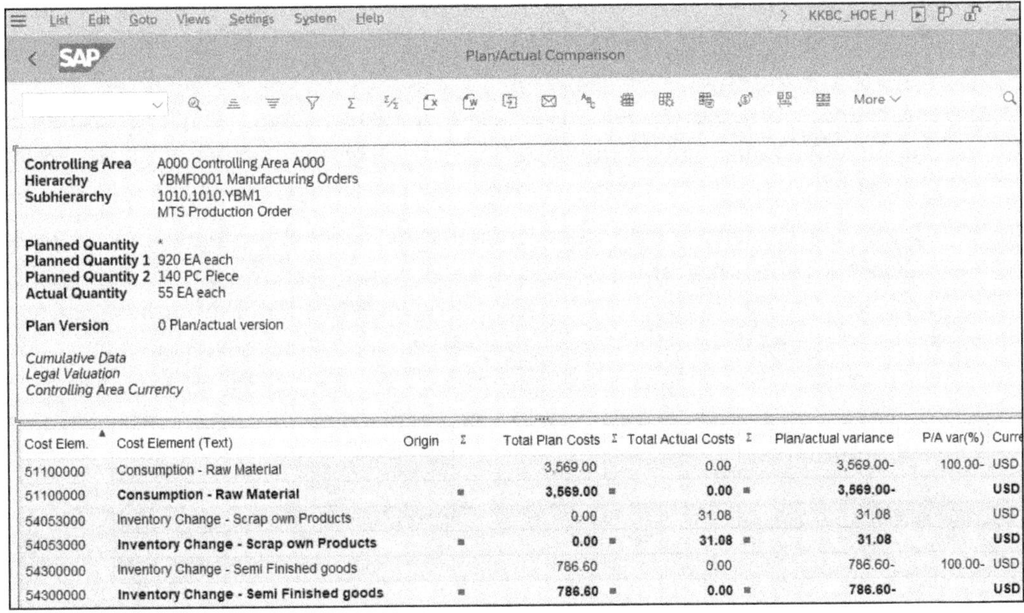

Figure 20.28 Summarization Cost Report on a Node or Object

Alternate Summarization Hierarchies

An advantage of summarization reporting is that you can create multiple hierarchies. 20
In the example hierarchy shown previously in Figure 20.21, we chose plant and order type as hierarchy levels. You can also create hierarchies with an additional level of detail.

You can create exception rules and display traffic light symbols in the results screen to help you quickly find nodes that require further analysis. To create exception rules, follow the IMG menu path **Controlling • Product Cost Controlling • Information System • Cost Object Controlling • Settings for Summarized Analysis/Order Selection • Define Exception Rules.** You include exception rules during data collection with Transaction KKRC and then select **Extras • Exception • Define Rule** from the menu bar. Double-click on an exception rule to attach it to the summarization hierarchy.

Summarization in SAP Fiori

SAP Fiori reports are built using a virtual data model comprising a series of CDS views that combine the information needed for reporting, such as transactional data, master data, and hierarchical data. There are no prebuilt summarization reports because all data is aggregated on the fly from the Universal Journal in line with the report selections. When we looked at the Analyze Costs by Work Center/Operation app in Chapter 10, the system automatically selected the data from the Universal Journal using the data from the selection part of the app. It aggregated the data by work and cost centers without requiring a data-collection step to summarize the data before reporting.

When you work with the Analyze Costs by Work Center/Operation app, the system automatically summarizes the costs for the work centers entered as selection parameters and displays them in the results area either graphically or as a list, as discussed in Chapter 10.

The selection parameters for the standard report include fiscal period, ledger, company code, work center, cost center, order, and order type, but there are many more fields available for selection. To reach these, choose **Adapt Filters**. In the dialog box shown in Figure 20.29, you first see the fields in the default selection screen and can remove any you don't need. You can then scroll down and add by choosing the relevant field, such as **Sales Order**, **Sales Order Item**, or **WBS Element** if you're working in a make-to-order environment.

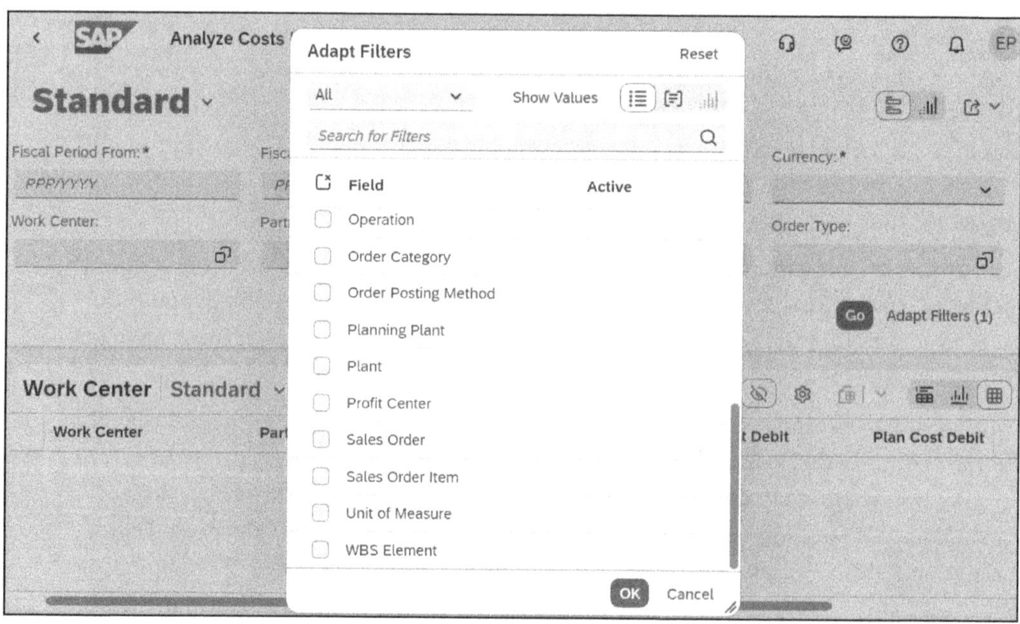

Figure 20.29 Extending the Selection Parameters

This behavior is dynamic, allowing you to aggregate on the fly to see the costs for many reporting dimensions. In addition to fields such as work center and cost center, which are shown by default, you can add many more fields by choosing the **Settings** icon and selecting additional fields, as shown in Figure 20.30.

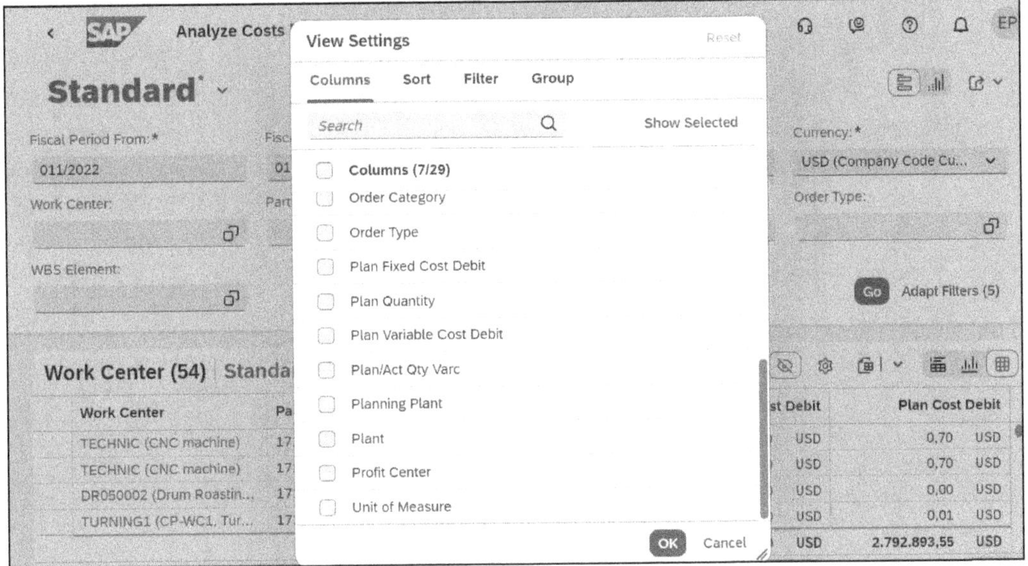

Figure 20.30 Extending the Results List

20.5 Detailed Reports

Now that we've discussed the standard summarization reports, let's look at standard detailed reports that allow you to examine the costs of individual production and process orders.

When you use summarized analysis, you'll most often drill down to detailed reports. You can also display detailed reports directly if you know the material or manufacturing order to analyze. You typically run a detailed report directly if you identify a material with large variances during variance analysis. Because you can't drill down to line-item details from the variance calculation output screen, you'll need to take note of the material number and run this report to drill down to line-item reports and source documents.

Detailed reports are useful during variance analysis because they provide cost element details by row and typically target, actual, and variance by column for an individual product cost collector or manufacturing order. You can display the costs for one or multiple periods or cumulatively for all periods. The cost element rows can be grouped by similar business transactions, such as confirmations, goods issues, and goods receipts, in the report.

This report is useful when analyzing variance for an order. You search for the cost element with the largest variance and drill down (double-click) to line-item details. You then sort the line-item list and double-click on the line item with the largest value to display the source document. The source document typically contains all the information needed to find the cause of the largest variances.

You can display and analyze target versus actual costs in detailed product cost collector reports with Transaction KKBC_PKO or via menu path **Accounting • Controlling • Product Cost Controlling • Cost Object Controlling • Product Cost by Period • Information System • Reports for Product Cost by Period • Detailed Reports.** A selection screen is displayed, as shown in Figure 20.31. A similar report is also available for production and process orders with Transaction KKBC_ORD.

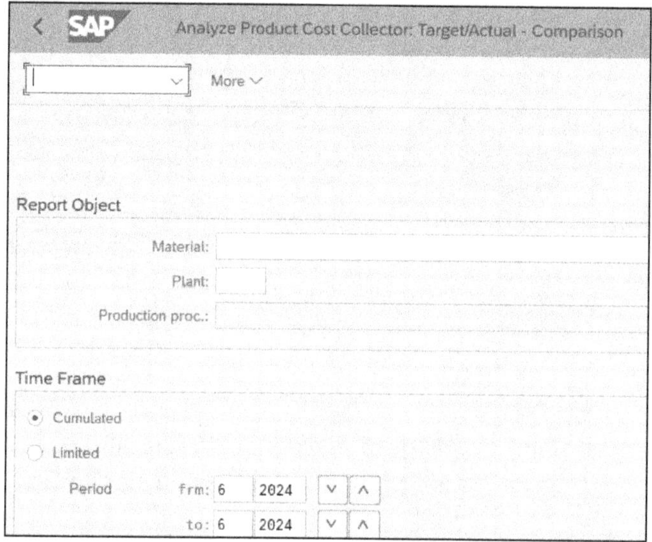

Figure 20.31 Analyze Product Cost Collector Selection Screen

Complete the **Material** and **Plant** fields. You can run the report for a **Limited** range of periods or **Cumulated** costs for all periods. Click the **Execute** button or press F8 to display the results shown in Figure 20.32.

Cost El...	Cost Element (Text)	Origin	Σ Total Actual Costs	Σ Target/actual var.	T/I var(%)	Crcy
420000	Direct labor costs		543.00	543.00		USD
841000	IS-RE Revenue from third-party rental		2,172.00-	2,172.00-		USD
416100	Electricity (fixed portion)		543.00	543.00		USD
400000	Consumption, raw material 1		543.00	543.00		USD

Figure 20.32 Analyze Product Cost Collector Results Screen

You see a detailed report with cost elements as rows and **Total Actual Costs**, **Target/actual var.**, and **T/I var(%) Ttl actual** as columns. Sort a column in descending order and double-click on the row containing the largest value to display line-item details, as shown in Figure 20.33.

Cost Elem.	Cost element name	Σ	Val.in RC	Total quantity	PUM	O	Offsetting Acct
400000	Consumptn. raw mat.1		543.00			S	830000
Order 400001		■	**543.00**				
		■■	**543.00**				

Figure 20.33 Line Items

You'll see line-item details of the summary line for cost element 400000 in Figure 20.33.

With SAP Fiori, the Production Cost Analysis app will provide a familiar view of your production costs. The initial screen shows a filtered view of all production orders that meet the selection criteria, as shown in Figure 20.34. As we discussed in Chapter 10, you can choose a production order and access detailed information on each one.

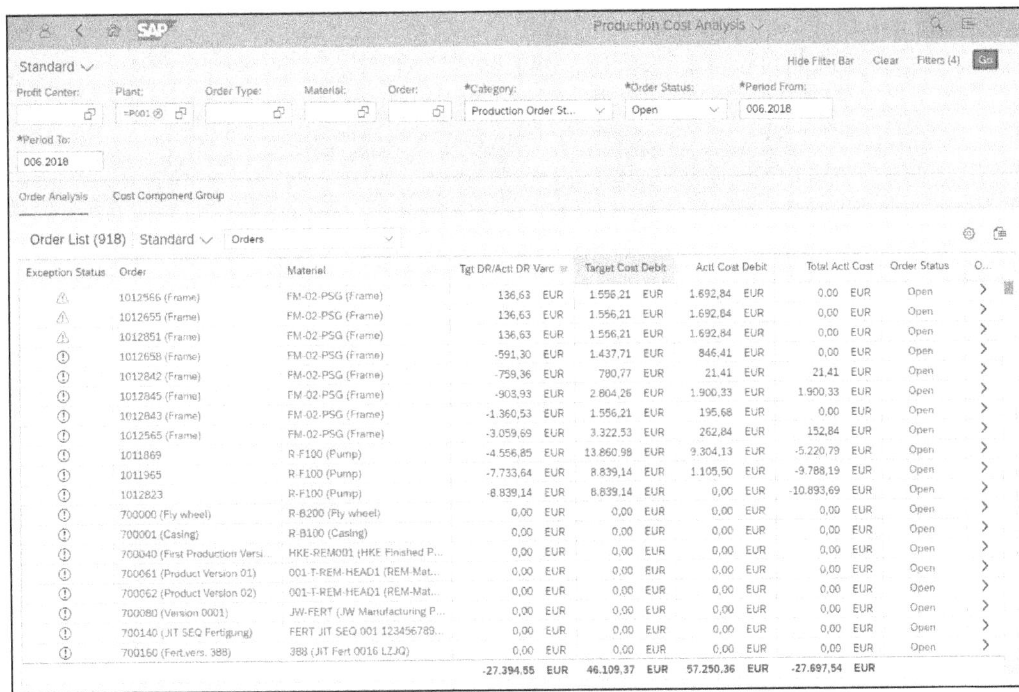

Exception Status	Order	Material	Tgt DR/Actl DR Varc	Target Cost Debit	Actl Cost Debit	Total Actl Cost	Order Status	O...
⚠	1012566 (Frame)	FM-02-PSG (Frame)	136,63 EUR	1.556,21 EUR	1.692,84 EUR	0,00 EUR	Open	>
⚠	1012655 (Frame)	FM-02-PSG (Frame)	136,63 EUR	1.556,21 EUR	1.692,84 EUR	0,00 EUR	Open	>
⚠	1012851 (Frame)	FM-02-PSG (Frame)	136,63 EUR	1.556,21 EUR	1.692,84 EUR	0,00 EUR	Open	>
ⓘ	1012658 (Frame)	FM-02-PSG (Frame)	-591,30 EUR	1.437,71 EUR	846,41 EUR	0,00 EUR	Open	>
ⓘ	1012842 (Frame)	FM-02-PSG (Frame)	-759,36 EUR	780,77 EUR	21,41 EUR	21,41 EUR	Open	>
ⓘ	1012845 (Frame)	FM-02-PSG (Frame)	-903,93 EUR	2.804,26 EUR	1.900,33 EUR	1.900,33 EUR	Open	>
ⓘ	1012843 (Frame)	FM-02-PSG (Frame)	-1.360,53 EUR	1.556,21 EUR	195,68 EUR	0,00 EUR	Open	>
ⓘ	1012565 (Frame)	FM-02-PSG (Frame)	-3.059,69 EUR	3.322,53 EUR	262,84 EUR	152,84 EUR	Open	>
ⓘ	1011869	R-F100 (Pump)	-4.556,85 EUR	13.860,98 EUR	9.304,13 EUR	-5.220,79 EUR	Open	>
ⓘ	1011965	R-F100 (Pump)	-7.733,64 EUR	8.839,14 EUR	1.105,50 EUR	-9.788,19 EUR	Open	>
ⓘ	1012823	R-F100 (Pump)	-8.839,14 EUR	8.839,14 EUR	0,00 EUR	-10.893,69 EUR	Open	>
ⓘ	700000 (Fly wheel)	R-B200 (Fly wheel)	0,00 EUR	0,00 EUR	0,00 EUR	0,00 EUR	Open	>
ⓘ	700001 (Casing)	R-B100 (Casing)	0,00 EUR	0,00 EUR	0,00 EUR	0,00 EUR	Open	>
ⓘ	700040 (First Production Versi...	HKE-REM001 (HKE Finished P...	0,00 EUR	0,00 EUR	0,00 EUR	0,00 EUR	Open	>
ⓘ	700061 (Product Version 01)	001-T-REM-HEAD1 (REM-Mat...	0,00 EUR	0,00 EUR	0,00 EUR	0,00 EUR	Open	>
ⓘ	700062 (Product Version 02)	001-T-REM-HEAD1 (REM-Mat...	0,00 EUR	0,00 EUR	0,00 EUR	0,00 EUR	Open	>
ⓘ	700080 (Version 0001)	JW-FERT (JW Manufacturing P...	0,00 EUR	0,00 EUR	0,00 EUR	0,00 EUR	Open	>
ⓘ	700140 (JIT SEQ Fertigung)	FERT JIT SEQ 001 123456789...	0,00 EUR	0,00 EUR	0,00 EUR	0,00 EUR	Open	>
ⓘ	700160 (Fert.vers. 388)	388 (JIT Fert 0016 LZJQ)	0,00 EUR	0,00 EUR	0,00 EUR	0,00 EUR	Open	>
			-27.394,55 EUR	**46.109,37 EUR**	**57.250,36 EUR**	**-27.697,54 EUR**		

Figure 20.34 Production Cost Analysis App

20.6 Line-Item Reports

You can drill down to line-item reports from detailed reports. This is useful because there can be many line items. Just as summarization reports group product cost collectors or production orders by characteristics for management variance reporting, detailed reports combine line items for production and management reporting.

Analyzing a detailed report for a product cost collector and drilling down on the cost element with the largest variance is a much more efficient variance analysis method than searching through many line items directly. However, if you are in a situation where you need to display line items directly, the procedure is described in the following sections.

20.6.1 Display Actual Costs Report

You display and analyze line-item reports with Transaction KRMI or via the menu path **Accounting • Controlling • Product Cost Controlling • Cost Object Controlling • Product Cost by Period • Information System • Reports for Product Cost by Period • Line Items • Product Cost Collectors • Actual Costs**. A selection screen is displayed, as shown in Figure 20.35. A similar report is also available for production and process orders with Transaction KOB1. These use compatibility views to access the data in the Universal Journal and display it as if using the old data structure.

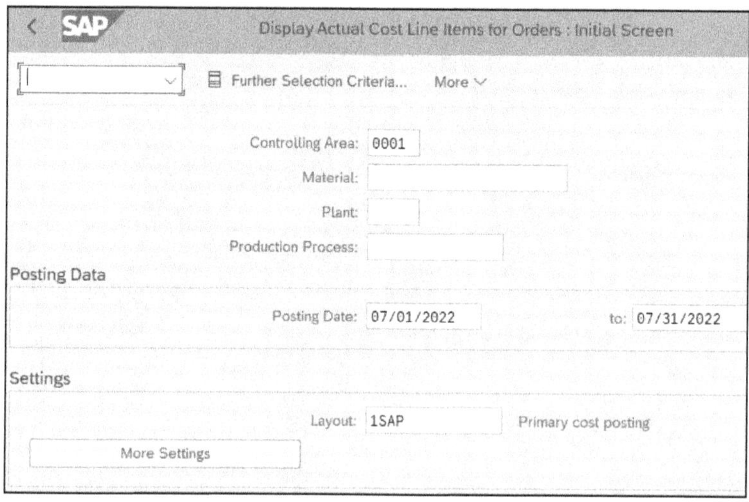

Figure 20.35 Line-item Report Selection Screen

When displaying line items directly, it's important to restrict the **Posting Date** range sufficiently to avoid long runtimes since there can be many line items. It's usually best to tightly restrict the posting date range initially, and then gradually increase it as required.

You typically sort line item lists by value or quantity and analyze the lines with the largest and smallest values by double-clicking to display the source documents. For confirmation line items, the source documents are activity confirmations. For goods receipts and goods issues, the source documents are material documents.

20.6.2 Display Line Items – Cost Accounting App

There are no specific apps for analyzing line items on production orders, but you can access the same information by choosing the Display Line Items—Cost Accounting app (SAP Fiori ID F4025). You can then either enter the order number or expand the selection parameters to include the fields that will allow you to select the relevant production order with the possible selection fields shown in Figure 20.36. SAP Fiori apps access the Universal Journal directly and do not use compatibility views.

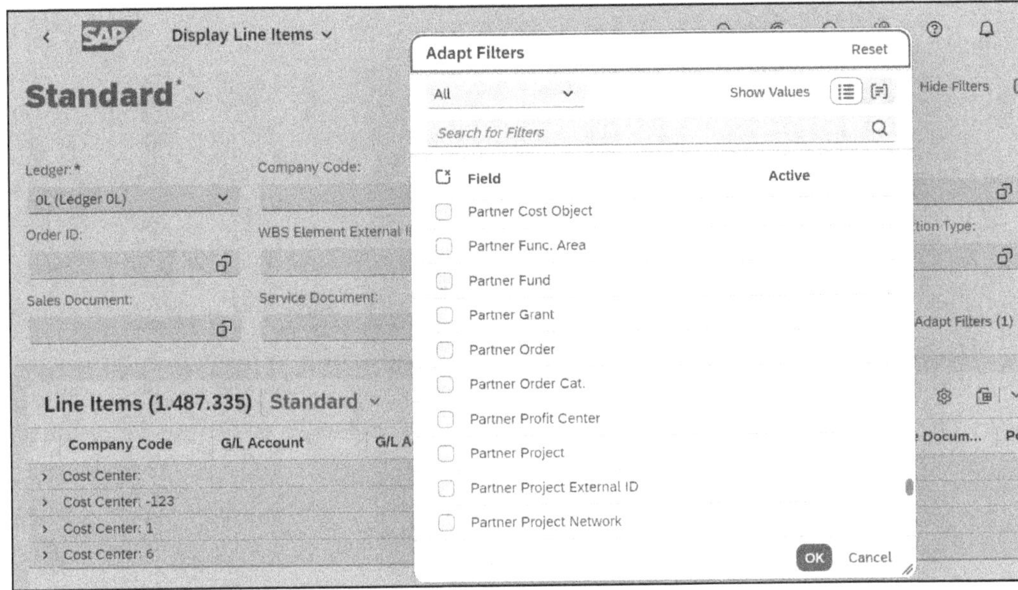

Figure 20.36 Selecting Line Items

You'll need to adjust the view settings to display the fields you require for your production order analysis, as shown in Figure 20.37. When you're happy with your view, save it to access your settings again.

We'll now return to the classic reports to discuss your options without SAP Fiori.

Consider whether your reports could benefit from including additional dimensions. For example, you might want to consider including information, such as the cost center or the activity type, that is available in the Controlling records but that the system didn't access during summarization for old reports. Be aware that adding dimensions to summarization reports is a modification.

SAP Notes

SAP Note 1523578 describes how to add a *profit center*.

SAP Note 1514091 describes how to add an *origin*.

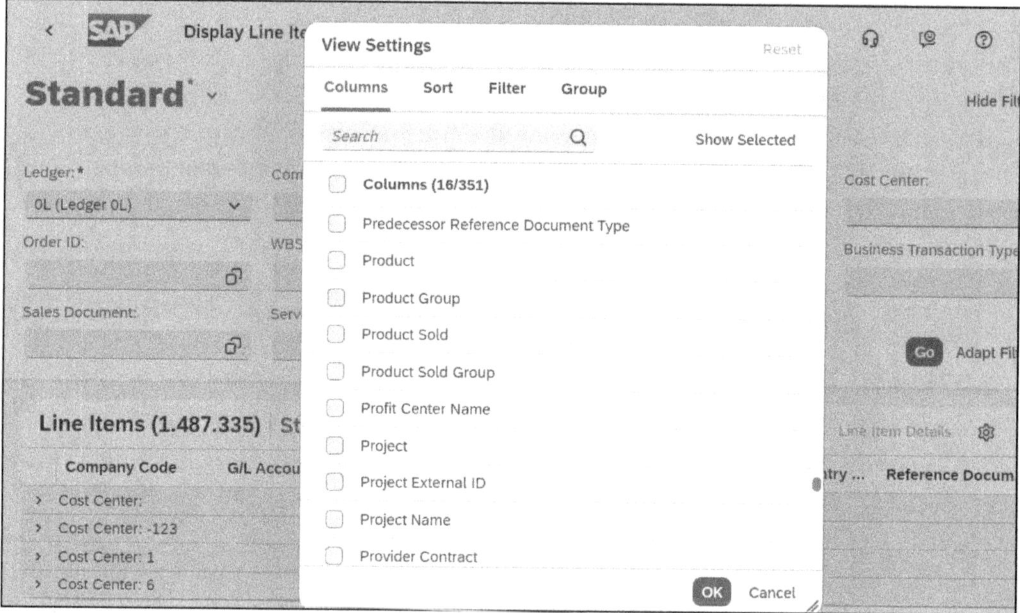

Figure 20.37 Adding Fields to the Report List

If you implement these notes, you must rebuild the drilldown tables for your historical data.

Now that we've examined summarization reports, let's examine production order reports.

20.7 Production Order Reports

Now that we've examined summarization, detailed, and line-item reports when analyzing production order costs and variances, let's look at two production order reports in the following sections. We'll first look at how you can display a simple list of production orders and then examine how to display an order list with cost information in columns.

20.7.1 Order Information System

The *order information system* is a tool for shop-floor control and process industries with reporting for production orders, planned orders, and process orders. You can, for example, display all production orders within a specific time frame and group the information according to order or operation. You can use this list to identify individual production orders, which may require separate variance analysis.

You access the order information system with Transaction COOIS or via the menu path **Logistics • Production • Shop Floor Control • Information System • Order Information System**. The system displays a selection screen, as shown in Figure 20.38.

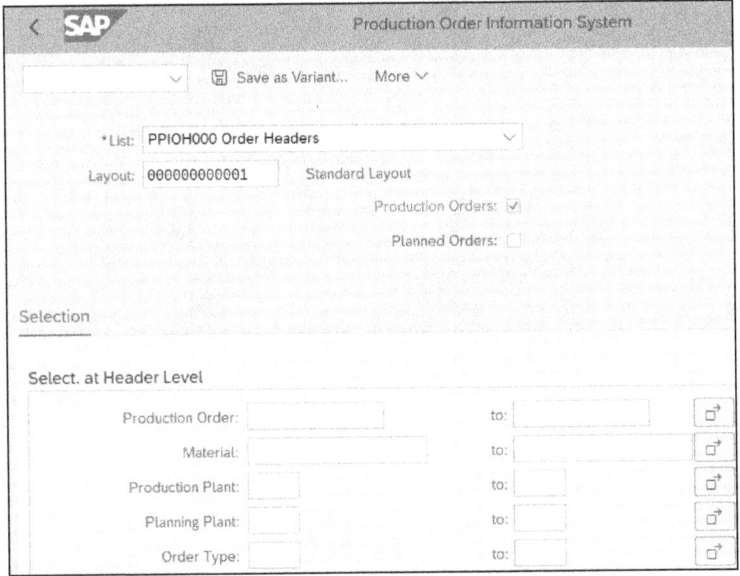

Figure 20.38 Production Order Information System Selection Screen

List Process Orders

You can display a list of process orders with Transaction COOISPI or via the menu path **Logistics • Production • Process • Process Order • Reporting • Order Information System • Process Order Information System**. These are part of the compatibility scope and are not recommended for new implementations. Instead, use CDS views.

The **Selection** tab provides many options for displaying a list of objects, such as order headers, status, or area of responsibility (e.g., MRP controller or production supervisor).

The **List** field at the top defaults to **Order Headers,** which you can change by clicking in the field and selecting from the dropdown list shown in Figure 20.39.

Scroll down the list of possible entries to view more objects. You can create and save your own variant by clicking the **Save as Variant** button.

Clicking in the **Layout** field allows you to choose any available standard or user-defined entries, determining how columns display information in the following report.

Complete the **Material** and **Production Plant** fields in Figure 20.38 and click the **Execute** button or press F8 to display a results list, as shown in Figure 20.40.

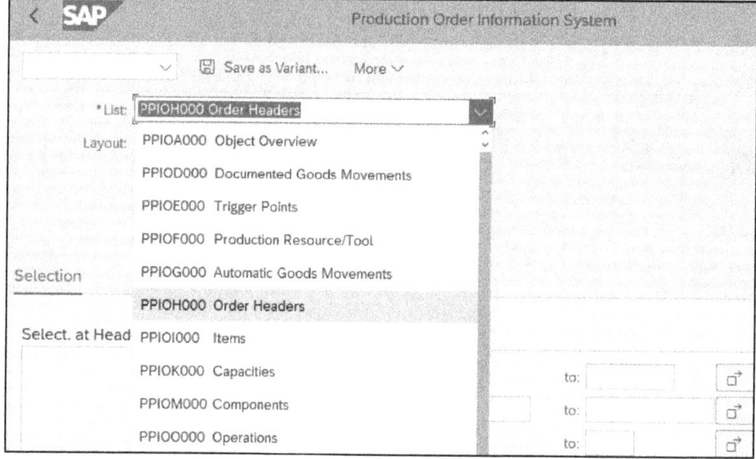

Figure 20.39 Possible Entries in Order Information System

Order	Material	Icon	Order Type	MRP ctrlr	Pr.Superv.	Plant	Target Qty	Unit	Bsc start	Basic fin. date	Type	System Status
1000037	MAT1		PP01	001	YB1	1710	1	PC	05/13/2022	05/16/2022		TECO MSPT PRC SETC
1000025			PP01	001			1	PC	05/16/2022	05/17/2022		TECO MSPT PRC SETC
1000027			PP01	001			1	PC	05/23/2022	05/24/2022		TECO MSPT PRC RELR
1000028			PP01	001			1	PC		05/24/2022		TECO MSPT PRC RELR
1000029			PP01	001			1	PC	05/24/2022	05/25/2022		TECO MSPT PRC RELR
1000023			PP01	001			1	PC	05/25/2022	05/26/2022		TECO MSPT PRC SETC
1000026			PP01	001			1	PC		05/26/2022		TECO MSPT PRC SETC

Figure 20.40 Production Order Results List

Click an item in the **Order** or **Material** field and click the **Edit** (pencil) or **Display** (glasses) icon to navigate directly to the corresponding edit or display screens. You can display a full list of standard ALV function icons, including sort, total, and filter, by clicking the right-pointing arrow icon at the left.

Tip

You can include production orders flagged for deletion by scrolling down the selection screen in Figure 20.38 and selecting the **With Deletion Flag/Indicator** checkbox shown in Figure 20.41.

Options

Maximum number of orders:

Displ. Compl. Collective Order:

With Deletion Flag/Indicator: ✓

Display Complete Rework:

Figure 20.41 Order Information System Deletion Flag/Indicator Checkbox

Production Orders with Status DLFL or DLT

Production orders with a system status of "DLFL" or "DLT," displayed in the last column of Figure 20.40, are included in the results list along with other orders.

20.7.2 Order Selection

Now that we've examined the order information system, let's examine another report that lists production orders and variance information. You can list production orders with costs displayed in columns with Transaction S_ALR_87013127 or via the menu path **Accounting • Controlling • Product Cost Controlling • Cost Object Controlling • Product Cost by Order • Information System • Reports for Product Cost by Order • Object List • Order Selection**. The selection screen in Figure 20.42 is shown.

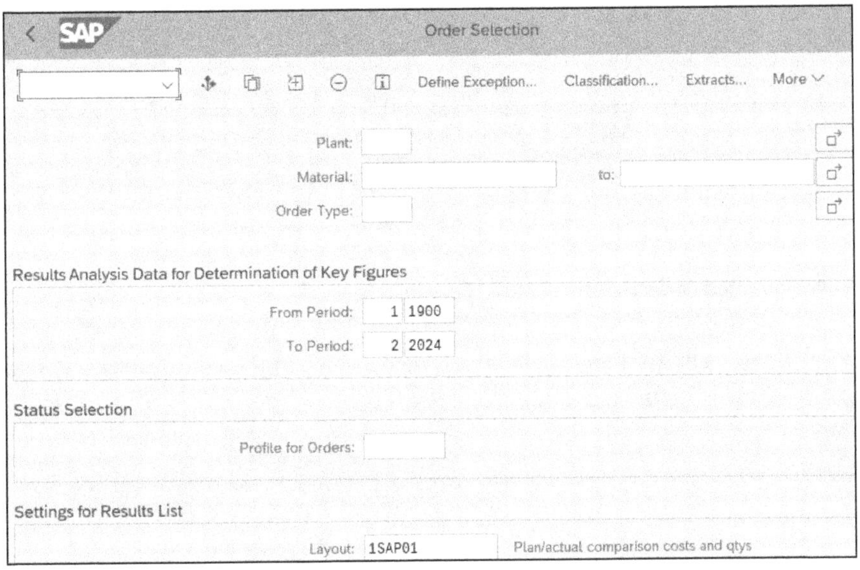

Order Selection

Define Exception... Classification... Extracts... More

Plant:

Material: to:

Order Type:

Results Analysis Data for Determination of Key Figures

From Period: 1 1900

To Period: 2 2024

Status Selection

Profile for Orders:

Settings for Results List

Layout: 1SAP01 Plan/actual comparison costs and qtys

Figure 20.42 Order Selection Initial Screen

Complete the selection fields as follows:

1. Complete the **Plant**, **Material**, and **Order Type** fields to narrow your costs displayed or leave them blank to see the costs for all orders within the timeframe.
2. Limiting the reporting periods with entries in the **From Period** and **To Period** fields specifies the periods from which key figures can be read. This limitation doesn't affect the orders selected.
3. For **Profile for Orders**, you can enter a profile to select production orders based on status. We discussed status selection profiles in Section 20.4.2
4. The **Layout** determines the columns in the cost report. Press F4 to display the possible entries, as shown in Figure 20.43.

Layout: Choose

Layout	Layout description
1SAP00	Plan/actual comparison
1SAP01	Plan/actual comparison costs and qtys
1SAP02	Target/actual comparison

Figure 20.43 Layout Possible Entries

Choose any of the three standard layouts and see which works for you. Many more selection fields are available than are shown in the initial screen shown earlier in Figure 20.42. Click the plus sign icon to display all selection fields, as shown in Figure 20.44.

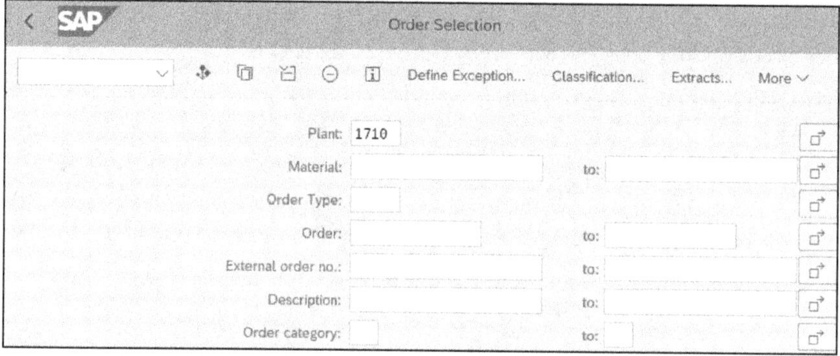

Figure 20.44 Expanded Order Selection Initial Screen

Press the [PageDown] key to display additional fields. Click in the **Order category** field and press [F4] to display the list of possible entries for the order category, as shown in Figure 20.45.

Order category (2) 14 Entries found

Order category	Short Description
01	Internal Order (Controlling)
02	Accrual Calculation Order (Controlling)
03	Model Order (Controlling)
04	CO Production Order
05	Product Cost Collector
06	QM Order
10	PP Production Order
20	Network
30	Maintenance order
40	Process Order
50	Inspection Lot
60	Personnel Order
70	Shipping deadlines
99	Master Planned order

Figure 20.45 Order Category Possible Entries

You can further restrict the results screen by order type. After you choose an **Order category** in Figure 20.45 for example, **PP Production Order**, the possible entries for **Order Type** in the selection screen in Figure 20.44 are restricted to order type by order category, for example, **PP: Production Order**. You can then type your order type in the first selection screen (shown earlier in Figure 20.42) without expanding the selection list and then saving it as a variant.

Limiting the reporting periods with entries in the **From Period** and **To Period** fields shown earlier in Figure 20.42 specifies the periods that key figures are read from. This limitation doesn't affect the orders selected. Type in your selection criteria and click the **Execute** button or press the F8 key to display the **Results List** in Figure 20.46.

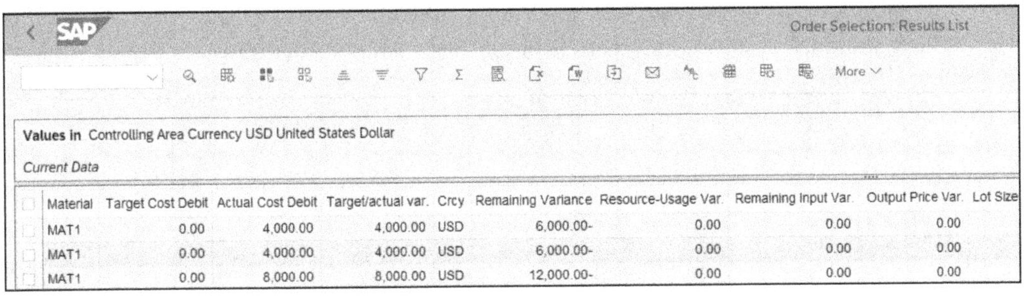

Order Selection: Results List

More

Values in Controlling Area Currency USD United States Dollar

Current Data

Material	Target Cost Debit	Actual Cost Debit	Target/actual var.	Crcy	Remaining Variance	Resource-Usage Var.	Remaining Input Var.	Output Price Var.	Lot Size
MAT1	0.00	4,000.00	4,000.00	USD	6,000.00-	0.00	0.00	0.00	
MAT1	0.00	4,000.00	4,000.00	USD	6,000.00-	0.00	0.00	0.00	
MAT1	0.00	8,000.00	8,000.00	USD	12,000.00-	0.00	0.00	0.00	

Figure 20.46 Order Selection Results List

You can display the variance columns shown in Figure 20.46 by clicking the **Change Layout** icon (third from right). Select the **variance categories** on the right of the **change layout** dialog box, click the left-pointing icon, and press Enter. You can sort, total, and filter the values in this report to highlight orders with the largest variance.

Even though the values displayed occurred during the periods specified earlier in Figure 20.42, all production orders outside the periods also appear with zero costs. To make the report more manageable, filter out the production orders with zero costs with the funnel icon. If you summarize the **Actual debit** column, you see the total manufacturing costs for the period, which isn't easily obtainable in other standard reports.

You can display a detailed report for individual orders by double-clicking an **Order** and then continuing to drill down to line items and source documents. You can display an individual production order by clicking an **Order** and selecting **Extras • Master Data** from the menu bar.

Now that we've examined how production order reports can assist during variance analysis, let's examine how cost center reports are useful.

20.8 Cost Center Reports

Standard cost center reports provide managers and users with a quick and efficient method to analyze planned and actual cost center costs and variances. The system posts costs in real time and reports at the cost-element level, as we'll discuss in this section. You can drill down directly to line items and source documents.

20.8.1 Cost Centers: Actual/Plan/Variance Report

You display and analyze cost center reports with Transaction S_ALR_87013611 or via the menu path **Accounting • Controlling • Cost Center Accounting • Information System • Reports for Cost Center Accounting • Plan/Actual Comparisons • Cost Centers: Actual/Plan/Variance.** A selection screen displays, as shown in Figure 20.47.

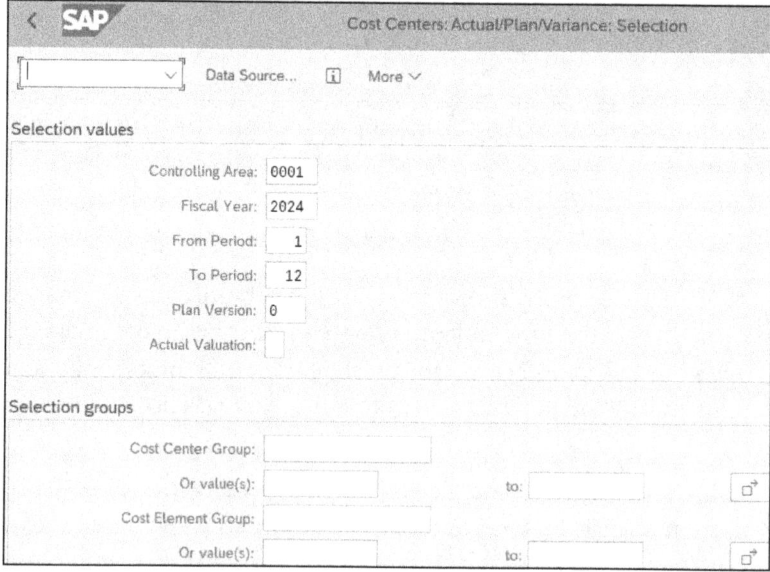

Figure 20.47 Cost Center Selection Screen

Complete the **Controlling Area**, **Fiscal Year**, **From Period** and **To Period**, **Plan Version**, and **Cost Center** fields. Enter **Cost Center Group** and optionally **Or value(s)** field entries and click the **Execute** button or press the F8 key to display the report shown in Figure 20.48.

> **Tip**
>
> Enter the standard hierarchy in the **Cost Center Group** field shown in Figure 20.47 to display the standard hierarchy structure on the left, as shown in Figure 20.48. You can determine the standard hierarchy with Transaction OKENN.

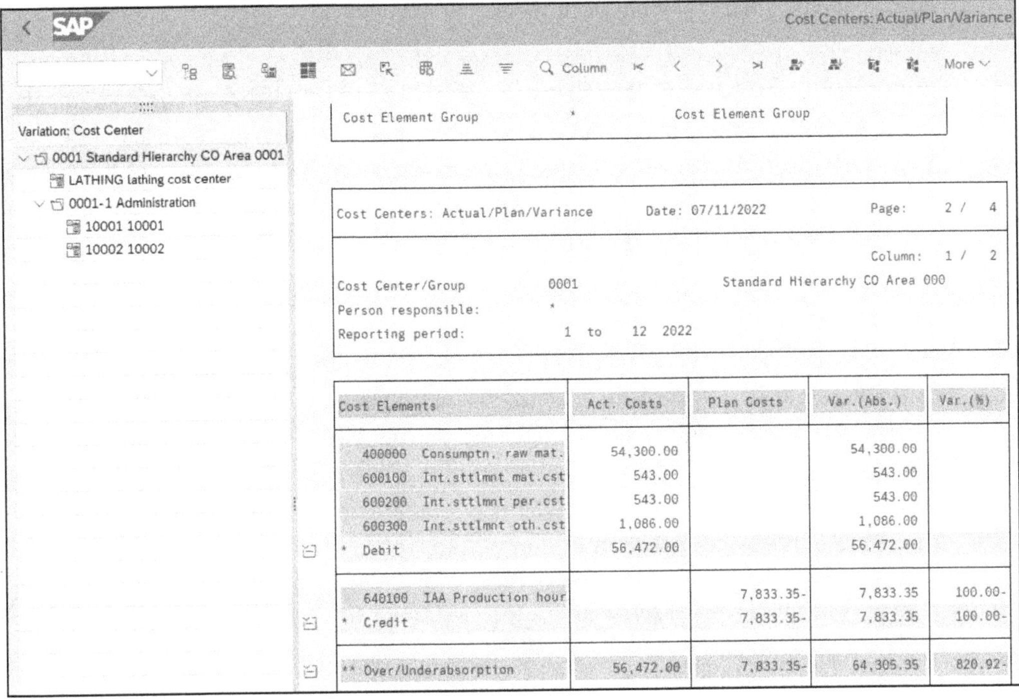

Figure 20.48 Cost Center Actual/Plan/Variance Report

You can expand and click any of the nodes on the left to display the corresponding information on the right. Double-click a cost in the **Act. costs** (actual) or **Plan costs** columns to display the corresponding line-item report. For example, double-click actual costs of **54,300.00** for cost element **400000** to display the corresponding line-item report shown in Figure 20.49.

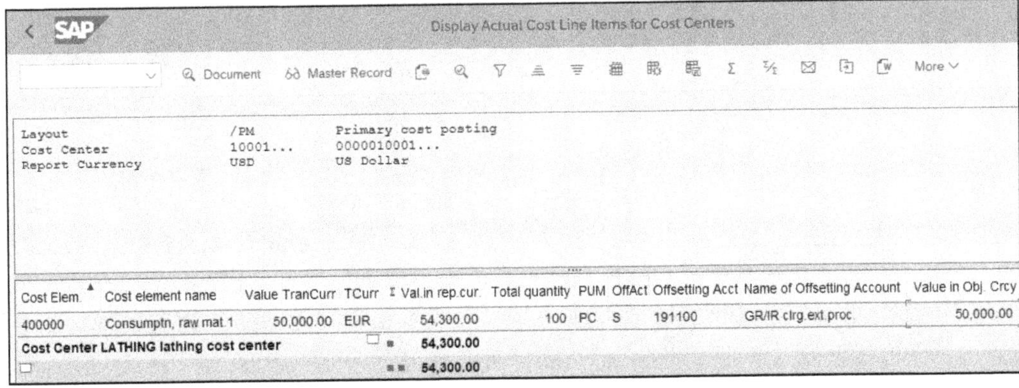

Figure 20.49 Cost Center Line-item Report

You can double-click any line item in Figure 20.49 to display the original source document.

20.8.2 Cost Centers – Plan/Actual App

Alternatively, you can use the Cost Centers—Plan/Actual app (SAP Fiori ID F0949) to display the cost centers and the planned and actual costs. The default view will display a flat list of cost centers and general ledger accounts, but if you place the cursor on the **Cost Center** field and choose **Hierarchy • Select Hierarchy**, you can select a hierarchy to group the cost centers in the list. You can do the same with the general ledger accounts to group your accounts by the financial statement version, the account group, or the cost element group.

The SAP Fiori reports are designed to use cross-application, so it's also worth understanding the role of the object types in your reports. Figure 20.50 and Figure 20.51 show the P&L – Plan/Actual app (SAP Fiori ID F0927), displaying the general ledger accounts and the object type. This provides a quick and simple way of looking at postings to margin analysis (object type **EO** for market segment), postings in cost center accounting (object type **KS** for cost center) and **KL** (for cost center and activity type), and in production (object type **OR** for order and **PS** for project).

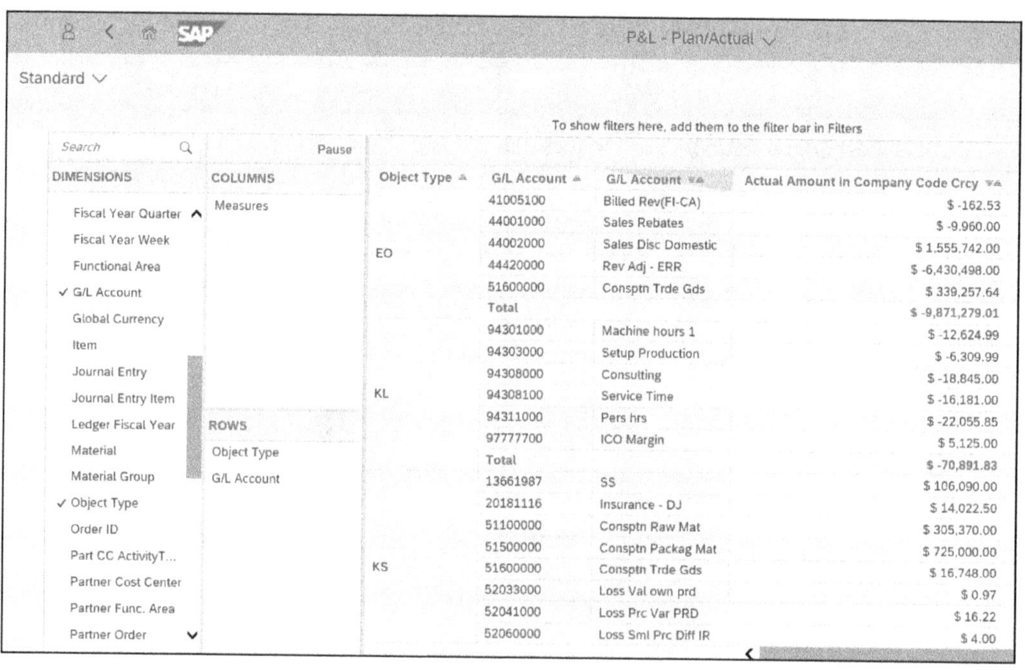

Figure 20.50 Plan /Actual App Showing Object Types and General Ledger Accounts (1/2)

There aren't yet any detailed variance reports showing target/actual costs and the variance categories for the cost centers. The list of available reports changes with each release as new reports are delivered. To stay abreast of changes, refer to SAP Note 2349297.

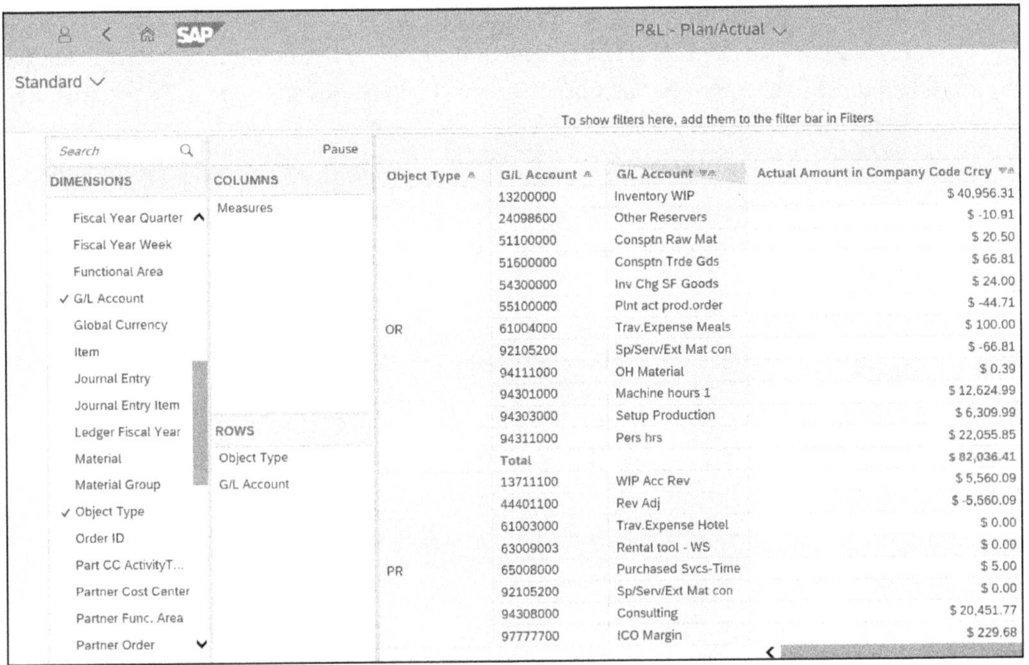

Figure 20.51 Plan /Actual App Showing Object Types and General Ledger Accounts (2/2)

20.9 Summary

In this chapter, we examined standard information system reports. For product cost planning, we examined how to analyze costing runs and create cost estimate list reports. We looked at how product drilldown reports have a predefined reporting hierarchy. We saw how to run data collection and the report. We also looked at summarization hierarchy reports, which allow you to create your own reporting hierarchies. And we saw how SAP Fiori reports summarize on the fly.

Summarized reports allow the selection of individual orders for further analysis. With detailed reports, you can analyze individual orders at a cost-element level. You can double-click a cost element to display a line-item report. You can also run line-item reports directly with a transaction without drilling down, although this is often unnecessary. We examined a method to display line-item reports directly, sort the rows based on value, and drill down on the largest values to examine the source documents, such as activity confirmations and material documents.

We also examined two reports that list production orders according to selection screen parameters. Order information system reports allow you to display a simple list of production orders within a period, while order selection reports display variance information directly in columns.

Cost center reports allow cost center managers and users to analyze cost center variances at a cost-element level and to drill down directly to line-item reports and source documents. We examined the Cost Centers—Plan/Actual app to display the cost centers and the planned and actual costs.

In conclusion, standard reports provide excellent reporting functionality, sufficient to analyze variances for most companies. You can use product drilldown reports with only a small amount of configuration and setup. Summarization hierarchy reports allow you to create your own multiple hierarchies. With detailed reports, you can view all postings by cost element and drill down to line-item details. Order list reports allow you to analyze variances by listing production orders, and cost center reports provide cost center managers with detailed analysis capability for cost center plan and actual costs.

Conclusion

Product cost controlling allows you to plan for future fiscal periods and years, post actual costs, compare plan, target, and actual costs, and analyze variances. This component receives costs from many other components through the Universal Journal, which you can analyze and use to report profitability with margin analysis.

Book Summary

We provide a comprehensive overview of the integration aspects of product cost controlling in this book. Let's first revisit the initial planning in Chapter 1.

Initial Planning

You can begin planning in several areas. We looked at a typical best practice flow, starting with entering sales plan data into either profitability analysis, margin analysis, or sales and operations planning. As in many other areas of SAP, these components allow you to enter different versions for scenario-based planning and budgeting. Sales managers test the profitability of different sales plan quantities for products or product groups and determine a preferred sales plan converted into a production plan. Sales and operations planning is gradually being replaced by SAP Integrated Business Planning for Supply Chain (SAP IBP).

You can transfer the production plan to demand management, which involves planning the requirement quantities and dates for finished products and important assemblies and defining the strategy for planning, producing, or procuring a finished product.

Long-term planning is crucial in accessing demand management's assembly and finished product requirements data and determining the requirements of lower-level components and activities by accessing bills of materials (BOMs) and routing master data. You can then transfer these planned requirements to the purchasing information system and cost center accounting (CCA).

You transfer data from long-term planning to the purchasing information system to determine requests for quotations (RFQ), which you send to suppliers to determine procurement prices for components to meet the sales plan. The preferred supplier quotation prices for components are stored as data in purchasing info records.

You analyze activity data from long-term planning on a work center capacity-planning basis and workload leveled across work centers. You then transfer this activity quantity data to cost center and activity type combinations, which you use to determine activity prices.

Chapter 2 through Chapter 9 examined product cost planning, which we'll now summarize.

Product Cost Planning

This module determines the planned cost to manufacture products based on the sales plan, master data, and configuration settings.

Chapter 2 examined Controlling master data, including cost elements, accounts, cost centers, and activity types. In Chapter 3, we discussed material master views relevant to product cost planning, including **MRP**, **Costing**, and **Accounting**, which are set up by the corresponding departments. In Chapter 4, we discussed logistics master data, including BOMs, work centers, and routings, which provide quantity structure information for cost estimates. Purchasing info records supply the material price information for cost estimates.

In Chapter 5 through Chapter 8, we examined configuration settings, starting with costing sheets containing overhead settings and then cost components, which identify costs of similar types, such as material, labor, and overhead, by grouping cost elements. In Chapter 7 and Chapter 8, we examined the configuration of costing variants, which are the primary source of information on how a cost estimate calculates the standard price.

In Chapter 9, we created standard, preliminary, and mixed cost estimates and examined how these determine planned procurement and manufacturing prices. We also examined how standard cost estimates update the standard price and revaluate inventory with costing runs. We also discussed how cost estimate user exits allow you to add your own functionality in a structured way to the standard program.

The next part of this book examined cost object controlling, as summarized in the next section.

Cost Object Controlling

Cost objects include cost centers, manufacturing orders, internal orders, and WBS elements, which collect planned and actual costs. Preliminary costing determines the planned costs for manufacturing orders. During simultaneous costing, discussed in Chapter 11, you collect actual costs while purchasing and manufacturing. During period-end processing, overhead (Chapter 12), work in process (Chapter 13), and variances (Chapter 14) are calculated, which are posted during settlement (Chapter 15).

In Chapter 16, we analyzed setting up and running actual costing, which valuates all goods movements within a period at the standard price. At the end of the period, the price and exchange rate differences are used to calculate an actual weighted average price for the period, called the periodic unit price (PUP).

In Chapter 17, we looked at sales order controlling, which involves costing sales order line items for individual customer requirements.

Chapter 18 analyzed special topics, including subcontracting, component delivery costs, and stock-in-transit functionality. Subcontracting is a purchasing process that involves sending components to an external supplier, who assembles them and returns the finished goods to inventory. We explained how to report component delivery costs as cost components. We also looked at accessing advanced stock-in-transit functionality by activating the business function LOG_MM_SIT (Cross-Company-Code Stock/Actual Costing).

In Chapter 19, we discussed the benefits of SAP S/4HANA, including universal parallel accounting and event-based processing.

Information System and Trends

In Chapter 20, we looked at the standards summarized, detailed, and line-item reports available to analyze costs and explain variances. The best practice method in period-end reporting is to analyze summarized costing reports to determine overall manufacturing plant performance. You then drill down from summarized reports to analyze specific objects further by cost element with detailed reports. You can obtain analysis in more detail by drilling down to line item reports, sorting on the largest values, and then accessing source documents to determine the cause of the variance. You can then take corrective action as required. We looked at the benefits of SAP S/4HANA to summarized and line item reporting.

Order information system and order selection reports allow you to list production orders based on various selection parameters. Order information system reports allow you to list orders based on a validity date range, while order selection reports list production orders with variance displayed directly in columns. You can sort, total, and filter this information in the user-friendly ABAP List Viewer (ALV) report format.

Cost center reports allow you to analyze production and overhead expenses per cost element. You can also drill down to actual and plan line item reports and source documents to explain the costs and variances from the plan and target.

Enhancement packages allow you to access the latest functionality contained in business functions. This allows you to activate and test only your required functionality without testing all your business processes.

Looking Ahead

The product cost controlling component integrates with many other components since most company activities involve costs. This book helps you understand integration aspects and new developments. Each new SAP S/4HANA release introduces many improvements.

> **Online Help Documentation**
>
> Access help documentation from any transaction by selecting **Help • Product Assistance** from the menu bar.

One of the aims of this book is to provide you with a detailed reference guide for the core functionality of product cost controlling, including basic overviews and concepts as well as detailed master data and configuration. You've received many notes, examples, and real-world case scenarios to help you understand the many product costing settings to improve your system.

SAP S/4HANA

SAP HANA is an in-memory column-based database and application platform. This technology has allowed a complete restructuring of the database and software to dramatically improve system performance. Three of the main drivers of this performance increase are:

- All data is held in the main memory of the server.
- Transaction data is organized by columns instead of rows.
- The software has been optimized for the new hardware architecture and uses parallel processing.

SAP HANA allows simplification of the table structure by:

- Removing aggregate and index tables
- Combining separate tables from individual modules into one table: the Universal Journal

This simplified table structure has several benefits, including the elimination of data reconciliation and locking issues, which allows the development of reports to take advantage of the Universal Journal such as:

- Trial balance with many drilldown functionalities
- Multidimensional income statement

All standard reports are still available as views, providing a seamless transition for the end user to SAP S/4HANA.

With SAP S/4HANA, the Universal Journal combines financial and management accounting, allowing users to understand the modules more easily. Secondary cost elements become SAP general ledger accounts, and the reconciliation ledger is unnecessary.

SAP continues developing SAP S/4HANA with universal parallel accounting, which provides a harmonized architecture for ledgers and currencies. This provides a foundation for calculating and posting values per ledger and currency. It also allows event-based processing, which provides the basis for real-time production accounting.

Keeping Up

One of the best ways to keep up with the constant improvements made to product cost controlling is to attend Controlling conferences run by SAP. With expert speakers from SAP and consulting firms, you can quickly get up to speed with the latest developments and SAP's future direction.

These events can also provide excellent opportunities to exchange information with colleagues from other companies about the different approaches to solving similar issues.

Appendices

Appendix A
Bibliography

Jordan, John, *Product Cost Controlling with SAP (3rd ed.)* Boston: SAP PRESS, 2016.

Jordan, John, Salmon, Janet *Production Variance Analysis in SAP S/4HANA* Boston: SAP PRESS, 2023.

Jordan, John, *100 Things You Should Know About Controlling with SAP (2nd ed.)* Boston: SAP PRESS, 2015.

Ovigele, Paul, *Material Ledger in SAP S/4HANA (2nd ed.)* Boston: SAP PRESS, 2023.

Reis, Vanda, *Actual Costing with The SAP Material Ledger (2nd ed.)* Boston: SAP PRESS, 2015.

Salmon, Janet, Walz, Stefan, *Controlling with SAPS/4HANA Business User Guide* Boston: SAP PRESS, 2021.

Appendix B
Glossary

We cover many terms in this book that are relevant to product costing. Let's review them.

Activity input planning Cost centers can provide planned output services based on activity quantities with Transaction KP26. You can also plan cost center activity input quantities from other cost centers with activity input planning using Transaction KP06.

Activity type An activity type identifies activities provided by a cost center to manufacturing orders. The secondary type general ledger account associated with an activity type identifies the activity costs on cost center and detailed reports.

Actual costing Actual costing determines what portion of the variance is debited to the next-highest level using material consumption. All purchasing and manufacturing difference postings are allocated upward through the BOM to assemblies and finished goods. Variances can be rolled up over multiple production levels and company codes to the finished product. We discuss actual costing in Chapter 16.

Actual costs Actual costs debit a product cost collector or manufacturing order during business transactions, such as general ledger account postings, inventory goods movements, internal activity allocations, and overhead calculations.

Additive cost estimate This is a material cost estimate in which you enter costs manually in the form of a unit cost estimate (spreadsheet format) so that manual costs can be added to an automatic cost estimate with a quantity structure. We discuss additive cost estimates in Chapter 9.

Allocation structure An allocation structure allocates the costs incurred for a sender by cost element or cost element group, and it is used for settlement and assessment. An assignment maps a source cost element group to a settlement general ledger account.

Alternative bill of materials There can be multiple methods of manufacturing an assembly and many possible bills of materials (BOMs). The alternative BOM allows you to identify one BOM in a BOM group.

Alternative hierarchy While there can only be one cost center standard hierarchy, you can create many alternative hierarchies. You create an alternative hierarchy by creating cost-center groups.

Alternative sequence An alternative sequence is a sequence of operations that can be used as an alternative to several consecutive operations from the standard sequence. Alternative sequences are used, for example, if the production process differs according to lot size.

Alternative unit of measure This is a unit of measure defined in addition to the base unit of measure. Examples of alternative units of measure include order unit (purchasing), sales unit, and unit of issue.

Apportionment method An apportionment method distributes the total costs of a joint production process to the primary products. Although the costs of the individual primary products may vary, they are apportioned using an apportionment structure.

Apportionment structure An apportionment structure defines how costs are distributed to co-products. The system uses the apportionment structure to create a settlement rule that distributes costs from an order header to the co-products. For each co-product, the system generates a further settlement rule that assigns the costs distributed to the order item to stock.

Assembly scrap Assembly scrap is the percentage of assembly quantity that does not meet required quality standards. Assembly scrap is an output scrap because it increases the planned output quantity of items in the production process. You plan assembly scrap in the **MRP 1** view. It can be ignored with the **Net ID** checkbox in the **Basic Data** tab of a BOM item.

Automatic account assignment Automatic account assignment allows you to enter a default cost center per general ledger account within a plant.

Auxiliary cost component split Only the main cost component split can update the results of the standard cost estimate to the material master. A second cost component split called the auxiliary cost component split, is used for statistical information purposes and can be used in parallel to the main cost component split. We discuss cost component splits in Chapter 6.
You define an auxiliary cost component split when assigning cost component structures to organizational units with Transaction OKTZ.

Backflush Backflushing is automatically posting a goods issue for components in an order during confirmation. It reduces the amount of work in warehouse management, especially for low-value parts. The material components from the BOM are assigned to operations in the routing.

Base quantity All component quantities in a BOM relate to the base quantity. Increasing the base quantity increases the accuracy of component quantities, similar in concept to the price unit.

Base unit of measure Material stocks are managed in the base unit of measure. The system converts all quantities you enter in other units of measure (alternative units of measure) to the base unit of measure.

Bill of materials A bill of materials (BOM) is a structured hierarchy of components required to manufacture an assembly. BOMs and purchasing info records allow cost estimates to determine the material costs of an assembly.

BOM application A BOM application is a costing variant component that allows for the automatic determination of alternative BOMs. We discuss costing variants in Chapter 7.

BOM group A BOM group is a collection of BOMs for a product or several similar products.

BOM item component quantity The quantity of a BOM item that is entered in relation to the base quantity of the product.

BOM item status The **Status/Long Text** tab of a BOM item contains six indicators, such as costing relevancy.

BOM status This controls the BOM's current processing status. For example, a BOM may have a default status of not active when initially created. It may be changed to active when used in material requirements planning (MRP) and released for planned orders.

BOM usage This determines a section of your company, such as production, engineering, or costing. You define which item statuses can be used in each BOM usage; for example, all items in BOMs with a certain usage may be relevant to production.

Bulk material Bulk materials are not relevant for costing in a cost estimate and are expensed directly to a cost center. The Bulk Material checkbox is maintained in the MRP 2 view and the BOM item. If a material is always used as a bulk material, set the indicator in the material master. If a material is only used as a bulk material in individual cases, set the indicator in the BOM item, which has a higher priority.

Business function Delivered by enhancement packages to an SAP system, you can activate business functions individually with Transaction SFW5 to access the new functionality.

Calculation base A calculation base is a group of cost elements to which overhead is applied in a costing sheet. The calculation base is a component of a costing sheet.

Capacity category The capacity category enables you to differentiate between machine and labor capacity. Machine capacity is the availability of a machine based on planned and unplanned outages and maintenance requirements. Labor capacity is the number of workers who can operate a machine at the same time.

Chart of accounts A chart of accounts is a group of general ledger accounts assigned to each company code. This chart of accounts is the operative chart of accounts used in both financial and cost accounting. All companies within the one controlling area must have the same operative chart of accounts. Other charts of accounts include the country-specific chart of accounts required by individual country legal requirements and the group chart of accounts required by consolidation reporting.

Company code A company code is the smallest organizational unit of financial accounting for which a complete self-contained chart of accounts can be drawn up for external reporting.

Component scrap Component scrap is the percentage of component quantity that does not meet required quality standards before being inserted in the production process. The plan quantity of components is increased. Component scrap is an input scrap because it is detected before use in the production process. You can plan component scrap in the **MRP 4** view and the **Basic Data** tab of the BOM item. An entry in the BOM item field takes priority over an entry in the material master **MRP 4** view.

Condition Conditions are stipulations agreed upon with vendors, such as prices, discounts, surcharges, freight, duty, and insurance. You maintain purchasing conditions in quotations, purchasing info records, outline agreements, and orders.

Condition technique The condition technique determines the purchase price by considering all the relevant pricing elements. This technique includes formulating rules and requirements.

Condition type A condition type is a key that identifies a condition. It indicates, for example, whether the system applies a price, a discount, a surcharge, or other pricing, such as freight costs and sales taxes.

Confirmation A confirmation documents the processing status of orders, operations, and individual capacities. You specify the operation yield, scrap and rework quantity, the activity quantity, work center, and who performed the operation.

Consignment material Consignment occurs when a vendor maintains a stock of materials at a customer site. The vendor retains ownership of the materials until they are withdrawn from the consignment stores.

Controlling level The controlling level determines the level of detail of procurement alternatives. The characteristics of the material/plant determine the standard setting of the controlling level. You specify which characteristics are updated for the production process by the controlling level.

Co-product You select the **Co-product** checkbox located in the MRP **2** and **Costing 1** views if a material is a valuated product produced simultaneously with one or more other products. Setting this checkbox allows you to assign the proportion of costs this material will receive in relation to other co-products within an apportionment structure.

Cost center A cost center is master data that identifies where the cost occurred. At the end of the period, a responsible person assigned to the cost center analyzes and explains cost center variances.

Cost component A cost component identifies costs of similar types, such as material, labor, and overhead costs, by grouping together cost elements in the cost component structure.

Cost component group Cost component groups allow you to display cost components in standard reports. In the simplest implementation, you create a cost component group for each component and assign each group to a corresponding cost component. You assign cost component groups as columns in cost estimate list reports and costed multi-level BOMs.

Cost component split The cost component split is the combination of cost components that makes up the total cost of a material. For example, if you need to view three cost components (material, labor, and overhead) for your reporting requirements, combining these three cost components represents the cost component split.

Cost component structure You define which cost components comprise a cost component split by assigning them to a cost component structure. Within the cost component structure, you assign cost elements and origin groups to cost components.

Cost component view Each cost component is assigned to a cost component view. When you display a cost estimate, you can choose a cost component view, which filters the cost components displayed in the cost estimate.

Cost element Cost elements are included in a general ledger account. Primary cost elements identify external costs, while secondary cost elements identify costs allocated within control, such as activity allocations from cost centers to manufacturing orders.

Cost estimate A cost estimate calculates the plan cost to manufacture a product or purchase a component. It determines material costs by multiplying BOM quantities by the standard price, labor costs by multiplying operation standard quantities by plan activity price, and overhead by costing sheet configuration

Costed multilevel BOM A costed multilevel BOM is a hierarchical overview of the values of all items of a costed material according to the material's costed quantity structure (BOM and routing). You display a costed multilevel BOM on the left side of a cost estimate screen. You can also view a costed multilevel BOM separately with Transaction CK86_99.

Costing BOM Costing BOMs are assigned a BOM usage of costing and are usually copied from BOMs with a production usage. You can adjust costing BOMs if required to differ from production BOMs. With system-supplied settings, standard cost estimates search for costing BOMs before production BOMs.

Costing lot size The costing lot size in the Costing 1 view determines the quantity on which the cost estimate calculations are based. To reduce lot size variance, the costing lot size should be set as close as possible to actual purchase and production quantities.

Costing run A costing run is a collective processing of cost estimates, which you maintain with Transaction CK40N.

Costing sheet A costing sheet summarizes the rules for allocating overhead from cost centers for cost estimates, product cost collectors, and manufacturing orders. The components of a costing sheet include the calculation base (group of cost elements), overhead rate (percentage rate applied to base), and credit key (cost center receiving credit).

Costing type The costing type determines if the cost estimate can update the standard price.

Costing variant The costing variant contains information on how a cost estimate calculates the standard price. For example, it determines whether the purchasing info record price is used for purchased materials or an estimated price is manually entered in the **Planned price 1** field of the material master **Costing 2** view. The purpose of the cost estimate is to set the standard price or commercial/tax prices.

Currency type The currency type identifies the role of the currency such as local or global.

Current cost estimate A current cost estimate is based on the current quantity structure and prices and is used to cost materials during the fiscal year and analyze cost changes and developments.

Demand management Demand management involves planning requirement quantities and dates for assemblies, as well as defining the strategy for planning and producing/procuring a finished product.

Demand planning Demand planning allows you to forecast market demand for a company's products and produce a demand plan. Supply network planning (SNP) can be integrated with SAP planning functionalities such as sales and operations planning.
Sales and operations planning is being replaced by SAP Integrated Business Planning for Supply Chain (SAP IBP), which supports all S&OP features. Sales and operations planning is an interim solution to allow a smooth transition from SAP ERP to on-premise SAP S/4HANA and SAP IBP. See SAP Note 2268064 for more details.

Dependent requirements Dependent requirements are caused by higher-level dependent and independent requirements when running MRP. Independent requirements, created by sales orders or manually planned independent requirement entries in demand management, determine lower-level dependent material requirements.

Detailed reports Detailed reports display cost element details of manufacturing orders and product cost collectors. During variance analysis, you can drill down on cost elements to display line-item reports.

Distribution rule You maintain distribution rules in settlement rules in manufacturing orders and product cost collectors.

Event-based processing As of SAP S/4HANA release 2022, event-based processing is available. Goods movements and confirmations represent events that trigger the overhead calculation according to the costing sheet. Depending on the order's status, this triggers either the posting of a journal entry for the work in process (WIP) or the cancellation of any existing WIP and the calculation of production variances.

External processing An external supplier performs external processing of a manufacturing order operation. This is distinct from subcontracting, which involves sending material parts to an external supplier, who manufactures the complete assembly via a purchase order.

Functional area A functional area allows you to create a profit and loss account in financial accounting using cost-of-sales (COS) accounting. This method compares the sales revenue for a given accounting period with the manufacturing costs of the activity. Expenses are allocated to the functional areas, such as production, sales and distribution, and administration.

Gross profit COS represents the expense related to labor, raw materials, and manufacturing overhead. This expense is deducted from the company's net revenue, which results in the first level of profit, or gross profit. It is used to analyze how efficiently a company uses raw materials, labor, and manufacturing-related fixed assets to generate profits.

Group counter A group counter identifies a unique routing within a task list (routing) group.

Initial cost split The initial cost split is based on a cost component structure for raw materials that contains separate cost components for all procurement costs such as purchase price, freight charges, insurance contributions, and administrative costs. With an additive cost estimate, you can enter a cost component split for costs such as freight and insurance for a material. These costs are added to the price from the material master.

Input variance Variances on the input side are based on goods issues, internal activity allocations, overhead allocation, and general ledger account postings. There are four input variances: input price, resource-usage, input quantity, and remaining input variance.

Internal order An internal order monitors an organization's costs and revenue for short—to medium-term jobs. You can carry out planning at a cost element and detailed level and budgeting at an overall level with availability control.

Inventory cost estimate An inventory cost estimate accesses tax-based and commercial prices in the Accounting 2 view for purchased parts, use these prices for valuation and then updates the costing results for finished and semi-finished products in the same fields. You can enter values such as determining the lowest value in the **Tax-based** and **Commercial price** fields of purchased parts.

Itemization Itemization provides detailed cost estimate data about the resources necessary to produce a product. The costing information for each item includes details such as the quantity, unit of measure, and value.

Line-item reports Line-item reports display a list of postings to a cost object, such as a cost center or order, within a time frame. During variance analysis, you can sort the value or quantity columns to find the largest postings.

Long-term planning Long-term planning allows you to enter medium- to longer-term production plans and simulate future production requirements with long-term MRP. You can determine future purchasing requirements for vendor requests for quotations (RFQs), update purchasing info records, and transfer planned activity requirements to cost center accounting.

Main cost component split The main cost component split is the principal cost component split used by a standard cost estimate to update the standard price. You define the main cost component split when assigning cost component structures to organizational units with Transaction OKTZ.

Manufacturing order Manufacturing order is an umbrella term for production and process orders.

Margin analysis Margin analysis involves categorizing costs depending on their nature, such as those directly incurred during manufacturing or collected as overhead. The categorized costs are then progressively subtracted from revenue to calculate profit margins, such as gross, operating, pre-tax, and net profit.

Mark standard cost estimate After a standard cost estimate is saved without errors, it can be marked. The cost estimate value populates the **Future** column in the **Costing 2** view. You can create and mark standard cost estimates many times before release. Within the same fiscal period, a new standard cost estimate overwrites the existing marked cost estimate.

Master data Master data is information that stays relatively constant over long periods of time. For example, purchasing info records contain vendor information such as a business name, which usually doesn't change.

Material assignment Material assignment determines which material is to be produced with a routing. Based on this, the routing can be used for sales and operations planning, MRP, production orders, and cost estimates for this material.

Material Ledger Allows inventory to be carried in up to three valuation approaches.

Material master A material master contains all the information required to manage a material. Information is stored in views, each corresponding to a department or area of business responsibility. Views conveniently group information together for users in different departments, such as sales and purchasing.

Material origin The **Material Origin** indicator in the **Costing 1** view determines if the material number is displayed in detailed reports. This is one of the most important indicators in providing visibility to the causes of variances. If you have already created material master records without the **Material Origin** indicator selected, you can use report RKHKMATO to select the indicator.

Material price determination Material price determination in the **Accounting 1** view is only applicable if the material ledger is active. Transaction-based material price determination (indicator 2) allows price control to be set at either moving average (V) or standard price.
Use the setting **S** if you are using multiple currencies and/or valuation approaches for transfer pricing but not the actual costing functionality. In a typical scenario, you are interested in reporting on global inventory valuation with and without mark-up, that is, internal company profit. Single-/multilevel price determination (indicator **3**) is only available for materials with standard price control (**S**), which remains unchanged during a period. A periodic unit price (PUP) is up-

dated for information during the period and used for material valuation in a closed period.

Material requirements planning (MRP) MRP guarantees material availability by monitoring stocks and generating planned orders for purchasing and production.

Material type A material type groups together materials with the same basic attributes such as raw materials, semifinished products, or finished products. Material types also determine whether materials are quantity and/or valuation relevant per plant with configuration Transaction OMS2.

Milestone confirmation The system automatically confirms all preceding operations up to the preceding milestone operation during the confirmation of a milestone operation. An operation is marked as a milestone operation in the **Confirmations** field of its control key. If several operations are marked as milestones, they must be confirmed in sequence.

Modified standard cost estimate A modified cost estimate is based on the latest quantity structure and planned prices. It is used to cost materials during the fiscal year and analyze developments.

Movement type This key indicates the type of material movement such as goods receipt, goods issue, and physical stock transfer. The movement type enables the system to find predefined posting rules, which determine how the stock and consumption general ledger accounts post and how the stock fields in the material master record are updated.

Moving average price The moving average price in the **Costing 2** view determines the inventory valuation price if price control is set at the moving average. It is updated during goods receipt.

Multiple BOM A multiple BOM is a group of BOM records with different combinations of materials (alternatives) for the same product.

Net ID indicator This indicator determines whether scrap for a BOM item is calculated on the basis of the net required quantity, that is, the required quantity without assembly scrap from the material master record. The following list details in which circumstances the **Net ID** indicator must be set:

- You must set this checkbox if you want assembly scrap to be ignored.
- You must set this checkbox if you enter operation scrap.

You can set this checkbox if you only enter component scrap to calculate scrap based on the net required quantity for the assembly.

Net profit Referred to as a company's profit margin or bottom line, you determine net profit by subtracting all expenses and tax from revenue.

Operating profit You subtract selling, general, and administrative (SG&A) expenses, or operating expenses, from gross profit to calculate the operating profit margin. Management has much more control over operating expenses than COS. Positive and negative trends in operating profit are, for the most part, directly attributable to management decisions.

Operating rate The operating rate percentage is defined as *(actual activity/plan activity) 100*. This calculation is used during cost center cost center target cost analysis.

Operation An operation is a work step in a plan or work order.

Operation scrap Operation scrap is the percentage of assembly quantity that does not meet required quality standards. Operation scrap is an output scrap because it reduces the planned output quantity in the production process. You can plan operation scrap in the routing **Operation Details** view and in the **Basic Data** tab of the BOM item.

Order type The order type categorizes orders according to their purpose and allows you to allocate number ranges and settlement profiles.

Organizational unit This represents an organizational structure such as a sales organization in sales and distribution, company code in financial accounting and asset accounting, and plant in materials management and sales and distribution.

Origin group An origin group separately identifies materials assigned to the same cost element, allowing them to be assigned to separate cost components. The origin group can also determine the calculation base for overhead in costing sheets.

Outline agreement This is a longer-term arrangement between a purchasing organization and a vendor for the supply of materials or provision of services over a certain period. The two types are contracts and scheduling agreements.

Output variance Variances on the output side result from too little or too much planned order quantity being delivered or because the delivered quantity was evaluated differently. Output variances are divided into the following categories during variance calculation: mixed price, output price, lot size, and remaining variance.

Overhead group An overhead group is used to apply different overhead percentages to individual materials or groups of materials. You assign an overhead group to an overhead key with Transaction OKZ2.

Overhead key An overhead key is used to apply overhead percentages to individual orders or groups of orders. You assign the overhead key in the overhead rate component of a costing sheet.

Overhead order Used for short- to medium-term monitoring of overhead costs such as marketing campaigns. You can monitor internal orders throughout their lifecycle from creation through the planning and posting of actual costs to settlement and archiving.

PA transfer structure A PA transfer structure allows you to assign costs and revenues from other functionalities to value and quantity fields in profitability analysis.

Phantom assembly A phantom assembly is a logical assembly created to efficiently maintain a single BOM, which is part of many higher-level assemblies. It is neither a physical assembly nor an inventory item. A phantom assembly is not included in a costing run; however, you can create an individual cost estimate.

Pipeline material Pipeline materials, such as oil or water, flow directly into the production process. Stock quantities are not changed during withdrawal.

Plan reconciliation This allows you to compare and overwrite the plan *activity quantity* that was manually entered in the second column of Transaction KP26 with the *scheduled activity* quantity that was automatically entered in the second last column. Scheduled activity quantities are transferred from S&OP, MRP, or long-term planning with Transaction KSPP.
You carry out plan reconciliation with Transaction KPSI. When planning activity prices and quantities with Transaction KP26, you can select the **Plan Quantity Set** checkbox in an activity type to default to being selected. This ensures that an activity quantity that was manually planned will not be overwritten during plan reconciliation.

Planned independent requirements Planned independent requirements are either created automatically by sales orders or created manually by entering quantities and dates in demand management. You may need to make manual entries to generate requirements for purchased components or assemblies with long delivery times to satisfy demand by sales orders that you expect but are not yet entered in the system. An example of this is long-term government defense contracts, which can specify several years before the first planned delivery to the customer.

Planned order MRP automatically creates planned orders when a material shortage occurs. A planner may also create a planned order manually. Planned orders are converted into production orders for in-house production and purchase requisitions for purchasing requirements.

Planned price 1, 2, and 3 You can manually enter prices in these fields in the **Costing 2** view. These are generally used to estimate the purchase price of components early in the lifecycle of a new or modified product.

Planning plant The plant in which the goods receipt takes place for the manufactured material. If the planning plant and production plant are identical, then you do not need to enter the planning plant as well. The production plant is copied automatically.

Planning variance Planning variance is a type of variance calculation based on the difference between costs on the preliminary cost estimate for the order and target costs based on the standard cost estimate and planned order quantity. You calculate planning variances with target cost version 2. Planning variances are for information only and are not relevant for settlement.

Preliminary cost estimate A preliminary cost estimate calculates the planned costs for a manufacturing order or product cost collector. There can be a preliminary cost estimate for every order or production version, while there can only be one released standard cost estimate for each material. The preliminary cost estimate can be used to valuate scrap and WIP in a WIP at target scenario.

Preliminary costing Preliminary costing is carried out when you create a cost estimate for a manufacturing order or product cost collector. It is generally based on a quantity structure, which consists of a BOM and routing.

Pre-tax profit A company has access to a variety of tax-management techniques that allow the company to manipulate the timing and magnitude of its taxable income.

Price control The **Price Control** field in the **Costing 2** view determines whether inventory is valuated at the standard or the moving average price.

Price indicator The activity type **Price indicator** field determines how the system calculates the price of an activity for a cost center.

Price unit The price unit is the number of units to which the price refers. Increasing the price unit can increase the accuracy of the price. To determine the unit price, divide the price by the price unit.

Primary cost component split The primary cost component split provides an alternative view of cost components based on cost center primary costs. By displaying each as a cost component, you can more easily analyze changes to your more significant primary costs, such as wages, energy, and depreciation. These costs normally are divided between cost components as activities are consumed during manufacturing.
This functionality is only required if you have significant primary costs that you need to analyze separately from the manufacturing process. To set this up, you need to create a primary cost component split in cost center accounting when calculating the activity price. You can only use the primary cost component split to determine your activity costs into components if you automatically calculate the activity price.

Process order A process order is a manufacturing order that is used in process industries. A master recipe and materials list are copied from the master data to the order. A process order contains operations that are divided into phases. A phase is a self-contained work-step that defines the detail of one part of the production process using a primary resource.
In process manufacturing, only phases are costed not operations. A phase is assigned to a subordinate operation and contains standard values for activities, which are used to determine dates, capacity requirements, and costs.

Procurement alternative A procurement alternative represents one of several different ways of procuring a material. You can control the level of detail in which the procurement alternatives are represented through the controlling level. Depending on the processing category, there are single-level and multilevel procurement alternatives. For example, a purchase order is a single-level procurement, while production is a multilevel procurement.

Procurement type The procurement type in the MRP 2 view defines the material as assembled in-house, purchased externally, or both:

- *In-house production* (E) means a cost estimate will search for a BOM and routing.
- *External procurement* (F) results in the system searching for a purchasing info record price.
- *Both* (X) means a planned order can be converted into either a production or purchase order.

The special procurement type can override the procurement type.

Product cost collector A product cost collector collects target and actual costs during the manufacture of an assembly. Product cost collectors are necessary for repetitive manufacturing and optional for order-related manufacturing.

Product drilldown reports Product drilldown reports allow you to slice and dice data based on characteristics such as product group, material, plant, cost component, and period. Product drilldown reports are based on predefined summarization levels.

Production line A production line is used in repetitive manufacturing and typically consists of one or more work centers.

Production order A production order is used for discrete manufacturing. A BOM and routing are copied from master data to the order. A sequence of operations is supplied by the routing, which describes how to carry out work-steps.
An operation can refer to a work center where it is to be performed. It contains planned activities required to carry out the operation. Costs are based on the material components and activity price multiplied by a standard value.

Production order information system This is a tool for shop floor control and process industries with reporting functions for production orders, planned orders, and process orders. You can, for example, display all production orders within a time frame and group the information according to order or operation.

Production process Product cost collectors are created regarding a production process that describes how a material is produced, that is, the quantity structure used. The quantity structure is taken from the production version, noted during the production process. The production process is determined by the following characteristics: material, production plant, and production version. One production process can be created for each production version.

Production resource/tool This tool is a moveable operating resource used in production or plant maintenance.

Production variance Production variance is a type of variance calculation based on the difference between net actual costs debited to the order and target costs based on the preliminary cost estimate and quantity delivered to inventory. You calculate production variance with target cost version 1. Production variances are for information only and are not relevant for settlement.

Production version A production version describes the types of production techniques that can be used for a material in a plant. It is a unique combination of BOM, routing, and production line maintained in the **MRP 4**, **Work Scheduling**, and **Costing 1** views of the material master.

Profit center A profit center receives postings made in parallel to cost centers and other master data, such as orders. Profit center accounting is integrated in the Universal Journal. You normally create profit centers based on areas in a company that generate revenue and have a responsible manager assigned.
If profit center accounting is active, you will receive a warning message if you do not specify a profit center, and all unassigned postings are made to a dummy profit center. You activate profit center accounting with configuration Transaction OKKP, which maintains the controlling area.

Profit margin Profit margin is a profit analysis technique that involves subtracting expenses and tax from revenue.

Profitability analysis Costing-based profitability analysis enables you to evaluate market segments, which can be classified according to products, customers, orders (or any combination of these), or strategic business units, such as sales organizations or business areas, with respect to your company's profit or contribution margin.

Purchase price variance When raw materials are valued at the standard price, a purchase price variance (PPV) will post during goods receipt if the goods receipt or invoice price differs from the material standard price.

Purchase requisition A purchase requisition is a request or instruction to Purchasing to procure a quantity of a material or service so that it is available at a certain point in time. Until a purchase order is created, there is no legal requirement to carry out the purchase. You can record the purchase requisition in commitment management because it may lead to actual expenditure in the future.

Purchasing info record A purchasing information record stores all the information relevant to procuring a material from a supplier. It contains the **Purchase Price** field, which the standard cost estimate searches for when determining the purchase price.

Purchasing organization A purchasing organization procures materials and services and negotiates conditions of purchase with suppliers.

Quantity structure A quantity structure consists of a BOM and a routing. In the process industries, a master recipe is used instead of routing, and in repetitive manufacturing, a rate routing is used instead of a routing. A quantity structure is used by a standard cost estimate to determine component and activity quantity.

Quantity structure control Quantity structure control is a costing variant component that automatically searches for alternatives if multiple BOMs and/or routings exist for a material when a cost estimate is created.

Quantity structure date The quantity structure date determines which BOM and routing are selected when initially creating a cost estimate. Because these can change over time, it is useful to select a particular BOM or routing by date when developing new products or changing existing products.

Real-time production accounting Real-time production accounting is based on event-based accounting instead of period-end-based accounting

Reference variant A reference variant is a costing variant component that allows you to create cost estimates or costing runs based on the same quantity structure to improve performance or make reliable comparisons.

Release standard cost estimate When you release a standard cost estimate, the results are written to the Costing 2 view as the current planned price and current standard price. Inventory is revalued, and accounting documents are posted. A standard cost estimate must be marked before it can be released, and it can be released only once per fiscal period.

Repetitive manufacturing Repetitive manufacturing eliminates the need for production or process orders in manufacturing environments with production lines and long production runs. It reduces the work involved in production control and simplifies confirmations and goods receipt postings.

Request for quotation A request for quotation (RFQ) is a request for a supplier to submit a quotation for materials or services.

Requirement This is the quantity of material needed in a plant at a certain time.

Results analysis key A product cost collector or order for which you want to create WIP must contain a results analysis key. A results analysis key means that the product cost collector or order is included in WIP calculation during period-end closing. It is also used to control event-based WIP and variance calculation, which we discuss in Chapter 19. You define a results analysis key with configuration Transaction OKG1.

Rework Assemblies or components that do not meet quality standards may either become scrap or require rework. Depending on the problem, cheaper items may become scrap, while more costly assemblies may justify rework.

Routing A routing is a list of tasks containing standard activity times required to perform operations to build an assembly. Routings, together with planned activity prices, provide cost estimates with the information necessary to calculate product labor and activity costs.

Routing header A routing header contains data that is valid for the entire routing. Select **Details • Header** from the menu bar when displaying routing operations to display the routing header.

Sales and operations planning Sales and operations planning allows you to enter a sales plan, convert it to a production plan, and transfer the plan to long-term planning.

Sales order A sales order is a customer's request to deliver goods or services at a certain time. If you are using non-valuated inventory, a sales order line item can be a real cost object.

SAP Fiori SAP Fiori is a web-based interface that can replace SAP GUI. SAP Fiori apps access the Universal Journal directly, taking advantage of additional fields like the work center and operation for improved variance reporting.
SAP Fiori is a user interface for modern business applications that can run on any device (desktop, tablet, smartphone).

SAP S/4HANA Cloud Private Edition SAP S/4HANA Cloud Private Edition is a managed cloud offering that enables a smooth, secure migration of an on-premise enterprise resource planning (ERP) system—including SAP ERP and SAP S/4HANA—to the cloud. This deployment option allows you to preserve your existing configuration, customization, and historical data from SAP ERP.

SAP S/4HANA Cloud Public Edition SAP S/4HANA Cloud Public Edition is a ready-to-run cloud ERP that delivers the latest industry best practices and continuous innovation. This deployment option is often abbreviated to SAP S/4HANA Cloud.

Scale A scale represents vendor quotations containing reduced prices for greater purchase quantities. Scales are maintained in purchasing info record conditions with Transaction ME12.

Scheduled activity Scheduled activity quantities are transferred from sales and operations planning, MRP, or long-term planning to cost center/ activity type planning with Transaction KSPP. The scheduled activity quantity appears in the second-to-last column when you are planning activity prices and quantities with Transaction KP26. You can compare and overwrite a manual plan activity quantity with the scheduled activity quantity during plan reconciliation.

Scheduling During scheduling, the system determines the start and finish dates of orders or of operations in an order. Scheduling is performed in MRP, capacity planning, and networks.

Scheduling agreement A scheduling agreement is a longer-term purchase arrangement with a supplier that covers the supply of materials according to predetermined conditions. These conditions apply for a predefined period and total purchase quantity.

Scope Items Scope items are the mechanism to deliver best practice content. They are self-contained building blocks that contain the configuration settings needed to run a particular feature. Some are automatically active (such as standard cost calculation), and some (such as country versions) must be activated during initial scoping.

Selection method The **Selection Method** field in the **MRP 4** view determines the method of selecting an alternate BOM.

Settlement WIP and variances are transferred to financial accounting, profit center accounting, and profitability analysis during settlement. Variance categories can also be transferred to value fields in profitability analysis.

Settlement profile A settlement profile, contained in the order type, contains the parameters necessary to create a settlement rule for manufacturing orders and product cost collectors.

Settlement rule A settlement rule determines which portions of a sender's costs are allocated to which receivers. A settlement rule is contained in a manufacturing order or product cost collector header data.

Simultaneous costing Simultaneous costing is the process of recording actual costs for cost objects, such as manufacturing orders and product cost collectors, in cost object control. Costs incurred typically include goods issues, receipts to and from an order, activity confirmations, and external service costs.

Source cost element Source cost elements identify costs that debit objects, such as manufacturing orders and product cost collectors.

Source list A source list is a list of available sources of supply for a material, which indicates the periods during which procurement is possible. Usually, a source list is a list of quotations for a material from different vendors.
You can specify a preferred vendor by selecting a fixed source of supply indicator. If you do not select this indicator for any source, a cost estimate will choose the lowest cost source as the cost of the component. You can also indicate which sources are relevant to MRP.

Source structure You define source structures when settling and costing joint products. A source structure contains several source assignments, each containing the individual cost elements or cost element intervals to be settled using the same distribution rules. You need a source structure for investment orders.

Special procurement type The special procurement type field found immediately below the procurement type in the MRP 2 view is used to more closely define the procurement type. For example, it may indicate if the item is produced in another plant and transferred to the plant you are analyzing.

Special procurement type for costing If you enter a special procurement type in the **Costing 1** view, it will be used by costing. If no entry is made in this field, the system will use the special procurement type in the **MRP 2** view.

Splitting rule A splitting rule determines how a splitting structure distributes the cost center costs of an individual cost element, range, or group over an activity type, range, or group.

Splitting structure You can allocate activity-independent cost center plan costs to activity types either with equivalence numbers or with a splitting structure. A splitting structure contains one or more assignments for which you assign splitting rules for corresponding cost elements and the activity types over which the costs are split. The plan price calculation splits the activity-independent costs automatically based on equivalence numbers or a splitting structure if assigned. You can also split the plan costs manually to see how the plan costs are distributed to the activity types.
A splitting structure can be used to view all cost center costs at the cost center/activity type level during cost center plan/target/actual comparison.

SSCUI Self-service configuration user interface. A simplified view of a configuration step allowing minor changes to, e.g., costing variants or costing sheets.

Standard cost estimate This is a material cost estimate used to calculate the standard price of a material. The cost estimate must be executed with a costing variant that updates the material master, and the cost estimate must be released. A standard cost estimate can be released only once per period and is typically created for each product at the beginning of a fiscal year or new season.

Standard hierarchy A standard hierarchy represents your company structure. A standard hierarchy is guaranteed to contain all cost centers or profit centers because a mandatory field in cost and profit center master data is a standard hierarchy node.

Standard price The standard price in the **Costing 2** view determines the inventory valuation price if price control is set at standard (**S**). The standard price is updated when a standard cost estimate is released. You normally value manufactured goods at standard price.

Standard value This is the planned value for executing an operation in a routing. For example, you may define that it normally takes five minutes to drill a hole in a metal plate or that it takes one hour of setup time to prepare to manufacture a batch of pharmaceuticals. The standard value is multiplied by the planned activity rate to determine the value of activities in cost estimates.

Standard value key The standard value key in the **Basic Data** tab of an operation defines and gives a dimension (e.g., time or area) to one of up to six standard values available in an operation.

Statistical key figure Statistical key figures define values describing cost centers, profit centers, and overhead orders, such as number of employees or minutes of long-distance phone calls. You can use statistical key figures as the tracing factor for periodic transactions such as cost center assessment. You can post both the plan and actual statistical key figures.

Subcontracting You supply component parts to an external vendor who manufactures the complete assembly. The vendor has previously supplied a quotation, which is entered in a purchasing info record with a category of subcontracting.

Summarization hierarchy reports Summarization reports are based on data collected at the nodes of a summarization hierarchy. A summarization hierarchy groups together manufacturing orders or product cost collectors at the lowest-level summarization nodes, which in turn are grouped together to create a pyramid structure. You create hierarchies with Transaction KKRO.

Target cost version A target cost version determines the basis for calculating target costs. It calculates total variance and is used to explain the difference between actual debits and credits on an order. It is the only target cost version that can be settled to financial accounting, profit center accounting, and profitability analysis.

Task list A task list (routing) is a list of tasks containing standard activity times required to perform operations to build an assembly. Task lists, together with planned activity prices, provide cost estimates with the information necessary to calculate labor costs of products.

Task list group A task list group identifies routings that have different production steps for one material.

Task list type A task list type classifies task lists according to their function. Typical task list types include routing, reference operation set, rate routing, and standard rate routing.

Total variance Total variance is a type of variance calculation based on the difference between actual costs debited to the order and credits from deliveries to inventory. You calculate total variance with target cost version 0, which determines the basis for calculation of target costs.

Tracing factor Tracing factors determine the cost portions received by each receiver from senders during periodic allocations, such as assessments and distributions.

Transfer control Transfer control is a costing variant component that requires a higher-level cost estimate to use recently created standard cost estimates for all lower-level materials. Preliminary cost estimates for product cost collectors use transfer control.

Transfer price This price is charged for transferring a material or product from one business unit (company code or profit center) to another. The amount of price markup is determined by tax authorities in each country involved in the transfer and is the basis of legal inventory valuation. Global companies are also interested in global inventory group valuation, excluding transfer pricing in consolidation reporting. This provides a view of inventory valuation based on the actual cost of purchase and manufacture for internal reporting and analysis.

Under/overabsorption Cost center balance, or under/overabsorption, represents the difference between cost center debits and credits during a period or range of periods.

Unit costing Unit costing is a method of costing that does not use BOMs or routings, typically when developing new products. You create a preliminary structure of materials and activities in a view similar to a spreadsheet.

Universal Journal The efficiency and speed of the SAP HANA in-memory database allowed the introduction of the Universal Journal single line-item tables ACDOCA (actual) and ACDOCP (plan). The Universal Journal allows all postings from the previous financial and controlling components to be combined into single items. The many benefits include the development of real-time accounting. In this book, we discuss both period-end and event-based processing.

Universal parallel accounting When you activate this business function, all ledgers and currencies configured for the universal journal will be relevant for the Material Ledger, Controlling, and asset accounting application components.

With universal parallel accounting, you can now transition tasks from the period close to real-time production accounting, with event-based processing. We discuss universal parallel accounting in Chapter 19.

User exit A user exit is a point in the standard program where you can call your own program. In contrast to customer exits, user exits allow developers to access and modify program components and data objects in the standard system.

Valuation approach A valuation approach describes the values that are stored in accounting as a combination of a currency type (such as the group currency) and a valuation view (such as the profit center valuation view). The combination of various valuation approaches is known as a currency and valuation profile.

Valuation category The valuation category in the **Accounting 1** and **Costing 2** views determines which criteria are used to group partial stocks of a material and value them separately. This category is part of split valuation.
Normally, you will have only one price per material per plant. Split valuation allows you to valuate, for example, batches separately. Moving average price (**V**) is the only price control setting available if you activate split valuation and enter a valuation category. You assign valuation types to valuation categories, and you assign valuation categories to plants in Transaction OMWC.

Valuation class The valuation class in the **Costing 2** view determines which general ledger accounts are updated as a result of inventory movement or settlement.

Valuation date The valuation date determines which material and activity prices are selected when you create a cost estimate. Purchasing info records can contain different vendor-quoted prices for different dates. Different plan activity rates can be entered per fiscal period.

Valuation grouping code The valuation grouping code allows you to assign the same general ledger account assignments across several plants with Transaction OMWD to minimize your work. The grouping code can represent one or a group of plants.

Valuation type You use valuation types in the split valuation process, which enables the same material in a plant to have different valuations based on criteria such as batch.
You assign valuation types to each valuation category, which specify its individual characteristics. For example, you can valuate stocks of a material produced in-house separately from stocks of the same material purchased externally from vendors. You then select procurement type as the valuation category and internal and external as the valuation types.

Valuation variant The valuation variant is a costing variant component that allows different search strategies for materials, activity types, subcontracting, and external processing. For example, the search strategy for purchased and raw materials typically searches first for a price from the purchasing info record.

Valuation variant for scrap and WIP This valuation variant allows a choice of cost estimates to valuate scrap and WIP in a WIP at target scenario. If the routing structure is changed after a costing run, WIP can still be valued with the valuation variant for scrap and WIP, resulting in a more accurate WIP valuation.

Valuation view In the context of multiple valuation and transfer prices, you can define the following views:

- Legal valuation view
- Group valuation view
- Profit center valuation view

Together with a currency type and a currency, the valuation view creates a valuation approach. You can maintain up to three different valuation approaches in financial accounting, and the Material Ledger.

Value field In costing-based profitability analysis, value fields store the base quantities and amounts for reporting. Value fields can either be highly summarized (representing a summary of cost element balances, for example) or highly detailed (representing just one part of a single cost element balance).

Variance calculation Variance calculation provides information to assist you during analysis of how the order balance occurred. In other words, it helps you determine the reason for the difference between order debits and credits. The three main types of variance calculation are total, production, and planning.

Variance categories During variance calculation, the order balance is divided into categories on the input and output sides. Variance categories provide reasons for the cause of the variance, which you can use when deciding corrective action.

Variance key Variances are calculated on manufacturing orders or product cost collectors containing a variance key. This key is defaulted from the **Costing 1** view when manufacturing orders or product cost collectors are created. The variance key also determines if the value of scrap is subtracted from actual costs before variances are determined.

Variance variant The variance variant (Transaction OKVG) determines which variance categories are calculated. If a variance category is not selected, variances of that category are assigned to remaining variances. Scrap variances are the only exception to this rule. If scrap variance is not selected, these variances enter all other variances on the input side.

Version Versions, formerly known as plan versions, enable you to have independent sets of planning and actual data.

WBS element See work breakdown structure.

Wildcard Wildcard characters allow you flexibility when searching for data in tables. You can use * to replace any string of characters and + to replace any single character.

Work in process at actual WIP at actual is valuated based on actual debits to a production order or product cost collector.

Work in process at target WIP at target is valuated based on a cost estimate.

Work breakdown structure A work breakdown structure (WBS) exists within Project System as part of a project hierarchy. WBS elements describe tasks in a project to be performed within a defined time period. A WBS element is a cost object.

Work center Operations are carried out at work centers representing, for example, machines, production lines, or employees. Work center master data contains a mandatory cost center field. A work center can only be linked to one cost center, while a cost center can be linked to many work centers.

Work in process (WIP) WIP represents production costs of incomplete assemblies. For balance sheet accounts to accurately reflect company assets at period end, WIP costs are temporarily moved to the WIP balance sheet and profit and loss accounts. WIP is canceled during period-end processing following the delivery of assemblies to inventory.

Appendix C
The Authors

John Jordan is the founder and principal consultant at ERP Corp, which provides expert SAP consulting. John runs the annual SAP Controlling Financials conference in San Diego, CA. John is also the author of *Production Variance Analysis in SAP S/4HANA* and *100 Things You Should Know About Controlling with SAP* (2nd ed., 2015), both SAP-PRESS bestsellers. He regularly speaks at conferences, publishes articles, and provides training sessions for clients. John is considered one of the leading experts in SAP ERP financials by clients and peers. You can reach John at *jjordan@erpcorp.com*.

Janet Salmon is the chief product owner for management accounting at SAP SE and has accompanied many developments to the controlling components of SAP ERP Financials as both a product and a solution manager. She regularly works with key customers and user groups in the United States and Germany to understand their controlling challenges and requirements. Her role is to design and implement innovative controlling solutions with SAP's development teams in Germany and China.

Index

B

C

N

O

P

S

Y

Z

- Master variance analysis with SAP S/4HANA to improve profitability and production
- Create cost estimates, assess actual costs, perform period-end close, and analyze scrap with step-by-step instructions
- Discover the new real-time accounting approach and key SAP Fiori apps

John Jordan, Janet Salmon

Production Variance Analysis in SAP S/4HANA

To effectively plan production costs, you know that guesswork and projections aren't enough. Optimize your production processes—and increase profitability—with this guide to production variance analysis in SAP S/4HANA. Start with the planning basics, and then perform costing runs and create cost estimates. See how to post actual costs with the Universal Journal, analyze variance during period-end processing, and assess scrap. Get details on SAP Fiori apps for reporting and the road ahead for variance analysis in this comprehensive resource!

301 pages, pub. 01/2023
E-Book: $94.99 | **Print:** $99.95 | **Bundle:** $109.99

www.sap-press.com/5629

- Master your core controlling tasks in SAP S/4HANA
- Assess overhead, manufacturing, sales, project, and investment costs
- Streamline your operations with both the new and classic user interfaces

Janet Salmon, Stefan Walz

Controlling with SAP S/4HANA: Business User Guide

SAP S/4HANA brings change to your routine controlling activities. Perform your key tasks in the new environment with this user guide! Get click-by-click instructions for your daily and monthly overhead controlling tasks, and then dive deeper into processes such as make-to-stock/make-to-order scenarios, margin analysis, and investment management. Finally, instructions for intercompany transactions and reporting make this your all-in-one resource!

593 pages, pub. 05/2021
E-Book: $74.99 | **Print:** $79.95 | **Bundle:** $89.99

www.sap-press.com/5282

Rheinwerk
Publishing

Interested in reading more?

Please visit our website for all new book and e-book releases from SAP PRESS.

www.sap-press.com